D0788596

Distribution, Biology, and Management of Exotic Fishes

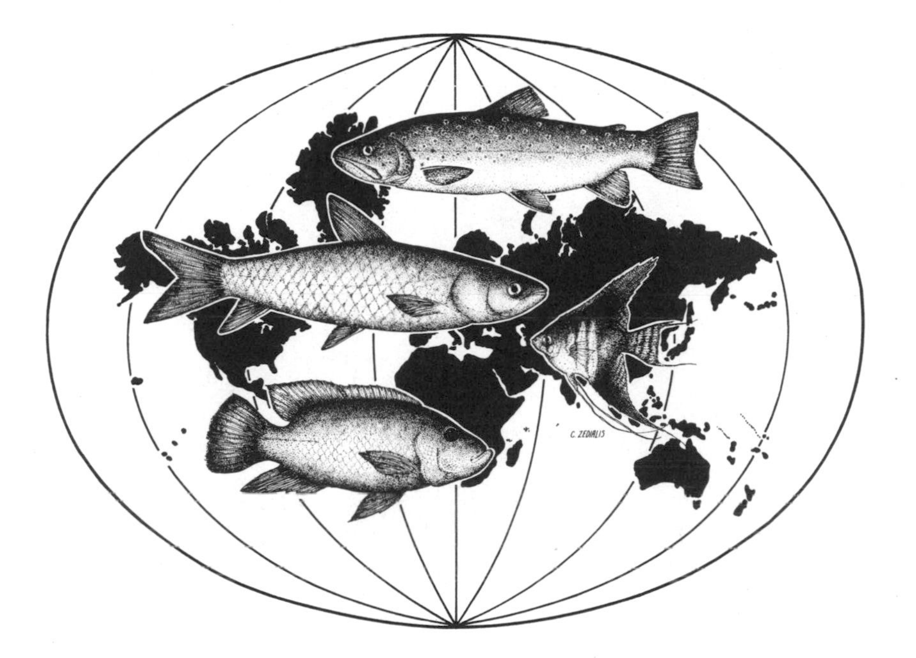

Distribution, Biology, and Management of Exotic Fishes

EDITED BY

Walter R. Courtenay, Jr.

Department of Biological Sciences, Florida Atlantic University

and Jay R. Stauffer, Jr.

Appalachian Environmental Laboratory, University of Maryland

THE JOHNS HOPKINS UNIVERSITY PRESS / Baltimore and London

Half-title page ornament: Logo of the Exotic Fish Section of the American Fisheries Society.
Drawing by C. Zedialis.

Chapter-opening ornament: *Tilapia mossambica* (Mozambique tilapia).
Drawing by Frances McKittrick Watkins.

Printed in the United States of America

The Johns Hopkins University Press, Baltimore, Maryland 21218
The Johns Hopkins Press Ltd., London

Library of Congress Cataloging in Publication Data
Main entry under title:

Distribution, biology, and management of exotic fishes.

Bibliography: p.
Includes index.
1. Fishes – North America – History. 2. Fishes – North America – Geographical distribution. 3. Animal introduction – North America – History. 4. Fish-culture – North America – History. 5. Fishes – Geographical distribution. 6. Animal introduction – History. 7. Fish-culture – History. I. Courtenay, Walter R. II. Stauffer, Jay R.
QL625.D57 1984 639.3'1 83-18723
ISBN 0-8018-3037-0

Contents

Foreword **vii**
Preface **ix**
Acknowledgments **xi**
Introduction **xiii**

CHAPTER 1. Involvement of the American Fisheries Society with Exotic Species, 1969–1982
James A. McCann **1**

CHAPTER 2. Colonization Theory Relative to Introduced Populations
Jay R. Stauffer, Jr. **8**

CHAPTER 3. International Transfers of Inland Fish Species
R. L. Welcomme **22**

CHAPTER 4. Distribution of Exotic Fishes in the Continental United States
Walter R. Courtenay, Jr., Dannie A. Hensley, Jeffrey N. Taylor, and James A. McCann **41**

CHAPTER 5. Introduction of Exotic Fishes into Canada
E. J. Crossman **78**

CHAPTER 6. Distribution and Known Impacts of Exotic Fishes in Mexico
Salvador Contreras-B. and Marco A. Escalante-C. **102**

CHAPTER 7. Exotic Fishes in Hawaii and Other Islands of Oceania
J. A. Maciolek **131**

CHAPTER 8. Exotic Fishes in Puerto Rico
Donald S. Erdman **162**

CHAPTER 9. Introductions of Exotic Fishes in Australia
Roland J. McKay **177**

CHAPTER 10. Exotic Fishes: The New Zealand Experience
R. M. McDowall **200**

CHAPTER 11. Bacteria, Parasites, and Viruses of Aquarium Fish and Their Shipping Waters
Emmett B. Shotts, Jr., and John B. Gratzek **215**

CHAPTER 12. Some Parasites of Exotic Fishes
Glenn L. Hoffman and Gottfried Schubert **233**

CHAPTER 13. Exotic Fishes in Warmwater Aquaculture
William L. Shelton and R. Oneal Smitherman **262**

CHAPTER 14. Control of Aquatic Weeds with Exotic Fishes
Jerome V. Shireman **302**

CHAPTER 15. Exotic Fishes and Sport Fishing
G. C. Radonski, N. S. Prosser, R. G. Martin, and R. H. Stroud **313**

CHAPTER 16. Known Impacts of Exotic Fishes in the Continental United States
Jeffrey N. Taylor, Walter R. Courtenay, Jr., and James A. McCann **322**

CHAPTER 17. Toward the Development of an Environmental Ethic for Exotic Fishes
Charles H. Hocutt **374**

CHAPTER 18. A Suggested Protocol for Evaluating Proposed Exotic Fish Introductions in the United States
Christopher C. Kohler and Jon G. Stanley **387**

Summary **407**
Contributors **409**
Index of Names **411**
Index of Fishes **423**

Foreword

Many books and articles deal with some aspect of the exotic animal and exotic plant question. Some describe a variety of events, or "horror stories," used by the author to promote an environmental message or ethic. Others, usually dealing with crops or insects brought in to control specific targets, tell a parallel series of anecdotes that make cheerier reading as they report successes in controlling aquatic weeds or in establishing new fisheries. In such instances, but especially in the first, the reader finds only the briefest documentation of the events chronicled and seldom any message about examples that depart from, much less contradict, the author's theme.

The introduction or transplantation of animals and plants is a powerful tool for wildlife and crop managers. That it must be employed cautiously, after careful research, becomes obvious from study of the chapters that follow.

Man has altered physically and chemically much of the freshwater habitat, creating ponds and lakes where none existed, creating warm havens in otherwise seasonally cold regimes, and completely changing the flow pattern, bottom type, and water clarity of creeks and rivers. Nonnative fishes offer the manager attractive opportunities for providing new food and game species, for controlling pests, and for solving quickly or reversing, so it would seem, embarrassing problems that past errors and environmental mismanagement helped create.

As will be clearly noted in this book, the seductiveness of a quick solution, the need to ease political pressure, and the seeming lack of fixed culpability when something goes amiss have led to the abuse of the process, often with results that are environmentally disastrous and irretrievably so.

The summary duly notes that man can shut off chemical abuse and undo physical damage to our freshwater environment, but, except in small isolated ponds, an aquatic organism, once established, is indeed a permanent resident, for better or worse. Decades of reclamation efforts on rivers and lakes to remove "trash" species and improve the lot of desirable species show that the improvements are at best temporary, and they become increasingly difficult to justify on a cost-effective basis. Such experiments point to the folly of believing that one can poison and remove undesirable nonnative organisms in a river or lake. Efforts to remove or eradicate aquatic animals quickly become control programs that usually become prohibitively expensive.

The information assembled here is, in part, found only in reports of scientific research in papers scattered among many journals, articles that mostly are of narrow focus. From a review of theories of colonization to up-to-date overviews of the current situation in fresh waters around the world, the authors have drawn upon those sources, augmented by their broader experience and unpublished information, and have pieced together a story that deserves the careful attention of environmentalists, wildlife managers, and those people who traffic in or rear nonnative aquatic organisms for profit or pleasure (aquaculturists and aquarists, for example).

This book is not simply a rewrite of past articles. It is broadly focused and the chapters are arranged so that the entire story unfolds. There are accounts of successes and failures – and many events described are to be lamented. But the avenue that the reader is advised to travel is not barricaded; it is abundantly marked with slow, caution, and danger signs that must be heeded if further damage to the fresh waters is to be avoided.

C. Richard Robins
Rosenstiel School of Marine and Atmospheric Science
University of Miami

Preface

The importation of exotic fishes into North America began in the late 1600s with the ornamental goldfish. Since then, at least 41 exotic fishes have become established as reproducing populations. Conflicts over the purposeful introduction of exotic species relate mostly to the degree of assurance needed so that the species would not become established and cause problems, not to whether the fish has demonstrated beneficial characteristics.

Importation of exotics and introductions that may result in establishment are not synonymous. Yet the importation of exotic fishes by federal, state, and private groups for purposes of aquaculture, sport fishing, aquatic weed control, insect control, or aquarium use has led directly to the release or escape of many imported fishes into our open waters. Now, however, investigations by many groups are underway to test and predict the potential and long-range effects of introduced fishes.

Few exotic fishes now established in U.S. waters have been beneficial, and some have had severe negative impacts on native fishes and their habitats. More importantly, none were adequately tested to determine their beneficial or negative characteristics before they were introduced.

In recent years, concerns have repeatedly been expressed about the introduction of exotic species. These are real and honest concerns, because beneficial characteristics (desirability as a new sport, food, or biological control species) are usually given only cursory study while potential harmful characteristics are often ignored or given only token consideration. Unfortunately, adequate study of potential negative effects can be more expensive than a study of potential beneficial effects. Once exotic fishes become established and expand into new territories, it is usually impossible or impractical to eradicate them; they will become, for better or for worse, a permanent addition to the fauna. Any costs attributable to a species' introduction will mount each year even if it does not spread and increase in number. Problems arise when the people reaping the benefits from the introduction – and who rightly should be prepared to pay the costs – are not the ones who will eventually have to pay them.

Communication among the various groups interested in using exotic fishes has been poor. Misinformation and distrust have resulted. Exotic fishes can and do provide legitimate and useful benefits. The pleasures that exotic fishes have provided to aquarium fish hobbyists are beyond measure; furthermore, that same hobby and import trade has brought many undescribed species to the attention of ichthyologists. Exotic fishes can also provide service in aquaculture, weed and insect control, and sport fishing. Nevertheless, individuals or groups who carefully seek an exotic because of some special trait have an added responsibility. The very characteristics they seek may be ones that cause the greatest problems once the species is freed from its native checks and balances.

This symposium was conceived at the first meeting of the Exotic Fish Section of the American Fisheries Society in 1980. The goals of the section and symposium were, and continue to be, the maximal use of beneficial exotic fishes with adequate protection of native aquatic ecosystems, along with cooperation and coordination of all interested persons and agencies and exchange of information. If we have achieved most of those goals through section activities and this symposium, we are indeed on a road to responsible progress.

James A. McCann
National Fisheries Center – Leetown
U.S. Fish and Wildlife Service
Kearneysville, West Virginia

Acknowledgments

Each chapter received peer review before publication. For their efforts, we sincerely thank the reviewers: R. M. Adams, G. R. Bell, H. A. Bullis, Jr., B. Campbell, J. T. Carlton, W. D. Davies, K. L. Dickson, B. L. Duncan, L. G. Eldredge, P. L. Eschmeyer, D. Gill, R. J. Goldstein, H. F. Henderson, D. A. Hensley, F. Iñigo, D. E. McAllister, R. Mann, R. R. Miller, J. D. Parrish, J. S. Ramsey, W. A. Rogers, D. B. Rouse, T. W. Rowell, R. H. Schaefer, P. L. Shafland, M. C. Swift, and T. L. Wellborn, Jr.

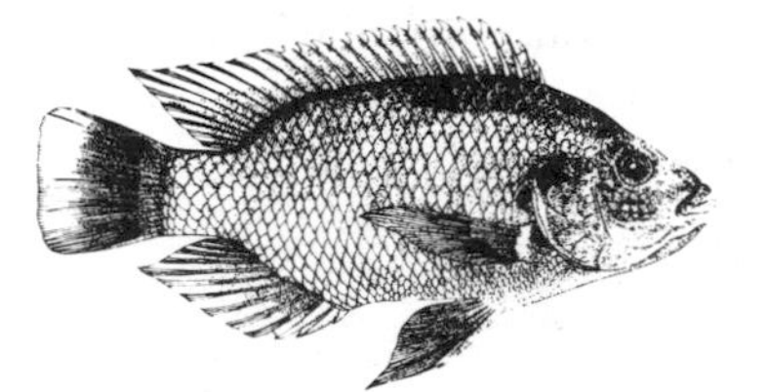

Introduction

In February 1969, the American Fisheries Society and the American Society of Ichthyologists and Herpetologists held an invitational Conference on Exotic Fishes and Related Problems in Washington, D.C. This conference sparked a great deal of interest in introductions of foreign fishes and their real and potential impacts on receiving ecosystems.

At the 1980 meeting of the American Fisheries Society (AFS), held in Louisville, Kentucky, James A. McCann, first president of the AFS Exotic Fish Section and, at that time, director of the National Fishery Research Laboratory (U.S. Fish and Wildlife Service, Gainesville, Florida), asked us to organize a symposium on exotic fishes for the 1981 AFS meeting to be held in Albuquerque, New Mexico. We recognized that the subject of exotic fishes is a complex one with many diverse aspects, and we had to decide what should be included in such a symposium. We also recognized that all aspects could not be treated.

We chose to concentrate on various aspects dealing with distribution, biology, and management of exotic fishes. We placed much emphasis on introductions using case histories from several geographic localities and, where known, some of the impacts of exotic fishes on native species or habitat. Aspects of microorganisms and parasites of exotic fishes, uses of exotic fishes, colonization theory, environmental ethics, and a proposed protocol for introductions were included to emphasize the complexity of this subject area.

It is our sincere hope that readers will recognize, through the contributions of the chapter authors, that many factors must be considered when one contemplates an introduction. An introduction is a faunal addition with potential to become a permanent member of the aquatic biomass. That concept of permanence demands careful study, evaluation, and caution in fisheries management with exotic fishes as well as transfers of native fishes beyond their historical ranges of distribution. If we, the symposium participants, alert fishery biologists and managers, as well as others working with exotic fishes, to these concerns, then we will have achieved our objectives.

Within the chapters we have chosen to refer to all mouthbrooding tilapias as *Tilapia,* thus following Robins et al. (1980). This was an editorial decision and does not reflect the views of some of the individual chapter authors.

Literature Cited

Robins, C. R., R. M. Bailey, C. E. Bond, J. R. Brooker, E. A. Lachner, R. N. Lee, and W. B. Scott. 1980. A list of common and scientific names of fishes from the United States and Canada. American Fisheries Society Special Publication 12. 174 pp.

Distribution, Biology, and Management of Exotic Fishes

CHAPTER 1

Involvement of the American Fisheries Society with Exotic Species, 1969–1982

James A. McCann

The beginning of the American Fisheries Society's organized involvement with, and concern about, the introduction of exotic species dates back only to February 1969, when the society cosponsored an invitational conference with the American Society of Ichthyologists and Herpetologists in Washington, D.C., entitled "Exotic Fishes and Related Problems" (Stroud 1969, King 1969, Anonymous 1970, Lachner et al. 1970). Richard H. Stroud, from the Sport Fishing Institute, opened that conference and outlined the group's two tasks: (1) "to bridge the many communication gaps among the various concerned interests that have generated unfortunate confusion and mutual suspicion, and (2) to develop possible guidelines looking toward rational management of troublesome conservation problems arising out of poorly controlled importation and release of exotic fishes and other aquatic organisms."

The fifty-seven people attending that meeting broadly represented the various concerned scientific, governmental, commercial, consumer, and conservation interests throughout North America. The participants reviewed the various interests and viewpoints relating to exotic species; the status and impact of exotic species already introduced into the nation's open water systems; the applicable laws and available statutes on importation and introduction of exotic fish; and the problems of the aquarium trade, fish importers, and fish farmers. The conferees, in an attempt to develop a conservation ethic concerning exotic species, recommended the establishment of a "Committee on Introduction of Exotic Fishes and Other Aquatic Organisms" to review all proposals for the introduction of new species of exotic fish into U.S. waters. Unfortunately, this committee was never officially developed, although the Exotic Fish committees of the American Fisheries Society have served in that capacity in the past and the Exotic Fish Section will provide that service in the future.

The conferees also attempted to standardize three terms frequently used in referring to exotic fish issues and problems: (1) *exotic*—any species introduced by man from a foreign land; (2) *transplanted*—native species moved by man into an ecosystem outside their native range but still within their country of origin; (3) *nonnative*—any species introduced by man into an ecosystem outside its original native range (includes exotic plus transplanted).

Since the 1969 conference, the American Fisheries Society has supported the activities of many successive Exotic Fish committees (table 1-1) to address some of the recommendations resulting from the conference. The activities of these committees have been many and varied. Some committees were productive but others suffered from internal conflict and outside interferences, and were hampered by nonattendance of members having difficulties in obtaining funds for travel to committee meetings.

In most years the committees found that they had too little time to initiate a work unit and produce a finished product before their one-year term of service had ended.

During this period of about twelve years the society has also supported six conferences or symposia dealing with exotic fishes:

1969—Conference on Use of Exotic Fishes and Related Problems. R. H. Stroud, Washington, D.C.
1971—Exotic and Transplanted Species—What Fishes, Where and When? J. E. Deacon, Salt Lake City, Utah
1973—Introduced Exotics in the Southeastern United States and Puerto Rico. W. R. Courtenay, Jr., Disney World, Florida
1976—Transplants and Introductions of Fishes in North America. W. R. Courtenay, Jr., Dearborn, Michigan
1978—Symposium on Culture of Exotic Fishes, Fish Culture Section. Atlanta, Georgia (see Smitherman et al. 1978)
1981—Distribution, Biology, and Management of Exotic Fishes. W. R. Courtenay, Jr., and J. R. Stauffer, Jr., Albuquerque, New Mexico

To address some of the shortcomings associated with one-year committee assignments, as well as other considerations, the Exotic Fish Section was formed in 1981. It is hoped that the establishment of this section will facilitate a long-term, coordinated, and integrated effort to be directed at major prob-

Table 1-1. EXOTIC FISH COMMITTEES OF THE AMERICAN FISHERIES SOCIETY

Years	Society president	Committee chairman	Name of committee
1968–1969	E. A. Seaman	J. E. Deacon	Exotic Fish Committee
1969–1970	C. J. D. Brown	J. E. Deacon[a]	Exotic Fish Committee
1970–1971	R. M. Jenkins	J. E. Deacon[b]	Exotic Fish Committee
1971–1972	E. L. Cooper	H. R. Axelrod	Exotic Fish Committee
1972–1973	C. J. Campbell	H. R. Axelrod	Exotic Fish Committee
1973–1974	R. E. Johnson	W. R. Courtenay, Jr.	Exotic Fish Committee
		R. B. Socolof	Ornamental Fish Committee
1974–1975	R. M. Bailey	W. R. Courtenay, Jr.	Exotic Fish Committee
1975–1976	J. C. Stevenson	Clark Hubbs	Exotic Fish Committee
1976–1977	R. F. Hutton	J. E. Thomerson	Exotic/Ornamental Fish Committee
1979–1980	R. H. Stroud	J. A. McCann	Exotic/Ornamental Fish Committee

[a] Activities reported by King (1971)
[b] Activities reported by Deacon (1972)

lems confronting the use of exotic species for beneficial purposes, while protecting the open water ecosystems of North America.

The use of grass carp for weed control and the interest of aquaculturists in other exotic species have increased concern about introduced exotic species.

Over the past thirty years a number of federal, state, and private organizations have been interested in and concerned with the use or control of exotic fishes. The U.S. Fish and Wildlife Service and state fish and game agencies have attempted to establish effective regulations to stop the importation and introduction of injurious species. The depth of these concerns is indicated by President Carter's 1977 Presidential Order No. 11987. With that order the president restricted all federal agencies from actively supporting projects that would increase the spread of exotic species, and encouraged the states, local governments, and private citizens to prevent the introduction of exotic species into the natural ecosystems of the United States. In addition to these actions, the recent passage of the National Aquaculture Act of 1980 in the United States, and the continuing multi-agency and private search for better aquaculture candidates, stimulated the U.S. Fish and Wildlife Service to establish the National Fishery Research Laboratory in Gainesville, Florida. The laboratory's primary mission is to oversee development of a research program that will support the use of exotic species for beneficial purposes while protecting the natural environment.

Although many obstacles have confronted the Exotic Fish committees, a number of significant contributions were completed. With the assistance of a contract from the National Fishery Research Laboratory—Gainesville, an updated national survey of the distribution, status, and impact of exotic fishes was completed by Courtenay et al. (chapter 4), and Taylor et al. (chapter 16). Major committee efforts were also expended to evaluate several exotic fish species being considered for introduction into open waters. A series of as yet unpublished evaluations or synopses were also prepared on a number of exotic species: Lake Ohrid trout (*Salmo letnica*), Nile perch (*Lates nilotica*), Amur pike (*Esox reicherti*), grass carp (*Ctenopharyngodon idella*), freshwater stingrays (Potamotrygonidae), and several species of snakeheads (Channidae). Three additional synopses on the milkfish (*Chanos chanos*), Mozambique tilapia (*Tilapia mossambica*), and the rohu (*Labeo rohita*) are nearing completion. These reports summarize the beneficial and potentially harmful characteristics of each species, and will enable professionals to develop a better understanding of selected exotic species before the fish are stocked into open water systems or into closed waters, from which the release or escape of fish into open waters is always possible.

In 1972 a survey of state regulations pertaining to the introduction of exotic species into U.S. open waters was compiled by one of these committees (Axelrod 1973). This analysis showed that some states prohibited the importation of species into the state and that all states had regulations prohibiting the unofficial release of nonnative species into their waters. An attempt to develop a "model" set of state regulations was begun but never completed. A list of exotic fishes considered undesirable for importation into North America was compiled and published (King 1971). Information for that list was taken largely from lists of fishes prohibited by various states.

The 1972 Exotic Fish Committee prepared a position paper for the American Fisheries Society's annual meeting in Hot Springs, Arkansas. This document was approved as a society policy at that meeting (American Fisheries Society 1973). It recommended that no exotic species be released into open waters unless a seven-step evaluation procedure (protocol) was followed: (1) develop a rationale for introduction; (2) search and evaluate all possible candidates, including native fishes; (3) conduct a preliminary assessment of the impacts on native ecosystems; (4) publicize the intent to introduce the species and solicit review; (5) conduct experimental research to test and evaluate the beneficial and potentially harmful characteristics; (6) encourage a full evaluation and review of all information available by all interested scientists; and (7) if all conditions are favorable, introduce the species under a closely monitored study, and widely circulate the results of the study. All new introductions should be conducted initially in areas where the study can be terminated and the species recovered or destroyed if potentially harmful effects are discovered. To facilitate these procedures several committees initiated (but never completed) a standard protocol to evaluate species for potential release into open waters. A preliminary report is presented later in this symposium (chapter 18). A detailed protocol is now being developed by the Exotic Fish Section with the assistance of the National Fishery Research Laboratory—Gainesville.

Several committees attempted to develop a mechanism that would facilitate the exchange of information between the society and the tropical fish industry. However, efforts to date have met with little success. It is hoped that the Exotic Fish Section can discover ways to include effectively contributions from that industry in future decisions by the society and the fishery profession. The committees in the past have (1) responded to numerous requests from federal, state, provincial, and private groups for assistance on exotic fish problems and issues; (2) prepared statements on proposed regulations about injurious fish introductions; and (3) commented on a Federal Disease Bill, a National Test Fish Production Center proposal, and several national aquaculture bills.

Other committees and sections of the society have worked on exotic fish problems and issues. A recent contract by the National Marine Fisheries Service and the U.S. Fish and Wildlife Service to the society's Committee on Names of Fishes assisted in the development of a "List of Exotic Fishes of Economic Importance to U.S. Interests." This list, to be published by the society, identified over 2,000 exotic fish species that are either now being imported into the United States or are likely to be in the near future. A review of this list indicates that it has indeed been fortunate that to date only 40 of 101 exotic species taken from open waters of the United States have become established, and that only a few of these have caused serious problems. Nevertheless, a number of species in the list of 2,000 will require careful evaluation before they are placed in areas where their release or escape into open waters might be possible. The function of the National Fishery Research Laboratory – Gainesville will be to help provide this needed research effort.

The Fish Culture Section of the American Fisheries Society has recently developed a "Tilapia Newsletter" which will aid in information and technology transfer among aquaculturists specializing in the raising of tilapias. In 1978 the section sponsored a "Symposium on the Culture of Exotic Fishes" (Smitherman et al. 1978) and in 1981 it conducted a symposium at the annual meeting of the American Fisheries Society at which up-to-date information was presented on the use of grass carp and hybrid bighead carp x grass carp for the control of vegetation.

It is hoped that the Exotic Fish Section and the other sections of the society, with the support of federal, state, and private groups, can develop a worldwide program that will promote the use of beneficial exotic fish and prevent releases of undesirable species.

The Exotic Fish Section can successfully serve as an information exchange center and as a sounding board and review body for proposals concerning introduction of exotic fish where the escape or release of the fish into open waters is likely. However, it will be up to federal, state, and private groups to conduct the necessary research and assist in the compilation and review of existing data if the beneficial and harmful characteristics of candidate species are to be evaluated.

The efforts to control the introduction of potentially harmful exotic species into the open waters of the continent must be supported by all interested parties if exotic species are to be successfully managed.

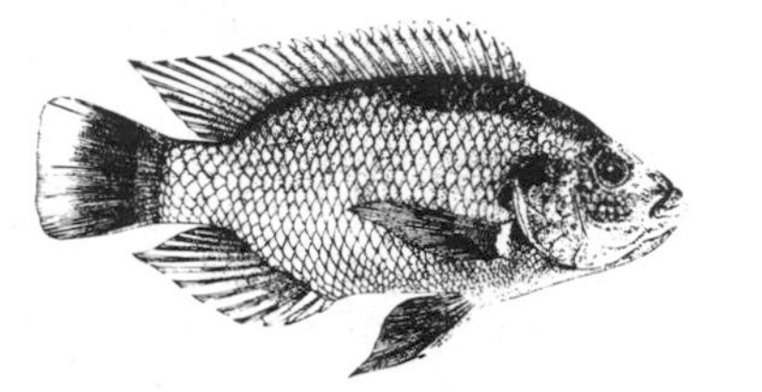

Literature Cited

American Fisheries Society. 1971. Resolution on the introduction of foreign fishes. Transactions of the American Fisheries Society 100(1):202. (Resolution No. 4, passed at 101st Annual Meeting, Salt Lake City, Utah, 15–18 September 1971.)

American Fisheries Society. 1973. Position of American Fisheries Society on Introduction of Exotic Aquatic Species. Transactions of the American Fisheries Society 102(1):268–269.

Anonymous. 1970. Report of the conference on use of exotic fishes and related problems. Transactions of the American Fisheries Society 99(1):291–295.

Axelrod, H. R. 1973. Report of Exotic Fishes Committee. Transactions of the American Fisheries Society 102(1):239–248.

Bailey, R. M. 1972. Report of Committee on Name of Fishes. Transactions of the American Fisheries Society 101(1):177–178.

Courtenay, W. R., Jr. 1979. Biological impacts of introduced species and management policy in Florida, 237–257. *In:* R. Mann (ed.). Exotic species in mariculture. MIT Press, Cambridge, Massachusetts.

Deacon, J. E. 1972. Report of Exotic Fish Committee. Transactions of the American Fisheries Society 101(1):186–187.

King, W. 1969. Bureau of Sport Fisheries and Wildlife, Department of the Interior, report on exotic fishes conference. American Fisheries Society Newsletter 13(59).

King, W. 1971. Report of Committee on Exotic Fishes. Transactions of the American Fisheries Society 100(1):183–185.

Lachner, E. A., C. R. Robins, and W. R. Courtenay, Jr. 1970. Exotic fishes and other aquatic organisms introduced into North America. Smithsonian Contributions in Zoology 59:1–29.

Martin, R. G. 1977. Exotic Fish in Southeast. Sport Fishing Institute Bulletin 288:6–7.

Smitherman, R. O., W. L. Shelton, and J. H. Grover (eds.). 1978. Culture of exotic fishes symposium proceedings. Fish Culture Section, American Fisheries Society, Auburn, Alabama. 257 pp. + 3.

Stroud, R. H. 1969. Conference on exotic fishes and related problems. Sport Fishing Institute Bulletin 203:1–4.

CHAPTER 2

Colonization Theory Relative to Introduced Populations

Jay R. Stauffer, Jr.

The introduction and subsequent establishment of nonnative fishes have concerned ichthyologists, fisheries managers, and sportsmen since before World War II. Attempts to introduce sport fish in certain areas have failed, while undesirable species have established reproducing populations despite efforts to prevent their success. Elsewhere in this symposium protocols for purposeful introductions and environmental ethics concerning exotic fishes are discussed. The purpose of this chapter is to review the constraints and the advantages that an introduced species faces relative to the establishment of a self-sustaining population.

While this symposium surveys the specific problem of exotic species (i.e., those species not native to the continent), the theory relative to the establishment of a new species in a particular habitat applies, regardless of the origin of that species.

Demographic Strategy

Theoretically, colonization by fishes can result from the dispersal or introduction of one breeding pair or in the case of livebearers, one pregnant individual. In reality, however, this is rarely the case. Yet single events can result in the establishment of populations. The introduction of the walking catfish in southern Florida can be traced to a single accidental release of several fish

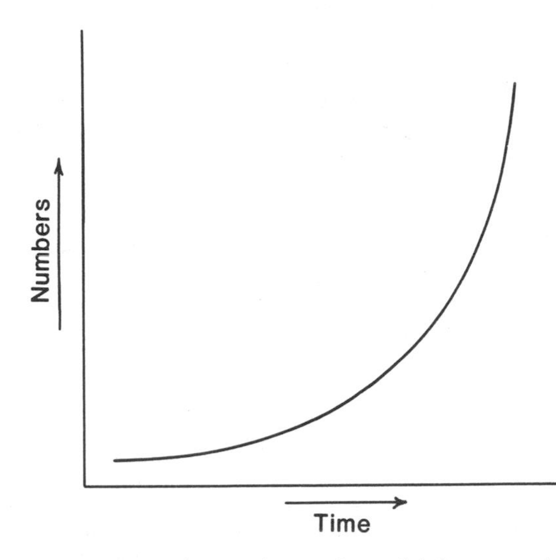

Figure 2–1. J-shaped growth model for natural populations, where N is equal to population size.

(W. R. Courtenay, Jr., personal communication). The establishment of other species is related to multiple releases. With certain common tropical fish (mollies, swordtails, etc.) there appears to be a constant trickle of fishes to the open waters (W. R. Courtenay, Jr., personal communication). Certainly multiple releases almost guarantee the successful introduction of a species. Finally, as documented by Courtenay et al. (1980), several exotics have been released as part of a planned managerial strategy.

Irrespective of how the initial population originated, growth in numbers is generated by the intrinsic rate of natural increase (r_{max}) and any environmental resistance; thus, the actual growth rate of the population is r_{max} minus environmental resistance. Growth curves of most populations take one of two general forms: J-shaped or S-shaped (figures 2–1 and 2–2). For those populations that exhibit a J-shaped curve, numbers increase rapidly and then stop as environmental resistance becomes effective; in the sigmoid form, environmental resistance acts on the population more gradually (Odum 1971) and the upper limit of the population is the carrying capacity (K). Population size (N) changes with time (t) based on either of the following:

$$\frac{dN}{dt} = rN \text{ or } \frac{dN}{dt} = rN \left(\frac{K-N}{K}\right)$$ (Whittaker and Goodman 1979).

Because of density dependent effects,* the probability that a particular

* Density dependent events are those parameters which influence population size as a function of the number of individuals per habitable space. For example, as density increases organisms are more vulnerable to disease, parasitism, predation, etc.

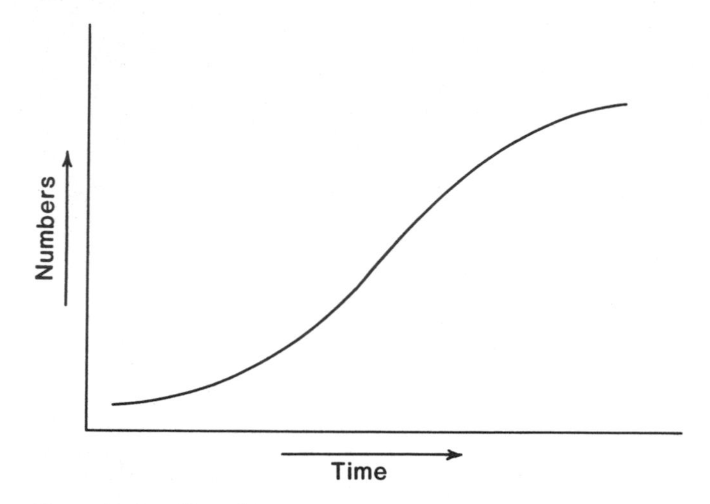

Figure 2–2. Sigmoid growth curve for natural populations, where N is equal to population size.

individual will survive or reproduce decreases as population density approaches the carrying capacity. Mechanisms that individuals utilize to minimize the subsequent debilitating effects of high density include (1) escape through dispersal, (2) escape through time (i.e., diapause), (3) tolerance for high densities, and (4) competitive superiority through interference mechanisms (Gill 1978).

Within the framework of the above population growth models, species can be said to employ two strategies. Those forms that maximize their reproductive potential are termed r-strategists, while those that maximize trophic level efficiency and competitive ability are K-strategists.

The r-K continuum has two extremes. At the r-endpoint it is assumed that an "ecological vacuum" exists where there are no density dependent effects so that there is essentially no competition (Pianka 1970). Conversely at the K-endpoint the environment is saturated with species, and density dependent effects operate at an extremely high level (Pianka 1970). It would appear that a species cannot excel at everything at once; therefore, Whittaker and Goodman (1979) suggested that the development of r-strategies occurs at the expense of K-strategies and vice versa. In reality, most species employ a strategy between the two extremes.

Successful demographic strategies in the tropics may be different from those in temperate areas (Pianka 1970, Dobzhansky 1950). Population fluctuations within the temperate zones appear to be related to phenomena that are independent of the genotype, while the relatively stable conditions which exist in the tropics appear to favor those species with highly developed competitive ability (Pianka 1970). This hypothesis, however, is currently being debated (D. E. Gill, personal communication). MacArthur (1962) suggested that, under constant environmental conditions, those genotypes that maximize a K-strategy will predominate over those that tend to maximize an r-strategy. This trend was further supported by the field observations of MacArthur and Wilson (1967), who noted that r-selected organisms are dominant in highly variable environments while K-selected organisms dominate stable environments. Thus, an r-strategy results from selection pressure caused by environmental variation, while a K-strategy results from selection pressures emanating from density dependent factors and interspecific competition (MacArthur 1972).

Since organisms tend toward unlimited increase (Crombie 1947), I will review those parameters that decrease population numbers. Irrespective of the strategy employed, populations are either resource limited or limited to a level below that set by their resources (Hairston et al. 1960). With respect to r-strategists, if they are located near the extreme of the r-K continuum, the resultant oscillations may increase the probability of extinction because of random fluctuations (Levins 1970). However, one cannot assume that those organisms that increase rapidly under optimum conditions will decrease rapidly under adverse ones (Whittaker and Goodman 1979). In other words, population increase and population decrease may not be controlled by the same factors (Whittaker and Goodman 1979). This phenomenon is especially true with r-strategists, because mortality is mostly governed by extrinsic environmental occurrences (i.e., density independent effects) rather than intrinsic biological parameters. Karlin (1966) provided a formula that estimates the probability of extinction for populations that exhibit geometric growth and have random birth and death rates (i.e., r-strategists):

$$\mu_i = \left(\frac{\mu}{\lambda}\right)^i \text{ if } \lambda > \mu,$$

where λ = probability of birth within a unit of time; μ = probability of death; μ_i = probability of extinction given an initial state of i individuals; and where $\lambda + \mu = 1$.

Conversely, K-strategists may adapt to adverse environmental conditions through physiological and/or behavioral changes (see section on Evolutionary Adaptations). As indicated by Whittaker and Goodman (1979), however, their adaptations do not necessarily provide a floor below which the population cannot fluctuate. They have suggested that the patchiness of the environment may serve to preserve some type of lower limit.

Interrelated with the intrinsic rate of natural increase and population strategies are the survivorship curves of a particular population. Deevey (1947) stated that there are basically three types of survivorship curves exhibited by populations (figure 2–3). In the negatively skewed curve (a), the probability of survival is good until a certain age is reached at which point mortality is high. Human populations in industrialized nations probably come closest to exhibiting this type of survivorship curve. The second curve (b) is

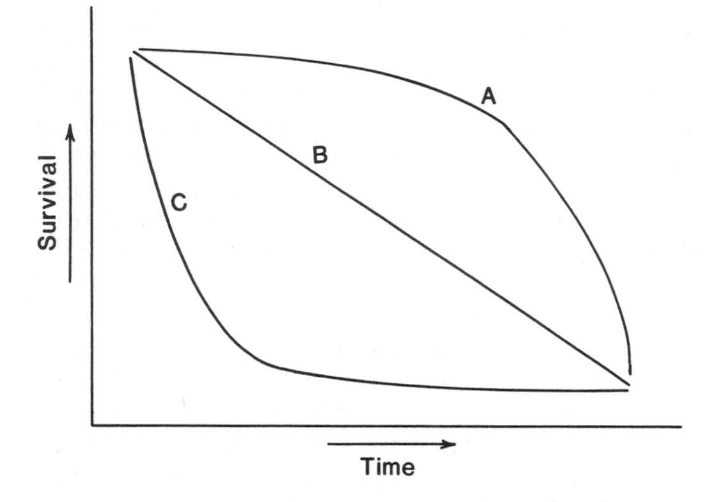

Figure 2–3. Survivorship curves, where S is equal to survival (Deevey 1947).

exhibited by certain microorganisms. Essentially, the probability of survival is constant throughout all life stages. Populations that have a survivorship curve that is positively skewed (c) exhibit extremely high mortality during the beginning and have a low expectation of death once a certain age is reached. A good example of this situation is the common carp (*Cyprinus carpio*).

Evolutionary Adaptations

Introduced populations are subjected to various ecological constraints and their ability to survive depends on adaptations to a specific environment and plasticity should that environment change.

By definition, an introduced organism has had no "experience" relative to its new habitat. If the new habitat is similar to its native one, the organism may survive and establish a self-sustaining population. If, on the other hand, the habitat is different from the native one, the ability to survive depends on whether the animal is preadapted or if the population is carrying remnant genes from some ancient ancestors that experienced a similar habitat (D. E. Gill, personal communication). Figure 2–4 depicts preadaptation as used in this context. The inner circle represents the situations which the population or species has experienced in its evolutionary history and to which it has adapted. The outer circle represents those situations to which the organism has never been exposed, but within which it can survive, thus to which it is preadapted. For example, water temperature of temperate streams did not exceed 30 C in pre-Columbian North America. Moreover, most temperate warmwater fishes can survive temperatures as high as 33–36 C and, in fact,

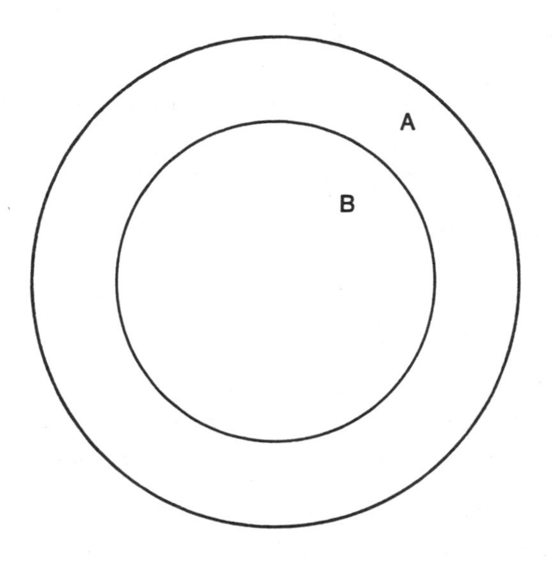

Figure 2–4. Venn diagram depicting preadaptation. Area A represents condition to which population is preadapted, while area B represents evolutionary experience of a population.

several species prefer temperatures greater than 30 C. One could argue that an evolutionary strategy that caused a poikilotherm always to prefer a warmer temperature than the one to which it is presently exposed could account for this phenomenon. These same species, however, will avoid water temperatures that are lethal (Stauffer et al. 1976) thus demonstrating a form of preadaptation to increased water temperatures caused by clearing of land and thermal outfalls, or the retention of traits that were of selective value from some ancestral form.

Organisms that are K-strategists adapt to changing environmental conditions through increased trophic efficiency, enzyme changes, and physiological adaptation. For example, the upper lethal temperature for fish changes with their acclimation temperature. Hochachka and Somero (1973) showed that there is an "on-off" production of different enzymes which occur seasonally in rainbow trout (*Salmo gairdneri*) relative to environmental temperatures, thus demonstrating a K-strategy with respect to temperature changes.

Conversely, r-strategists adapt to changing environmental conditions by exposing a multitude of genotypes to the situation, thereby allowing natural selection to take its course. With respect to introduced populations, an r-strategist may be adversely influenced by the founder effect. According to Koslov (1979; 2), "in the genomes of all the cells of organisms of a given species all the genes coding for the essential characters of this species are conserved." The founder effect states that because the introduced population often represents a small segment of the original population, its gene pool and, therefore, the population's inherent ability to adapt, may be substantially decreased. Moreover, the probability of the founder population to be preadapted is also decreased because genetic diversity is decreased. For example, Allendorf and Phelps (1980) reported a significant loss of genetic variation in hatchery stock of cutthroat trout (*Salmo clarki*) when compared to the wild stock from which it was derived some fourteen years earlier. They attributed this loss to the founder effect and genetic drift.

Ecological Constraints of Dispersing Fishes

Once a population of introduced organisms has become established, it has the opportunity to disperse and invade new areas. Udvardy (1969) observed that individuals are responsible for dispersal and that, if a new area is invaded, it is achieved in most cases by the multiple interactions between individuals and the environment. For the most part, introduced fishes are subjected to the same constraints as indigenous forms with respect to avenues of dispersal. One obvious exception is the walking catfish (*Clarias batrachus*) which can travel via terrestrial routes. Dispersal is a dynamic event and the conditions that govern it change with time; that is, the effectiveness of all barriers is limited in space and temporary in time (Udvardy 1969).

The environment offers resistance to dispersal through extrinsic barriers. Extrinsic dispersal barriers can be classified as physical barriers, ecological barriers, or time and distance barriers (Udvardy 1969). With respect to most fishes, any land mass is a physical barrier, as are dams, dikes, and other man-made structures. Ecological barriers would include any area in which an organism cannot live indefinitely. Salt water may be an effective ecological barrier for many freshwater fishes. Other examples are areas with low dissolved oxygen and/or temperature extremes. Any water body that has physicochemical conditions outside the zone of tolerance for a particular organism (Fry 1947) is an ecological barrier. Time and distance become barriers if a species has not invaded a particular area that is separated by its present location from habitable grounds.

Intrinsic constraints on dispersal include structural, physiological, behavioral, and life history attributes. While structural or anatomical characteristics may be important, Udvardy (1969) noted that the speed or ease of locomotion is not directly correlated with vagility (i.e., the ability to disperse). Vagility is governed by the interrelationships with extrinsic barriers and the intrinsic attributes of the dispersing organisms. For example, those organisms that can withstand exposure to saline water certainly can cross saltwater barriers successfully. On the other hand, many organisms do not invade areas that are apparently separated by habitable areas. For example, the orangethroat darter (*Etheostoma spectabile*) occurs only in the western tributaries of the Ohio River, while the sharpnose darter (*Percina oxyrhyn-*

cha) inhabits only the eastern tributaries. Such distributional anomalies may be the result of a time and distance barrier or may relate to interactions between behavioral characteristics of the species and the environment. Finally, particular stages of the life cycle may have greater vagility than others.

Colonization

Irrespective of the demographic strategy employed, or ability to adapt to various environmental conditions, the success of a potentially introduced population is dependent to some extent on the biological components that presently inhabit the area in question. Certain environments are saturated with species while other areas are considered to be depauperate. Hocutt and Hambrick (1973) hypothesized that the Roanoke darter (*Percina roanoka*) was a recent introduction to the New River in Virginia, which exhibited a population explosion and subsequent rapid dispersal throughout portions of the drainage. Part of the success of *P. roanoka* can be directly attributed to the scarcity of indigenous species to the New River. Conversely, certain species do not occur in areas that are potentially habitable in terms of physical attributes, even though they have reached such areas in suitable numbers for successful colonization to occur (DeBach 1966, Colwell and Fuentes 1975). Such incidents are probably related to niche space and species packing.

For the remainder of this discussion a niche will be defined as a multidimensional hypervolume that depicts the conditions under which the species can exist (Hutchinson 1957).* It will be assumed, based on the Volterra-Gause principle, that no two species can occupy the same niche. This principle assumes that (1) for two species to coexist they must segregate themselves along at least one dimension of the niche and (2) there is an eventual incompressibility of any one dimension (Schoener 1974). In other words, only a set number of species can separate along one dimension and once this limit has been reached a new dimension must be added to the environment if more species are to be successful.

* According to Fry (1947) all organisms have a zone of tolerance relative to environmental parameters within which the organism can live indefinitely. The Mozambique tilapia's (*Tilapia mossambica*) zone of tolerance to temperature ranges between 13 and 37 C. Outside of this zone of tolerance is the zone of resistance. Survival within this zone of resistance is a function of both time and the intensity of the stress. If one could plot the upper and lower limits of the zones of tolerance relative to all chemical, physical, and biological parameters in multidimensional space, a graphical representation of the organism's niche would result. These environmental parameters are referred to as dimensions. If two organisms differ with respect to one dimension, then their niches are theoretically different. It should be noted that as an organism approaches the limit for one dimension, the limits may change for others.

MacArthur and Levins (1967) expanded on this concept and stated that there is a limit to the similarity among species that can coexist and that the total number of species is proportional to the variety of the environment divided by the mean niche breadth of the coexisting species. They further stated that the number of coexisting species may be decreased by unequal quantity of resources but increased by adding important niche dimensions. In this respect the seasonality and number of resources limit the number of coexisting species (MacArthur 1969).

Relative to the population attributes discussed earlier, the diversity of coexisting species may be partially set by the lower limit of abundance of each species (Schoener 1965, in MacArthur and Levins 1967). On the other hand, the upper limit of abundance, which is set by predators and parasites, may increase the number of species that can coexist. For example, let us assume that two species, A and B, are occupying similar niches in the environment and that species A is slowly displacing species B. If a third species, C, is introduced into the environment and preys exclusively on species A, then it would be possible for species A and B to coexist. In effect, a new dimension (i.e., predation) has been added along which species A and B can segregate.

At one time it was assumed that the predominant factor that controls the organization of natural communities was interspecific competition (Dunham et al. 1979). This assumption was inferred from niche shifts and morphological changes in the form of character displacement which are correlated with geographic variation (Dunham et al. 1979). Simberloff and Wilson (1969), however, indicated that competition may not be so important as previously thought, at least for insular areas. Pianka (1981), based on work by several authors, stated that the outcome of interspecific competition depends upon (1) the initial population densities (Neyman et al. 1956), (2) environmental conditions (Park 1954), and (3) the genetic constitution of competing populations (Park et al. 1965). Nevertheless, initial numbers may not be important in terms of which species "wins" (DeBach 1966). The above discussion assumes that competition is effected by the improvement of a species' own intrinsic capabilities through either an r or K strategy. However, species can also successfully compete by developing means to impair its competitors (i.e., α-selection; Gill 1974). Irrespective of the above theoretical discussion, the outcome of two ecological homologs was probably best described by DeBach (1966:189) when he stated: "If two ecological homologues simultaneously happen to invade a new area by chance — an unlikely possibility — or by purposeful manipulation by man — which is more probable — then the winner will be the species which produces the most female progeny which survive to reproduce per parental female per unit time."

Despite the fact that most authors and mathematical treatments predict that ecological homologs cannot coexist, there are instances when homologs do appear to do so. DeBach (1966) offers the following explanations for cases in which there is apparent coexistence: (1) both of the species occur at such low densities that they do not compete; (2) the two species have different controlling factors and, therefore, are not ecological equivalents; (3) the advantage of one species is continually reversed by environmental fluctuations, or (4) the superior utilization of a resource by one species is offset by the superior ability of the other to discover and exploit alternate resources. Horn and MacArthur (1972) hypothesized that coexistence of similar species can also occur in patchy environments if local extinction, dispersal, and colonization occur.

DeBach (1966) further states that two species do not have to be homologs for one to displace the other. This phenomenon can occur if the broad niche of one species completely overlaps the narrow niche of the other, or if one in utilizing a particular resource coincidently destroys the resource of another. For example, if Dutch elm disease kills all of the American elm trees, then all insects that are specific to American elms would also be destroyed.

Applicability to Exotic Fishes

Courtenay and Hensley (1980) listed 15 exotic fishes that have expanding ranges (table 2–1). A quick survey of the demographic characteristics of these species indicated that 13 are traditional "K-strategists" which inhabit tropical or subtropical waters. The 2 r-strategists, goldfish (*Carassius auratus*) and grass carp (*Ctenopharyngodon idella*), inhabit temperate waters. Moreover, they have established populations in states that have few other exotic species. The same facts apply to the only 2 species which Courtenay and Hensley (1980) list as having broad and stable distributions, brown trout (*Salmo trutta*) and common carp (i.e., they are r-strategists that inhabit temperate waters).

The tropical fish industry has been active in this country since about 1930. Many of the aquarium fishes that were popular before 1960 included cyprinids and characins. Certainly these species had ample opportunity to escape to the natural waters of this continent. Nevertheless, it was not until the cichlids started to increase in popularity that exotic aquarium fish started to colonize successfully.

The success of many exotic fishes that presently have expanding populations must be related to their physiological attributes and demographic strategies. Moreover, they must be able to acclimate to a wide range of environmental conditions. Some of them have unique attributes that permit them to disperse via routes not available to indigenous fishes (i.e., walking catfish). Furthermore, personal observations from laboratory experiments have shown that several *Tilapia* species can live in water ranging from 0–30 o/oo salinity. Certainly these species are not restricted by the same environmental constraints that govern many native forms. In this sense, they are preadapted for dispersal via seawater.

Table 2–1. EXOTIC FISHES WITH EXPANDING RANGES

Species	Common name
Acanthogobius flavimanus (Temminck and Schlegel)	yellowfin goby
Astronotus ocellatus (Cuvier)	oscar
Carassius auratus (Linnaeus)	goldfish
Cichlasoma bimaculatum (Linnaeus)	black acara
Cichlasoma meeki (Brind)	firemouth cichlid
Cichlasoma nigrofasciatum (Günther)	convict cichlid
Cichlasoma octofasciatum (Regan)	Jack Dempsey
Clarias batrachus (Linnaeus)	walking catfish
Ctenopharyngodon idella (Valenciennes)	grass carp
Hemichromis bimaculatus (Gill)	jewelfish
Poecilia "sphenops complex" (*P. mexicana* Steindachner component)	shortfin molly
Tilapia aurea (Steindachner)	blue tilapia
Tilapia mariae (Boulenger)	spotted tilapia
Tilapia mossambica (Peters)	Mozambique tilapia
Tilapia zilli (Gervais)	redbelly tilapia

Source: Courtenay and Hensley 1980.

Thirteen of the species listed in table 2–1 exhibit some form of parental care. The successful establishment of these species when compared to other potential colonizers suggests that some form of postnatal care is advantageous to establish a reproducing population in the southern portion of North America. It can be assumed that the energy channeled into reproduction is divided between courtship and parental investment (Trivers 1972). Parental investment can be further divided between the development and fertilization of eggs, and postnatal care. The evolution of postnatal care is related to (1) the difference in ability between the sexes to provide parental investment after the eggs are laid and (2) the availability of mates (Perrone and Zaret 1979). In most animals, the parental investment of the females exceeds that of the males (Emlen and Oring 1977). Emlen and Oring (1977) further state that male incubation lessens the metabolic load of the female. If it is assumed that more energy is needed to produce ova than sperm, then male incubation (i.e., the blackchin tilapia, *Tilapia melanotheron*) would appear to have a selective advantage, especially if multiple broods occur and the sex ratios are nearly equal. If, on the other hand, the sex ratios are skewed, then it might be more advantageous for a male to fertilize several females (i.e., *Tilapia mossambica*) that subsequently brood the eggs. In any case, postnatal parental care appears to be a distinct advantage to potential colonizers of southern North America.

The success of the majority of the species listed in table 2–1 may be related to their ability to maximize both r and K strategies at the same time. Most of these species evolved in species-rich tropical areas where the development of a K-strategy was essential to their survival. When compared to the indigenous

North American fauna, however, these same species employ an r-strategy. For example, the blackchin tilapia and blue tilapia (*Tilapia aurea*) spawn throughout the year, thus producing many more young than indigenous centrarchids with which they compete. Centrarchids spawn at most twice during the year (W. R. Courtenay, Jr., personal communication). The result is an organism that has a superior K-strategy (i.e., mouthbrooding versus a nest) and a superior r-strategy (i.e., more progeny produced). Couple these strategies with a wider tolerance to environmental conditions, and the outcome must be increased survival of the exotic and the subsequent decline of indigenous populations.

Conclusion

Exotic fishes pose two distinct dilemmas, depending upon certain value judgments. First, there is the failure of desirable introductions and, second, there is the successful colonization of undesirable ones. If a species successfully colonizes a particular area, then the new habitat must obviously contain conditions that are within the physiological tolerances of the organism and possess the essential requirements for its survival and propagation. Moreover, the organism must be able to coexist with existing biological components of its new environment. In general, it would appear that r-strategists are successful in temperate areas that are not saturated with competing species, while K-strategists are much more successful in subtropical and tropical areas.

The colonization of fishes recently associated with the tropical fish industry results from the type of fishes presently being imported and cultured. The cichlids appear to be preadapted to habitats found in southern North America. Moreover, they have a wide tolerance to many environmental parameters that allows them to inhabit and disperse through areas that are barriers to indigenous fishes. Finally, it is hypothesized that they can implement both r and K strategies to an extent that they are replacing native species. Current management strategies emphasize the eradication of unwanted species (i.e., pests) rather than the introduction of desired species. Conway (1981) lists five major techniques that are used to control pests: (1) pesticides, (2) biological controls (natural enemies), (3) cultural practices that change the habitat, (4) plant and animal resistance, and (5) sterile matings. The first three techniques probably are most applicable to the control of exotic fishes. In reality, however, very few fish species have been successfully eradicated once they have become established. Therefore, emphasis must be placed on preventing escape of these fishes into our natural waters.

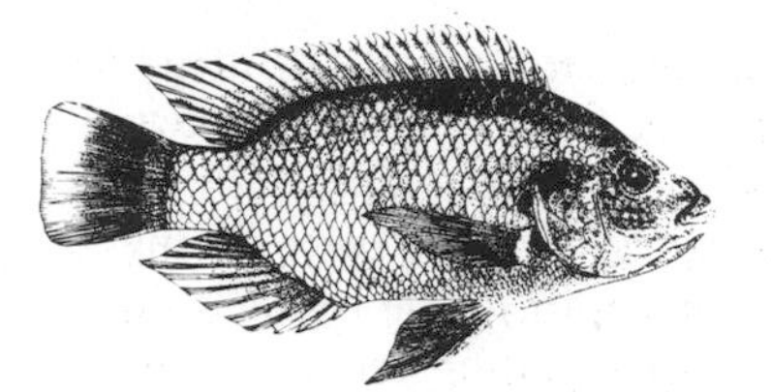

Literature Cited

Allendorf, F. W., and S. R. Phelps. 1980. Loss of genetic variation in a hatchery stock of cutthroat trout. Transactions of the American Fisheries Society 109:537–543.

Colwell, R. K., and E. R. Fuentes. 1975. Experimental studies of the niche. Annual Review of Ecology and Systematics 6:281–310.

Conway, G. 1981. Man versus pests, 356–386. *In:* R. M. May (ed.). Theoretical ecology, principles and applications. Sinauer Associates, Sunderland, Massachusetts.

Courtenay, W. R., Jr., and D. A. Hensley. 1980. Special problems associated with monitoring exotic species, 281–307. *In:* C. H. Hocutt and J. R. Stauffer, Jr. (eds.). Biological monitoring of fish. Lexington Books, Lexington, Massachusetts.

Crombie, A. C. 1947. Interspecific competition. Journal of Animal Ecology 16:44–73.

DeBach, P. 1966. The competitive displacement and co-existence principles. Annual Review of Entomology 11:183–212.

Deevey, E. S. 1947. Life tables for natural populations of animals. Quarterly Review of Biology 22:283–314.

Dobzhansky, T. 1950. Evolution in the tropics. American Scientist 38:209–221.

Dunham, A. E., G. R. Smith, and J. N. Taylor. 1979. Evidence for ecological character displacement in western American catostomid fishes. Evolution 33(3):877–896.

Emlen, S. T., and L. W. Oring. 1977. Ecology, sexual selection, and the evolution of mating systems. Science 197:215–223.

Fry, F.E.J. 1947. Effects of the environment on animal activity. University of Toronto Studies, Biological Series no. 55. Publications of Ontario Fisheries Research Laboratory 68:1–62.

Gill, D. E. 1974. Intrinsic rate of increase, saturation density, and competitive ability. II. The evolution of competitive ability. American Naturalist 108:103–113.

Gill, D. E. 1978. On selection at high population density. Ecology 59(6):1289–1291.

Hairston, N. G., F. E. Smith, and L. B. Slobodkin. 1960. Community structure, population control and competition. American Naturalist 94:421–425.

Hochachka, P. W., and G. N. Somero. 1973. Strategies of biochemical adaptation. W. B. Saunders Company, Philadelphia.

Hocutt, C. H., and P. S. Hambrick. 1973. Hybridization between the darters *Percina crassa roanoka* and *Percina oxyrhyncha* (Percidae, Etheostomatini), with comments on the distribution of *Percina crassa roanoka* in New River. American Midland Naturalist 90(2):397–405.

Horn, H. S., and R. H. MacArthur. 1972. Competition among fugitive species in a harlequin environment. Ecology 53:749–752.

Hutchinson, G. E. 1957. Concluding remarks. Cold Spring Harbor Symposium of Quantitative Biology 22:415–427.

Karlin, S. 1966. A first course in stochastic processes. Academic Press, New York.

Kozlov, A. P. 1979. Evolution of living organisms as a multilevel process. Journal of Theoretical Biology 81:1–17.

Levins, R. 1970. Extinction, 75–107. *In:* Some mathematical questions in biology. American Mathematical Society.

MacArthur, R. H. 1962. Some generalized theorems of natural selection. Proceedings of the National Academy of Sciences, U.S. 48:1893–1897.

MacArthur, R. H. 1969. Species packing and what competition minimizes. Proceedings of the Natural Academy of Sciences, U.S. 64:1369–1371.

MacArthur, R. H. 1972. Geographical ecology. Harper and Row, New York.

MacArthur, R. H., and R. Levins. 1967. The limiting similarity, convergence and divergence of co-existing species. American Naturalist 101:377–385.

MacArthur, R. H., and E. O. Wilson. 1967. The theory of island biogeography. Princeton University Press, Princeton.

Neyman, J., T. Park, and E. L. Scott. 1956. Struggle for existence. The *Tribolium* model: Biological and statistical aspects, 41–79. *In:* Proceedings of the 3d Berkeley Symposium on Mathematical Statistics and Probability. Vol. 4. University of California Press, Berkeley.

Odum, E. P. 1971. Fundamentals of ecology. W. B. Saunders Company, Philadelphia. 574 pp.

Park, T. 1954. Experimental studies of interspecies competition. II. Temperature, humidity, and competition in two species of *Tribolium.* Physiological Zoology 27:177–238.

Park, T., P. L. Leslie, and D. B. Metz. 1964. Genetic strains and competition in populations of *Tribolium.* Physiological Zoology 37:97–162.

Perrone, M., Jr., and T. M. Zaret. 1979. Parental care patterns of fishes. American Naturalist 113(3):351–361.

Pianka, E. 1970. On r and K selection. American Naturalist 104:592–597.

Pianka, E. R. 1981. Competition and niche theory, 167–196. *In:* R. M. May (ed.). Theoretical ecology. Sinauer Associates, Sunderland, Massachusetts.

Schoener, T. W. 1965. Evolution of bill size differences among sympatric congeneric species of birds. Evolution 19:189–213.

Schoener, T. W. 1974. Resource partitioning in ecological communities. Science 185: 27–39.

Simberloff, D. S., and E. O. Wilson. 1969. Experimental zoogeography of islands: The colonization of empty islands. Ecology 50:278–295.

Stauffer, J. R., Jr., K. L. Dickson, J. Cairns, Jr., and D. S. Cherry. 1976. The potential and realized influences of temperature on the distribution of fishes in the New River, Glen Lyn, Virginia. Wildlife Monographs 50:1–40.

Trivers, R. 1972. Parental investment and sexual selection, 136–179. *In:* B. Campell (ed.). Sexual selection and the descent of man. Aldine, Chicago.

Udvardy, M.D.F. 1969. Dynamic zoogeography. Van Nostrand Reinhold Company, New York.

Whittaker, R. H., and D. Goodman. 1979. Classifying species according to their demographic strategy. I. Population fluctuations and environmental heterogeneity. American Naturalist 113(2):185–200.

CHAPTER 3

International Transfers of Inland Fish Species

R. L. Welcomme

If Theinemann (1950) and Balon (1974) are correct, transfers of fish species in Europe through human activities may date as far back as Roman times when common carp (*Cyprinus carpio*) from the Danube were used for pond culture in Greece and Italy. Whatever the earliest history of the common carp, there is little doubt that it was diffused throughout much of Europe during the Middle Ages, when it was cultured widely in monastic or village ponds and inevitably escaped into adjacent rivers and lakes. This diffusion occurred sufficiently long ago for the species to have become incorporated into the native faunas and, as such, it would be difficult to unravel its impact. From about the middle of the nineteenth century until 1940, transfers of fish species, most ostensibly in the interests of sport, intensified but in many cases, one suspects, primarily for the nostalgia for familiar things associated with the displacement of peoples and with the colonial experience. During this period, many salmonid species were exchanged between Europe and North America, or were introduced widely around the world. Many of these transfers were highly inappropriate and did not withstand the rigors of their new habitats, although much of the considerable spread of rainbow trout (*Salmo gairdneri*) dates from this period. At the same time some odd introductions were attempted, often with complete disregard of the biology of the species concerned. Such was the attempted establishment of the tench (*Tinca tinca*) in the tropical waters of Africa and Asia, although, ironically, some populations of this species still exist in India, Indonesia, and southern Africa.

Since 1945 the numbers and scope of transfers of fish species have increased and, with the development of advanced techniques for artificial breeding, species that are incapable of reproducing in their new country have become common objects of introduction. Some of these transfers have had disastrous consequences and, as a result, an attitude of caution has grown up. Potential candidates for introduction must now be more rigorously examined for their desirability or for any environmental impacts. There is also a greater awareness of the hazards of acquisition of unwanted disease organisms along with the new transfer. Pressure continues, however, for additional transfers and the location of suitable candidate species for a wide range of purposes. In order to document these transfers more closely, the Food and Agriculture Organization (FAO) of the United Nations was requested by its member countries to prepare and maintain a register of international transfers. A first version of this has been published (Welcomme 1981), and many of the observations made in this chapter are based on personal communications received during its preparation. The register lists movements of over 160 species (table 3-1) to some 120 countries in over 800 entries. The register is to be updated periodically to reflect new transfers and additional species.

Purposes of Transfers

Aside from the often ill-defined motivations for transfer by private individuals, a number of major objectives have emerged as legitimate reasons for moving fish species about the world.

SPORT OR RECREATION

Many of the earlier transfers of fish species were to establish suitable sporting species in waters from which they were absent. The nineteenth-century diffusion of salmonids or the early-twentieth-century introductions of largemouth bass (*Micropterus salmoides*) were to this end. Sports fisheries were fairly conservative but with the expansion of fishing as a recreational pastime for the urban populations in the latter half of the twentieth century, the search for improved or novel species continues. The role of introduced species for sports fisheries is discussed in detail in chapter 15.

AQUACULTURE

The growth and spread of aquaculture throughout the world have been based on relatively few species which have been transferred to countries or continents from which they were previously absent. Thus, for instance, the success of European aquaculture in the sixties and seventies was founded primarily on rainbow trout, a species that first appeared in Europe in about 1880 (MacCrimmon 1971). Similarly, attempts to introduce aquaculture into the tropics have relied much on a relatively small number of *Tilapia* species. This trend has had the advantage that culture procedures based on known

Table 3–1. TAXONOMIC LIST OF SPECIES SUBJECT TO INTERNATIONAL TRANSFERS

Acipenseridae:	*Acipenser baeri*
	A. ruthensis
Anguillidae:	*Anguilla anguilla*
	A. australis
	A. japonicus
Clupeidae:	*Dorosoma petenense*
	Limnothrissa miodon
Osteoglossidae:	*Arapaima gigas*
	Heterotis niloticus
Osmeridae:	*Hypomesus transpacificus*
Salmonidae:	*Hucho hucho*
	Oncorhynchus gorbuscha
	O. keta
	O. kisutch
	O. masou
	O. nerka
	O. rhodurus
	O. tshawytscha
	Salmo clarki
	S. gairdneri
	S. marmorata
	S. salar
	S. sebago
	S. trutta
	Salvelinus alpinus
	S. fontinalis
	S. leucomaenius
	S. namaycush
Coregonidae:	*Coregonus spp.*
	C. laevaratus
	C. peled
Esocidae:	*Esox lucius*
	E. reicherti
Umbridae:	*Umbra krameri*
	U. pygmaea
Chanidae:	*Chanos chanos*
Erythrinidae:	*Hoplias malabaricus*
Cyprinidae:	*Alburnus alburnus*
	Aristichthys nobilis
	Barbus barbus
	B. conchonius
	B. holubi
	B. kimberleyensis
	B. natalensis
	B. schwanfelti
	B. tetrazona
	Blicca bjoerkna
	Carassius auratus
	C. carassius
	Catla catla
	Cirrhina molitorella
	C. mrigala

Table 3–1. TAXONOMIC LIST OF SPECIES SUBJECT TO INTERNATIONAL TRANSFERS *(continued)*

	Ctenopharyngodon idella
	Cyprinus carpio
	Hemibarbus maculatus
	Hypophthalmichthys molitrix
	Hemiculter eigenmanni
	H. leucisculus
	Labeo molitorella
	L. rohita
	Leuciscus idus
	L. leuciscus
	Ospariichthys uncirostris
	Pimephales promelas
	Mylopharyngodon aetiops
	M. piceus
	Pseudogobio rivularis
	Pseudorasbora parva
	Puntius gonionotus
	Rhodeus sericeus
	Rutilus rutilus
	Scardinius erythrophthalmus
	Tinca tinca
Cobitidae:	*Misgurnus anguillicaudatus*
Bagridae:	*Bagrus meridionalis*
Siluridae:	*Silurus glanis*
Ictaluridae:	*Ictalurus catus*
	I. melas
	I. nebulosus
	I. punctatus
Loricariidae:	*Hypostomus* spp.
Clariidae:	*Clarias batrachus*
	C. lazera
Cyprinodontidae:	*Aplochielus panchax*
Poeciliidae:	*Belonesox decemmaculatus*
	Gambusia affinis
	Phalloceros caudomaculatus
	Poecilia reticulata
	P. latipinna
	P. mexicana
	Poeciliopsis gracilis
	Xiphophorus helleri
	X. maculatus
	X. variatus
Oryziatidae:	*Oryzias latipes*
Atherinidae:	*Basilichthys bonariensis*
Centropomidae:	*Lates nilotica*
Centrarchidae:	*Ambloplites rupestris*
	Lepomis auritus
	L. cyanellus
	L. gibbosus
	L. gulosus
	L. macrochirus
	L. microlophus

Table 3–1. TAXONOMIC LIST OF SPECIES SUBJECT TO INTERNATIONAL TRANSFERS *(continued)*

	Micropterus coosae
	M. dolomieui
	M. punctulatus
	M. salmoides
	Pomoxis annularis
Sciaenidae	*Aplodinotus grunniens*
Percichthyidae:	*Morone saxatilis*
Percidae:	*Perca fluviatilis*
	Stizostedion lucioperca
	S. vitreum
Cichlidae:	*Astatoreochromis alluaudi*
	Astronotus ocellatus
	Cichla ocellaris
	C. temensis
	Cichlasoma bimaculatum
	C. facetum
	C. guttulatum
	C. macracanthus
	C. managuense
	C. meeki
	C. motaguense
	C. nigrofasciatum
	C. octofasciatum
	C. salvini
	C. severum
	Etroplus suratensis
	Hemichromis bimaculatus
	Pterophyllum scalare
	Serranochromis robustus
	Tilapia andersoni
	T. aurea
	T. esculenta
	T. galilaea
	T. hornorum
	T. leucosticta
	T. macrochir
	T. melanotheron
	T. mortimeri
	T. mossambica
	T. nilotica
	T. shiranum
	T. spilura
	T. mariae
	T. rendalli
	T. sparmanni
	T. zilli
Eleotridae:	*Hypoeleotris swinhonis*
	Perccottus glehni
Gobiidae:	*Rhinogobius similis*
	Acanthogobius flavimanus
Anabantidae:	*Anabas testudineus*
	Betta splendens
	Colisa lalia

Table 3–1. TAXONOMIC LIST OF SPECIES SUBJECT TO INTERNATIONAL TRANSFERS *(continued)*

	Ctenopoma nigropannosum
	Helostoma temmincki
	Macropodus opercularis
	Trichopsis vittata
Osphronemidae:	*Osphronemus gouramy*
	Trichogaster leeri
	T. pectoralis
	T. trichopterus
Channidae:	*Channa striatus*

techniques have been adapted to new environments, but at the same time have tended to suppress interest in local species of possibly equal or greater value. As such species are discovered, a potentially greater number of fishes may be transferred for culture. This and other aspects of introduced species in aquaculture are treated in chapter 13.

ECOLOGICAL MANIPULATION

A number of introductions are made to fill apparently vacant niches in fish communities or to substitute for species judged inferior. In some cases these may be in existing lakes with established fish communities such as in Lake Kivu (Zaire-Rwanda), where an open-water planktonophage was lacking despite the existence of a rich planktonic community. To remedy this, the clupeid *Limnothrissa miodon* was introduced and now forms the basis for a growing fishery. The same species was also introduced into a reservoir, Lake Kariba (Zambia-Zimbabwe), where it has been equally successful, and additional transfers into other large lakes are being contemplated. Less successful are those attempts to introduce forage species which often accompany the introduction of a major predator and where, as we shall see, the forage species have upset the ecological balance of the preexisting communities. More often the water bodies into which this type of introduction is made are man-made lakes, where fish communities are new or unstable. In particular, there may be an apparent lack of species that are able to adapt to the lacustrine conditions (Fernando 1980). Such would appear to apply in parts of Southeast Asia where the introduction of various tilapias has often resulted in great increases in harvest. For instance, lakes in Thailand or in India that customarily produced only 10–15 kg/ha/year of native carps, stocked each year from downstream, now yield 40–80 kg/ha/year of Nile tilapia (*Tilapia nilotica*) on a sustained basis. Fernando's (1980) hypothesis is perhaps less convincing in Latin American waters where large populations of piranhas colonize new reservoirs, although these are regarded as of only dubious utility. Sometimes valuable additions to local faunas are made through accidental escapes from aquaculture; thus gourami (*Osphronemus gouramy*), introduced into Madagascar in an abortive attempt to farm the species, now contributes much to the catch of the coastal Pangalanes. Similarly, the barb *Puntius gonionotus,* introduced into rivers of islands east of the Wallace Line (the demarcation between the Oriental and Australian zoogeographic regions), has given rise to a successful commercial fishery in Celebes, and oriental weatherfish (*Misgurnus anguillicaudatus*) has occupied highland streams in the Philippines where it forms the basis of a moderately important food fishery.

CONTROL OF UNWANTED ORGANISMS

The use of fishes for the control of aquatic pests has been widespread for two rather diverse purposes. First, the control of vector organisms, principally the larvae of malarial mosquitoes, whereby the poeciliids guppy (*Poecilia reticulata*) and mosquitofish (*Gambusia affinis*) have achieved almost worldwide distribution. Attempts have also been made to control the alternative hosts of bilharzia in the same way by the use of mollusc-eating cichlids such as *Astatoreochromis alluaudi* with varying degrees of success. Second, fishes have been used to control aquatic vegetation in lakes and irrigation canals, an aspect that will be treated in more detail in chapter 14. Here, the major species involved has been grass carp (*Ctenopharyngodon idella*), which has often proved successful for its purpose in many parts of the world. Happily, so it was thought, *C. idella* would not breed outside of its native rivers and stocks could thus be controlled and maintained either by artificial breeding or by continuous import. This impression has since been dispelled by the appearance of naturally reproducing populations in certain Russian rivers, the Danube system, and parts of the Mississippi. Less successful has been the experience with redbreast tilapia (*Tilapia rendalli*) and redbelly tilapia (*Tilapia zilli*), which for some reason have gained an undeserved reputation as weed eaters. True, these species will eat weeds, but they will eat anything else as well from detritus to young fish, and resulting population explosions have not generally endeared the species to those in whose waters it has been introduced.

ORNAMENT

One species in particular, goldfish (*Carassius auratus*) in its golden variety, has acquired an almost global distribution because of its value for ornament, but Conroy (1975) estimates that about 1,000 species may be shipped around the world at any one time for the aquarium fish trade, and Lachner et al. (1970) have indicated that over 6,000 species may eventually be of interest to aquarists (see also International Trade Center, 1979, for analysis of aquarium trade). The ecological impacts of this movement have not yet been fully evaluated, nor in many cases are they immediately apparent. Evidently, the transfer of species from the tropics to the colder waters of the temperate zone bears little risk of the establishment of such fishes in their new environment, except in cases such as the populations of the guppy and redbelly tilapia reported by Wheeler and Maitland (1973) as living in the warmwater discharges of power stations in England. However, in warmer water areas such introductions can create havoc; for instance, Courtenay and Hensley (1980) list 18 species as being established in Florida with an additional 31 as possible candidates for that distinction. The aquarium fish breeders of Hong Kong, Singapore, Malaysia, or Brazil have for some years now been raising a gamut of species, native or otherwise, for export. Many of these undoubtedly escape and eventually become established in the wild with totally unknown effects on the local fish communities, and almost certainly an enormous number of small ornamental species will become diffused around the tropical world in this manner.

NONPURPOSEFUL INTRODUCTIONS

While the original transport of most fishes between countries is presumably purposeful, many of the actual introductions, whereby the species has become established within natural waters, have taken place unintentionally. Over ten percent of the transfers listed by Welcomme (1981) are reported as entering natural waters by accident, and many of the appearances of exotic fishes to which no reason could be ascribed were doubtless of similar origin. Furthermore, a great number of species which were originally imported for aquaculture have escaped from their ponds to become resident in the wild. Of the true accidental introductions, many may have occurred when small species, usually cyprinids, have been transferred along with the juveniles of common carp or grass carp (*Ctenopharyngodon idella*); the appearance of a Japanese cyprinid, *Pseudorasbora parva,* in the Danube, bleak (*Alburnus alburnus*) and silver bream (*Blicca bjoerkna*) in Cyprus, and no fewer than ten species of cyprinids and eleotrids in the waters of Tashkent are typical examples. Yet other "accidents" have taken place when unwanted stocks of ornamental fish have been released from aquaria or by breeders from ponds. Other accidental introductions occur by diffusion. Courtenay and Hensley (1979) provided an example of the rapidity with which a species, spotted tilapia (*Tilapia mariae*), may spread through a system that is initially foreign to it. Similar introductions by diffusion have occurred many times as, for example, with eastern mudminnow (*Umbra pygmaea*) and North American catfishes (*Ictalurus* sp.) in the lowland waters of northern Europe, or the various introductions into the Danube River. Certainly, where open or communicating systems exist there is a real and continuing danger that species introduced into one region will spread rapidly to other areas outside the scope of the original transfer. Nowhere is this more dangerous than in the case of euryhaline species such as Mozambique tilapia (*Tilapia mossambica*), which can spread along the coast and even occupy saltwater habitats.

Recognition of the ease and frequency with which fishes can escape from confinement necessitates careful consideration of the advisability of effecting any transfer before the candidate species arrives in its new country.

Effects of Transfers

Experience throughout the world has shown that, in addition to the undoubtedly beneficial effects of some transfers, introductions of fish species into waters from which they were previously absent can be accompanied by a number of undesirable, if not downright disastrous, consequences. These may be produced directly on the fish community within the waters, although in some cases marked effects on the environment also have been noted.

PREDATION AND COMPETITION

Introduction of new fish species into a water body frequently reduces the numbers of those species already there. It is sometimes difficult to distinguish whether such reductions are caused by competition for common resources or by predation and, on occasions, both have been found to operate together. The most evident cases are those where a newly introduced predator has reduced disastrously or even eliminated endemic fishes. Many examples exist; for instance, tucanare (*Cichla ocellaris*), introduced into the Chagres River, Panama, eventually found its way into Gatun Lake and there progressively eliminated local species including the atherinid *Melaniris chagrensis,* four species of characins, and two species of poeciliids (Zaret and Paine 1973). Understandably, in the process the food webs became vastly simplified (figure 3-1) in those areas occupied by *Cichla.* Similarly, the introduction of largemouth bass has disrupted local communities in a Mexican lake, Lago de Patzcuaro, where populations of a native food fish, blanco de Patzcuaro (*Chirostoma ester*), have been greatly reduced along with several species of Goodeidae (Preciado 1955); in Lake Naivasha in Africa where several cichlid species, themselves earlier introductions, were the victims; in the glacial lakes of northern Italy where *Alburnus alborella,* northern pike (*Esox lucius*), perch (*Perca fluviatilis*), and introduced sunfishes (*Lepomis* spp.) have been supplanted; and in some Zimbabwe waters Rhodesian mountain catfish (*Amphilius platychir*) has been locally exterminated. One of the most consistent records for damaging stocks of endemic fish species is held by the salmonids, particularly rainbow trout and brown trout (*S. trutta*), which, by reason of their worldwide distribution, have had more opportunity to do so. Rainbow trout have been implicated in the decline of endemic salmonids in Lake Ohrid, Yugoslavia, the rare *Oreodaemion quathlambae* in Lesotho, *Trichomycterus* sp. in Colombia, and together with brown trout, the endemic galaxiids and New Zealand grayling (*Protroctes oxyrhynchus*) from New

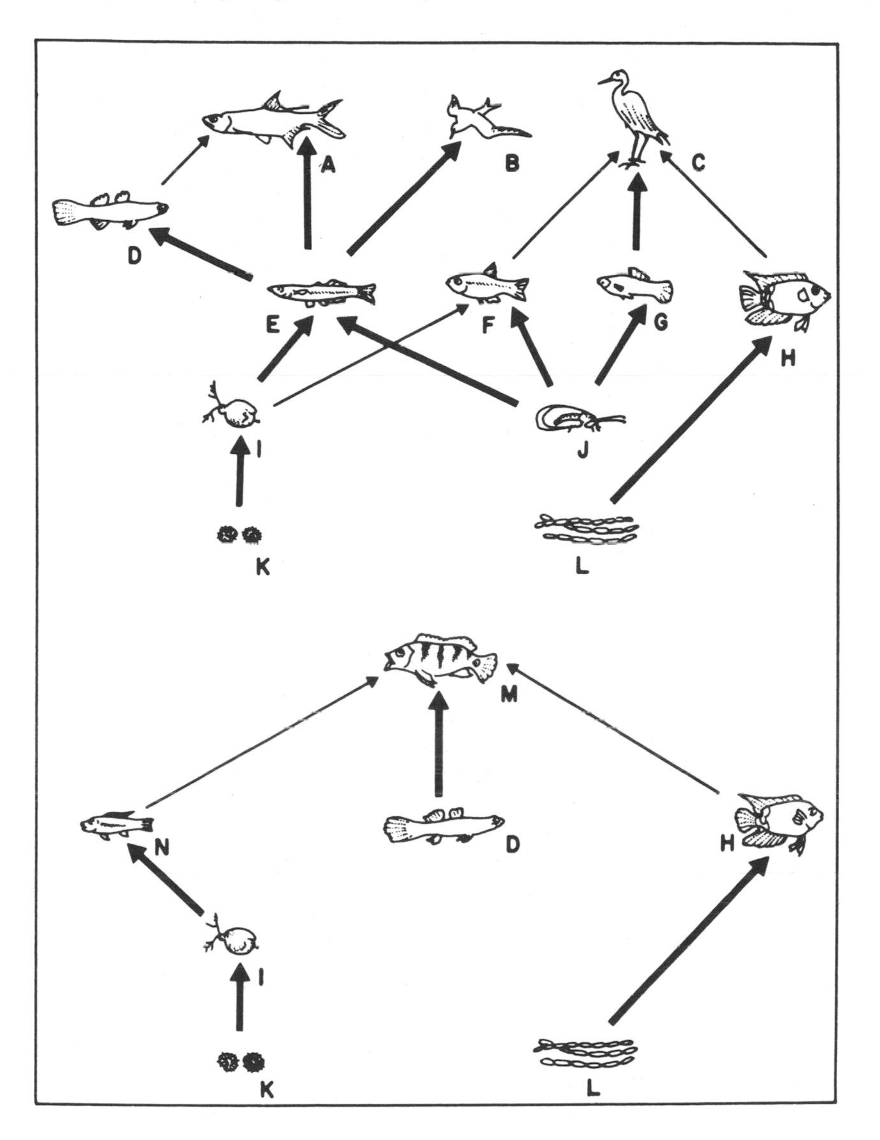

Figure 3–1. Food web before (top) and after (bottom) introduction of *Cichla* sp.

Zealand. Jackson (1960) reported that trout and largemouth bass had eliminated haarder (*Trachyistoma euronotus*) and cape kurper (*Sandelia capensis*) in some South African lakes. In Lake Titicaca the trout, principally rainbow trout, combined with *Basilichthys bonariensis* to damage a species flock of some 19 species of *Orestias* and *Trichomycterus rivulatus*. In this case, after reducing the local fishes through predation, their numbers were kept low by subsequent competition for the benthic invertebrates which were the common source of food (Campbell 1976). Direct competition for food probably accounts for the reduction in numbers of the cyprinodont White River springfish (*Crenichthys baileyi*) after the introduction of the guppy into the southwestern United States (Deacon et al. 1964). Competition for similar nursery beaches for young fish apparently caused a decline in numbers and catches of *Tilapia variabilis* after the insertion of redbelly tilapia into the Lake Victoria fauna (Welcomme 1967).

HYBRIDIZATION

When introducing closely related species whose specific integrity is maintained by geographic barriers, there is always a risk of hybridization. This has been observed on several occasions; for example, the identity of many stocks is now in doubt as a consequence of the dissemination of various species of whitefishes and ciscoes (*Coregonus*) among the glacial lakes of Europe. However, few families of fishes appear to possess the extraordinary genetic plasticity of the cichlids, a family renowned for its production of species flocks in many lakes in Africa. Of the cichlids, it is species of the tilapias that have been most extensively transported around the tropical world. The extent to which they hybridize can be measured by table 3-2, taken from Wohlfarth and Hulata (1981). Although in many cases disturbed sex ratios among the progeny are common, apparently self-breeding stocks of hydrids do appear to occur in nature, and thus new genotypes are apt to form in those parts of the world in which more than one of these species have become established. A good example of this occurred in Lake Itasy, Madagascar, where parental stocks of Nile tilapia and longfin tilapia (*T. macrochir*) virtually disappeared in favor of a hybrid called "tilapia trois quarts." Later an equilibrium between Nile tilapia and the new hybrid was established, which was more favorable to Nile tilapia, but longfin tilapia has not recovered.

INTRODUCTION OF DISEASE AND PARASITIC ORGANISMS

Rosenthal (1976) and chapters 11 and 12 of this book examine in some detail the many cases where disease and parasitic organisms have been or could be introduced along with fish species into waters from which they were hitherto absent. Most notably, perhaps furunculosis was probably native only to rainbow trout of western North America. After rainbow trout had been introduced, however, the disease became a serious problem in European brown trout and is now present almost everywhere salmonids are cultured. Another case arose from infectious dropsy of cyprinids, which first appeared in Czechoslovakia in 1929 after common carp were imported from Yugoslavia. The disease subsequently spread rapidly throughout continental Europe after stock carp were transferred without adequate sanitary controls

(Havelka 1973). Other diseases that have apparently spread in a similar manner are infectious haematopoietic necrosis and whirling disease of salmonids.

The scale of transfer of parasitic organisms is indicated by Hoffman (1970) who listed at least 48 species that have become established on continents other than those of their origin through the transfer of infected fishes. Typical of these are the 5 monogenetic trematode species which accompanied various tilapias from Africa to Israel. More recently Courtenay and Robins (1975) have cited grass carp as a particularly bad carrier of parasites, many of which may infect native cyprinids where they occur in its introduced range.

CROWDING AND STUNTING

One of the most common complaints directed at introduced species is that they undergo rapid, often explosive expansion in numbers but, at the same time, they reduce the age and size at which the fish mature. In this way, the water bodies into which such fishes are introduced become filled with numerous small or stunted individuals which may even supplant native species through sheer pressure on living space. Various species of *Tilapia* sometimes show a similar pattern of such stunting in waters to which they are native. This behavior on the part of *Tilapia mossambica,* which was transferred around much of the tropical world in the 1950s for aquaculture, renders it virtually useless for this purpose and, combined with resistance to attempts at its elimination, makes it nearly impossible to carry out additional introductions with more suitable species. The cichlids are by no means the only fishes to behave in this manner and similar cases have been reported for many other species including bleak (*Alburnus alburnus*) and silver bream (*Blicca bjoerkna*) in Cyprus, redbreast sunfish (*Lepomis auritus*), pumpkinseed (*L. gibbosus*) and catfishes (*Ictalurus* spp.) in France, Italy, and the Netherlands, *Pseudorasbora parva* in the Danube basin and in parts of the U.S.S.R., and particularly common carp. The last has become notorious over much of its introduced range; Australia, India, South Africa, and the United States all report bad experiences with it, although in all fairness in several of these countries the species also has its admirers, for example, McLaren (1981) who has discussed its advantages in Australia.

ENVIRONMENTAL EFFECTS

The common carp has also been implicated in environmental changes that followed its introduction in many areas. These principally led to eutrophication of the water through an increase in turbidity, which has been ascribed to the fish's habit of rooting or digging in the bottom. While it is true that carp do turn over the bottom and, in so doing, increase the turbidity of the water, it is also thought that the fish, by ingesting organisms from the phosphate-

Table 3–2. HYBRIDS BETWEEN DIFFERENT TILAPIAS

Breeding type	Species	In nature	Deliberate crosses carried out: Observations		Sex of hybrid progeny in deliberate crosses
			In ponds or tanks	Under lab conditions	
Maternal mouthbrooder	*T. nilotica* x *T. spilura niger*	x	x		Surplus of males
x	*T. nilotica* x *T. macrochir*	x			Only males when *T. nilotica* female parent
Maternal mouthbrooder	*T. nilotica* x *T. variabilis*	x	x		Only males when *T. nilotica* female parent
	T. nilotica x *T. leucosticta*		x		Surplus of males
	T. nilotica x *T. hornorum*		x		Only males when *T. nilotica* female parent
	T. nilotica x *T. mossambica*		x		Surplus of males
	T. mossambica x *T. hornorum*		x		Only males when *T. mossambica* female parent
	T. mossambica x *T. andersoni*	x			
	T. mossambica x *T. spilura niger*		x		Surplus of males
	T. mossambica x *T. aurea*		x		Surplus of males (*T. aurea* misidentified as *T.nilotica*)
	T. vulcani x *T. hornorum*		x		Large surplus of males

	T. spilura niger x *T. hornorum*		x		Only males when *T. spilura niger* female parent
	T. spilura niger x *T. leucosticta*	x			Surplus of males
	T. amiphetes x *T. esculenta*	x			
	T. vulcani x *T. aurea*		x		Large surplus of males
	T. hornorum x *T. aurea*		x		Surplus of males
Maternal mouthbrooder x Paternal mouthbrooder	*T. melanotheron* x *T. mossambica*			x	Only females when *T. melanotheron* female parent
	T. melanotheron x *T. nilotica*			x	
Maternal mouthbrooder x Substrate breeder	*T. tholloni* x *T. mossambica*			x	Only females when *T. tholloni* female parent
	T. tholloni x *T. nilotica*			x	Only females when *T. tholloni* female parent
	T. zilli x *T. mossambica*			x	
	T. zilli x *T. spilura niger*	x	x		Only males; sex of parents not given
Paternal mouthbrooder x Substrate breeder	*T. tholloni* x *T. melanotheron*			x	
Substrate breeder x Substrate breeder	*T. zilli* x *T. rendalli*	x	x		Sex ratio 1:1

Source: Wohlfarth and Hulata 1981.

rich substratum and later excreting phosphate, mobilize this nutrient in a more soluble form. At the same time, aquatic vegetation is likely to suffer both mechanically through the digging activities and through shading by consequent planktonic blooms. The altered environment becomes less suitable for native species which favor clear waters and these subsequently disappear; for example, the fishery for schizothoracids in India (Jhingran and Sehgal 1978). Although elimination of aquatic vegetation is sometimes desirable and has been a major objective for the transfer of such weed-eating species as grass carp or redbreast tilapia, the successful accomplishment of this aim has a certain environmental cost. Beds of submersed vegetation are used as spawning grounds by other species of fish which may be threatened by such elimination, and outside of the fish community water birds or amphibians (Van Leeuwen 1979) may also be adversely affected.

In addition to the effects of introduced species on the environment, there appears to be some effect of environment on species introductions even where climatic and hydrological factors are otherwise suitable to the introduction. As a general principle, it would appear that the more complex the environment and the endemic fish community into which a species is introduced, the less its immediate impact. There are few records, for instance, of introduced species taking over the species-rich environment in the tropics. Indeed, in some such regions apparently highly successful transfers have, after a first bloom, withered and disappeared as with Nile tilapia in some reservoirs in Thailand or with common carp when faced with competition from tilapias in Madagascar. Records of disastrous introductions usually come from regions with either simple (i.e., of low specific diversity) or relatively fragile fish communities or which are under pressure for other reasons such as excessive fishing or environmental modification.

Environmental modification would also account for the success of certain introductions, notably the tilapiine cichlids and common carp in artificial lakes or in rivers below dams where native fish communities have been unable to adapt to the new conditions. In fact, it might be fair to argue that, in many areas where common carp have flourished, it is because of man-made changes to the ecosystem rather than because the carp have modified and simplified the ecosystem in their own favor. This characteristic implies that certain introductions may lie dormant as a relatively insignificant part of a host community, later to become a dominant element should that community be stressed by overfishing, pollution, or other forms of environmental change. It would appear that in many cases where the introduction of a species has been followed by explosive expansion, it was because the species has been able to tap some food source hitherto little exploited; but, as the fish come into trophic equilibrium, the populations become less dense and eventually the species becomes incorporated as a normal component of the population.

Conclusion

Of the 163 species listed by Welcomme (1981), 96 (58 percent) were either unsuccessful in that they failed to become established or have been so localized in their new country as to be insignificant. One suspects that the percentage is far higher in reality, because many unsuccessful attempts almost certainly fail to be reported. Most of the many hundreds of ornamental species introduced for the aquarium fish trade, for instance, are not listed except insofar as they have been positively reported as breeding in the wild in their new environment. Four species have not bred naturally in the majority of countries to which they have been introduced. These species are generally highly desirable for aquaculture or to exploit specialist niches, such as feeding on plankton or macrophytes, and interest in them is sufficient to justify continued shipments of young from the countries of origin or the establishment of artificial propagation facilities. Of the remaining species, 65 have become established in the wild and may give some indication of the effectiveness and dangers of such introductions. Three categories may be distinguished.

Twenty-seven (table 3–3) of the 63 remaining species (43 percent) have been widely classified as pests. Almost without exception, these are small species, or large species that stunt readily, whose main nuisance value lies in their high fecundity, rapid proliferation, and general uselessness for any human purpose.

Clarias batrachus and *Tilapia mossambica* (table 3–3) have enjoyed some success in aquaculture and will probably continue to be introduced either for culture or on their own; the walking catfish (*C. batrachus*) has recently been found to be successful in Kwangtung Province in China, and Mozambique tilapia have been used for polyculture or for hybridization. Additional proposals to transfer any of the other species on this list should probably be treated with extreme caution.

Sixteen species (25 percent) (table 3–4) have been found to be generally satisfactory where introduced in that they have either fulfilled their intended function or have given rise to important fisheries outside the original scope of their introduction. At the same time, no noticeable side effects have yet been detected from these species. Some of the species will doubtlessly be considered for additional introduction.

Twenty species (32 percent) are viewed with mixed feelings because they either have been highly successful in some areas of the world and less successful in others or, while being highly desirable for those objectives for which they were originally selected, have posed serious problems of one type or another. This group (table 3–5) includes some of the most widely introduced species, and a small number of them, common carp and rainbow trout in particular, have already been transferred so widely that additional transfers will not greatly influence their history.

Table 3–3. SPECIES WIDELY CLASSIFIED AS PESTS WHERE INTRODUCED

Alburnus alburnus	*Opsarichthys uncirostris*
Blicca bjoerkna	*Percottus glehni*
Clarias batrachus	*Pseudogobio rivulatus*
Hemibarbus maculatus	*Pseudorasbora parva*
Hemiculter eigenmanni	*Rhinogobius similis*
Hemiculter leucisculus	*Rutilus rutilus*
Hypseleotris swinhoris	*Tilapia mossambica*
Ictalurus melas	*Tilapia mariae*
Ictalurus nebulosus	*Tilapia rendalli*
Lepomis auritus	*Tilapia zilli*
Lepomis gibbosus	*Umbra krameri*
Leuciscus leuciscus	*Xiphophorus helleri*
Poecilia latipinna	*Xiphophorus maculatus*
Poecilia reticulata	

Table 3–4. SPECIES FOUND TO BE ON THE WHOLE BENEFICIAL WHERE INTRODUCED

Aristichthys nobilis[a]	*Labeo rohita*
Cichlasoma managuense	*Limnothrissa miodon*
Coregonus laeveraetus	*Misgurnus anguillicaudatus*
Coregonus peled	*Onchorynchus tshawytscha*
Dorosoma petenense	*Puntius gonionotus*
Etroplus suratensis	*Stizostedion lucioperca*
Heterotis niloticus	*Trichogaster leeri*
Hypophthalmichthys molitrix[a]	*Trichogaster pectoralis*

[a] Species needing artificial propagation

These lists, and the last particularly, provoke some reflection on the criteria for determining whether or not any particular species is of value. Clearly there is a pronounced difference between industrialized societies, which place an emphasis on protection of the environment and the species in it, and those primarily poor rural nations where prime concern is for the production of sufficient food. The two societies are clearly apt to view introduction of the same species in quite different ways. An industrial society, located mainly in the temperate zones, is apt to approve only those species which enhance its environment and to be minutely critical of even slight disturbances in the quality of its natural waters. Conversely, the rural societies of the tropics may tolerate quite considerable environmental change in favor of an immediate solution to greater food production or perhaps more important to food stability. A second factor influencing decisions is the argument, advanced earlier in this chapter, that simple and more sensitive fish communities of the temperate zones seemingly respond in a greater degree to a new introduction than the more complex and generally more resistant fish communities of tropical waters. While it remains difficult to predict the outcome

Table 3–5. SPECIES VIEWED WITH MIXED FEELINGS

Basilichthys bonariensis	*Perca fluviatilis*
Carassius auratus	*Salmo gairdneri*
Ctenopharyngodon idella	*Salmo trutta*
Cyprinus carpio	*Salvelinus fontinalis*
Esox lucius	*Tilapia aurea*
Gambusia affinis	*Tilapia galilaea*
Lates nilotica	*Tilapia leucosticta*
Micropterus salmoides	*Tilapia melanotheron*
Oncorhynchus gorbuschka	*Tilapia nilotica*
Osphronemus gouramy	*Tinca tinca*

of species transfers, it should be evident that haphazard introductions should not be undertaken, and that those considered desirable should be subjected to the judgment of disinterested experts. However, many of the current uncertainties regarding particular introductions could be reduced in the case of future proposals for transfers if fundamental ecological research can be undertaken. Here, more careful documentation of community ecology in both the original habitats of species and in communities into which introductions have occurred is important to an understanding of the types of adjustment that will occur as a new species is absorbed into its new ecosystem.

In the meantime, there is a great lack of information upon which to base judgments on species of genuine utility for the most pressing problems of the present – the development of local (regional) species for aquaculture and the selection of species for maximizing of yield from artificial impoundments. Doubtless other species will be suggested for those roles.

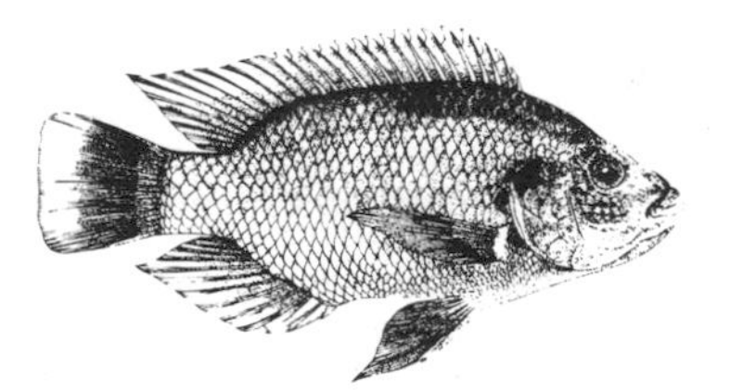

Literature Cited

Balon, E. K. 1974. Domestication of the carp *Cyprinus carpio* L. Royal Ontario Museum, Life Sciences Miscellaneous Publications. 37 pp.

Campbell, H. J. 1976. Reconocimiento de las pesquerías en aguas continentales del Peru. Food and Agriculture Organization: FI:DP/PER/71/012/2. 32 pp.

Conroy, D. A. 1975. An evaluation of the present state of world trade in ornamental fish. Food and Agriculture Organization Fisheries Circular 335, 120 pp.

Courtenay, W. R., and D. A. Hensley. 1979. Range expansion in southern Florida of the introduced spotted tilapia, with comments on its environmental impress. Environmental Conservation 6(2):149–151.

Courtenay, W. R., and D. A. Hensley. 1980. Special problems associated with monitoring exotic species, 281–307. *In:* C. H. Hocutt and J. R. Stauffer, Jr. (eds.). Biological monitoring of fish. Lexington Books, Lexington, Massachusetts.

Courtenay, W. R., and C. R. Robins. 1975. Exotic organisms: An unsolved complex problem. BioScience 25(5):306–313.

Deacon, J. E., C. Hubbs, and B. J. Zahuranec. 1964. Some effects of introduced fishes on the native fish fauna of southern Nevada. Copeia 1964(2):384–388.

Fernando, C. H. 1980. The fishery potential of man-made lakes in Southeast Asia and some strategies for its optimization, 25–38. Biotrop Anniversary Publication.

Havelka, J. 1973. Infectious dropsy of cyprinids. Food and Agriculture Organization (IDC) in Czechoslovakia. EIFAC Technical Paper 17, supplement 2:124–132.

Hoffman, G. L. 1970. Intercontinental and transcontinental dissemination and transformation of fish parasites with emphasis on whirling disease (*Myxosoma cerebralis*). Special publication of the American Fisheries Society 5:69–81.

Jackson, P.B.N. 1960. On the desirability or otherwise of introducing fishes to waters that are foreign to them, 157–164. Proceedings of the 4th CCTA/CSA: Hydrobiology and Inland Fish Symposium.

Jhingran, V. G., and K. L. Sehgal. 1978. Coldwater fisheries of India. India, West Bengal, Inland Fisheries Society of India. 239 pp.

Lachner, E. A., C. R. Robins, and W. R. Courtenay, Jr. 1970. Exotic fishes and other aquatic organisms introduced into North America. Smithsonian Contributions in Zoology 59:1–29.

MacCrimmon, H. R. 1971. World distribution of rainbow trout (*Salmo gairdneri*). Journal of the Fisheries Research Board of Canada 28(5):663–704.

McLaren, P. 1981. Is carp an established asset? Fisheries 5(6):31–32.

Preciado, A. S. 1955. La pesca en el Lago de Patzcuaro, Mich. y su importancia económica regional. Mexico, Secretaría de la Marina, Dirección General de Pesca. 58 pp.

Rosenthal, H. 1976. Implications of transplantations to aquaculture and to ecosystems. Food and Agriculture Organization Technical Conference on Aquaculture, FIR:AQ/Conf.76/E.67. 19 pp.

Thienemann, A. 1950. Die Binnengewässer. Band 18, Vebreitungsgeschichte der Süsswassertierwelt Europas. Schweizerbart, Stuttgart. 809 pp.

Van Leeuwen, B. H. 1979. Grass carp, a threat to our amphibia? Environmental Conservation 6(4):264.

Welcomme, R. L. 1967. Observations in the biology of the introduced species of Tilapia in L. Victoria. Revue de Zoologie et de Botanique africaine 76:249–279.

Welcomme, R. L. 1981. Register of international transfers of inland fish species. Food and Agriculture Organization Fisheries Technical Paper no. 213. 120 pp.

Wheeler, A., and P. S. Maitland. 1973. The scarcer freshwater fishes of the British Isles. Part 1. Introduced species. Journal of Fish Biology 5:49–68.

Wohlfarth, G. W., and G. J. Hulata. 1981. Applied genetics of *Tilapias.* ICLARM Studies and Reviews 6:1–26.

Zaret, T. M., and R. T. Paine. 1973. Species introduction in a tropical lake. Science 182:449–455.

CHAPTER 4

Distribution of Exotic Fishes in the Continental United States

Walter R. Courtenay, Jr.,
Dannie A. Hensley,
Jeffrey N. Taylor, and
James A. McCann

This interesting note is important, as establishing the practicability of introducing foreign fishes into our waters, and as recording an important fact in Ichthyology. We invite other patriotic individuals to make similar experiments with other species, which are now limited to the other side of the Atlantic.

DeKay (1842:189) made these comments on a letter received from "Henry Robinson, Esq." of Newburgh, Orange County, New York, reporting successful establishment (a breeding population) of the common carp in the Hudson River in the early 1830s. While we doubt that patriotism has been a factor in subsequent introductions of foreign (or exotic) fishes, we record 41 alien species as established in the forty-eight contiguous United States (table 4-1). We also record 63 nonestablished exotic fishes from this same area, including formerly established species and others that have been collected or reported (tables 4-2 and 4-3).

Introductions and distributions of exotic fishes in North America are changing continually. New species are added, some already established are

Table 4–1. SUMMARY OF THE STATUS OF ESTABLISHED EXOTIC FISHES (IN ALPHABETICAL ORDER) IN THE CONTIGUOUS UNITED STATES

Species with expanding populations		
Acanthogobius flavimanus	yellowfin goby	California[a]
Astronotus ocellatus	oscar	Florida[a]
Cichlasoma bimaculatum	black acara	Florida[a]
Cichlasoma citrinellum	Midas cichlid	Florida[a]
Cichlasoma meeki	firemouth	Florida[a]
Cichlasoma nigrofasciatum	convict cichlid	Nevada[a]
Cichlasoma octofasciatum	Jack Dempsey	Florida[a]
Clarias batrachus	walking catfish	Florida[a]
Ctenopharyngodon idella	grass carp	Arkansas,[a] Louisiana,[a] Mississippi,[a] Tennessee[a]
Hemichromis bimaculatus	jewelfish	Florida[a]
Poecilia mexicana	shortfin molly	California, Montana, Nevada,[a] Texas
Tilapia aurea	blue tilapia	Arizona,[a] California,[a] Florida,[a] Georgia,[b] North Carolina, Oklahoma, Texas[a]
Tilapia mariae	spotted tilapia	Florida,[a] Nevada
Tilapia melanotheron	blackchin tilapia	Florida[a]
Tilapia mossambica	Mozambique tilapia	Arizona, California,[a] Florida,[a] Georgia,[b] Texas
Tilapia zilli	redbelly tilapia	Arizona,[a] California,[a] Nevada, North Carolina, Texas
Species with localized distributions (= not expanding)		
Bairdiella icistia	bairdiella	California
Belonesox belizanus	pike killifish	Florida
Cynoscion xanthulus	orangemouth corvina	California
Hypostomus sp.[c]	suckermouth catfish	Florida, Nevada, Texas
Misgurnus anguillicaudatus	oriental weatherfish	California, Michigan
Poeciliopsis gracilis	porthole livebearer	California
Trichopsis vittata	croaking gourami	Florida
Tridentiger trigonocephalus	chameleon goby	California
Xiphophorus helleri	green swordtail	Florida, Montana, Nevada
Xiphophorus maculatus	southern platyfish	Florida, Nevada
Xiphophorus variatus	variable platyfish	Florida, Montana
Species with apparent broad and stable distributional limits		
Carassius auratus	goldfish	Alabama,[b] Arizona,[b] Arkansas,[b] California, Colorado,[b] Connecticut,[b] Delaware, District of Columbia, Georgia, Idaho, Illinois, Indiana, Iowa, Kansas,[b] Kentucky, Maryland, Massachusetts, Michigan, Minnesota,[b] Mississippi,[b] Missouri,[b] Montana,[b] Nebraska, Nevada, New Hampshire, New Jersey,[b] New Mexico,[b] New York, North Carolina,[b] North Dakota, Ohio, Oklahoma, Oregon,[b] Pennsylvania, Rhode Island, South Carolina, South Dakota, Tennessee, Texas, Virginia, Washington, Wisconsin, Wyoming[b]

Table 4–1. SUMMARY OF THE STATUS OF ESTABLISHED EXOTIC FISHES (IN ALPHABETICAL ORDER) IN THE CONTIGUOUS UNITED STATES *(cont.)*

Cyprinus carpio	common carp	all continental states except Alaska
Salmo trutta	brown trout	Arizona, Arkansas, California, Colorado, Connecticut, Delaware, Georgia, Idaho, Illinois, Indiana, Iowa, Kentucky, Maine, Maryland, Massachusetts, Michigan, Minnesota, Missouri, Montana, Nebraska, Nevada, New Hampshire, New Jersey, New Mexico, New York, North Carolina, North Dakota, Oregon, Pennsylvania, Rhode Island, South Carolina, South Dakota, Tennessee, Utah, Vermont, Virginia, Washington, West Virginia, Wisconsin, Wyoming
Species with apparently declining populations		
Leuciscus idus	ide	Connecticut, Pennsylvania (?)
Rhodeus sericeus	bitterling	New York
Scardinius erythrophthalmus	rudd	Maine (?), New York
Tinca tinca	tench	California, Colorado, Connecticut, Idaho, New Mexico, Washington
Insufficient information available to determine populational status		
Pterygoplichthys sp.		Florida
Geophagus surinamensis	redstriped eartheater	Florida
Hypomesus nipponensis	wakasagi [b]	California
Poecilia reticulata	guppy	Arizona, California,[b] Florida, Idaho, Nevada, Texas, Wyoming
Tilapia hornorum	Wami tilapia	California

[a] Indicates states where populations are expanding or can be expected to expand
[b] Indicates current status unknown (= still established?)
[c] Three morphologically distinct species

expanding their ranges naturally or via a major vector, *Homo sapiens,* and some are dying out or being purposefully eradicated. As a result, when this book is published, some species accounts in our chapter will probably already be out of date.

The major purposes of this chapter are (1) to present all the information we have on distributions of exotic fishes in North America, gleaned from personal and sponsored research, journals, reports, books, and personal communications as of July 1983, and (2) to urge input from readers on records we missed and reports we did not uncover and especially to solicit records of future findings of exotic fishes in North America.

Family entries in this chapter are arranged in the phyletic sequence followed by Robins et al. (1980), with generic and specific names in alphabetical order. In addition to the native and known distributional ranges in North

Table 4–2. SUMMARY OF THE EXOTIC FISHES KNOWN TO HAVE BEEN ESTABLISHED IN THE 48 CONTIGUOUS UNITED STATES

Scientific name	Common name	Native range	Where established	Reason for extirpation	Reference(s)
Serrasalmus humeralis	pirambeba	Amazon, Peru, Bolivia, and Guyana	Dade Co., Florida	purposeful eradication 1981	Shafland and Foote 1979
Hoplias malabaricus	trahira	Amazon, Guyana, Venezuela, Colombia, Paraguay, and Argentina	Hillsborough Co., Florida	cold 1977	personal observation
Oryzias latipes	medaka	Japan, Korea, southern Manchuria, eastern China to Hainan	Santa Clara Co., California; Suffolk Co., New York	unknown in California; cold 1978 in New York	Shapovalov et al. 1981; D. E. Rosen, personal communication
Cynolebias bellottii[a]	Argentine pearlfish	Río de la Plata basin	Riverside Co., California	unknown	Moyle 1976, Shapovalov et al. 1981
Rivulus harti	Trinidad rivulus	Trinidad, Guyana?, northern Brazil	Riverside Co., California	unknown	St. Amant 1970, Moyle 1976; Shapovalov et al. 1981
Aequidens pulcher[b]	blue acara	Trinidad, northern Venezuela	Manatee Co., Florida	unknown	R. B. Socolof, personal communication
Cichlasoma beani	green guapote	Pacific slope of Mexico from Jalisco north to Sonora	Solano Co., California	unknown	Shapovalov et al. 1981

Cichlasoma salvini	yellowbelly cichlid	Veracruz, Mexico, to Puerto Barrios, Guatemala	Broward Co., Florida	purposeful eradication 1981	P. L. Shafland, personal communication; personal observation (WRC)
Cichlasoma severum	banded cichlid	Río Negro, Brazil	Clark Co., Nevada	purposeful eradication 1963	Courtenay and Deacon 1983
Cichlasoma trimaculatum	threespot cichlid	Pacific slope of southern Mexico to Guatemala	Manatee Co., Florida	purposeful eradication 1975	Shafland 1976
Anabas testudineus[b]	climbing perch	southern China, Indo-China, Malaya, India, Philippines, and Indo-Australian archipelago	Manatee Co., Florida	unknown	R. B. Socolof, personal communication
Betta splendens	Siamese fighting-fish	Thailand	Broward Co., Florida	cold 1977	personal observation (WRC)
Ctenopoma nigropannosum	twospot ctenopoma	Niger delta to Congo River	Manatee Co., Florida	unknown	personal observation (DAH)
Macropodus opercularis	paradisefish	Korea, China, Vietnam, Taiwan	"Everglades," Florida	unknown	Myers 1940

[a] Stocks included *Cynolebias nigripinnis* and *C. whitei* (Shapovalov et al. 1981), stocked in Butte County, California; these species failed to reproduce.
[b] Reported by R. B. Socolof as formerly established. No specimens known to have been collected.

Table 4–3. NONESTABLISHED EXOTIC FISHES KNOWN FROM THE WATERS OF THE UNITED STATES

Anguilla anguilla	European eel	California
Osteoglossum bicirrhosum	aruana	California, Nevada
Esox reicherti	Amur pike	Pennsylvania
Plecoglossus altivelis	ayu	California
Coregonus maraena	German whitefish	Michigan
Salmo letnica	Ohrid trout	Colorado, Montana, Tennessee, Wyoming
Chanos chanos	milkfish	California
Colossoma spp.	pacu	Arizona, California, Florida, Missouri, Ohio
Colossoma bidens		Florida, Georgia
Colossoma nigripinnis	blackfin pacu	Florida
Gymnocorymbus ternetzi	black tetra	Florida
Metynnis lippincottianus		Florida
Metynnis roosevelti		Kentucky
Serrasalmus sp.	piranha	Illinois, Kentucky, Michigan, Pennsylvania
Serrasalmus nattereri	red piranha	Florida, Massachusetts, Michigan, Pennsylvania
Barbus sp.	tinfoil barb	Florida
Barbus conchonius	rosy barb	Florida
Barbus gelius	golden barb	Florida
Barbus tetrazona	tiger barb	California, Florida
Brachydanio rerio	zebra danio	California, Florida
Danio malabaricus	giant danio	Florida
Hypophthalmichthys molitrix	silver carp	Arkansas
Labeo chrysophekadion	black sharkminnow	Florida
Oxydoras niger		Florida
Platydoras costatus		Florida
Pterodoras sp.		Florida
Pterodoras granulosus		Florida
Callichthys sp.	callichthys	Florida
Corydoras sp.	corydoras	Florida
Poecilia hybrids		Florida, Nevada
Chirostoma jordani		Texas
Channa micropeltes	giant snakehead	Maine, Rhode Island
Lates nilotica	Nile perch	Texas
Stizostedion lucioperca[a]	European pike-perch	New York
Ameca splendens	butterfly goodeid	Nevada
Cichla ocellaris	tucanari	Florida
Cichlasoma labiatum	red devil	Florida
Geophagus brasiliensis	pearl eartheater	Florida
Melanochromis auratus	gold mbuna	Nevada
Melanochromis johanni		Nevada
Pseudotropheus zebra	zebra mbuna	Nevada
Pterophyllum sp.		Florida
Tilapia sparmanni	banded tilapia	Florida
Colisa fasciata	giant gourami	Pennsylvania
Colisa labiosa	thicklip gourami	Florida
Colisa lalia	dwarf gourami	Florida
Helostoma temmincki	kissing gourami	Florida
Macropodus opercularis	paradisefish	Florida
Trichogaster leeri	pearl gourami	Florida
Trichogaster trichopterus	blue gourami	Florida

[a] Unconfirmed report

America, we have attempted to include information on the dates, sources, and reasons for these introductions wherever possible.

SALMONIDAE – SALMONS AND TROUTS

***Salmo trutta* Linnaeus brown trout**

The native range of the brown trout includes Iceland, Ireland, Great Britain, North Africa (northern Morocco, Algeria, Tunisia), and Eurasia (Spain to northern Scandinavia, eastward to the Timanskiy-Ural mountain ranges, southward into Lebanon, headwaters of the Tigris and Euphrates, tributaries of the Caspian Sea in Turkey, southeastward to the Amu Darya drainage of the Aral Sea, through Afghanistan to West Pakistan; Corsica, Sardinia, and Sicily [MacCrimmon and Marshall 1968, MacCrimmon et al. 1970]).

In the United States, populations of brown trout are established in Arizona, Arkansas, California, Colorado, Connecticut, Delaware, Georgia, Idaho, Illinois, Indiana, Iowa, Kentucky (perhaps), Maine, Maryland, Massachusetts, Michigan, Minnesota, Missouri, Montana, Nebraska, Nevada, New Hampshire, New Jersey, New Mexico, New York, North Carolina, North Dakota, Oregon, Pennsylvania, Rhode Island, South Carolina, South Dakota, Tennessee, Utah, Vermont, Virginia, Washington, West Virginia, Wisconsin, and Wyoming (MacCrimmon and Marshall 1968). Attempts to establish the species in Florida, Kansas, Ohio, and Oklahoma failed. Of the states with established populations, this fish was not being stocked as of 1968 in Arkansas and Kentucky (MacCrimmon and Marshall 1968).

The first introduction into the United States was made in April 1883 in the Pere Marquette River, Michigan, by the U.S. Fish Commission (Mather 1889, Goode 1903, Laycock 1966). The eggs from which the introduced fry hatched were shipped from Germany in the winter of 1883.

The brown trout was introduced as a food and game species. It is stocked annually in many states by state and federal agencies.

OSMERIDAE – SMELTS

***Hypomesus nipponensis* McAllister wakasagi**

The wakasagi is native to Hokkaido and possibly Honshu, Japan (McAllister 1963).

This fish was introduced in 1959 to six California reservoirs, Dodge Reservoir, Lassen County; Dwinnel Reservoir, Siskiyou County; Freshwater Lagoon, Humboldt County; Spalding Reservoir, Nevada County; Sly Park Reservoir, El Dorado County; and Big Bear Lake, San Bernardino County. There have been subsequent releases into other reservoirs (Moyle 1976). The current status of some of these populations is unknown, but Moyle (1976) suggested that this fish can be expected in the lower Klamath and Sacramento river systems and possibly in other drainages.

The wakasagi was introduced by the California Department of Fish and Game as a forage fish for trout (Wales 1962).

CYPRINIDAE – CARPS AND MINNOWS

Carassius auratus (Linnaeus) goldfish

The goldfish is native to the People's Republic of China, Taiwan, southern Manchuria, Korea, Japan, Hainan, the Lena River of eastern Europe to the Amur Basin (Berg 1949a) and the Tym and Poronai rivers of Sakhalin. Okada (1966) said that China introduced the goldfish into Japan, suggesting the same for this fish on Hainan and Taiwan.

The goldfish was the first exotic fish to be introduced into North America (Courtenay and Hensley 1980). DeKay (1842) recorded the first releases as being in the late 1600s. This fish has been collected in the wild in every state except Alaska. It is established in California, Delaware, District of Columbia, Georgia, Idaho, Illinois, Indiana, Iowa, Kentucky, Maryland, Massachusetts, Michigan, Nebraska, Nevada, New Hampshire, New York, Ohio, Oklahoma, Pennsylvania, Rhode Island, South Carolina, South Dakota, Tennessee, Texas, Virginia, Washington, and Wisconsin. Its establishment is uncertain in Alabama, Arizona, Arkansas, Colorado, Connecticut, Kansas, Louisiana, Minnesota, Mississippi, Missouri, Montana, New Jersey, New Mexico, North Carolina, North Dakota, Oregon, West Virginia, and Wyoming. Although specimens are collected periodically, self-sustaining populations appear to be lacking in Florida, Maine, Utah, Vermont, and West Virginia.

Releases appear to have been made by aquarists, ornamental pondfish hobbyists, and fishermen (as excess bait fish); escapes from culture facilities have also occurred. This species has been used in state and federal hatcheries as forage for largemouth bass.

Ctenopharyngodon idella (Valenciennes) grass carp

The grass carp is native to the middle and lower Amur River (as far north as Blagoveshchensk, U.S.S.R.), the Sungari and Ussuri rivers and Lake Khanka, in eastward flowing rivers of the People's Republic of China south to Guangzhou, Kwangtung Province (Berg 1949a).

This species is recorded as established in the lower Mississippi River at three locations: near Eudora, Chicot County, Arkansas; near Simmesport, Avoyelles Parish, Louisiana; and near St. Francisville, West Feliciana Parish, Louisiana (Conner et al. 1980). Grass carp larvae have been collected in the Atchafalaya River near Butte la Rose, St. Martin Parish, Louisiana (R. P. Gallager, personal communication). Juvenile grass carp less than 40 mm TL were collected in a Mississippi River backwater near Memphis, Shelby County, Tennessee (Conner et al. 1980; L. Koch, personal communication). Because of the finding of larval grass carp in open waters, Conner et al. (1980) believed this species to be established in the lower Mississippi River.

As of 1977, specimens of grass carp had been collected in the wild in Alabama, Arkansas, Florida, Georgia, Michigan, Missouri, New York, Pennsylvania, and West Virginia. Unconfirmed reports indicate the presence of this fish in the Mississippi River in Wisconsin and Minnesota and in the Missouri River in Kansas and Nebraska, as far upstream as Gavins Point Dam near Yankton, South Dakota.

The grass carp was first imported by the U.S. Fish and Wildlife Service Fish Farming Experimental Station (Stuttgart, Arkansas) from Malaysia and by Auburn University (Auburn, Alabama) from Taiwan in 1963 (Guillory and Gasaway 1978). It was subsequently distributed to research agencies or companies with research capabilities in eleven states (Provine 1975). The first release was in Arkansas, the result of escapes from the Fish Farming Experimental Station, followed by intentional releases made by the Arkansas Game and Fish Commission (Sneed 1972, Greenfield 1973).

The rationale for importation and introduction of the grass carp was biological control of aquatic weeds.

Cyprinus carpio **Linnaeus** **common carp**

Berg (1949a) described the native range of the common carp as the basins of the Black, Caspian, and Aral seas, perhaps western Europe, Volga River, and the rivers flowing into the Pacific and in eastern Asia from the Amur River southward to Burma. Mišik (1958), Balon (1974), and others suggested that the species first appeared in Asia Minor and the basin of the Caspian Sea and spread from there into western Europe and eastward to China. Transfer into other waters of Europe outside the Danube River (where this fish is native) as early as the Roman Empire obscured the geographical origins of the common carp, leading to assumptions that it was native to China and introduced into Europe (Balon 1974).

Self-sustaining populations of common carp exist in the forty-eight contiguous states (Allen 1980). The greatest population densities of this fish are in the midwestern states.

The first introduction in North America was into the Hudson River in New York in 1831 and was made by an individual citizen (DeKay 1842). In 1872, 5 carp imported from Germany were stocked in a pond in California's Sonoma Valley (Moyle 1976). In May 1877, the U.S. Fish Commission successfully imported 338 common carp from Germany and distributed them for culture and introductions to applicants throughout the United States and Canada until 1896 (Baird 1879, Laycock 1966, Scott and Crossman 1973). The reason for these importations was the culture and introduction of a food fish (DeKay 1842, Baird 1879).

Leuciscus idus **(Linnaeus)** **ide**

The ide is native to Europe from the Rhine River eastward, the rivers of the Danube basin, northward to 61°10′ N in Norway, to the rivers at the northern end of the Gulf of Bothnia in Sweden, in Finland to 68°30′ N, Baltic

Sea basin, eastward in the basins of the Arctic Ocean, White Sea, and Volga, and rivers of the northern coast of the Black Sea to the Danube (Berg 1949a).

This fish was established in one pond in East Lyme Township, New London County, Connecticut (Whitworth et al. 1968) but has since been eradicated (R. A. Jones, J. C. Moulton, personal communications). It was formerly established in ponds in Delaware and York counties, Pennsylvania, and may exist in other private waters of that state (E. L. Cooper, C. N. Shiffer, personal communications). Bean (1901, 1904) reported the ide as established on Long Island, New York; this population is no longer extant. It was collected in the Chanango River, a tributary of the Susquehanna River, between Hamilton and Norwich, New York, in 1950–51 (C. R. Robins, personal communication). Smith and Bean (1899), Bean and Fowler (1929), and Schwartz (1963) reported it as established in the Potomac River but Musick (1972) doubted that it was still extant there.

An isolated population was found in a private pond at Holden, Penobscot County, Maine, in late March 1983 (P. G. Walker, personal communication). That population was eradicated in May 1983 by the Maine Department of Inland Fisheries and Wildlife. Recently (July 1983), this fish was being utilized as a bait species in Tennessee (L. B. Starnes, personal communication).

Although we presently are unaware of any established population of the ide in the United States, the likelihood of finding such a population is great because of its past history and because sources in New Jersey and Florida are distributing the fish. Moreover this species is often misidentified as goldfish.

The ide was imported by the U.S. Fish Commission in 1877 for distribution (Baird 1879). Baird (1893) reported a flood in the fish ponds in Washington, D.C., that washed most of the fishes being held there, including the ide, into the Potomac River. Although no specific reason was given for importing and distributing this species, one can assume that it was to be used as both an ornamental and food fish. It is popular with European anglers (Holčík and Mihálik 1968).

***Rhodeus sericeus* Pallas bitterling**

The bitterling is native in Europe from the Seine eastward, the basins of the Black and Caspian seas, rivers entering the Baltic Sea from the south to the Neva basin, Transcaucasia, Asia Minor, rivers entering the Aegean Sea, the Amur basin, Sungari, Ussuri, and Uda rivers, the Tym and Poronai rivers on Sakhalin, southward into the northern People's Republic of China; it is absent in central Asia and Siberia (Berg 1949a).

This species is established in the Bronx River at Bronxville, Westchester County, New York (Greeley 1937, Schmidt et al. 1981). It was formerly abundant in the Sawmill River at Tarrytown, Westchester County, New York (Dence 1925, Myers 1925, Bade 1926) but is no longer considered extant there (Schmidt et al. 1981).

The bitterling was probably an aquarium fish release (Myers 1925, Bade 1926, Schmidt et al. 1981).

***Scardinius erythrophthalmus* (Linnaeus) rudd**

The native range of the rudd is Great Britain (except as noted below), Europe eastward to the Ural and Emba rivers, Asia Minor, Caucasia, southern coast of the Caspian Sea, Aral Sea basin, the Bay of Volkhov in Lake Ozero, Gulf of Finland, in Finland north to 63°20′ N, north Caucasus, western and eastern Transcaucasia, Aral Sea basin, and Chu, Syr Darya, and Sary Su rivers; absent from Greece, Ireland, Spain, northern Scotland, northern and central Scandinavia (Berg 1949a).

The rudd is established in Cobbosseeconte Lake near Augusta, Kennebec County, Maine (P. G. Walker, personal communication), in Copake Lake in the Taghkanic Creek drainage, and below the dam at the outlet of Robinson Pond in the Roeliff-Jansen Kill, Columbia County, New York (C. L. Smith, personal communication). Greeley (1937) reported this species from the Roeliff-Jansen Kill. It was reported as established in Central Park Lake, New York City, and a lake in Hudson County Park, Jersey City, Hudson County, New Jersey (Myers 1925), but apparently is no longer extant at these sites (C. L. Smith and R. W. Hastings, personal communications). C. R. Robins (personal communication) collected the rudd in the early 1950s in Cascadilla Creek near Ithaca, Tompkins County, New York; we are unaware of any recent records from that area. It was also established in Oconomowoc Lake, Waukesha County, Wisconsin (Cahn 1927, Greene 1935); it is now considered extinct in Wisconsin by G. C. Becker (personal communication via H. E. Booke) and J. L. McNelly and D. Fago (personal communication), although we are unaware of any recent studies to confirm its presence or absence.

There is no information available as to the source or rationale for the introductions of this species in New York or New Jersey. Cahn (1927) and Greene (1935) stated that the rudd was introduced into Wisconsin in 1917 by the Wisconsin Conservation Department; the rationale was probably the introduction of a food and sport fish, as it is caught by anglers in Europe (Holčík and Mihálik 1968).

***Tinca tinca* (Linnaeus) tench**

Berg (1949a) cited the native range of the tench as western Europe (except northern Norway and Sweden), England, Scotland, Ireland, rivers of the Baltic, Caspian and Black Sea basins eastward to the Ural and Emba rivers, Sicily, southward flowing rivers entering the Aegean Sea, near Istanbul, Caucasus, western and eastern Transcaucasia, middle and lower Kura River eastward to the Ob and Yenisei rivers and Lake Baikal.

In North America, the tench is established in California (Moyle 1976), Colorado (Beckman 1974), Connecticut (one small lake in Litchfield County; C. W. Wilde, personal communication), Idaho (Simpson and Wallace 1978), New Mexico (Koster 1957), and Washington (Wydoski and Whitney 1979). It is possibly established in Delaware, Maryland, and New York (Schwartz 1963). C. R. Robins (personal communication) collected this species in the early 1950s in Cascadilla Creek near Ithaca, Tompkins County, New York. J. E. Johnson (personal communication) has not collected this fish in the central Rio Grande in New Mexico in recent years. This species was recorded established in Oregon by Bond (1973) but he (personal communication) stated that it has not been collected there for more than two decades. The tench has been collected but is not known to be established in Arizona (Minckley 1973) and Missouri (Baughman 1947).

The tench was imported by the U.S. Fish Commission from Germany in 1877 (Baird 1879) and subsequently was distributed to applicants in several states. Baird (1893) listed Colorado, Indian Territory, Indiana, Kansas, Maryland, Michigan, and Missouri as places where the tench was delivered in 1890–91. In 1889, tench were washed into the Potomac River when federal fish ponds in Washington, D.C., were flooded (Baird 1893). The rationale for introductions of this species was food and sport.

COBITIDAE – LOACHES

***Misgurnus anguillicaudatus* (Cantor) oriental weatherfish**

The native range of the oriental weatherfish is eastern Asia including the Tugur and Amur river basins, the Tym and Poronai rivers of Sakhalin, the Sedanka River near Vladivostok, U.S.S.R., the Tumen-Ula River in North Korea, Hokkaido and Kyushu in Japan, Taiwan, the People's Republic of China from the Liao River south to Guangzhou, Kwangtung Province, and inland to Yunnan Province, Hainan, headwaters of the Irrawaddy River in Burma, and Tomkin and Annam provinces of North Vietnam (Berg 1949a).

In North America this fish is established in several flood control channels in Huntington Beach and Westminster, Orange County, California (St. Amant and Hoover 1969; M. H. Horn, personal communication), and in the headwaters of the Shiawassee River, Oakland County, Michigan (Schultz 1960; M. L. Smith, personal communication).

Imported as an aquarium fish since at least the late 1930s, the oriental weatherfish is believed to have escaped from an aquarium fish culture facility in Westminster, California (St. Amant and Hoover 1969). Escapes in Michigan were from an aquarium supply company which imported the fish from Kobe, Japan, in 1939. They were stocked in Sunset Pond east of Holly, Michigan, and escaped during periods of high water (Schultz 1960).

CLARIIDAE – AIR-BREATHING CATFISHES

***Clarias batrachus* (Linnaeus) walking catfish**

The walking catfish is native to fresh and brackish waters from Sri Lanka, eastern India, Bangladesh, Burma, and the Malay Archipelago (Mookerjee and Mazumdar 1950, Sterba 1966).

This species is established in Brevard, Broward, Charlotte, Collier, Dade, De Soto, Glades, Hendry, Highlands, Hillsborough, Indian River, Lee, Manatee, Martin, Monroe, Okeechobee, Palm Beach, Polk, Sarasota, and

St. Lucie counties in Florida (Courtenay 1978, 1979a). It is possibly established in Orange and Osceola counties. Specimens have been collected in the All American Canal, Riverside County (Minckley 1973), and the San Joaquin River, Sacramento County (M. R. Brittan, personal communication) in California; the Flint River in Georgia (Courtenay and Miley 1975); Rogers Spring above the Overton Arm of Lake Mead in Clark County, Nevada (J. E. Deacon, personal communication); and in Waldo Lake, Brockton, Norfolk, and Plymouth counties, Massachusetts (Halliwell 1979). There is no evidence of establishment outside of Florida.

Albino juvenile walking catfish were imported from Bangkok, Thailand, in the early 1960s for sale in the aquarium fish trade. Adults, subsequently imported as brood stock, either escaped from culture facilities or from a truck transporting brood fish between Miami, Dade County, and Parkland, Broward County, Florida (Courtenay 1979b) in the mid-1960s. Releases in the Tampa Bay area, Hillsborough County, in about 1968 resulted in at least one population being established there (Courtenay et al. 1974, Courtenay and Miley 1975). As of 1968, this exotic fish was confined to three counties of Florida; by 1978 it had spread to twenty counties in the southern half of peninsular Florida – a dramatic feat for a fish.

LORICARIIDAE – SUCKERMOUTH CATFISHES

***Hypostomus* spp.** **suckermouth catfishes**

The native range of this genus is from the Río de la Plata system northward throughout South America (except Pacific slope drainages in Chile and Peru), Panama, and Costa Rica (Fowler 1954, Bussing 1966).

At least three morphologically distinct but unidentified species of *Hypostomus* are established in the United States. One is in Six Mile Creek near Eureka Springs, Hillsborough County, Florida (Courtenay et al. 1974). A second species is established in Indian Spring, Clark County, Nevada (Minckley 1973, Courtenay and Deacon 1982). The third species is established in the San Antonio River, Bexar County, Texas (Barron 1964, Hubbs et al. 1978, Hubbs 1982).

Suckermouth catfishes are popular aquarium fishes. The Hillsborough County, Florida, population was an alleged escape from a culture facility (Burgess 1958). Releases elsewhere in Florida may have been from aquarium fish farms, or, more likely, were made by aquarium hobbyists. Inasmuch as the only fishes in Indian Spring, Nevada, are popular aquarium fishes, the release of aquarium fishes there appears to have been the source of that *Hypostomus*. The Texas population was an escape from the San Antonio Zoological Gardens (Barron 1964, Hubbs et al. 1978).

***Pterygoplichthys multiradiatus* (Hancock)**

This genus is native to the Río Magdalena, Colombia; Guyana; throughout the Amazon and Río San Francisco, Brazil; upper Amazon tributaries in Bolivia, Peru, and Venezuela; Río de la Plata, Paraguay and Argentina (Gosline 1945, Fowler 1954, Isbrücker 1980).

In June 1983, biologists of the Non-Native Fish Research Laboratory, Florida Game and Fresh Water Fish Commission, found an established population of *Pterygoplichthys multiradiatus* in a canal in northeastern Dade County, Florida (P. L. Shafland, personal communication). Single individuals of this catfish genus have been collected at separate localities in Dade County since 1971.

POECILIIDAE – LIVEBEARERS

***Belonesox belizanus* Kner** **pike killifish**

The native range of the pike killifish is the Atlantic slope of Middle America from Laguna San Julian, northwest of Ciudad Veracruz, Mexico, to Costa Rica (Caldwell et al. 1959, Miller 1966).

This species is established in canal systems and saline cooling canals (40 ppt) in southeastern Dade County, Florida (Belshe 1961, Rivas 1965, Lachner et al. 1970, Courtenay and Robins 1973, Courtenay et al. 1974, Miley 1978; C. R. Robins, personal communication).

The pike killifish was released into a canal along SW Eighty-seventh Avenue, Dade County, in November 1957, after a research grant in which the fish was being used experimentally was terminated (Belshe 1961, Miley 1978). Whether the release was made by researchers or importers holding excess stocks is unknown.

***Poecilia mexicana* (Steindachner)** **shortfin molly**

In a 1979 unpublished report on exotic fishes in the United States to the National Fishery Research Laboratory, U.S. Fish and Wildlife Service, Gainesville, Florida, Courtenay and Hensley treated the shortfin molly as a "species complex," following Schultz and Miller (1971). They recognized that records of other members of the complex in North America (*P. latipunctata, P. petenensis,* and *P. sphenops*) could have contained misidentifications. Because the identifications of *P. mexicana* in North America are now fairly certain, we confine our discussion to this species.

The native range of the shortfin molly includes the Atlantic slope of Central America from the Río San Juan (Rio Grande basin), Nuevo León, Mexico, and the Pacific slope from Río del Fuerte basin, Sonora, Mexico, southward to the Caribbean slope of Colombia, the Pacific slope of eastern Panama (Río Tuira), and the Netherlands and Colombian West Indies (Rosen and Bailey 1963, Miller 1966).

North of the Rio Grande basin, the shortfin molly was introduced and has become established in a drainage canal south of Mecca, Riverside County, California (St. Amant 1966, 1970, St. Amant and Sharp 1971, Hubbs et al. 1979, personal observation [WRC]); Trudau Pond, Madison County, Montana (Brown 1971); in several springs and most of the Moapa River, Clark County, and in Lincoln County, Nevada (Deacon et al. 1964, Hubbs and

Deacon 1965, Deacon 1979, Courtenay and Deacon 1982). Hahn (1966) reported a "large population" in hot springs and associated drainages at the ghost town of Valley View, Saguache County, Colorado, but the current status of this population is unknown. Fishes belonging to the species complex of which *P. mexicana* is a member have been reported from Dade County, Florida (Courtenay and Robins 1973, Courtenay et al. 1974; J. M. Pestrak, personal communication); although Courtenay and Robins (1973) and Courtenay et al. (1974) considered some of these fishes as established in Florida, their current status is unknown. Minckley and Deacon (1968) and Minckley (1973) collected the shortfin molly once in the metropolitan canal system near Phoenix, Arizona, but this species is apparently not established in Arizona.

The introductions of the shortfin molly, like those of the guppy, probably resulted from the dumping of unwanted pet fishes. St. Amant (1966, 1970) and St. Amant and Sharp (1971) suggested that the release of this fish in Riverside County, California, was the result of an escape from a nearby aquarium fish farm.

***Poecilia reticulata* Peters guppy**

The guppy occurs naturally in the Netherlands Antilles and Venezuelan islands, Trinidad, Windward (Barbados) and Leeward (St. Thomas and Antigua) islands, and from western Venezuela to Guyana. Rosen and Bailey (1963) suggested that records from the Lesser Antilles may be introductions. Endler (1980) cited northeastern Venezuela, Margarita, and Tobago as part of its natural range.

In North America, the guppy is established in Arizona (Minckley 1973), Florida (F. W. King, personal communication), Idaho (Simpson and Wallace 1978), Nevada (Deacon et al. 1964, Williams et al. 1980, Courtenay and Deacon 1982), Texas (Hubbs et al. 1978, Hubbs 1982), and Wyoming (Baxter and Simon 1970). It may be established locally in sewage treatment ponds in California (Moyle 1976). Courtenay et al. (1974) listed the guppy as established in Hillsborough and Palm Beach counties, Florida. The populations from Hillsborough County are probably not self-sustaining (escapes from local aquarium fish farms) and the formerly established local population in Boca Raton, Palm Beach County, was exterminated when its habitat dried in the late 1970s. F. W. King (personal communication to J. A. McCann) reported an established population near High Springs, Alachua County, Florida.

It appears that all releases of guppies in North America were made to dispose of unwanted pet fish.

***Poeciliopsis gracilis* Heckel porthole livebearer**

The porthole livebearer is native to the Atlantic and Pacific slopes of Central America from southern Mexico to Honduras (Rosen and Bailey 1963, Miller 1966).

This species was established before 1965 in an agricultural drain south of

Mecca, Riverside County, California (Mearns 1975, Moyle 1976, Hubbs et al. 1979; J. A. St. Amant, personal communication; personal observation [WRC]) and remains extant there. This site is adjacent to an aquarium fish farm, the presumed source for this self-sustaining population.

Xiphophorus helleri **Heckel green swordtail**

The native range of the green swordtail is the Atlantic slope of Middle America from the Río Nautla, Veracruz, Mexico, to northwestern Honduras (Rosen 1960, Rosen and Kallman 1969).

This species is established in a pond at the Satellite Beach Civic Center, Brevard County, and in certain canals and roadside ditches near Ruskin, Hillsborough County, Florida (Courtenay et al. 1974, Dial and Wainright 1983) and Trudau Pond, Madison County, Montana (Brown 1971). St. Amant and Hoover (1969) collected green swordtails near an aquarium fish farm in Westminster, Orange County, California. This species has also been collected in a drainage canal near Oasis, Mono County, and near Mecca, Riverside County, but neither population appears to be established. Moyle (1976) was uncertain whether permanent breeding populations occur in California. This species was established at Rock Springs, Maricopa County, Arizona, but the population was destroyed by flooding in 1965 (Minckley 1973). LaRivers (1962) reported the green swordtail as established in Rogers Spring, Clark County, Nevada, but Deacon et al. (1964) believed this record applied to nearby Blue Point Spring; this fish was not found in either spring in October 1980 (Courtenay and Deacon 1982, 1983). It was established in Indian Spring, Clark County, Nevada, in 1975 (J. E. Deacon, personal communication) but the only *Xiphophorus* present there as of October 1980 was a hybrid of what appears to be the green swordtail and southern platyfish (*Xiphophorus maculatus*).

The established populations in Florida appear to be releases of pet fish or escapes from local aquarium fish farms. The population in Montana and the hybrids in Indian Spring, Nevada, originated from aquarium releases.

Xiphophorus maculatus **(Günther) southern platyfish**

Miller (1966) described the native range of this fish as the Atlantic slope from just south of Ciudad Veracruz, Mexico, to northern Belize.

This species is established in canals at Satellite Beach, Brevard County (Dial and Wainright 1983), and in some roadside ditches near Ruskin and perhaps elsewhere in Hillsborough County, Florida (Courtenay et al. 1974, Dial and Wainright 1983; personal observation [WRC]). It is common only in isolated localities and never abundant or dominant. What appears to be a hybrid of this fish and the green swordtail is established in Indian Springs, Clark County, Nevada (Courtenay and Deacon 1982). It has been collected near a fish farm in Westminster, Orange County, California (St. Amant and Hoover 1969), but was not found there in collections made in December 1980 (M. H. Horn, personal communication). Hubbs et al. (1979) listed the species as introduced in California but not established. Hubbs (1972) cited the southern platyfish as known or suspected of having been released in Texas without indicating establishment.

The population at Satellite Beach, Florida, and the hybrid in Indian Springs, Nevada, are aquarium releases. The population near Ruskin, Florida, apparently escaped from nearby fish farms.

***Xiphophorus variatus* (Meek) variable platyfish**

The variable platyfish is endemic to Mexico, occurring from southern Tamaulipas, to eastern San Luis Potosí and northern Veracruz (Rosen 1960).

This species has been reported as established in canals and roadside ditches of the eastern shore of Tampa Bay, Hillsborough County, Florida (Courtenay and Robins 1973, Courtenay et al. 1974); the current status of these populations is uncertain. Burgess et al. (1977) and J. A. McCann found established populations in Gainesville, Alachua County, Florida. Brown (1971) reported established populations in thermal springs, in Beaverhead, Granite, and Madison counties, Montana. A population was established in the Salt River at Tempe, Maricopa County, Arizona, in 1963–65, but the population was destroyed by a flood in 1965 (Minckley and Deacon 1968; Minckley 1973). Minckley (1973) reported this fish from Yuma without evidence of establishment. St. Amant and Hoover (1969) and St. Amant and Sharp (1971) recorded the possible establishment of this species in Orange and Riverside counties, California, but it is doubtful that the variable platyfish still exists there (Moyle 1976, Hubbs et al. 1979; J. A. St. Amant, personal communication); Shapovalov et al. (1981) reported those populations as extinct.

Introductions of variable platyfish in the United States are probably due to dumping of unwanted aquarium fish and occasional escapes from aquarium fish farms. In at least two localities (Arizona State University, Tempe, and the University of Florida, Gainesville), college students may be the source of local introductions (W. L. Minckley, personal communication; personal observation [JAM]).

SCIAENIDAE – DRUMS

***Bairdiella icistia* (Jordan and Gilbert) bairdiella**

The bairdiella is native to the Pacific coast of Mexico to Almejas Bay (24°28′ N; 111°24′ W), just east of Isla Santa Margarita, Baja California Sur (Miller and Lea 1972). Berdegue (1956) reported this fish as abundant in the Gulf of California and near Mazatlán.

This fish was introduced and became established in the Salton Sea, Imperial and Riverside counties, California, where 67 individuals were released beginning in October 1950 (Walker et al. 1961, Shapovalov et al. 1981) by personnel of the California Department of Fish and Game. Walker et al. (1961) wrote that it was present in the millions in 1953, and Moyle (1976) listed this species as one of the three dominant fishes in the Salton Sea.

The bairdiella was introduced as a game species and a forage fish for another exotic fish, the orangemouth corvina (*Cynoscion xanthulus*).

***Cynoscion xanthulus* Jordan and Gilbert orangemouth corvina**
Miller and Lea (1972) recorded the native range of this fish from Acapulco, Mexico, northward into the Gulf of California. Berdegue (1956) cited it as abundant at Mazatlán.

Between 1950 and 1955, about 272 individuals were introduced as a game fish by personnel of the California Department of Fish and Game into the Salton Sea, Imperial and Riverside counties, California. Walker et al. (1961) reported millions there in 1953.

CICHLIDAE – CICHLIDS

***Astronotus ocellatus* (Cuvier) oscar**
The oscar is native to the Orinoco, Amazon, and La Plata river systems of South America.

This popular aquarium species first appeared in canals in Miami, Dade County, Florida, in the late 1950s after a deliberate release from an aquarium fish farm (Courtenay et al. 1974). It is now established in Broward, Dade, Glades, Hendry (probably), and Palm Beach counties. It has also been collected but is not established in Massachusetts (Halliwell 1979), Mississippi (Anonymous 1979), Pennsylvania (C. N. Shiffer, personal communication), and Rhode Island (W. H. Krueger, personal communication). These were aquarium fish releases.

***Cichlasoma bimaculatum* (Linnaeus) black acara**
The native range of the black acara is eastern Venezuela and Trinidad, Guyana, Surinam, French Guiana, Brazil, Ecuador (possibly), Bolivia, Paraguay, Uruguay, and northern Argentina (Regan 1905, Fowler 1954, Ringuelet et al. 1967).

This species is established in Broward, Dade, Collier, Hendry, Glades (probably), Monroe, and Palm Beach counties of Florida (Rivas 1965, Lachner et al. 1970, Kushlan 1972, Courtenay et al. 1974, Hogg 1976a, b, personal observation [DAH]). The first specimens in open waters of Florida were found in the early 1960s (Rivas 1965) and probably escaped or were released from aquarium fish farms.

***Cichlasoma citrinellum* (Günther) Midas cichlid**
The Midas cichlid is native to the Atlantic slope of Nicaragua (including the Great Lakes basin) and Costa Rica (Miller 1966).

This fish was found by biologists of the Non-Native Fish Research Laboratory, Florida Game and Fresh Water Fish Commission, established in Black Creek Canal and adjoining canals northeast of Homestead, Dade County, Florida, in late May 1981. As of 12 June 1981, its range was known to encompass 4.8 km of that canal system. Its introduction could have been from the release of unwanted pet fish or from an aquarium fish farm.

***Cichlasoma meeki* (Brind) firemouth**

The native range of the firemouth is Atlantic slope drainages from the Río Tonala in Veracruz and Tabasco, Mexico, to southern Belize, including the Yucatán Peninsula and the upper Usumacinta basin in Guatemala.

Courtenay et al. (1974) reported the firemouth as established in a rockpit on a private estate in South Miami, Dade County, and in a canal near Lantana, Palm Beach County, Florida. The current status of the population in the rockpit is unknown, but the Palm Beach County population is extinct. This species is established in the Comfort and probably in the connecting Tamiami and Snapper Creek canals in Dade County. It has been collected in a canal in southeastern Dade County (C. R. Robins, personal communication). A well-established population in an isolated borrow pit in Broward County was eradicated by personnel of the Florida Game and Fresh Water Fish Commission in late July 1981.

Minckley (1973) reported one specimen caught in a canal in Mesa, Maricopa County, Arizona; this species is not established there.

The populations in Dade County appear to have been the result of escapes or releases from aquarium fish farms. The population formerly established in Broward County, along with a recently eradicated (July 1981) population of the yellowbelly cichlid (*Cichlasoma salvini*), appears to have been the result of a deliberate release as the borrow pit is on the grounds of a former amusement park.

***Cichlasoma nigrofasciatum* (Günther) convict cichlid**

The convict cichlid is native to Pacific slope drainages from Guatemala to Costa Rica and Atlantic slope drainages of Costa Rica (Miller 1966).

This species is established in Rogers Spring, Clark County, and in the outflows of Crystal and Ash springs, Lincoln County, Nevada (Deacon et al. 1964, Hubbs and Deacon 1965, Deacon 1979, Courtenay and Deacon 1982, 1983). Minckley (1973) reported two previously established populations from Arizona: one in Mesa that did not survive the winter of 1970–71 and another in a borrow pit near eastern Phoenix, Maricopa County, that was destroyed by a flood in 1973. Rivas (1965) reported the convict cichlid from a rockpit in northwest Miami, Dade County, Florida, but there are no recent records of this fish from Florida.

Introductions appear to have been from releases of unwanted pet fish. In Nevada, it was apparently moved from one site of establishment into other waters where it quickly became established (Hubbs and Deacon 1964, Courtenay and Deacon 1982, 1983).

***Cichlasoma octofasciatum* (Regan) Jack Dempsey**

This species is native to Atlantic slope drainages from the Río Chachalacas basin, Veracruz, Mexico, to the Río Ulua basin in Honduras including the Yucatán Peninsula (Miller 1966).

The Jack Dempsey is established in four Florida counties: in ditches on the University of Florida campus, Gainesville, Alachua County; in canals near the Satellite Beach Civic Center and other canals from Satellite Beach to Canova Beach, Brevard County (Dial and Wainright, 1983); in Black Creek and Snapper Creek canals, Dade County (Courtenay et al. 1974; Hogg 1976a, b); and in a roadside ditch in Ruskin, Hillsborough County (Courtenay et al. 1974). An established population of this fish was eradicated from a rockpit in Levy County by the Florida Game and Fresh Water Fish Commission (Levine et al. 1979). Another population was established in a canal near an aquarium fish farm west of Lantana, Palm Beach County (Courtenay et al. 1974), but appears to have been extirpated. This species has been collected near Micco, Brevard County, but does not appear to be established there (R. G. Gilmore, personal communication).

The populations in Dade and Hillsborough counties appear to have resulted from escapes or releases from aquarium fish farms; those in Brevard County probably resulted from the releases of pet fish.

***Geophagus surinamensis* Bloch redstriped eartheater**

The native range of the redstriped eartheater includes the Guianas and the Amazon basin of Bolivia, Brazil, Colombia, and Peru (Gosse 1975). A reproducing population of this species was found by personnel of the Non-Native Fish Research Laboratory, Florida Game and Fresh Water Fish Commission, in 1982 in Snapper Creek, Dade County, Florida (P. L. Shafland, personal communication). Its presence there may be the result of an escape from a fish farm.

***Hemichromis bimaculatus* Gill jewelfish**

The jewelfish is native to rivers and lakes throughout western Africa, in the Chad basin, Nile River, and south to the Congo River (Daget and Iltis 1965).

This species is established in Dade County, Florida, in the Hialeah Canal and connecting canals to the west and south of Miami International Airport and in the Comfort Canal, the channelized South Fork of the Miami River (Rivas 1965; Courtenay and Robins 1973; Courtenay et al. 1974; Hogg 1976a, b). C. R. Robins (personal communication) reported its probable establishment in a canal east of Goulds. It may also be established in Snapper Creek, north of the Tamiami Canal (P. L. Shafland and R. J. Wattendorf, personal communications). The jewelfish has also been collected in a canal near an aquarium fish farm near Micco, Brevard County, Florida, but there is no evidence of its establishment there (R. G. Gilmore, personal communication).

Sources for the introduction of this fish in Dade County are unknown. Its distribution is largely centered in canals around Miami International Airport. Possibly, the jewelfish was released at the airport or from aquarium fish farms northward along the Hialeah Canal.

***Tilapia aurea* (Steindachner) blue tilapia**

The blue tilapia is native to the Senegal River, the Middle Niger River as far south as Bussa (not recorded from the lower Niger or from the Volta

River), Lake Chad, pools and lagoons of the lower Chari and Logone rivers, the lower Nile from near Cairo to the Delta lakes (but apparently only in fresh water), the Jordan River system, the Na'aman and Yarkon rivers in Israel, and the Asraq marshes and hot pools at Ein Fashka, Jordan (Trewavas 1965).

The blue tilapia is stocked annually in ponds at Auburn University, Alabama (Smith-Vaniz 1968). Although this fish overwintered outdoors in 1971, it usually would be expected to die if not brought into heated buildings. This species (or a hybrid with *T. nilotica,* the Nile tilapia) is established and locally dominant in the lower Colorado River in extreme southwestern Arizona and southeastern California below Laguna Dam and in the Gila River north of Yuma (W. L. Minckley and J. N. Rinne, personal communications; personal observation [WRC]). A. R. Essbach (personal communication) reported that this fish was used several years ago for algal control in an irrigation district near Gila Bend, Arizona. R. J. Behnke (personal communication) reports that this or a closely related tilapia is raised commercially on a catfish farm and on a hog farm near Alamosa, San Luis Valley, Colorado; its success there apparently is due to thermal springs.

In Florida the blue tilapia is established in eighteen counties: Alachua, Brevard, Dade, De Soto, Hardee, Hernando, Hillsborough, Lake, Manatee, Marion, Orange, Osceola, Palm Beach, Pinellas, Polk, Sarasota, Seminole, and Volusia. Foote (1977) recorded this species from Broward, Charlotte, Glades, and Pasco counties, but we are unaware of any specimens from these counties. This species is also reproducing in the saline waters of Tampa Bay.

It has been reported (but not confirmed) that *Tilapia aurea* is established in golf course ponds at Sea Island and St. Simons Island, Glynn County, Georgia. The blue tilapia was stocked in ponds in Iowa to test growth potential; although it reproduced there, it did not overwinter (Pelren and Carlander 1971). It was introduced in Lake Julian, a heated reservoir of Carolina Power and Light Company, Buncombe County, North Carolina, in 1965, and a self-sustaining population is extant (W. H. Tarplee, Jr., personal communication). Pigg (1978) reported the blue tilapia as established in 1977 in the North Canadian River, northwest of Harrah, Oklahoma County, Oklahoma, with a confirmed range of 383 km from Lake Overholser to Lake Eufaula. This population was believed to have died in late 1977 and early 1978, but additional specimens were collected there in 1979 (J. Pigg, personal communication via C. Hubbs) suggesting that the population persists.

In Texas the blue tilapia was reported as established in Braunig Reservoir, Bexar County; Canyon Reservoir, Comal County (Hubbs et al. 1978); Lake Nasworthy, Tom Green County; Amistad Reservoir, Val Verde County (Hubbs 1976); Tradinghouse Creek Reservoir, McLennan County; Lake Colorado City, Mitchell County; and Lake Fairfield, Freestone County (R. L. Bounds, personal communication via C. Hubbs). C. Hubbs (personal communication, 1981) reported this fish as dominant for 60 km in the Rio Grande below Falcon Reservoir, Zapata County. It was established in Trinidad Lake,

Henderson County (Noble et al. 1975, Germany 1977) but has been extirpated there (Noble 1977, Hubbs et al. 1978).

The sources and reasons for introduction of established populations of the blue tilapia are varied and in some cases only suggested. Introductions in Alabama were made by Auburn University for research purposes (Smith-Vaniz 1968); in Arizona by the Arizona Department of Game and Fish for algal control (A. R. Essbach, personal communication); in Colorado by private individuals for aquaculture (R. J. Behnke, personal communication); in Florida, initially by the Florida Game and Fresh Water Fish Commission for research purposes (mostly biological control) and subsequently by individuals (Crittenden 1962; Buntz and Manooch 1968; Courtenay and Robins 1973, 1975; Courtenay et al. 1974; Harris 1978); in Georgia possibly by private individuals for aquatic vegetation control; in North Carolina as a potential gamefish (J. H. Cornell in letter to R. M. Bailey of 25 September 1972); in Oklahoma possibly by Oklahoma Gas and Electric Company as an aquaculture experiment (Pigg 1978); and in Texas for aquaculture (Stickney 1979), "inadvertently introduced" (Noble et al. 1975), and the release of bait fish (C. Hubbs, personal communication). Its present distribution in the lower Colorado River in Arizona, California, and perhaps southern Nevada, is expected to expand because juveniles are being used as bait fish.

***Tilapia hornorum* Trewavas Wami tilapia**

The Wami tilapia is native to the Wami River basin, Tanzania (Trewavas 1966).

This fish, if its identity is correct, is established in California in the Bolsa Chica Flood Control Channel in Huntington Beach, Orange County. It, or a hybrid with *Tilapia mossambica,* is established in the Cerritos Flood Control Channel, Cerritos Lagoon, and the Coyote Creek–San Gabriel River drainage, Long Beach, Los Angeles County (Knaggs 1977, personal observation [WRC]). The Wami tilapia was reported as having been introduced and established in drainage channels in Imperial and Riverside counties and in Coyote Creek, Los Angeles County (Hauser et al. 1976, Legner and Pelsue 1977, Legner 1979, Legner et al. 1980).

Releases in California were made for aquatic plant control and to reduce mosquito and chironomid midge populations. The introductions were made by biologists at the University of California – Riverside, and by the Southeast and Orange County Mosquito Abatement districts (Legner and Pelsue 1977; E. H. Knaggs, personal communication).

***Tilapia mariae* (Boulenger) spotted tilapia**

Tilapia mariae is native to coastal lowlands in fresh water from the middle Ivory Coast to southwestern Ghana and from southeastern Benin to southwestern Cameroon (Thys van den Audenaerde 1966, Trewavas 1974).

The spotted tilapia is established in Florida throughout most of eastern and central Dade County northward into southeastern Broward County (Hogg 1974, 1976a, b; Courtenay and Hensley 1979, 1980). It has been reported from a pond south of Copeland, in southern Collier County.

An established population was discovered in Rogers Spring above the Overton Arm of Lake Mead, Clark County, Nevada, in October 1980 (Courtenay and Deacon 1982, 1983). One specimen was collected from nearby Blue Point Spring, but there was no evidence of a reproducing population at that site. Minckley (1973) stated that the Nile tilapia was introduced experimentally in southern Arizona. The photograph of this fish and the description of the juveniles, however, match the spotted tilapia. Nevertheless, there is no evidence that this tilapia is established in Arizona.

Hogg (1974; 1976a, b) suggested that the Florida populations originated from escapes or releases from aquarium fish farms. Courtenay and Hensley (1980) suggested possible purposeful release of this fish near Miami. The Nevada population probably resulted from the release of pet fish (Courtenay and Deacon 1982, 1983).

***Tilapia melanotheron* (Rüppell)** **blackchin tilapia**

This species is native to river delta lagoons from middle Liberia to south Cameroon (Thys van den Audenaerde 1971).

The blackchin tilapia is established in Florida in Hillsborough County from Lithia Springs to the mouth of the Alafia River, southward along the eastern shore of Tampa Bay to Cockroach Bay, Manatee County (Springer and Finucane 1963, Finucane and Rinckey 1964, Buntz and Manooch 1969, Lachner et al. 1970, Courtenay et al. 1974). It is established in Brevard County in canals near Satellite Beach and in the Indian and Banana rivers from Merritt Island southward to below Canova Beach, a distance of 27 km (Dial and Wainright 1983).

Springer and Finucane (1963) suggested that this species either escaped or was released from an aquarium fish farm on the eastern shore of Tampa Bay. The Brevard County population may have resulted from aquarium fish releases into the reflecting pool at the Satellite Beach Civic Center (Dial and Wainright 1983).

***Tilapia mossambica* (Peters)** **Mozambique tilapia**

The native range of the Mozambique tilapia is the eastward-flowing rivers of Africa, from the lower Zambezi and Shire systems in Mozambique southward in coastal drainages to Algoa Bay, South Africa (Jubb 1967, Thys van den Audenaerde 1968).

This fish has been established in agricultural drains and mitigation ponds near Yuma, Yuma County, Arizona, since the early 1960s (Minckley 1973, personal communication; personal observation [WRC]). *Tilapia aurea* now coexists with *T. mossambica* in these same mitigation ponds (personal observation [WRC], 1981). It also occurred in drains and in various portions of the

Gila River from Phoenix to just north of Yuma but appears to have been recently replaced in several locations by blue tilapia or a hybrid with the Nile tilapia (personal observation [WRC]). Flooding destroyed other populations established at Warm Springs on the San Carlos River, Gila and Graham counties, and in the Salt River in Tempe, Maricopa County (Minckley 1973).

In California the Mozambique tilapia is established in Imperial, Los Angeles, Orange, and Riverside counties. It was released recently (January 1983) into High Rock Spring, Lassen County (J. E. Williams, personal communication). Before 1976 it was locally established in a pond and tributary on the eastern side of the Salton Sea, in drainage canals near Bard and in the lower Colorado River (St. Amant 1966, Hoover and St. Amant 1970, Moyle 1976). It was subsequently introduced into other agricultural drains after 1976 and began invading the Salton Sea in 1978 (G. F. Black, F. G. Hoover, and J. A. St. Amant, personal communications); it, or a hybrid with an unknown congener, perhaps blue tilapia, is now the dominant fish in the Salton Sea in terms of biomass. This species was introduced with *Tilapia hornorum* (and possibly with hybrids of these species) into Coyote Creek, a tributary of the San Gabriel River (Legner et al. 1980) and waters in Long Beach, Los Angeles County, and the Santa Ana River, Orange County, in 1973. Specimens have been collected from marine waters at Seal Beach, Orange County, and Cerritos Lagoon, Long Beach, Los Angeles County (Knaggs 1977). Knaggs (1977) suggested that this fish moved along the coast from the San Gabriel River into Cerritos Lagoon and from there into Cerritos Flood Control Channel; it now appears that youngsters may have transplanted the Mozambique tilapia from the San Gabriel River into Cerritos Flood Control Channel to create a local sport fishery (E. H. Knaggs, personal communication on site to WRC).

The Mozambique tilapia is established in Florida in the saline Banana River near Cocoa Beach, Brevard County (Dial and Wainright 1983), and at five locations in Dade County: in ponds of the Aventura condominium community, Snapper Creek canal for about 9 km above its juncture with the Tamiami Canal, Tamiami Canal at U.S. highways 41 and 27, the Comfort Canal, and in two canals east of Goulds.

This fish was established in Lake Julian, Buncombe County, North Carolina. Introduced in 1965 (J. H. Cornell in letter to R. M. Bailey of 25 September 1972), it failed to survive beyond the early 1970s (W. H. Tarplee, Jr., personal communication).

Brown (1961) first recorded this fish in Texas from springs in the San Marcos River, Hays County, and in headwater springs of the San Antonio River, Bexar County; it remains at these two localities (Hubbs et al. 1978).

Smith-Vaniz (1968) stated that this species is stocked annually in numerous farm ponds and a few state-owned lakes in Alabama; there is no evidence of overwintering there. A commercial catfish farm at Hooper, San Luis Valley,

Colorado, maintains this fish in its raceways for aquatic plant control (R. J. Behnke, personal communication). It may be established in golf course ponds on Sea Island and St. Simons Island, Glynn County, Georgia. Childers and Bennett (1967) reported the stocking of this fish for experimental purposes in Arrowhead Pond, Allerton Park, of the University of Illinois near Monticello, during the spring of 1962–65; although it reproduced, it was moved indoors during winter months. Brown and Fox (1966) reported the collection of four specimens in 1962–63 from a spring-fed pond in Bearmouth, Granite County, Montana; this pond was later destroyed by highway construction (Brown 1971).

The Mozambique tilapia was introduced into Arizona by the Arizona Game and Fish Department and the Arizona Cooperative Fishery Research Unit, U.S. Fish and Wildlife Service, for aquatic plant control (Minckley 1973). It was introduced in California for aquatic plant, mosquito, and chironomid midge control by many different agencies, including the Orange County and Southeast Mosquito Abatement districts (Knaggs 1977, Legner and Pelsue 1977); the Salton Sea population, however, appears to have originated from stocks released into irrigation drains after 1976 by unknown persons (G. L. Black, F. G. Hoover, and J. A. St. Amant, personal communications).

Populations in Brevard County, Florida, appear to have originated from stocks of pet fish released at Satellite Beach (Courtenay et al. 1974). Most of those in Dade County probably escaped from aquarium fish farms; at one locality, establishment apparently resulted from an introduction for aquatic plant control by a developer.

In Lake Julian, North Carolina, the Mozambique tilapia was introduced as a sport fish by the North Carolina Wildlife Resources Commission (J. H. Cornell in a letter to R. M. Bailey of 25 September 1972). Brown (1961) listed the source of the San Antonio River, Texas, population as an escape from the San Antonio Aquarium in 1956; Hubbs et al. (1978) cited the A. E. Wood State Fish Hatchery and the Fish Culture Station of the U.S. Fish and Wildlife Service as the sources for escapes of this fish into the San Marcos River during 1958–59 and 1960–61, respectively.

***Tilapia zilli* (Gervais) redbelly tilapia**

The redbelly tilapia occurs in Africa in the Senegal, Sassandra, Bandama, Volta, Niger, Benue, Chari, Ubangi, Uele, and Ituri rivers, lakes Chad, Albert, and George, and in the Near East in the Jordan River (Thys van den Audenaerde 1968).

This species, or perhaps a hybrid with *T. guineensis* (P. V. Loiselle, personal communication), is established in ponds in Papago Park, Scottsdale, and in other waters near Phoenix, Maricopa County, and perhaps elsewhere in Arizona (Minckley 1973; personal communication, personal observation

[WRC]). It, or the hybrid, is also established in irrigation drains in the Coachella, Imperial, and Palo Verde valleys of California (Pelzman 1973, Moyle 1976). It was introduced into twenty ponds, lakes, and creeks in Kern, Los Angeles, Orange, Riverside, and Santa Clara counties but the status of these introductions is unknown (Moyle 1976). It was introduced recently (January 1983) in High Rock Spring, Lassen County (J. E. Williams, personal communication). Legner and Pelsue (1977) reported the species as no longer extant in Los Angeles County. It has been collected from marine coastal waters in Orange County but is not established there (Knaggs 1977). Pelzman (1973) reported introduced populations as having survived two winters in Napa County before becoming extirpated.

The redbelly tilapia was established in ponds on a golf course in Pahrump Valley, Nye County, Nevada (Courtenay and Deacon 1982), but is no longer extant there (personal observation [WRC]). It is established in headwater springs of the San Antonio River, Bexar County, Texas (C. Hubbs, personal communication).

This species is stocked annually in farm ponds and lakes in Alabama (Smith-Vaniz 1968) but is not established there. It was established in a rockpit near Perrine, Dade County, Florida, but was eradicated by personnel of the Florida Game and Fresh Water Fish Commission (P. L. Shafland, personal communication). Unconfirmed reports indicate that the redbelly tilapia is cultured as a food fish in Idaho. It is stocked for aquatic vegetation control by Texasgulf, Inc., near Aurora, Beaufort County, and by Carolina Power and Light Company in a heated reservoir at Wilmington, Brunswick County, North Carolina (W. H. Tarplee, Jr., personal communication). It is also cultured for aquatic vegetation control and a potential food fish in a heated portion of the Santee-Cooper Reservoir in South Carolina.

Introductions in Arizona were made by personnel of the University of Arizona and the Arizona Department of Game and Fish for vegetation control (Minckley 1973, personal communication). California released this fish for the same purpose and to reduce mosquito and chironomid midge populations (Legner and Pelsue 1977); there, introductions were conducted by many agencies, including the Goleta, Orange County, Northwest and Southwest Mosquito Abatement districts, the University of California, Riverside (E. H. Knaggs, personal communication), and a private individual (J. E. Williams, personal communication). The release in Nevada was made illegally by a private developer (Courtenay and Deacon 1982). The source of release of this fish in Texas appears to be the San Antonio Zoo (C. Hubbs, personal communication). Annual stockings in Alabama are conducted by personnel from the Alabama Department of Conservation and Auburn University (W. L. Shelton, personal communication).

The population that was established in southern Dade County, Florida, appears to have resulted from a release of pet fish or a release from a culture facility (Hogg 1976a, b). The Florida Game and Fresh Water Fish Commission experimented with this fish or a closely related species (probably *Tilapia rendalli* [Boulenger]) as a potential vegetation control during the late 1960s; these studies were terminated and the fish stock was destroyed in the early 1970s when it was determined the fish could have more negative than positive impacts in Florida waters (C. L. Philippy, personal communication).

GOBIIDAE – GOBIES

***Acanthogobius flavimanus* (Temminck and Schlegel) yellowfin goby**

The yellowfin goby is native to marine, brackish, and freshwater rivers of southern Japan, South Korea, and China (Tomiyama 1936, Okada 1955, Fowler 1961).

This fish is established in California in the San Joaquin River and Stockton Deepwater Channel, San Joaquin County (Brittan et al. 1963); the San Francisco Bay area, Sacramento Delta, Delta-Mendota Canal, San Luis Reservoir, Suisun Bay, and Bolinas Lagoon, Alameda, Contra Costa, Marin, Napa (possibly), San Francisco, San Mateo, Santa Clara, Solano, and Sonoma counties (Brittan et al. 1970); Elkhorn Slough near Moss Landing, Monterey County (Kukowski 1972); Tomales Bay, Marin County (Miller and Lea 1972); Newport Bay, San Gabriel River and Bolsa Chica Flood Control Channel, Orange County; Los Angeles and Long Beach harbors and the mouth of the Los Angeles River, Los Angeles County (Haacker 1979; Usui 1981). C. A. Usui (personal communication) reports this species as far south as San Diego and perhaps to Baja California Norte, Mexico, in 1980.

The source of the initial and perhaps subsequent introductions may have been the ballast pumped from transoceanic ships (Brittan et al.1963).

***Tridentiger trigonocephalus* (Gill) chameleon goby**

This species is native to the southern Amur River, fresh water in the Suifen River, mouths of rivers emptying into Peter the Great Bay, Tumen-ula River, eastern coast of Korea south to Pusan, Chemulpo, Liao River, Lusbun (Port Arthur), China south to at least Guangzhou (Canton), and Japan (Berg 1949b).

The chameleon goby is established in California in San Francisco Bay and Los Angeles Harbor (Brittan et al. 1970, Miller and Lea 1972, Moyle 1976, Haacker 1979).

Hubbs and Miller (1965) theorized that the initial introduction into San Francisco Bay may have been as fertilized eggs on the introduced Japanese oyster, *Crassostrea gigas.*

ANABANTIDAE – GOURAMIES

***Trichopsis vittata* (Kuhl and Van Hasselt) croaking gourami**

The croaking gourami is native to Java, Borneo, Sumatra, Malaya, Thailand, Laos, Cambodia, and Vietnam (Smith 1945).

It is established in a localized area on the south side of Lake Worth Drainage District canal L-36, Delray Beach, Palm Beach County, Florida. It probably escaped from one of the nearby aquarium fish farms.

Acknowledgments

We are indebted to the many persons who provided information, specimens, references, identifications, and advice that made much of this chapter possible. In particular, we thank D. F. Austin, J. R. Bailey, R. M. Bailey, F. G. Banks, B. Barnett, J. N. Baskin, R. J. Behnke, T. R. Bender, R. D. Bishop, G. Black, the late J. E. Böhlke, J. C. Briggs, P. T. Briggs, M. R. Brittan, G. H. Burgess, R. L. Butler, J. S. Carter, L. Chako, P. G. Chapman, N. R. Clark, E. L. Cooper, D. T. Cox, E. J. Crossman, J. E. Deacon, H. DeWitt, S. Dobkin, C. F. Duggins, Jr., R. England, P. H. Eschmeyer, W. N. Eschmeyer, A. R. Essbach, W. L. Fink, L. Fishelson, S. Gebhards, C. R. Gilbert, R. G. Gilmore, C. D. Goforth, R. J. Goldstein, V. Guillory, J. Hall, D. B. Halliwell, D. Hendrickson, C. H. Hocutt, F. J. Hoover, the late C. L. Hubbs, C. Hubbs, R. E. Jenkins, R. A. Jones, P. Krieger, W. H. Krueger, J. Kushlan, F. H. Langford, J. N. Layne, R. N. Lea, D. Levine, D. S. Lee, E. F. Legner, P. V. Loiselle, A. D. Linder, H. Loyacano, D. L. McAllister, W. J. McConnell, S. McKenney, G. A. Marsh, W. W. Miley II, R. R. Miller, W. L. Minckley, D. P. Moody, F. A. Morello, P. B. Moyle, J. G. Nickum, W. L. Pflieger, J. S. Ramsey, E. C. Raney, H. M. Ratledge, G. K. Reid, K. Relyea, R. E. Roberts, C. R. Robins, D. E. Rosen, R. H. Rosenblatt, J. A. St. Amant, R. E. Schmidt, W. B. Scott, P. L. Shafland, C. N. Shiffer, C. L. Smith, R. B. Socolof, J. G. Stanley, G. Staples, L. B. Starnes, R. H. Stasiak, J. R. Stauffer, Jr., J. A. Stolgitis, R. H. Stroud, C. Swift, R. E. Thomas, J. E. Thomerson, D. A. Thomson, D.F.E. Thys van den Audenaerde, J. C. Underhill, P. G. Walker, and F. J. Ware.

For assistance in field studies, we acknowledge P. and R. M. Adams, K. Aasen, M. Belfit, G. Black, L. M. Boyer, W. R. Courtenay III, R. S. Dial, T. Grail, P. Greger, T. Hardy, B. C. Hartig, V. Hensley, M. H. Horn, F. J. Hoover, S. Jackle, M. M. Leiby, G. Lattin, P. C. Marsh, J. McDonald, D. P. Moody, J. Y. Remus, J. N. Rinne, W. Rinne, D. B. Snyder, L. Stanaland, C. A. Usui, and S. A. Wainright.

We owe special thanks to J. L. Lane and B. J. Rice for their sincere interest in this project which extended far beyond the typewriter or word processor support provided.

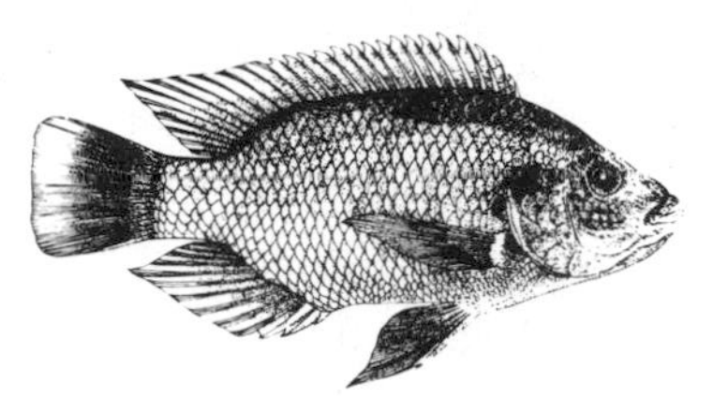

Literature Cited

Allen, A. W. 1980. *Cyprinus carpio* Linnaeus, 152. *In:* D. S. Lee, C. R. Gilbert, C. H. Hocutt, R. E. Jenkins, D. E. McAllister, and J. R. Stauffer, Jr. (eds.). Atlas of North American freshwater fishes. North Carolina Biological Survey Publication 1980–12.

Anonymous. 1979. Lateral line. Newsletter of the North American Native Fish Association.

Bade, E. 1926. The central European bitterling found in the States. Bulletin of the New York Zoological Society 29(6):188, 205–206.

Baird, S. F. 1879. The carp, 40–44. *In:* Report of the Commissioner. Report of the United States Fish Commission for 1876–77. U.S. Government Printing Office, Washington, D.C.

Baird, S. F. 1893. Report of the Commissioner, 1–96. *In:* Report of the United States Fish Commission for 1890–91. U.S. Government Printing Office, Washington, D.C.

Balon, E. K. 1974. Domestication of the carp *Cyprinus carpio* L. Royal Ontario Museum, Life Sciences Miscellaneous Publications. 37 pp.

Barron, J. C. 1964. Reproduction and apparent overwinter survival of the suckermouth catfish, *Plecostomus* sp., in the headwaters of the San Antonio River. Texas Journal of Science 16:449–450.

Baughman, J. L. 1947. The tench in America. Journal of Wildlife Management 11(3):197–204.

Baxter, G. R., and J. R. Simon. 1970. Wyoming fishes. Bulletin of the Wyoming Game and Fish Department 4:1–168.

Bean, B. A., and H. W. Fowler. 1929. The fishes of Maryland. Maryland Conservation Bulletin 3:1–120.

Bean, T. H. 1901. Catalogue of the fishes of Long Island, 373–478. Sixth Annual Report of the Forest, Fish and Game Commission of New York.

Bean, T. H. 1904. Catalogue of the fishes of New York. New York State Museum 60, Zoology 9, Appendix 5:1–784.

Beckman, W. C. 1974. Guide to the fishes of Colorado. University of Colorado Museum, Boulder, Colorado. 110 pp.

Belshe, J. F. 1961. Observations of an introduced tropical fish (*Belonesox belizanus*) in southern Florida. M.S. thesis, University of Miami, Coral Gables, Florida. 71 pp.

Berdegue, A. J. 1956. Peces de importancía comercial en la costa nor-occidental de México. Secretaría de Marina, Direccíon General de Pesca e Industrías Conexas, México. 347 pp.

Berg, L. S. 1949a. Freshwater fishes of the U.S.S.R. and adjacent countries. Vol. 2. Academy of Sciences of the U.S.S.R. Zoological Institute. Guide to the fauna of the U.S.S.R. 29. 1328 pp. (1964 translation by Israel Program for Scientific Translations. 496 pp.)

Berg, L. S. 1949b. Freshwater fishes of the U.S.S.R. and adjacent countries. Vol. 3. Academy of Sciences of the U.S.S.R. Zoological Institute. Guide to the fauna of the U.S.S.R. 30. 1331 pp. (1964 translation by Israel Program for Scientific Translation. 510 pp.)

Bond, C. E. 1973. Keys to Oregon freshwater fishes. Agricultural Experimental Station, Oregon State University, Technical Bulletin 58:1–42.

Brittan, M. R., A. B. Albrecht, and J. D. Hopkirk. 1963. An oriental goby collected in the San Joaquin River Delta near Stockton, California. California Fish and Game 49:302–304.

Brittan, M. R., J. D. Hopkirk, J. D. Conners, and M. Martin. 1970. Explosive spread of the oriental goby *Acanthogobius flavimanus* in the San Francisco Bay-Delta region of California. Proceedings of the California Academy of Sciences 38:207–214.

Brown, C.J.D. 1971. Fishes of Montana. Big Sky Books, Montana State University, Bozeman, Montana. 941 pp.

Brown, C.J.D., and A. C. Fox. 1966. Mosquito fish (*Gambusia affinis*) in a Montana pond. Copeia 1966(3):614–616.

Brown, W. H. 1961. First record of the African mouthbreeder *Tilapia mossambica* Peters in Texas. Texas Journal of Science 13:352–354.

Buntz, J., and C. S. Manooch III. 1968. *Tilapia aurea* (Steindachner), a rapidly spreading exotic in south central Florida. Proceedings of the Annual Conference of the Southeastern Association of Game and Fish Commissioners 22:495–501.

Buntz, J., and C. S. Manooch III. 1969. A brief summary of the cichlids in the south Florida region. Mimeographed report, Florida Game and Fresh Water Fish Commission. 3 pp.

Burgess, G. H., C. R. Gilbert, V. Guillory, and D. C. Taphorn. 1977. Distributional notes on some north Florida freshwater fishes. Florida Scientist 40(1):33–41.

Burgess, J. E. 1958. The fishes of Six Mile Creek, Hillsborough County, Florida, with particular reference to the presence of exotic species. Proceedings of the Annual Conference of the Southeastern Association of Game and Fish Commissioners 12:1–8.

Bussing, W. A. 1966. New species and records of Costa Rican freshwater fishes with a tentative list of species. Revista de Biología trópicale 14(2):205–249.

Cahn, A. R. 1927. An ecological study of southern Wisconsin fishes, the brook silversides (*Labidesthes sicculus*) and the cisco (*Leucichthys artedi*) in their relations to the region. Illinois Biological Monograph 2:1–151.

Caldwell, D. K., L. H. Ogren, and L. Giovannoli. 1959. Systematic and ecological notes on some fishes collected in the vicinity of Tortuguero, Caribbean coast of Costa Rica. Revista de Biología Trópicale 7(1):7–33.

Childers, W. F., and G. W. Bennett. 1967. Experimental vegetation control by largemouth bass–tilapia combinations. Journal of Wildlife Management 31:401–407.

Conner, J. V., R. P. Gallagher, and M. F. Chatry. 1980. Larval evidence for natural reproduction of the grass carp (*Ctenopharyngodon idella*) in the lower Mississippi River. Proceedings of the Fourth Annual Larval Fish Conference. Biological Services Program, National Power Plant Team, Ann Arbor, Michigan, FWS/OBS-80/43:1–19.

Courtenay, W. R., Jr. 1978. Additional range expansion in Florida of the introduced walking catfish. Environmental Conservation 5(4):273–276.

Courtenay, W. R., Jr. 1979a. Continued range expansion in Florida of the walking catfish. Environmental Conservation 6(1):20.

Courtenay, W. R., Jr. 1979b. The introduction of exotic organisms, 237-252. *In:* H. P. Brokaw (ed.). Wildlife and America. U.S. Government Printing Office, Washington, D.C.

Courtenay, W. R., Jr., and J. E. Deacon. 1982. The status of introduced fishes in certain spring systems in southern Nevada. Great Basin Naturalist 42(3):361-366.

Courtenay, W. R., Jr., and J. E. Deacon. 1983. Fish introductions in the American southwest: A case history of Rogers Spring, Nevada. Southwestern Naturalist 28(2):221–224.

Courtenay, W. R., Jr., and D. A. Hensley. 1979. Range expansion in southern Florida of the introduced spotted tilapia, with comments on its environmental impress. Environmental Conservation 6(2):149–151.

Courtenay, W. R., Jr., and D. A. Hensley. 1980. Special problems associated with monitoring exotic species, 281-307. *In:* C. H. Hocutt and J. R. Stauffer, Jr. (eds.). Biological monitoring of fish. Lexington Books, Lexington, Massachusetts.

Courtenay, W. R., Jr., and W. W. Miley II. 1975. Range expansion and environmental impress of the introduced catfish in the United States. Environmental Conservation 2(2):145–148.

Courtenay, W. R., Jr., and C. R. Robins. 1973. Exotic aquatic organisms in Florida with emphasis on fishes: A review and recommendations. Transactions of the American Fisheries Society 102(1):1–12.

Courtenay, W. R., Jr., and C. R. Robins. 1975. Exotic organisms: An unsolved, complex problem. BioScience 25(5):306–313.

Courtenay, W. R., Jr., H. F. Sahlman, W. W. Miley II, and D. J. Herrema. 1974. Exotic fishes in fresh and brackish waters of Florida. Biological Conservation 6(4):292–302.

Crittenden, E. 1962. Status of *Tilapia nilotica* in Florida. Proceedings of the Annual Conference of the Southeastern Association of Game and Fish Commissioners 16:257–262.

Daget, J., and A. Iltis. 1965. Poissons de Côte d'Ivorie. Mémoires de l'Institut français d'Afrique noire 74:1–385.

Deacon, J. E. 1979. Endangered and threatened fishes of the West. *In:* The endangered species: A symposium. Great Basin Naturalist Memoirs 3:41–64.

Deacon, J. E., C. Hubbs, and B. J. Zahuranec. 1964. Some effects of introduced fishes on the native fish fauna of southern Nevada. Copeia 1964(2):384–388.

DeKay, J. E. 1842. Zoology of New York – IV: Fishes. W. and A. White and J. Visscher, Albany, New York.

Dence, W. A. 1925. Bitter carp (*Rhodeus amarus*) from New York State waters. Copeia 142:33.

Dial, R. S., and S. C. Wainright. 1983. New distributional records for non-native fishes in Florida. Florida Scientist 46(1):8–15.

Endler, J. A. 1980. Natural selection on color patterns in *Poecilia reticulata*. Evolution 34(1):76–91.

Finucane, J. H., and G. R. Rinckey. 1964. A study of the African cichlid *Tilapia heudeloti* Dumeril, in Tampa Bay, Florida. Proceedings of the Annual Conference of the Southeastern Association of Game and Fish Commissioners 18:259–269.

Foote, K. J. 1977. Blue tilapia investigations. Study I: Preliminary status investigations. Mimeographed report, Florida Game and Fresh Water Fish Commission. 71 pp.

Fowler, H. W. 1954. Os peixes de água doce do Brasil. Archivos de Zoología do Estado São Paulo 9:203–404.

Fowler, H. W. 1961. A synopsis of the fishes of China. Part 9. The gobioid fishes. Quarterly Journal of the Taiwan Museum 14:203–250.

Germany, R. D. 1977. Population dynamics of the blue tilapia and its effects on fish populations of Trinidad Lake, Texas. Ph.D. dissertation, Texas A & M University, College Station, Texas. 85 pp.

Goode, G. B. 1903. American fishes. Dana Estes and Co., Boston, Massachusetts. 496 pp.

Gosline, W. A. 1945. Catálogo dos nematognatos de água-doce da América do Sul e Central. Boletim do Museau Nacional (Brasil), Zoologia 33:1–138.

Gosse, J. P. 1975. Révision du genre *Geophagus* (Pisces, Cichlidae). Mémoire Académie Royale de Sciences d'Outre-Mer. Classe de Sciences Naturelles et Médicales, N.S. 19-3:1–172.

Greeley, J. R. 1937. Fishes of the area with annotated list. *In:* A biological survey of the lower Hudson watershed. Supplement to the 26th Annual Report of the New York State Conservation Department 1936 (2):45–103.

Greene, C. W. 1935. The distribution of Wisconsin fishes. Wisconsin Conservation Commission, Madison, Wisconsin. 235 pp.

Greenfield, D. W. 1973. An evaluation of the advisability of the release of the grass carp, *Ctenopharyngodon idella,* into the natural waters of the United States. Transactions of the Illinois State Academy of Sciences 66(1/2):47–53.

Guillory, V., and R. D. Gasaway. 1978. Zoogeography of the grass carp in the United States. Transactions of the American Fisheries Society 107(1):105–112.

Haacker, P. L. 1979. Two Asiatic gobiid fishes, *Tridentiger trigonocephalus* and *Acanthogobius flavimanus,* in southern California. Bulletin of the Southern California Academy of Sciences 78(1):56–61.

Hahn, D. E. 1966. An introduction of *Poecilia mexicana* (Osteichthyes: Poeciliidae) into Colorado. Southwestern Naturalist 11(2):296–312.

Halliwell, D. B. 1979. Massachusetts fish list. Fauna of Massachusetts, ser. 4:1–13.

Harris, C. 1978. Tilapia: Florida's alarming foreign menace. Florida Sportsman 9(11):12, 15, 17–19.

Hauser, W. J., E. F. Legner, R. A. Medved, and S. Platt. 1976. *Tilapia*—a management tool for biological control of aquatic weeds and insects. Fisheries (Bethesda) 1(2):15–16.

Hogg, R. G. 1974. Environmental hazards posed by cichlid fish species newly established in Florida. Environmental Conservation 1(1):176.

Hogg, R. G. 1976a. Established exotic cichlid fishes in Dade County, Florida. Florida Scientist 39(2):97–103.

Hogg, R. G. 1976b. Ecology of fishes of the family Cichlidae introduced into the fresh waters of Dade County, Florida. Ph.D. dissertation, University of Miami, Coral Gables, Florida. 142 pp.

Holčík, J., and J. Mihálik. 1968. Fresh-water fishes. Hamlyn Publishing Group, Limited, New York. 128 pp.

Hoover, F. G., and J. A. St. Amant. 1970. Establishment of *Tilapia mossambica* Peters in Bard Valley, Imperial County, California. California Fish and Game 56(1):70.

Hubbs, C. L., W. I. Follett, and L. J. Dempster. 1979. List of the fishes of California. Occasional Papers of the California Academy of Sciences 133:1–51.

Hubbs, C. L., and R. R. Miller. 1965. Studies of cyprinodont fishes. Part 22. Variation in *Lucania parva,* its establishment in western United States, and description of a new species from an interior basin in Coahuila, Mexico. Miscellaneous Publications of the Museum of Zoology, University of Michigan 127:1–111.

Hubbs, C. 1972. A checklist of Texas freshwater fishes. Texas Parks and Wildlife Department, Technical Series 11:1–11.

Hubbs, C. 1976. A revised checklist of Texas freshwater fishes. Texas Parks and Wildlife Department, Technical Series 2:1–11.

Hubbs, C. 1982. Occurrence of exotic fishes in Texas waters. Pearce-Sellards Series, Texas Memorial Museum 36:1–19.

Hubbs, C., and J. E. Deacon. 1965. Additional introductions of tropical fishes into southern Nevada. Southwestern Naturalist 9(4):249–251.

Hubbs, C., T. Lucier, G. P. Garrett, R. J. Edwards, S. M. Dean, and E. Marsh. 1978. Survival and abundance of introduced fishes near San Antonio, Texas. Texas Journal of Science 30(4):369–376.

Isbrücker, I.J.H. 1980. Classification and catalog of the mailed Loricariidae. Verslagen en Technische Gegevens, Instituut voor Taxonomische Zoölogie (Zoölogisch Museum), Universiteit van Amsterdam 22:1–181.

Jubb, R. A. 1967. Freshwater fishes of southern Africa. A. A. Balkema, Cape Town. 248 pp.

Knaggs, E. H. 1977. Status of the genus *Tilapia* in California's estuarine and marine waters. California-Nevada Wildlife Transactions 1977:60–67.

Koster, W. J. 1957. Guide to the fishes of New Mexico. University of New Mexico Press, Albuquerque, New Mexico. 116 pp.

Kukowski, G. E. 1972. Southern range extension for the yellowfin goby, *Acanthogobius flavimanus* (Temminck and Schlegel). California Fish and Game 58:326–327.

Kushlan, J. A. 1972. The exotic fish (*Aequidens portalegrensis*) in the Big Cypress Swamp. Florida Naturalist 45:29.

Lachner, E. A., C. R. Robins, and W. R. Courtenay, Jr. 1970. Exotic fishes and other aquatic organisms introduced into North America. Smithsonian Contributions in Zoology 59:1–29.

LaRivers, I. 1962. Fishes and fisheries of Nevada. Nevada State Fish and Game Commission, Reno, Nevada. 782 pp.

Laycock, G. 1966. The alien animals. Natural History Press, Garden City, New York. 240 pp.

Legner, E. F. 1979. Considerations in the management of *Tilapia* for biological aquatic weed control. Proceedings of the California Mosquito Control Association 47:44–45.

Legner, E. F., R. A. Medved, and F. Pelsue. 1980. Changes in chironomid breeding patterns in a paved river channel following adaptation of cichlids of the *Tilapia mossambica-hornorum* complex. Annals of the Entomological Society of America 73:293–299.

Legner, E. F., and F. W. Pelsue. 1977. Adaptations of *Tilapia* to *Culex* and chironomid midge ecosystems in south California. Proceedings of the California Mosquito Control Association 45:95–97.

Levine, D. S., J. T. Krummrich, and P. L. Shafland. 1979. Renovation of a borrow pit in Levy County, Florida containing Jack Dempseys (*Cichlasoma octofasciatum*). Mimeographed report, Non-Native Fish Research Laboratory, Florida Game and Fresh Water Fish Commission, Boca Raton, Florida. 6 pp.

McAllister, D. E. 1963. Revision of the smelt family, Osmeridae. Bulletin of the National Museum of Canada 191:1–53.

MacCrimmon, H. R., and T. L. Marshall. 1968. World distribution of brown trout, *Salmo trutta*. Journal of the Fisheries Research Board of Canada 25(12):2527–2548.

MacCrimmon, H. R., T. L. Marshall, and B. L. Gots. 1970. World distribution of brown trout, *Salmo trutta:* Further observations. Journal of the Fisheries Research Board of Canada 27(4):811–818.

Mather, F. 1889. Brown trout in America. Bulletin of the U.S. Fish Commission 7(1887):21–22.

Mearns, A. J. 1975. *Poeciliopsis gracilis* (Heckel), a newly introduced poecillid fish in California. California Fish and Game 61(4):251–253.

Miley, W. W., II. 1978. Ecological impact of the pike killifish, *Belonesox belizanus* Kner (Poeciliidae), in southern Florida. M.S. thesis, Florida Atlantic University, Boca Raton, Florida. 55 pp.

Miller, D. L., and R. N. Lea. 1972. Guide to the coastal marine fishes of California. California Department of Fish and Game, Fisheries Bulletin 157:1–235.

Miller, R. R. 1966. Geographical distribution of Central American freshwater fishes. Copeia 1966(4):773–802.

Minckley, W. L. 1973. The fishes of Arizona. Arizona Game and Fish Department, Phoenix. 293 pp.

Minckley, W. L., and J. E. Deacon. 1968. Southwestern fishes and the enigma of "endangered species." Science 159(3822):1424–1432.

Mišik, V. 1958. Biometrika dunajskeho kapra (*Cyprinus carpio carpio* L.) z dunajskeho systemu na Slovensku (Biometry of the Danube wild carp [*Cyprinus carpio carpio* L.] of the Danube basin in Slovakia). Biologicke prace Slovenskej Akademie Vied. 4(6):55–125.

Mookerjee, H. K., and S. R. Mazumdar. 1950. Some aspects of the life history of *Clarias batrachus* (Linn.). Proceedings of the Zoological Society of Calcutta 3:71–79.

Moyle, P. B. 1976. Inland fishes of California. University of California Press, Berkeley, California. 405 pp.

Musick, J. A. 1972. Fishes of Chesapeake Bay and the adjacent coastal plain, 175–212. *In:* M. L. Wass (ed.). A checklist of the biota of lower Chesapeake Bay. Special Scientific Report, Virginia Institute of Marine Science 65.

Myers, G. S. 1925. Introduction of the European bitterling (*Rhodeus*) in New York and of the rudd (*Scardinius*) in New Jersey. Copeia 140:20–21.

Myers, G. S. 1940. An American cyprinodont fish, *Jordanella floridae,* reported from

Borneo, with notes on the possible widespread introduction of foreign aquarium fishes. Copeia 1940(4):267–268.
Noble, R. L. 1977. Response of reservoir fish populations to *Tilapia* reduction. Mimeographed Progress Report to the Sport Fishery Research Foundation. 4 pp.
Noble, R. L., R. D. Germany, and C. R. Hall. 1975. Interactions of blue tilapia and largemouth bass in a power plant cooling reservoir. Proceedings of the Annual Conference of the Southeastern Association of Game and Fish Commissioners 29:247–251.
Okada, Y. 1955. Fishes of Japan. Maruzen Company, Limited, Tokyo, Japan. 434 pp.
Okada, Y. 1966. Fishes of Japan. Uno Shoten Company, Limited, Tokyo, Japan. 474 pp.
Pelren, D. W., and K. D. Carlander. 1971. Growth and reproduction of yearling *Tilapia aurea* in Iowa ponds. Proceedings of the Iowa Academy of Sciences 78:27–29.
Pelzman, R. J. 1973. A review of the life history of *Tilapia zillii* with a reassessment of its desirability in California. California Department of Fish and Game, Inland Fisheries Administration Report 74-1:1–9.
Pigg, J. 1978. The tilapia *Sarotherodon aurea* (Steindachner) in the North Canadian River in central Oklahoma. Proceedings of the Oklahoma Academy of Sciences 58:111–112.
Provine, W. C. 1975. The grass carp. Mimeographed special report, Texas Parks and Wildlife Department, Inland Fisheries Research. 51 pp.
Regan, C. T. 1905. A revision of the fishes of the American cichlid genus *Cichlasoma* and of the allied genera. Annals and Magazine of Natural History (ser. 7) 16:60–77, 225–243, 316–340, 433–445.
Ringuelet, R. A., R. H. Aramburu, and A. A. de Aramburu. 1967. Los peces Argentinos de agua dulce. Comisión de Investigación Científica, B. A., La Plata. 602 pp.
Rivas, L. R. 1965. Florida freshwater fishes and conservation. Quarterly Journal of the Florida Academy of Sciences 28(3):255–258.
Robins, C. R., R. M. Bailey, C. E. Bond, J. R. Brooker, E. A. Lachner, R. N. Lea, and W. B. Scott. 1980. A list of common and scientific names of fishes from the United States and Canada. American Fisheries Society Special Publication 12. Bethesda, Maryland. 174 pp.
Rosen, D. E. 1960. Middle-American poeciliid fishes of the genus *Xiphophorus*. Bulletin of the Florida State Museum, Biological Sciences 5(4):57–242.
Rosen, D. E., and R. M. Bailey. 1963. The poeciliid fishes (Cyprinodontiformes), their structure, zoogeography, and systematics. Bulletin of the American Museum of Natural History 126(1):1–176.
Rosen, D. E., and K. D. Kallman. 1969. A new fish of the genus *Xiphophorus* from Guatemala, with remarks on the taxonomy of endemic forms. American Museum Novitates 2379:1–29.
St. Amant, J. A. 1966. Addition of *Tilapia mossambica* Peters to the California fauna. California Fish and Game 52(1):54–55.
St. Amant, J. A. 1970. Addition of Hart's rivulus, *Rivulus harti* (Boulenger), to the California fauna. California Fish and Game 56(2):138.
St. Amant, J. A., and F. G. Hoover. 1969. Addition of *Misgurnus anguillicaudatus* (Cantor) to the California fauna. California Fish and Game 55(4):330–331.

St. Amant, J. A., and I. Sharp. 1971. Addition of *Xiphophorus variatus* (Meek) to the California fauna. California Fish and Game 57(2):128–129.

Schmidt, R. E., J. M. Samaritan, and A. Pappantoniou. 1981. Status of the bitterling, *Rhodeus sericeus,* in southeastern New York. Copeia 1981(2):481–482.

Schultz, E. E. 1960. Establishment and early dispersal of a loach, *Misgurnus anguillicaudatus* (Cantor), in Michigan. Transactions of the American Fisheries Society 89(4):376–377.

Schultz, R. J., and R. R. Miller. 1971. Species of the *Poecilia sphenops* complex (Pisces: Poeciliidae). Copeia 1971(2):282–290.

Schwartz, F. J. 1963. The fresh-water minnows of Maryland. Maryland Conservationist 40(2):19–29.

Scott, W. B., and E. J. Crossman. 1973. Freshwater fishes of Canada. Bulletin of the Fisheries Research Board of Canada 184:1–966.

Shafland, P. L. 1976. The continuing problem of non-native fishes in Florida. Fisheries 1(6):25.

Shafland, P. L., and K. J. Foote. 1979. A reproducing population of *Serrasalmus humeralis* Valenciennes in southern Florida. Florida Scientist 42(4):206–214.

Shapovalov, L., A. J. Cordone, and W. A. Dill. 1981. A list of the freshwater and anadromous fishes of California. California Fish and Game 61(1):4–38.

Simpson, J. C., and R. L. Wallace. 1978. Fishes of Idaho. University Press, Idaho, Moscow, Idaho. 237 pp.

Smith, H. M. 1945. The fresh-water fishes of Siam, or Thailand. Bulletin of the United States National Museum 188:1–622.

Smith, H. M., and B. A. Bean. 1899. Fishes known to inhabit the waters of the District of Columbia and vicinity. Bulletin of the United States Fish Commission 1898:179–187.

Smith-Vaniz, W. F. 1968. Freshwater fishes of Alabama. Auburn University, Auburn, Alabama. 211 pp.

Sneed, K. 1972. The history of introduction and distribution of grass carp in the United States. Bureau of Sport Fisheries and Wildlife. Mimeographed report. 5 pp.

Springer, V. G., and J. H. Finucane. 1963. The African cichlid, *Tilapia heudeloti* Dumeril, in the commercial fish catch of Florida. Transactions of the American Fisheries Society 92(3):317–318.

Sterba, G. 1966. Freshwater fishes of the world. Pet Library, Limited, New York. 877 pp.

Stickney, R. R. 1979. Principles of warmwater aquaculture. Wiley-Interscience, New York. 375 pp.

Thys van den Audenaerde, D.F.E. 1966. Les *Tilapia* (Pisces, Cichlidae) du Sud-Cameroun et du Gabon; étude systématique. Annales, Musée royale de l'Afrique centrale, Sciences zoologiques 153:1–98.

Thys van den Audenaerde, D.F.E. 1968. An annotated bibliography of *Tilapia* (Pisces, Cichlidae). Musée royale de l'Afrique centrale, Documentation zoologique 14:1–406.

Thys van den Audenaerde, D.F.E. 1971. Some new data concerning the *Tilapia* species of the subgenus *Sarotherodon* (Pisces, Cichlidae). Revue de Zoologie et de Botanique africaine 84(3-4):203–216.

Tomiyama, I. 1936. Gobiidae of Japan. Japanese Journal of Zoology 7:37–112.

Trewavas, E. 1965. *Tilapia aurea* (Steindachner) and the status of *Tilapia nilotica exul,*

T. monodi, and *T. lemassoni* (Pisces, Cichlidae). Israel Journal of Zoology 14:258–276.

Trewavas, E. 1966. A preliminary review of fishes of the genus *Tilapia* in the eastward-flowing rivers of Africa, with proposals of two new specific names. Revue de Zoologie et de Botanique africaine 74(3-4):394–424.

Trewavas, E. 1974. The freshwater fishes of rivers Mungo and Meme and lakes Kotto, Mboandong and Soden, West Cameroon. Bulletin of the British Museum (Natural History), Zoology 26(5):331–419.

Usui, C. A. 1981. Behavioral, metabolic, and seasonal size comparisons of an introduced gobiid fish, *Acanthogobius flavimanus,* and a native cottid, *Leptocottus armatus,* from upper Newport Bay, California. M.A. thesis, California State University, Fullerton, California. 52 pp.

Wales, J. B. 1962. Introduction of pond smelt from Japan into California. California Fish and Game 48(2):141–142.

Walker, B. W., R. R. Whitney, and G. W. Barlow. 1961. The fishes of the Salton Sea, 77–91. *In:* B. W. Walker (ed.). The ecology of the Salton Sea, California, in relation to the sportfishery. California Department of Fish and Game, Fishery Bulletin 113.

Whitworth, W. R., P. L. Berrien, and W. T. Keller. 1968. Freshwater fishes of Connecticut. Bulletin of the State Geological and Natural History Survey, Connecticut 101:1–134.

Williams, J. E., C. D. Williams, and C. E. Bond. 1980. Survey of fishes, amphibians, and reptiles on the Sheldon National Wildlife Refuge, Nevada, 1. Fishes of the Sheldon National Wildlife Refuge. Mimeographed report, Contract 14-16-0001-78025, U.S. Fish and Wildlife Service. Oregon State University, Corvallis, Oregon, 58 pp.

Wydoski, R. S., and R. R. Whitney. 1979. Inland fishes of Washington. University of Washington Press, Seattle. 220 pp.

CHAPTER 5

Introduction of Exotic Fishes into Canada

E. J. Crossman

Man has long "managed" fish faunas by manipulating the exploited fauna, the exploiter, or both simultaneously. In a 1968 symposium on the introduction of exotic species, I suggested (Crossman 1968) we might well divide fisheries management into periods on the basis of which tool we relied on for salvation. I further suggested that we were then in the age of eradication and had passed through the ages of law enforcement, regulations, hatcheries, and in 1850–1900s the age of introductions. In the late 1800s Canada quickly followed the United States' lead in the importation of foreign fishes. Most of these, like the brown trout (*Salmo trutta*), were sport fishes innocently imported with the idea of increasing the number of species available to the angler. For the common carp (*Cyprinus carpio*), which was one of the earliest to be tried, the idea was different. Carp were imported into Ontario with the idea of "furnishing in the future a cheap article of food" (Crossman 1968). Samuel Wilmott, the father of fish culture in Canada, failed in his efforts to rehabilitate the Atlantic salmon in Lake Ontario. He then turned to carp, saying that he hoped "soon to see those waters deserted by the salmon now, stocked with a fish that will be welcomed to the poor man's table" (McCrimmon 1968:20). However, within nineteen years of their introduction, the Ontario Department of Fish and Game reported that "it is generally conceded that the promiscuous introduction of carp on this continent had been attended with nothing but evil results" (Crossman 1968:8).

Here, early in the twentieth century, we already have the first attempt in Canada to compensate for man's extirpation of a population of one fish by the introduction of an exotic. The high hopes were dashed in nineteen short years, but the exotic is with us still.

It is interesting that in the 1960s exotics were once again looked on as a potential salvation of another disastrous period of environmental degradation. Foreign species, longer subjected to pollution, were investigated as possibly better able to withstand the degraded environmental conditions in the Great Lakes (Christie 1968, 1970a).

That interest, and another attempt to introduce various Pacific salmons (*Oncorhynchus* spp.) in the East, could well be called Canada's Second Age of Exotics.

As will be seen below, Canada's history of introductions of exotics, as defined in this book, has been long and not always attended with success. The number of species involved (18) is, however, small. Included are four salmonids, one umbrid, two cyprinids, one ictalurid, three cichlids, two anabantids, and a pleuronectid, as follows: Danube salmon (*Hucho hucho*), cherry salmon (*Oncorhynchus masou*), golden trout (*Salmo aguabonita* *), brown trout (*Salmo trutta* *), Alaska blackfish (*Dallia pectoralis*), goldfish (*Carassius auratus* *), common carp (*Cyprinus carpio* *), margined madtom (*Noturus insignis* *), mosquitofish (*Gambusia affinis* *), sailfin molly (*Poecilia latipinna* *), guppy (*Poecilia recticulata*), green swordtail (*Xiphophorus helleri*), convict cichlid (*Cichlasoma nigrofasciatum*), jewelfish (*Hemichromis bimaculatus* *), angelfish (?*Pterophyllum scalare*), Siamese fighting fish (*Betta splendens*), blue gourami (*Trichogaster trichopterus*), and European flounder (*Platichthys flesus* *). Of these, only nine (marked *) are known or thought to be still in existence in at least some part of the area into which they were introduced.

Only the province of Prince Edward Island and the Northwest and Yukon territories have neither introduced, nor added by immigration, exotics as defined. Newfoundland and Nova Scotia introduced only *Salmo trutta*. If we include exotics introduced and those immigrating into the area as a result of nearby introductions, then New Brunswick has had 3, Manitoba 4, Quebec and Ontario 5 each, British Columbia 6, and Alberta 13(?) or 10. The term "introduced" as used in this chapter will include both official (purposeful) and unofficial (illegal or accidental) introductions.

We are dealing here with exotics as defined by the Exotic Fish Section AFS which technically excludes rainbow trout, since that species is native to western Canada. It actually does qualify, however, since the fish used to introduce it outside its native Canadian range were steelheads from the McCloud River in California, liberated into the waters of Lake Superior near Sault Sainte Marie, Ontario, in 1883. That story is extensive, and is well documented by MacCrimmon (1971), MacCrimmon and Gots (1972), and Scott and Crossman (1973), so it will not be included here.

The records of introductions into Canada are not as good as they should be. The details of many of the lesser and early introductions are not now well known. Subsequent reports are often contradictory, and the fact that many did not succeed and are now of no consequence is not a satisfactory excuse for the lack of adequate, persistent documentation. In the early years when the Dominion government was responsible for all fish culture, the records were in the reports of a single agency. As each province took over the responsibility for freshwater fishes in its area, the number of experiments proliferated and the location of records became diffuse. The interest in documenting these introductions, before details disappear, varies drastically across the country. Only two provincial publications actually summarize the history of introductions. These document the situation for the provinces of Newfoundland (Andrews 1965, 1966) and Saskatchewan (Marshall and Johnson 1971). For British Columbia, there is a summary for species of the genus *Oncorhynchus* (Aro 1979) and a massive computer printout that cryptically summarizes the introduction of exotics and the stocking of native species. A report for Manitoba (Wright and Sopuk 1979) summarizes introductions and stocking in at least the northern part of the province, and good records are available for the remainder of that province also. Some provinces said good records exist, but were too time-consuming to summarize. In some provinces, like Ontario, records now appear to be nonexistent or so scattered as to be unavailable. The existing publications dealing with the fish faunas of individual provinces or regions generally have a short section, separate from the general text, that concerns exotic or introduced fishes. These have been cited where appropriate and listed in the references.

Apparently the only person who attempted to summarize introductions across the whole of Canada was Dymond (1955). He included all species, native or otherwise, "introduced" into areas of Canada outside the native range of the species. Scott and Crossman (1973 and reprints) included references and brief notes on various introduced species. Dymond (1955) gave information on four exotics as defined for this exercise—*Salmo trutta, Carassius auratus, Cyprinus carpio,* and *Tinca tinca*. Since that time the list has increased to the 15 species and a number of hybrids included here.

This present summary was prepared by searching the literature, preparing lists of the information found on each species for each province or territory, submitting the appropriate list to one or more agencies or individuals in each area, and requesting that they correct or add only as required and return the list. The responses varied, as would be expected, from excellent to no return. Responsibility for introductions in Canada now rests with two federal agencies (Fisheries and Oceans, National Parks), and eight provincial and two territorial agencies. These were all contacted. A considerable part of the information was communicated in letters. The designation (personal communication) was not used in all cases as its frequency would have been disruptive. All those who contributed information are acknowledged.

This discussion includes not only the main section on exotics as defined but also shorter sections on exotics that appeared in Canada after spreading from U.S. introductions, introductions that can be said to have contributed "exotic genes," and a brief comparison of the extent of introduction of exotics with that of transfer of native species.

Exotic Species Currently Recognized by the Exotic Fish Section, AFS

SALMONIDAE – SALMONS AND TROUTS

Hucho hucho (Linnaeus) Danube salmon or huchen

This elongate, large (to 52 kg), predatory salmonid is native to the Danube and is a permanent river resident. In order to benefit from a sport fish with these characteristics, the Province of Quebec introduced this species in 1968 and 1969 (or between 1968 and 1971 according to Mongeau 1979). It obtained 30,000 eggs of a Danube River population from Czechoslovakia. These were hatchery-reared in Quebec and the young released in the St. Lawrence River, Richelieu River, Rivière du Nord, and Lake Mandeville in the Laurentians (Legendre 1980).

Two specimens are recorded as caught by anglers in Lake Mandeville, but a self-reproducing population apparently was never established in any of the waters. No subsequent introductions were made, and the species is now considered extinct in Quebec. A map provided by Mongeau (1979) indicated the location of the releases in the Richelieu River.

Oncorhynchus masou (Brevoort) cherry salmon

This is one of the most trout-like of the Pacific salmons (Christie 1970a), which, like *Oncorhynchus nerka,* is known to form natural landlocked populations. It occurs in tributaries to the seas of Japan and Okhotsk in Korea, the U.S.S.R., and Japan.

This species was considered for introduction in a search for "new species with characteristics making them better suited to the new environmental circumstances than the historical species" (Christie 1968, 1970a:1) "with a potential to restore the depreciated fisheries of the Great Lakes" (Christie 1970b:378).

The source was eggs of an anadromous population in the Shiribetsu River, Japan. These were shipped to Ontario in October 1965, and hatched in the Glenora hatchery. In July 1966, 5,500 young were liberated in Westward Lake, in Algonquin Provincial Park. It was intended to study the success of this species in this semiremote lake from which fish could not escape into public waters.

In 1967 and 1968, a total of 13 fish were taken, all of which were sexually mature. No young were ever seen. Other cherry salmon were liberated in the same lake in order to study aspects of the life history in nonnative fresh waters before any introduction elsewhere (Christie 1970b). Netting in Westward Lake in 1970 did not result in the capture of any specimens. No cherry salmon were ever released in what could be considered public waters of the province.

***Salmo aguabonita* Jordan golden trout**

This southern Pacific mountain species has been introduced in Alberta and British Columbia. Introductions apparently started in 1959 with eggs from Wyoming. Young were released into South Fork, Three Isle, Gap and Galatea lakes, high altitude lakes in the Alberta Rocky Mountains. Later the species was introduced in the drainage of the Red Deer River in Alberta. In 1960, eyed eggs from California were placed in Golden Lake in Jasper National Park, Alberta, and Kaufman Lake in Kootenay National Park, British Columbia.

In all cases the intent was to establish sport fisheries, in some cases in barren alpine lakes. In Alberta, the species has survived in various locations, and is considered to be of great success in the South Fork lakes where it reproduces naturally. One measure of its success is the fact that this species is treated in the main section of *The Fishes of Alberta* (Paetz and Nelson 1970), rather than in the section on exotics. The 1960 introduction did not establish a permanent population in the two national parks and no additional introductions were made there.

This introduction has added a species to the sport fishery without apparently causing problems with native species at this time.

Some confusion arose when fish called "golden trout" by the anglers appeared in catches in Ontario in 1974. There were, in fact, palomino or "golden" rainbow trout, an artificially selected strain of *Salmo gairdneri,* released in the Great Lakes system by the Pennsylvania Fish Commission (Craig and Crossman 1977).

***Salmo trutta* Linnaeus brown trout**

Historically, this species is the second exotic introduced into "Canada," if I apologize to Newfoundlanders for pushing back the date of Confederation of the two Dominions. Brown trout are also the most widely introduced of the exotics. MacCrimmon and Marshall (1968:2533), summarizing world distribution, said that "although the dissemination of brown trout in Canada was slow and dependent on neighboring American states, all provinces but Prince Edward Island and Manitoba and the Northwest and Yukon territories ultimately experienced successful introductions. In spite of naturalization in nine of ten provinces, Nova Scotia and Alberta remain as the only two presently conducting stocking programs." Apparently Nova Scotia stopped releasing fish sometime after 1968 and Manitoba is still releasing hatchery fish.

Dymond (1955) discussed the reasons for success or failure of this species in various parts of Canada. He cited eastern waters made too warm for *Salvelinus fontinalis* by man's activities and difference in spawning time compared to that of *Salmo* in the West.

Because the importation and stocking of exotics have largely been in the hands of individual provinces, the history and impact are most readily treated by dealing with the political areas one at a time, at least for those species with broad present-day distributions and extensive histories.

Newfoundland: Brown trout were brought to Newfoundland in either 1884 (Frost 1938, Scott and Crossman 1964) or 1886 (Andrews 1965, 1966). These were Lochleven trout from the private hatchery of Sir James Maitland in Howietoun, Scotland (McNeily 1909). More Lochleven eggs were imported from Scotland in 1886. In 1892 eggs of so-called German brown trout were imported, reared, and released. In 1905–6, the variety was increased by importing brown trout from England. The stocking in the years 1886–89 was apparently based on 100,000 eggs received annually (Andrews 1965) but introductions apparently ended in the early 1900s.

The species is moderately widely distributed on the Avalon, Burin, and Bonavista peninsulas and around Trinity Bay. The natural spread of the species was slow and thought to be the result of lower than optimal environmental temperatures. New territory has probably been colonized by the spread of sea-run populations that developed soon after the original introductions.

The brown trout provides an added recreational fishery for both landlocked and sea-run forms. It does, however, seem to be displacing native salmonids as the dominant species in a number of lakes near Saint John's.

Nova Scotia: This species was first introduced into Nova Scotia (supposedly both Lochleven and German strains) in 1925 from stocks already in New Brunswick and New Hampshire. These original fish were released in three rivers in Guysborough County. In 1929–30 both strains were again imported and released in Yarmouth, Queens, and Lunenburg counties. Further introductions were made in 1934, and later they were extended to Pictou, Cumberland, and Annapolis counties.

This species is no longer cultured or released in Nova Scotia but is established in many watersheds. The introduction had the positive result of increasing the diversity of the anglers' catch, providing potential trophy-sized trout, and providing populations that survived in areas not well suited to native salmonids. Negative aspects include competition with native salmonids, low angler catch success, and the opinion that they are considered not as good to eat as native salmonids (Catt 1950, Gilhen 1974).

New Brunswick: Although this species was listed for New Brunswick as early as 1852, that record more likely refers to *Salvelinus fontinalis* (Scott and Crossman 1959), since brown trout were apparently not introduced until 1921. The original releases in New Brunswick were in the Loch Lomond system but later releases included waters in Saint John and Kings counties (Catt 1950).

The species is still extant. There are confirmed annual reports (including 1981) of captures in the Nashwaak River, as well as unconfirmed reports from other waters also. No introductions or fish culture activity support these existing populations. The numbers of brown trout in New Brunswick would seem to be too small at present to create a significant positive or negative impact, although they do provide both freshwater and saltwater angling.

Quebec: There are published references to the importation of brown trout into Quebec from Europe about the end of the nineteenth century. Recent records from Departement de Chasse et Pêche and Catt (1950) suggest that the earliest authentic record is 1890 for 25,000 fry from the Caledonia hatchery, New York. These were introduced into Lac Brule near Ste. Agathe, probably in North River, a tributary of the lake. Another introduction, into the Chateauguay River, involved 3,000 fingerlings from Johns' River, Vermont (tributary of Lake Memphremagog), in 1951. The most recent introduction reported consisted of *Salmo trutta macrostigma* from lakes in the Atlas Mountains of Morocco. These were released in Lake Bernard, Mont Tremblant Park, and in Ruisseau Tremblant (a tributary to Lake Tremblant), or Ruisseau des Cascades, Terrebonne County.

This last-mentioned, quite exotic stock did not establish a self-reproducing population, and apparently disappeared. The Ste. Agathe populations exist, are intensively fished, and are supported by regular stocking from provincial hatcheries. The eastern township populations are also regularly supported from provincial hatcheries and provide good angling from the U.S. border to tributaries of the lower St. Lawrence River beyond Quebec City.

These were purposeful introductions, to provide sport fishing, and no negative aspects were mentioned.

Ontario: The original introduction apparently consisted of fingerlings put in the Speed River near Hespler, Waterloo County, in other streams in that county, and in Norfolk and Perth counties in 1913. It has been suggested that unauthorized introductions by private citizens may have occurred before that time (MacKay 1963) and later as well. This trout was extensively stocked in Ontario between 1913 and 1918. According to the 1938 correspondence of the late Dr. J. R. Dymond, no introductions occurred between 1919 and 1928, but in 1929 brown trout were released in Peterborough, Frontenac County's Muskoka District, and near Sudbury. In 1930 they were released in seven lakes near Kenora. That correspondence suggests that the original stock for these introductions was derived mostly from Michigan hatcheries, and indicates that the "greatest care has been taken to maintain a clean bill of health for generations." Since plantings were made mostly in agricultural areas of southern Ontario not particularly suited to brook trout, negative impact was thought to be minimal. However, anglers regularly comment that they are difficult to catch, and it was thought (possibly incorrectly) that brown trout populations gradually limited brook trout to the upper reaches of rivers.

Release of hatchery fish ended in 1962 but populations exist in rivers tributary to lakes Ontario, Erie, and Huron. Large brown trout are taken in small numbers from the lakes themselves.

More recent consideration of the management of the sport fishery has been given to the possible benefit of reintroducing more suitable lake-resident strains of this species.

Manitoba: Release of brown trout apparently began in 1943 (MacCrimmon and Marshall 1968) or 1944 (provincial records). West Blue Lake, in what is now Duck Mountain Provincial Park, received 20,000 fingerlings in 1944 and others in 1951–53. That population apparently did not reproduce. The following year, Child's Lake, in the same area, was also stocked. Stocking of brown trout has continued over the years to 1981, especially in the provincial parks. A provincial map distributed to anglers in 1978 listed 313 lakes as some of the favorite angling lakes. These included all the lakes presently stocked with trout as well as 9 lakes said to contain brown trout.

A number of the attempts to sustain brown trout populations with hatchery fish have failed as a result of unsuitability of the waters, regular winterkill, and dominance of northern pike (*Esox lucius*). Angling for brown trout is said to be good at least in southeast Manitoba, in the Birch River and nearby waters. No significant negative impact has been mentioned.

Saskatchewan: According to Marshall and Johnson (1971:10), "brown trout were introduced from Banff [National Park] Alberta in 1924" into the Cypress Hills region. More fish were obtained later from Montana, Wisconsin, and Alberta. Early stocking was successful but they were later restricted because *Salmo trutta* was a "less desirable angling species than brook trout" (*Salvelinus fontinalis*). Brown trout were introduced into only thirty-seven waters in the province. Thirteen introductions in Cypress Hills between 1924 and 1930 resulted in self-sustaining populations in seven tributaries of the Saskatchewan and Missouri rivers. Attempts between 1931 and 1955 in the Hudson Bay and North Battleford–Meadow Lake areas failed probably as a result of adverse environmental temperatures. Other plantings between 1924 and 1967 were unsuccessful but apparently Reid and Cypress lakes served as refuges for a few large individuals.

Suitable situations for trout in Saskatchewan are limited. It is unlikely that the brown trout could adversely affect native species, and introduced populations provide angling opportunities. Any impact seems of a positive or neutral character.

Alberta: Brown trout were first introduced into Alberta in 1924 when the species was placed in the Raven River (Red Deer system) and in Lake Annette, Jasper National Park. Lochleven Trout (?) entered the Bow River system, in Carrot Creek in 1925, when a hatchery truck taking fish from Banff Hatchery to eastern Alberta broke down there. From this chance introduction, brown trout eventually spread 141 km downstream (Nelson 1965). The Lake Annette stock was later chemically eradicated when considered unpopular with anglers. This species was later successfully introduced and established throughout much of western Alberta in the Athabasca, North Saskatchewan, Red Deer, Bow, and Milk drainages.

These introduced populations provide angling in lower elevation streams but the trout are said to be difficult to catch. No negative impact has apparently been recorded.

British Columbia: The first introduction of brown trout into British

Columbia was apparently in 1932, when eyed eggs were brought from Lodge, Wisconsin, hatched in British Columbia, and the fish placed in Cowichan Lake, Vancouver Island. In the same year, brown trout were brought from Manitoba. Those introductions came as a direct result of requests by anglers (Neave 1949). Between 1932 and 1935, fry, fingerlings, and older fish were planted in the Cowichan River, Cowichan Lake, and Little Qualicum River. A spawning run took place in 1937 and the population was established. This stock was used in 1959 for introductions into the Kettle River, Kootenay region, but none has been reported recently. The same stock was introduced more recently into the Adam River, 75 km north of Campbell River, Vancouver Island. Sea-run forms developed (Carl et al. 1967). In 1935, eggs from Montana were used to introduce the species in nine more locations on Vancouver Island and in later years stock was also imported from Washington.

The species remains, although stocking ended in the province about 1961 and the only known established population is the one in the Cowichan-Qualicum location. This self-sustaining population is heavily angled and annually yields fish three to nine pounds in weight. No significant negative impact has been recorded.

UMBRIDAE – MUDMINNOWS

***Dallia pectoralis* Bean Alaska blackfish**

In spite of the useful tendency of authors to treat fauna on a nonpolitical basis and include *Dallia* in accounts of freshwater fishes in Canada (Scott and Crossman 1973), this species is exotic as defined for this book (imported from the United States and not native to Canada).

Blackfish from the Kuskokwim River, near Bethel, Alaska, were brought to southern Ontario in 1956 and introduced in a few farm ponds. It was hoped they would survive the drastic oxygen depletion in winter and provide recreation from shallow ponds. The original animals were apparently never seen after their introduction and it seems they did not survive the first winter.

CYPRINIDAE – CARPS AND MINNOWS

***Carassius auratus* (Linnaeus) goldfish**

This species constitutes an example of an exotic that entered the wild Canadian fauna, not as a result of considered and purposeful introduction, but as an "escape" from the activities and marketing associated with private aquaria and garden ponds.

New Brunswick: Goldfish were not known before 1972 and were not listed by Scott and Crossman (1959). Two self-sustaining populations were discovered around 1972, one in Killarney Lake within the limits of the city of Fredericton, and the other in a farm pond just upstream of the Mactaquac Dam on the St. John River. Attempts to eradicate these populations were successful only in the case of the farm pond and goldfish still existed in Killarney

Lake in 1981. Both populations are presumed to result from deliberate, illegal introductions.

Quebec: The populations of this species in Quebec are thought now to be extinct. There was a 1974 record from the Richelieu River (Mongeau et al. 1974). No legal introductions were made, so it has been assumed any goldfish were escapes from private aquaria or had moved eastward from Ontario. The latter seems unlikely since no nearby population is presently known. The species was apparently unknown in Quebec in 1967 but discovered at least by 1974. They were eventually recorded in ponds in the eastern townships (southeast of the St. Lawrence River); on Mount Royal and St. Helen's Island, Montreal; a municipal pond in St. Bruno, Lac des Sables, and near Ste. Agathe, and the Richelieu record mentioned above.

Ontario: Goldfish were reported by Radforth (1944) as in lakes Erie and Ontario without record of official introduction. Dymond (1947) added the Detroit River and Scott (1967) documented that after 1954 they were known from Lake St. Clair; Grenadier Pond; Toronto; Musselman's Lake north of Toronto; and Gillies Lake near Timmins. There is no knowledge of any extensive increase in distribution since that time. Impact would seem to be negligible except that hybridization with carp in Lake Erie seems to be reducing populations of carp.

Manitoba: Although reported as not present in 1974 (Fedoruk 1974), this species was taken in 1975, south of the Pas, and in Cooks Creek near East Selkiv in 1976. These must have resulted from the escape of specimens from a private aquarium or garden pond.

Saskatchewan: According to Marshall and Johnson (1971:8), "The only wild goldfish population in Saskatchewan was found in a reservoir tributary to Moose Jaw Creek in 1958." These had been released by a local citizen. This population reproduced but was later eradicated.

Alberta: According to Paetz and Nelson (1970), several releases have been made in Alberta. Since goldfish occur there in many garden ponds they are the likely source of the wild fish. Henderson Lake in Lethbridge had a population, but that lake was rehabilitated for trout, and goldfish may no longer exist there.

British Columbia: Carl et al. (1967) stated that an established population of goldfish was first noted in a pond at Salmon Arm in 1935, and that it occurred also in Lac du Bois near Kamloops. No more recent information was received.

***Cyprinus carpio* Linneaus common carp**

McCrimmon (1968) has summarized extensive information on biology, the fisheries and regulations, the distribution and intentional introductions in Canada, and the effects on native fauna for this species. Pertinent points derived from that publication and from Scott and Crossman (1973) are given here, or modified where required by more recent information. Data are sparse, and it is likely that not all introductions were recorded. Although carp were kept at the government hatchery at Newcastle, Ontario, there are no records of carp released from a government hatchery, and the government of Canada assumed no responsibility for the establishment of carp in Canada. At least until 1893 private importation was, however, not only approved but also assisted by government agencies. As early as 1880 carp were sent by the U.S. Fish Commission to private individuals at Cedar Grove (near Markham), York County, Ontario, and Andover, New Brunswick. Carp are now widely distributed in Quebec, Ontario, Manitoba, Saskatchewan, and British Columbia. This species is present in Saskatchewan by extension of range, not by introduction, and its presence in the past in Alberta is debated.

New Brunswick: If the report of an 1880 shipment is correct, that attempt failed and either no subsequent effort was made or no subsequent effort was successful.

Quebec: Carp have been present in the Montreal section of the St. Lawrence River since 1910 but probably constitute natural spread downstream from fish introduced in Ontario, and from fish introduced in Lake Champlain in the United States. There are no records of official, intentional introductions in Quebec. Dates of first knowledge in various parts of Quebec were given by McCrimmon (1968).

Ontario: Carp were first introduced from the United States in 1880 to ponds at Cedar Grove near Markham, York County, by two men called Reesor. In 1896, a dam broke at Dykes Pond, Newmarket (on water tributary to Lake Simcoe), and it is assumed by many that carp first escaped into public waters that way. Nevertheless, the first documentation of a resident population is often said to be one in the Grand River (Lake Erie) in 1910 resulting from introduction started there in 1891. Dominion archives record that carp were established by the early 1890s at least in a pond draining into Turkey Point marsh, Lake Erie, and resident by then in Long Point Bay and other locations. At least by 1920 carp had spread to include the shores of lakes Erie and Ontario, all of southern Ontario east to the Trent Canal system, the St. Lawrence River, and the shore of Lake Huron as far north as Sault Sainte Marie. This spread is considered to have resulted from local introductions and from the northward movement of older U.S. populations. The only significant area of extension since that time is an area of the shore of Lake Superior from the Nipigon River southwest (first recorded 1953, Ryder 1956).

The commercial fishery was exploiting this species by 1900, and by 1915 there was a commercial fishery even in the waters of the Trent Canal system. In 1980, 268,174 pounds were marketed in Ontario at a landed value of $62,487. Carp also enter into a marginal sport fishery at least in and near urban areas. In contrast to this is the negative impact of destruction of vegetation, and of competition for food, space, and spawning grounds.

Manitoba: Carp from adjacent U.S. populations were planted in Manitoba in 1885, 1886, and 1889. In 1885, 100 carp, six months old, from a hatchery at St. Paul, Minnesota, were planted in ponds near Springfield, Portage la Prairie, and Minnesota. In 1889, carp received from Washington (D.C.?) were released in Lake Minnewawa (or Minnewasta) and a mill pond near Rapid City on the Assiniboine River. No populations were established, however, and carp were unknown until 1938 when they were discovered in the Red River near the international border. Atton (1959) suggested their presence there represented downstream invasion from Minnesota and North Dakota. By 1956, the carp had penetrated north to the Nelson River itself, and west via the Assiniboine into Saskatchewan. By 1974, the species was known as far downstream (north) as York Factory at Hudson Bay.

Commercial production of carp has been recorded at least since 1942, and in 1965–66 landed value for the whole province was $18,878.

A more recent summary of the distribution and impact on flora and fauna in Manitoba was given by Swain (1979).

British Columbia: The only populations now linked to actual introduction would be those on Vancouver Island. A population in Glen Lake is believed to have resulted from fish that escaped from a nearby pond to which carp from Oregon may have been introduced (Carl et al. 1967). That population is thought to have been removed with rotenone in 1961. More recently, however, carp have been illegally introduced into farm ponds in the Courtenay-Oyster rivers area of Vancouver Island. The populations in the south central mainland, particularly those in waters of the Columbia and Fraser rivers, are thought to have arisen not from introductions but from natural spread of populations established in the Columbia River, Washington, during the 1880s. Carp have been reported in the outfall of the Fraser River as far as Saturna Island. The history of this spread was well described by McCrimmon (1968).

ICTALURIDAE – BULLHEAD CATFISHES

***Noturus insignis* (Richardson) margined madtom**

In 1971, four specimens of this species were collected from the stream tributary to Lac la Pêche, Quebec (Rubec and Coad 1974). The original report suggested this constituted an introduction, possibly via a bait bucket. Subsequent collections that indicated other populations and slightly wider distribution may require a reevaluation of that origin.

POECILIIDAE – LIVEBEARERS

***Gambusia affinis* (Baird and Girard) mosquitofish**

This fish, an unlikely candidate for release in Canada, was introduced in Manitoba, Alberta, and possibly in British Columbia.

The mosquitofish was planted in a Manitoba pond in 1958 for experimental purposes. The fish lived through two winters only to be killed when the pond dried up (McAllister 1969).

In Alberta apparently the form *G. a. affinis,* introduced to and moved from California (Nelson 1984), was released in 1924 in the marshy outflow area below Cave and Basin hot springs in Banff National Park. It was hoped it would provide mosquito control. The population established itself and survives today (Nelson 1984). They are also known from the sulphur hot springs area of third Vermillion Lake (McAllister 1969, Paetz and Nelson 1970).

In 1928 (Carl et al. 1967) or 1929 (in lit.), "top minnows" were introduced into a pond (on the golf course?) at Kelowna, British Columbia, for mosquito control. These fish were quite active, and apparently effective, in the first summer. None were seen the following summer, however, and they were thought to have been winterkilled. There have been no subsequent reports.

This record has usually been given as *Gambusia* sp.? (Carl et al. 1967, McAllister 1969) but the writer of the letter containing the above information used the name guppy. It is, therefore, equally likely that the record refers to *Poecilia reticulata.*

***Poecilia latipinna* (Lesueur) sailfin molly**

This species was discovered in the marshy outflow of Cave and Basin hot springs, Banff National Park, Alberta, in 1968. It was presumed that it was released there from local aquaria as early as 1960 (McAllister 1969). The population was extant there in 1981 (Nelson 1984).

***Poecilia reticulata* Peters guppy**

This South American fish was introduced in Alberta. Guppies were apparently introduced by aquarists into Cave and Basin hot springs in the 1960s and perhaps also in the 1970s. They were present in large numbers in 1968. A report of their presence there in 1976 is now considered to be erroneous; they disappeared about that time; none were captured in 1981 (Nelson 1984).

For a possible record in British Columbia, see *Gambusia affinis* above.

***Xiphophorus helleri* Heckel green swordtail**

This middle American species also was first noted in 1968 (McAllister 1969) in the same Alberta location as that for the other poeciliids and is also assumed to have been introduced from local aquaria. It was always rare, has not been observed there in several years, and was not seen in collections made in 1981 (Nelson 1984).

CICHLIDAE – CICHLIDS

***Cichlasoma nigrofasciatum* (Günther) convict cichlid**

The information for this middle American species introduced into Alberta is the same as for the poeciliids above. It has been recorded in the past as zebra cichlid. This species has not been observed in several years and none was taken in 1981 collections (Nelson 1984); the population probably no longer exists.

Hemichromis bimaculatus **Gill** **jewelfish**
This cichlid was reported from the same Banff, Alberta, location only in 1976 in a MS report by W. D. Reynolds (J. S. Nelson, personal communication). It constitutes an added species to these unique introductions into national parks in Alberta, and collections in 1981 (Nelson 1984) indicated its survival to that date.

?*Pterophyllum scalare* **Castelnau** **angelfish**
Under this name, Nelson (1984) recorded for the first time the rumor that another cichlid had at some time been introduced in the Banff, Alberta, hot springs. Nelson saw none in his 1981 survey.

ANABANTIDAE – GOURAMIS

Betta splendens **Regan** **Siamese fightingfish**
Nelson (1984) recorded for the first time a local report that this tropical species had been released into waters east of Cave Springs in the Banff, Alberta, location. There is apparently no verification, and Nelson saw none in 1981.

Trichogaster trichopterus **(Pallas)** **blue gourami**
This anabantid was also reported to have been introduced in waters east of Cave Springs, Banff, Alberta. It was said to have reproduced, but later died out when water flow was interrupted. It was not seen in the 1981 survey (Nelson 1984).

PLEURONECTIDAE – RIGHTEYE FLOUNDERS

Platichthys flesus **(Linnaeus)** **European flounder**
One specimen of this flatfish was taken in Lake Erie in 1974 and one in 1976. Emery and Teleki (1978) suggested that they had been introduced via ballast water from European vessels. Nothing more was seen of this species until another specimen was captured in 1981 in Lake Superior. It must, at present, be assumed the same agency was responsible. Basically marine, this species regularly penetrates fresh waters in Europe.

This completes the list of those fishes introduced in Canada which strictly meets the definition set for this book.

Natural Immigration into Canada of Species Introduced in the United States

The definition of exotics for this book includes the words "introduced from another country." It is, therefore, necessary to insert a section to deal with those species that are exotic but were not directly introduced by man into Canada.

The only truly exotic fish in this category is *Tinca tinca,* the tench. This Eurasian cyprinid has regularly been recorded as an exotic in Christina, Tugulnuit, and Osoyoos lakes on the Columbia River system of British Columbia. It was also said to have been reported from Vancouver Island, but this has apparently never been verified. There is no record of the actual introduction of this species in British Columbia. It has always been assumed that its presence resulted from natural movement upriver in the Columbia River system of fish introduced in 1895 into small lakes in Spokane County, Washington, and Washington County, Oregon (Carl et al. 1967). I received no word as to whether or not this species has been seen for many years and it may no longer exist in those lakes. In 1954, a specimen was entered into the collection of the University of British Columbia (BC54–90) from Lake Tugulnuit near Osoyoos, British Columbia. See Dymond (1955) and Wydoski and Whitney (1979) for concepts of its introduction in Washington.

A number of spiny-rayed fishes foreign to British Columbia but native to other parts of Canada occur in various locations in mainland British Columbia. There is no record of the purposeful introduction of these species and it has been assumed (Carl et al. 1967) that they result from natural upstream spread in the Columbia River of animals introduced in the 1890s in Washington. These include black bullhead (*Ictalurus melas*), yellow perch (*Perca flavescens*), pumpkinseed (*Lepomis gibbosus*), black crappie (*Pomoxis nigromaculatus*), and largemouth bass (*Micropterus salmoides*). Brown bullhead (*Ictalurus nebulosus*) in the Fraser River system may result from a natural dispersal or a combination of immigration and introduction. American shad (*Alosa sapidissima*) occur in British Columbia as a result of introductions in the United States and striped bass (*Morone saxatilis*), infrequently taken in salt water off British Columbia, are also probably strays of fish introduced in the United States.

Common carp in Quebec, Ontario, and Saskatchewan are at least in part the results of introductions in the United States.

Chain pickerel (*Esox niger*), now widespread in New Brunswick, may have originated in parts of that province from the immigration of chain pickerel introduced into Maine near the border of New Brunswick (Scott and Crossman 1959).

Coho salmon (*Oncorhynchus kisutch*) in Nova Scotia, New Brunswick, and Quebec could also be strays from introductions in the United States since none have been introduced in those provinces. In certain parts of Ontario, many salmonids, resulting from fish culture programs, originated in part from introductions in the United States.

Salmo gairdneri **rainbow trout**

Palomino rainbow trout captured in Ontario were strays from introductions of this color-variant into Lake Erie by the state of Pennsylvania (Craig and Crossman 1977).

Genetic Exotics

Although the terms of this book define exotic precisely, it is appropriate to mention in passing the concept of the introduction of exotic gene pools. The species involved may not *per se* be exotic, but the impact on local fishes, although insidious, may equal or exceed the introductions of foreign fishes.

An extreme example is the rearing and release of young muskellunge (*Esox masquinongy*) in Ontario. The total fish-culture output for the province comes from one hatchery which uses a single, wild, parent stock from Stony Lake. In the past, young of this source have been stocked widely in the waters of the province, including the St. Lawrence and Niagara rivers. Muskellunge in these two rivers have life-history characteristics somewhat different from muskellunge in the Kawartha Lakes. These characteristics may have been built up over long periods of time and may cause the fish to adapt better not only to living in riverine conditions but also to doing well in sympatry with the northern pike that also occur in those rivers (Harrison 1978, and Harrison and Hadley 1979). Any such adaptations could be lost by long-term introductions of the lacustrine genes into the riverine populations. To add to this complication is the fact that New York State is now considering introducing tiger muskellunge (the hybrid, *Esox lucius* x *E. masquinongy*) into New York waters tributary to Lake Ontario and the St. Lawrence River.

A similar complication could already be taking place as a result of the release in Quebec of the Ohio River muskellunge. This semidiscrete race, occurring from Chautauqua Lake, New York, to Kentucky and Tennessee, is different in some aspects from the race in Quebec. From 1950 to 1962 this form was imported from the Chautauqua Lake hatchery and released in the St. Lawrence River as an added sport-fishing attraction.

***Salvelinus alpinus* Linnaeus French alpine char**

Although the Arctic char is native to large areas of Canada, exotic char, called here for convenience French alpine char, have been imported and experimentally introduced. Char from Lake Geneva were imported into Ontario in 1962 or 1963. These were never released into any water. Char from the same source in France were imported into Ontario between 1963 and 1966. These were liberated in Westward Lake, Algonquin Park (see *Oncorhynchus masou*) in 1966 or 1967.

Single individuals were caught in 1968 and 1969. In 1970, nets were set to see if any char persisted but none was caught. It is presumed none exists there now.

This species was introduced also into Saskatchewan. Eggs were obtained from France in 1963 and 1964, hatched, and 27,500 fingerlings released in three lakes near Lac la Ronge. Only one char was recovered, from Downtown Lake. No permanent population was established, and failure was attributed to incomplete detoxification after rehabilitation and to reentry of predators (Marshall and Johnson 1971).

***Salmo salar* Linnaeus Atlantic salmon**

This species was imported to British Columbia in 1934 and introduced into Cowichan Lake. It is now difficult to tell how many of the Atlantic salmon imported into British Columbia from 1905 to 1933 were from Canadian or Scottish sources.

Salmonids

Hybrids have been widely introduced in Canada. As early as 1889 there are records of hybrids being imported to Newfoundland or produced there and released. It would appear that the 1889 hybrids were possibly *Salmo trutta* x *Salmo salar* (Scott and Crossman 1964). In addition to being a hybrid genetic mix, the *Salmo trutta* genes would have been exotic. The splake, *Salvelinus fontinalis* x *Salvelinus namaycush,* has been developed in both the West and the East and liberated within areas populated by various salmonid species. The "Cranbrook trout," *Salmo clarki lewisi* x *Salmo gairdneri,* was cultured and has been released in eastern British Columbia. The hybrid *Salvelinus fontinalis* x *Salvelinus malma* was produced in the Maligne River fish hatchery, Jasper National Park, and in 1955 released in Emerald Lake in Yoho National Park, British Columbia. The lake was chemically rehabilitated in 1959 before any assessment of this hybrid was made. The same facts pertain to the hybrid *Salvelinus namaycush* x *Salvelinus malma* (or *S. confluentus*); this form was cultured and introduced as a predator for coarse fish control. The unusual hybrid, *Salvelinus fontinalis* x *Salvelinus alpinus* (*marstoni*), was apparently released in 1962 in lakes in Banff and Jasper National Park in Alberta (Paetz and Nelson 1970).

Conclusion

The introduction of exotic fishes into Canada began in the 1880s and has continued into the 1980s. Some of the areas could have barely been occupied by Europeans when requests for introductions began. Significant settlement of Saskatchewan only followed the completion of the Canadian Pacific Railroad in 1885. Marshall and Johnson (1971) said requests for introduction came from there as early as 1887. The low total for the number of species introduced (18), and the even lower number of introduced species that have persisted (9) can be looked upon as a blessing. Even of greater benefit is the fact that probably only one exotic, common carp, can definitely be said to have had a negative impact on the environment and on indigenous fishes. The impact of exotics has, in some way, been proportional to the degree of "exoticness" and their ability to spread after naturalization. In 1955, Dymond recorded that the temperate common carp was unknown in Manitoba until 1938, had shortly thereafter immigrated into the province naturally, and fifteen years later had spread 320 km downstream in the Nelson River. In contrast are the unplanned introductions of poeciliids and cichlids from Central and South America into hot springs in national parks in Alberta. These have been unable to invade even the nearby native waters as a result of the greater degree of their exotic nature. They do, however, threaten the survival of a possible hot spring endemic, *Rhinichthys cataractae smithi.*

The exotics are virtually all freshwater fishes. The only marine component of Canada's exotic species consists of populations derived from freshwater introductions; brown trout, strays from U.S. introductions (shad and striped bass), and the European flounder, one of the few flatfishes highly tolerant of fresh water. Certain native salmonids, which have been transferred from one coast to another for trials of mariculture, have escaped into Canadian marine waters. These freshwater exotics fall largely into two categories: (1) sport fishes purposely introduced, supposedly to increase recreational potential, and (2) aquarium fishes released illegally.

Of the 18 exotic species known, or thought, to have been introduced into Canada, 11—*Noturus insignis, Carassius auratus, Poecilia latipinna, P. reticulata, Xiphophorus helleri, Cichlasoma nigrofasciatum, Hemichromus bimaculatus,* ?*Pterophyllum scalare, Betta splendens, Trichogaster trichopterus,* and *Platichthys flesus*—can be said to have been unplanned (i.e., unofficial, illegal, or simply accidental). It is fortunate that across such a vast territory there have been so few. Many readers will probably immediately assume that there have been others but the harsh Canadian winters have eradicated them; that could be true. In many areas, however, winters are not that harsh and surface water temperatures under the ice reach a winter minimum, which varies little thereafter. In addition, in urban areas where illegal introductions on aquarium species are most likely, heated effluents become more available yearly.

Isolated captures of several species exotic to Canada, but well known in the United States, have been reported in Canada in recent years. These are stoneroller (*Campostoma anomalum*), silver shiner (*Notropis photogenis*), black buffalo (*Ictiobus niger*), flathead catfish (*Pylodictis olivaris*), black striped topminnow (*Fundulus notatus*), and warmouth (*Lepomis gulosus*) (1979 reprint of Scott and Crossman, 1973). Each of these might appear to be an unofficial (bait pail) introduction; however, on the basis of the species involved and the distance to other populations, these records would seem to constitute natural, pioneering immigrants, or strays.

The 18 exotics introduced constitute fewer than half the 37 indigenous species that have been moved into areas of Canada where they are not native. The indigenous transplants involve almost all large Canadian groups except the suckers, Catostomidae. They include such unusual choices as American eel (*Anguilla rostrata*), clupeids, gasterosteids, and percichthyids.

A separate section above listed as genetic exotics a group of introductions less qualified for the chosen definition of exotic. This category was included to convey the idea of the foreign genes introduced by those species. The impact of foreign genes (diluting the adaptive suitability of the indigenous populations) of the 37 native species transferred within Canada could be equal to that of the exotics. Concern for this problem is included in the guidelines for future introductions mentioned below.

One thinks today of national parks as wilderness areas, or areas where man-made changes are held to a minimum. Sadly enough, in the mountain national parks in Alberta and British Columbia a policy of maximizing the enjoyment of park visitors included the purposeful introduction of exotics and unique salmonid hybrids, and the transfer of nonindigenous native salmonids. The presence in these same parks of hot springs led to the purposeful introduction of 1 tropical species and the illegal introduction of 8 others. I hope that introductions into national parks of exotics of any nature are a thing of the past.

Simple access or changes in environmental conditions that could be brought about by damming or diverting waters, especially over international boundaries, can unknowingly transfer exotic species. The recent controversy over the Garrison Dam involved, in part, the potential introduction into Manitoba of an exotic fish.

Changed environmental conditions have, in the recent past, been the basis for suggesting and accommodating the introduction of exotics (Canada's second age of introductions). A more difficult and costly, but more responsible and suitable, solution is habitat improvement and careful rehabilitation of indigenous fishes. Regier (1968), discussing introduction of exotics in Canada from the ecological standpoint, stated that "we choose exotics for what they can do for us not what they can do for the non-human system" (p. 97). He considered that we should address ourselves to counteracting the destruction of the natural environment. He told Canadians that he considered it retreating when exotics are proposed as a solution to the problem of ecosystems that have changed sufficiently "that the valued earlier members of the system can no longer maintain their numbers in it" (p. 110). Regier urged both caution and the implementation of rigid guidelines to evaluate adequately candidates for introduction. He warned also that "any manager who believes that simply the act of injecting another species or two into a community will provide a lasting answer . . . is deluding himself" (p. 195).

Canada has profited, it is hoped, from its history and from the caution expressed in the past from several quarters. Processes for ample prior consideration of all proposed introductions are imperative. Provinces adjacent to the one proposing to make an introduction must be taken into consideration since they should not have to put up with an exotic they don't want, but may receive via stream connections. The need for caution concerning possible emigration of exotics across the international boundary is even greater.

Basic control of introductions has long existed in Canada, in that "the transport or introduction of live fish or fish eggs without specific written permission of the Minister of Fisheries is prohibited by law" (Federal Fisheries

Act). What is required, however, are federal-provincial agreements providing criteria for decisions on whether or not a proposed introduction is dangerous, appropriate, or practical. Such decisions need to be backed by authority to prevent the introduction if necessary. It seems that agencies on the East and West coasts of Canada are much further ahead in policies and controls than are most of the inland agencies concerned with fresh waters. In inland situations, it would appear that the Fish Health Protection Regulations are the only real guidelines presently in place requiring that a potential import be adjudicated before it is carried out.

In British Columbia, new regulations have been drafted to prevent and control the importation of pet or ornamental exotic fishes. A list of such fishes to be prohibited has also been developed. It might seem the threat from imported ornamentals is not so great in Canada as it is in Florida. This little quirk of nature does not justify the attitude that regulatory guidelines are not needed.

Canada has developed, or is party to, a number of agreements that can be used to bring about an appropriate kind of deliberation before an introduction. Some of these are: the Introductions Working Group of the International Council for the Exploration of the Seas (ICES); the 1977 Fish Health Protection Regulations, which apply to all interprovincial as well as international imports of salmonids; new Canadian fisheries regulations that will supersede those presently in place and give force of law to a number of guidelines currently used to control stocking of all aquatic species; the Advisory Committee on the Introduction of Non-indigenous Species of the Scotia-Fundy Region, Department of Fisheries and Oceans (T. W. Rowell, personal communication), and the British Columbia Interagency Committee on Transplants and Introductions of Fish and Aquatic Invertebrates.

The last group stated:

> Both federal and provincial fisheries management agencies have real concern for the potentially disastrous effects of introduction of non-indigenous fish and invertebrates on native fauna in both fresh- and seawater environments. With increasing capabilities for semi-natural or artificial fish culture production and the diversity of responsibility for species by both agencies [Province of British Columbia, Department of Fisheries and Oceans], it is apparent that some form of review of proposed transplants or introductions should be instituted. Problems of concern include, but are not restricted to: potential disease introduction, interspecific competition and genetics of native stocks, including suitability of the environment, racial problems, timing and quality of runs and methods of harvest or utilization. (G. R. Bell, personal communication)

With these sorts of concern, and with the helpful guidelines suggested as long ago as 1949 (Turner 1949), more recently by Courtenay and Robins (1973) and Mann (1979a, b), Canada should be able to become *Maître chez nous* in regard to introductions of exotics and the transfer of indigenous fishes. It is interesting that the workers associated with shellfish seemed to have led the way in both the recognition of the needs for controls and the implementation of such controls.

As has been stated above, the available records for past Canadian introductions often leave much to be desired. Records no longer exist, records have never been drawn together, and the records of two agencies operating in the same area disagree on the details concerning what we must suppose is the same introduction. Common names are often used to record introductions or transfers, and these refer in different parts of Canada to different things (e.g., *Osmerus mordax* was introduced into Newfoundland as frostfish but in Ontario this name is applied to *Prosopium cylindraceum*). It is now difficult to tell whether an introduction in Newfoundland under the name salmon trout was *Salvelinus namaycush* or the hydrid, *Salmo salar* x *S. trutta*). The documentation of introductions and discoveries of species new to an area would be most available in the published literature where they should be recorded by abstracting agencies and in the computer literature surveys. There is a trend, however, for some Canadian journals to look upon such records as not worthy of publication.

It should also be remembered that the transfer of a species to a new environment can lead to rapid changes in the phenotype if not the genotype. Therefore, appropriate documentation of each introduction should include deposition in a reference collection of a suitable sample of specimens from the donor population at the time of the transfer. It would, in that way, be possible to trace more completely the lasting effects of Canada's experience with the introduction of exotics.

Acknowledgments

In order to draw together the information across the whole of Canada, I had to depend on the help of a large number of individuals. I sincerely thank them for this help and acknowledge them simply in alphabetical order as follows: D. Aggett, F. M. Atton, G. R. Bell, J. F. Bergeron, B. F. Bidgood, M. Borodacz, W. J. Bruce, M. Burridge, K. Campbell, T. Carey, J. T. Carlton, W. J. Christie, D. Cucin, A. J. Derksen, D. G. Dodge, P. M. Etherton, G. A. Goodchild, Y. Gravel, I. Hagenson, C. Horkey, W. N. Howard, L. Johnson, R. Kingsley, S. Kirkpartrick, W. H. Lear, C. C. Lindsey, V. Legendre, D. E. McAllister, J. McClean, A. O. MacPhee, R. Mann, N. V. Martin, J. F. Nelson, A. E. Peden, J.H.C. Pippy, M. H. Prime, M. A. Redmond, M. M. Roberge, T. W. Rowell, N. M. Simmons, A. Smith, L. A. Sunde, R. B. Tinling, W. C. Turnbull, H. Valiant, and T. H. Whillans.

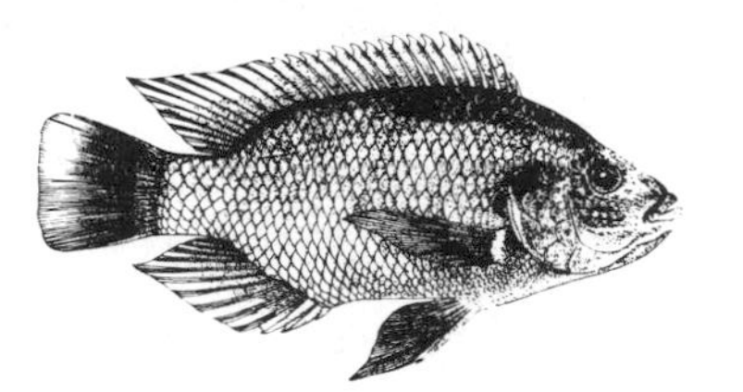

Literature Cited

Andrews, C. W. MS 1965. Early importation and distribution of exotic fresh water fish species in Newfoundland. Newfoundland Game Fish Protection Society.

Andrews, C. W. 1966. Early importation and distribution of exotic fresh water fish species in Newfoundland. Canadian Fish-Culturist 36:35–36.

Aro, K. V. 1979. Transfer of eggs and young of Pacific salmon within British Columbia. Canada Department of Fisheries and Oceans. Fisheries and Marine Service Technical Report no. 861. 147 pp.

Atton, F. M. 1959. The invasion of Manitoba and Saskatchewan by carp. Transactions of the American Fisheries Society 88(3):203–205.

Carl, G. C., W. A. Clemens, and C. C. Lindsay. 1967. The fresh-water fishes of British Columbia. British Columbia Provincial Museum Handbook 5. 132 pp.

Catt, J. 1950. Some notes on brown trout with particular reference to their status in New Brunswick and Nova Scotia. Canadian Fish-Culturist 4(5):15–18.

Christie, W. J. 1968. The potential of exotic fishes in the Great Lakes, 73–91. *In:* A Symposium on Introduction of Exotic Species. Ontario Department of Lands and Forests, Research Report no. 82. 111 pp.

Christie, W. J. 1970a. A review of the Japanese salmons *Oncorhynchus masou* and *Oncorhynchus rhodurus* with particular reference to their potential for introduction into Ontario waters. Ontario Department Lands and Forests, Research Information Paper (Fisheries) no. 37. 45 pp.

Christie, W. J. 1970b. Introduction of the cherry salmon *Oncorhynchus masou* in Algonquin Park, Ontario. Copeia 2:378–379.

Courtenay, W. R., Jr., and C. R. Robins. 1973. Exotic aquatic organisms in Florida with emphasis on fishes: A review and recommendations. Transactions of the American Fisheries Society 102(1):1–12.

Craig, E. G., and E. J. Crossman. 1977. Genetic variants in Canada of the rainbow trout *Salmo gairdneri,* called golden trout and palomino trout. Canadian Field-Naturalist 91(3):93–94.

Crossman, E. J. 1968. Changes in the Canadian freshwater fish fauna, 1–20 *In:* A Symposium on Introductions of Exotic Species. Ontario Department Lands and Forests. Research Report no. 82. 111 pp.

Dymond, J. R. 1947. A list of the freshwater fishes of Canada east of the Rocky Mountains. Royal Ontario Museum Zoology, Miscellaneous Publication no. 1. 36 pp.

Dymond, J. R. 1955. The introduction of foreign fishes in Canada. Proceedings of the International Association of Theoretical and Applied Limnology 12:543–553.

Emery, A. R., and G. Teleki. 1978. European flounder (*Platichthys flesus*) captured in Lake Erie, Ontario, Canada. Canadian Field-Naturalist 92(1):89–91.

Fedoruk, A. N. 1974. Freshwater fishes of Manitoba (1971, reprinted 1974). Manitoba Department of Mines, Resources and Environmental Management, Development and Extension Service. 130 pp.

Frost, N. 1938. Trout and their conservation. Newfoundland Department of Natural Resources, Service Bulletin 6 (Fisheries):1–16.

Gilhen, J. 1974. The fishes of Nova Scotia lakes and streams. Nova Scotia Museum. 49 pp.

Harrison, E. J. MS 1978. Comparative ecological histories of sympatric populations of *Esox lucius* and *Esox masquinongy* of the upper Niagara River and its local watershed. Ph.D. dissertation, State University of New York at Buffalo. 303 pp.

Harrison, E. J., and W. F. Hadley. 1979. Biology of muskellunge (*Esox masquinongy*) in the upper Niagara River. Transactions of the American Fisheries Society 108(5):444–451.

Legendre, V. 1980. Les salmonidae au Québec; une sommaire de la famille. Ministère du Loisir, de la Chasse et de la Pêche. 4 pp.

McAllister, D. E. 1969. Introduction of tropical fishes into a hotspring near Banff, Alberta. Canadian Field-Naturalist 83(1):31–35.

McCrimmon, H. R. 1969. Carp in Canada. Fisheries Research Board of Canada, Bulletin 165. 93 pp.

MacCrimmon, H. R. 1971. World distribution of rainbow trout (*Salmo gairdneri*). Journal of the Fisheries Research Board of Canada 28(5):663–704.

MacCrimmon, H. R., and B. L. Gots. 1972. Rainbow trout in the Great Lakes. Ontario Ministry of Natural Resources. 66 pp.

MacCrimmon, H. R., and T. L. Marshall. 1968. World distribution of brown trout *Salmo trutta*. Journal of the Fisheries Research Board of Canada 25(12):2527–2548.

MacKay, H. H. 1963. Fishes of Ontario. Department of Lands and Forests. 300 pp.

McNeily, A.J.W. 1909. Some old-time anglers. Newfoundland Quarterly 8(4):5–8.

Mann, R. (ed.). 1979a. Exotic species in mariculture: Proceedings of a Symposium on Exotic Species in Mariculture. Case histories of the Japanese oyster, *Crassostrea gigas* (Thunberg), with implications for other fisheries, held at Woods Hole Oceanographic Institution, Woods Hole, Massachusetts, 18–20 September 1978. MIT Press, Cambridge. 363 pp.

Mann, R. 1979b. Exotic species in aquaculture. Oceanus 22(1):29–35.

Marshall, T. L., and R. P. Johnson. 1971. History and results of fish introductions in Saskatchewan 1900–1969. Department of Natural Resources, Fisheries and Wildlife Branch, Fisheries Report no. 8. 31 pp.

Mongeau, J.-R. 1979. Dossiers des poissons du bassin versant de la Baie Missisquoi et de la Rivière Richelieu, 1954 à 1977. Québec Ministère du Tourisme de la Chasse et de la Pêche. Service de l'Aménagement et de l'Exploration de la Faune. Rapport Technique. 251 pp.

Mongeau, J.-R., A Courtemanche, G. Masse, et B. Vincent. 1974. Cartes de répartition géographique des espèces de poissons au sud du Québec, d'après les inventaires ichthyologiques effectués de 1963 à 1972. Québec, Ministère du Tourisme, de la Chasse et de la Pêche. Service de l'Aménagement de la Faune, Rapport Special no. 4. 92 pp.

Neave, F. 1949. Game fish populations of the Cowichan River. Fisheries Research Board Canada Bulletin 84. 32 pp.

Nelson, J. S. 1965. Effects of fish introductions and hydro-electric development on fishes in the Kananaskis River system, Alberta. Journal of the Fisheries Research Board of Canada 22(3):721–753.

Nelson, J. S. 1984. The tropical fish fauna in Cave and Basin Hotsprings drainage, Banff National Park. Canadian Field-Naturalist 97(3):255–261.

Paetz, M. J., and J. S. Nelson. 1970. The fishes of Alberta. Government of Alberta. 282 pp.

Radforth, I. 1944. Some considerations of the distribution of fishes in Ontario. Royal Ontario Museum of Zoology. Contribution no. 25. 116 pp.

Regier, H. A. 1968. The potential misuse of exotic fish as introductions, 92–111 *In:* A Symposium on introductions of exotic species. Ontario Department of Lands and Forests, Research Report no. 82. 111 pp.

Rubec, P. J., and B. W. Coad. 1974. First record of the margined madtom (*Noturus insignis*) from Canada. Journal of the Fisheries Research Board of Canada 31(8):1430–1431.

Ryder, R. A. MS 1956. Occurrence of carp on the north shore of Lake Superior, Port Arthur and Geraldton Districts. Ontario Department of Lands and Forests. Fish and Wildlife Management Report no. 31. 2 pp.

Scott, W. B. 1967. Freshwater fishes of eastern Canada. 2d ed. University of Toronto Press. 137 pp.

Scott, W. B., and E. J. Crossman. 1959. The freshwater fishes of New Brunswick: A checklist with distributional notes. Royal Ontario Museum, Division of Zoology and Palaeontology, Contribution no. 51. 37 pp.

Scott, W. B., and E. J. Crossman. 1964. Fishes occurring in the fresh waters of insular Newfoundland Canada. Department of Fisheries. 124 pp.

Scott, W. B., and E. J. Crossman. 1973. Freshwater fishes of Canada. Fisheries Research Board of Canada. Bulletin 184. 966 pp.

Swain, D. P. MS 1979. Biology of the carp (*Cyprinus carpio* L.) in North America and its distribution in Manitoba, North Dakota and neighboring U.S.A. waters. Manitoba Department of Natural Resources, Report no. 79-3. 35 pp.

Turner, H. J. 1949. Report on investigations of methods of improving the shellfish resources of Massachusetts. Massachusetts Department of Conservation, Division of Marine Fisheries: 7–28.

Wright, B. H., and R. D. Sopuk. MS 1979. A history of fish stocking in northern Manitoba. Manitoba Department of Mines, Natural Resources, and Environment. Report 79-6. 69 pp.

Wydoski, R. S., and R. R. Whitney. 1979. Inland fisheries of Washington. University of Washington Press, Seattle. 220 pp.

CHAPTER 6

Distribution and Known Impacts of Exotic Fishes in Mexico

Salvador Contreras-B. and
Marco A. Escalante-C.

The distribution of exotic fishes in Mexico is extensive for most common species. Nevertheless, there have been few attempts to document the species that have been introduced, when and why they were released, and their known status and impacts. Arredondo (1973) and Escalante and Contreras (in press) treated 19 introduced food and ornamental (aquarium) species; while their data were mostly from official records, the second reference included numerous collection data.

Introductions of some exotic fishes date back to the last century when Cházari (1884) discussed culture methods for rainbow and brown trouts and carps, presumably already or soon to become established, as well as for some native species. Meek (1904) mentioned goldfish and carp as present in lakes of the Valley of Mexico and that a trout hatchery was in operation in Lerma.

De Buen (1941) first discussed impacts of exotic fishes in Mexico in his evaluation of the effects of introduced largemouth bass (*Micropterus salmoides*) on pescado blanco (*Chirostoma* spp.) in Lago de Pátzcuaro. Other papers that dealt with declines and extinctions of endemic fishes because of introductions of exotic species and other factors include those of Miller (1961, 1963), Contreras (1969, 1975, 1978), Contreras et al. (1976), Escalante and Contreras (in press), and Smith and Miller (in press).

We use the term *exotic* to mean introduced or nonnative fishes occurring beyond their historical ranges of distribution as a result of movement by man. Parasitologists use the term transfaunated for such organisms. We feel, however, that it is useful to subdivide exotic into fishes of foreign origin (introduced from beyond the borders of Mexico) and transplants (fishes introduced outside of their native ranges within Mexico and those stocked from a foreign part of a range that extends into Mexico). Such subdivision is useful when communicating with decision-makers who are in charge of stocking and transfers.

The preparation of this chapter gave us the opportunity to review and synthesize the available information on introductions and distributions of exotic fishes in Mexico. Impacts, however, were difficult to determine in most cases; evidence for negative impacts was typically circumstantial and often (and understandably) synergistic with other detrimental factors. Moreover, there has been little direct study of the interactions of exotic and native fishes.

We have compiled a list of 55 exotic fishes (table 6–1) of which 26 are of foreign origin and 29 are transplants. We categorized the status of each species as follows: reproductive when known to breed in at least one site; nonreproductive when failing to establish at least one breeding population; unsuccessful when a species became established and later died out; and unknown when data were unavailable on reproductive success. Also included are the four primary reasons for introductions as treated by Hubbs (1982): (1) release of bait fish; (2) disposal of ornamental (aquarium) species; (3) aquaculture (including releases of potential food fishes into open waters, stocking of forage fishes, and aquatic weed control); and (4) sport. We have added two other categories: (5) species protection and (6) mosquito control.

In Mexico as elsewhere, most of the introduced fishes are purported to fill a so-called empty or vacant niche which, in our view, is a mistaken concept. The role a fish plays in a community is its niche; it is the sum of its activities and responses. Therefore, if a species is absent from a community, there can be no vacant niche; there may be one or more vacant trophic levels, but *never* a vacant niche.

Our list of exotic fishes in Mexico is probably incomplete and open to question concerning some species. Reasons for this include our having to use references containing only a fish name and incomplete data (Rosas 1976c, FIDEFA 1975), other reports referring to species groups such as charal (any small *Chirostoma*) or blanco (any large *Chirostoma*) from one or several areas, a generic name such as *Tilapia* or *Lepomis* without species designations, or newspapers mentioning aquarium fishes (see our entries for *Barbus*). Reports containing references only to a genus, as we found in the files of the Departamento de Pesca, or others that mentioned "tres especies mas" for unknown cichlids (Delgadillo 1976) were of minimal value.

Additional questions arise concerning the ranges of some introduced species when we disagree with other authors as to whether or not a species is

Table 6–1. EXOTIC FISHES IN MEXICO, WITH INDICATION OF BEING FOREIGN (F), TRANSPLANTS (T), REPRODUCTIVE (R), NONREPRODUCTIVE (N), UNSUCCESSFUL (U), OR UNKNOWN (?), AQUACULTURAL (A), BAIT (B), ORNAMENTAL (O), SPORTS (S), PROTECTION OF SPECIES (P), OR MOSQUITO CONTROL (M)

Scientific name	Common name	Origin	Status	Purpose
Clupeidae, herrings				
Dorosoma petenense (Günther)	threadfin shad	T	R	A
Salmonidae, trouts and salmons				
Salmo gairdneri Richardson	rainbow trout	F	R	S
Salvelinus fortinalis (Mitchill)	brook trout	F	?	S
Cyprinidae, carps, and minnows				
Algansea lacustris Steindachner	acumara	T	?	A
Barbus conchonius (Hamilton-Buchanan)	rosy barb	F	U	O
Barbus titteya (Deraniyagala)	cherry barb	F	U	O
Carassius auratus (Linnaeus)	goldfish	F	R	O
Ctenopharyngodon idella (Valenciennes)	grass carp	F	R	A
Cyprinus carpio Linnaeus	common carp	F	R	A
Gila bicolor mohavensis (Snyder)	Mohave tui chub	F	?	P
Gila orcutti (Eigenmann and Eigenmann)	arroyo chub	F	?	P
Hypophthalmichthys molitrix (Valenciennes)	silver carp	F	N	A
Notemigonus crysoleucas (Mitchill)	golden shiner	F	U	B
Notropis lutrensis (Baird and Girard)	red shiner	T	R	B
Pimephales promelas (Rafinesque)	fathead minnow	T	R	B
Pimephales vigilax (Baird and Girard)	bullhead minnow	T	R	B
Catostomidae				
Carpiodes carpio (Rafinesque)	river carpsucker	T	R	A
Cobitidae				
Misgurnus anguillicaudatus (Cantor)	oriental weatherfish	F	?	A
Ictaluridae				
Ictalurus furcatus (Lesueur)	blue catfish	T	R	A
Ictalurus melas (Rafinesque)	black bullhead	F	R	A
Ictalurus punctatus (Rafinesque)	channel catfish	T	R	A
Cyprinodontidae				
Fundulus zebrinus Jordan and Gilbert	plains killifish	F	R	B
Poeciliidae				
Gambusia affinis (Baird and Girard)	mosquitofish	T	R	M

Gambusia panuco Hubbs	Pánuco gambusia	T	?	?
Poecilia latipinna (Lesueur)	sailfin molly	T	R	O
Poecilia reticulata (Peters)	guppy	F	R	O
Xiphophorus helleri Heckel	green swordtail	T	R	O
Xiphophorus maculatus (Günther)	southern platyfish	T	R	O
Xiphophorus variatus (Meek)	variable platyfish	T	R	O
Atherinidae				
Chirostoma estor Jordon	blanco de Pátzcuaro	T	R	A
Chirostoma grandocule (Steindachner)	charal blanco	T	R	A
Chirostoma jordani Woolman	charal común	T	R	A
Chirostoma labarcae Meek	charal de Chapala	T	R	A
Chirostoma sphyraena Boulenger	blanco de Chapala	T	R	A
Menidia beryllina (Cope)	tidewater silverside	T	R	B
Percichthyidae				
Morone chrysops (Rafinesque)	white bass	T	R	A
Morone saxatilis (Walbaum)	striped bass	T	R	A
Centrarchidae				
Ambloplites rupestris (Rafinesque)	rock bass	F	?	A
Lepomis auritus (Linnaeus)	redbreast sunfish	F	?	A
Lepomis cyanellus (Rafinesque)	green sunfish	T	R	A
Lepomis gulosus (Cuvier)	warmouth	F	R	A
Lepomis macrochirus (Rafinesque)	bluegill	T	R	A
Lepomis megalotis (Rafinesque)	longear sunfish	T	R	A
Lepomis microlophus (Günther)	redear sunfish	F	?	A
Lepomis punctatus (Valenciennes)	spotted sunfish	F	?	A
Micropterus dolomieui Lacepède	smallmouth bass	F	?	S
Micropterus salmoides (Lacepède)	largemouth bass	T	R	S
Pomoxis annularis Rafinesque	white crappie	F	R	A
Pomoxis nigromaculatus (Lesueur)	black crappie	F	R	A
Cichlidae				
Cichlasoma cyanoguttatum (Baird and Girard)	Río Grande cichlid	T	R	?
Cichlasoma "urophthalmus" (Günther)	mojarra criolla	T	R	A
Petenia splendida Günther	tenhuayaca	T	R	A
Tilapia aurea (Steindachner)	blue tilapia	F	R	A
Tilapia mossambica (Peters)	Mozambique tilapia	F	R	A
Tilapia zilli (Gervais)	redbelly tilapia	F	R	A

native and within its historical range of distribution. When such records were recent, initially indicating the species as rare or scarce followed by a large population increase, or when similar records appeared following reservoir construction, we chose to regard these fishes as exotic.

The following accounts for exotic fishes in Mexico are in phylogenetic order after Robins et al. (1980) with generic and specific names in alphabetical order. Each account contains summary information on native range, records in Mexico, rationale for introduction, and status and impacts where known.

CLUPEDIAE – HERRINGS

***Dorosoma petenense* (Günther) threadfin shad**

Native range: Ohio River of Kentucky and southern Indiana, west and south to Oklahoma, Texas, and Florida, along the Gulf coast to northern Guatemala and Belize (Burgess 1980)

Records in Mexico: Río Colorado near Yuma and San Luis, Arizona/Sonora; Río Colorado near the delta in Sonora and Baja California Norte (Burgess 1980); upper Río Grande at El Paso, Texas/Ciudad Juarez to Ysleta, Chihuahua

Date of release: unknown

Source: United States

Purpose: a forage species (aquaculture)

Status: established

Impact: an associated factor with the disappearance of native fishes in the lower Río Colorado

Taxonomic note: several nominal subspecies; needs revision

SALMONIDAE – SALMONS AND TROUTS

***Salmo gairdneri* Richardson rainbow trout**

Native range: Kuskokwim River, Alaska to Río Presidio, Durango, including some interior areas (Behnke 1980)

Records in Mexico: western Sierras (Behnke 1980), Michoacán (Rosas 1976c), upper Río Yaqui (Hendrickson et al. 1980), Valley of Mexico (Alvares and Navarro 1957), general (FIDEFA 1975), Sierra Volcánica Transversal (Arrendondo 1973); Zacapú, Michoacán, and Montebello, Chiapas. Numerous federal stockings.

Date of release: 1870s

Source: United States

Purpose: sport and food (aquaculture)

Status: established

Impact: potential hybridization with native and introduced trouts

Taxonomic note: Mexican forms need reevaluation since *S. g. nelsoni* from Sierra San Pedro Mártir, Baja California Norte, is questionable and there appear to be undescribed forms from Chihuahua and Durango (Miller 1978, Uyeno and Miller 1979).

***Salvelinus fontinalis* (Mitchill) brook trout**

Native range: eastern North America from Hudson Bay, Canada, south to the Appalachian Mountains (Hendricks 1980)

Records in Mexico: presence assumed in the Valley of Mexico (Alvarez and Navarro 1957), Río Yaqui (Hendrickson et al. 1980), Cuitzitán, Michoacán (Rosas 1976c), and elsewhere (FIDEFA 1975). Numerous federal stockings.

Date of release: possibly mid-1800s

Source: United States

Purpose: sport and food (aquaculture)

Status: unknown

Impact: unknown

CYPRINIDAE – CARPS AND MINNOWS

***Algansea lacustris* Steindachner acumara**

Native range: Lago de Pátzcuaro (Alvarez 1970)

Records in Mexico: widely distributed but no specific localities mentioned (FIDEFA 1975, Rosas 1976c)

Date of release: since 1968 (Rosas 1976c) as transplants

Source: Lago de Pátzcuaro

Purpose: food (aquaculture)

Status: unknown

Impact: unknown, but hybridization is likely with congeners common in central Mexico

***Barbus conchonius* (Hamilton-Buchanan) rosy barb**

Native range: northern India and Assam

Records in Mexico: Río Santa Catarina, Monterrey (Contreras 1978)

Date of release: 1967

Source: an ornamental (aquarium) fish release

Purpose: to attract public attention to the ceremonial opening of the Municipal Aquarium in Monterrey. The officers of the aquarium asked children to release aquarium fishes into the river, which was done effectively. It was impossible to obtain a full list of all species released.

Status: unsuccessful

Impact: unknown

***Barbus titteya* (Deraniyagala) cherry barb**

Native range: Sri Lanka

Records in Mexico: Río Santa Catarina, Monterrey (Contreras 1978)

Date of release: 1967

Source: an ornamental (aquarium) fish release

Purpose: same as for *Barbus conchonius*

Status: unsuccessful

Impact: unknown

***Carassius auratus* (Linnaeus) goldfish**

Native range: Eurasian region (Banarescu 1964); primarily Chinese (Hubbs 1982)

Records in Mexico: Valley of Mexico (Meek 1904, Alvarez and Navarro 1957), upper Río Lerma (Romero 1967), Nuevo León (Contreras 1967), said to be rare in Mexico (Hensley and Courtenay 1980). Stocked in several localities in central Mexico (Departamento de Pesca).

Date of release: unknown

Source: unknown

Purpose: an ornamental (aquarium) fish release; perhaps as a food fish (aquaculture)

Status: established in some localities

Impact: refer to common carp. Goldfish and common carp are usually found together in the same localities. Occurs in Potosí, Nuevo León (Contreras 1978), and El Chorro, Coahuila (Contreras 1969). May not be so detrimental as common carp.

Taxonomic note: several recognizable subspecies and many cultivated stocks that do not merit naming (Banarescu 1964)

***Ctenopharyngodon idella* (Valenciennes) grass carp**

Native range: Pacific Asia from the Amur River of China and Siberia to the West River in China and Thailand (Guillory 1980)

Records in Mexico: Río Cupatitzio, Michoacán (Rosas 1976a, 1976c) and Temazcal, Oaxaca (Delgadillo 1976); also in Presa de la Boca, Nuevo León. Numerous federal stockings.

Date of release: 1964 or 1965

Source: People's Republic of China

Purpose: aquatic weed control and as a food fish (aquaculture)

Status: established only at Cupatitzio, Michoacán since 1974–75 (Rosas 1976a, 1976c)

Impact: unknown

***Cyprinus carpio* Linnaeus common carp**

Native range: Eurasia (Banarescu 1964)

Records in Mexico: Valley of Mexico lakes (Meek 1904), Arizona/Sonora border in the Río Colorado and in Chihuahua (Allen 1980), middle Río Grande (Hubbs et al. 1977), lower Río Grande (Treviño-Robinson 1959), lower Río Casas Grandes (Contreras et al. 1976, Miller and Chernoff 1980), Río Santa María (Miller and Chernoff 1980), Baja California (Follett 1961), Río Conchos (Contreras et al. 1976, Contreras 1978), upper Río Mezquital (Contreras et al. 1976), Parras basin (Contreras 1969, 1978), Valley of Mexico (Alvarez and Navarro 1957), upper Río Yaqui (Hendrickson et al. 1980), lower Río Yaqui (Branson et al. 1960), Michoacán (Alvarez and Cortes 1962), Nuevo León (Contreras 1967), San Luis Potosí (Alvarez 1959), and Aguascalientes (Contreras and Contreras, in press).

Date of release: 1872–73
Source: Haiti (Escalante and Contreras, in press); Israeli carp imported in 1956 (Obregón 1960)
Purpose: a food fish (aquaculture)
Status: established
Impact: associated with the disappearance of native fishes in Casas Grandes, Bustillos, Camargo in Chihuahua, Peña del Aguila and Tunal in Durango, Parras in Coahuila, and San Juan del Río in Querétaro (Contreras 1969, 1975, 1978; Contreras et al. 1976).
Taxonomic note: several subspecies recognized; varieties and cultivated stocks also named but not recognized by taxonomists (Banarescu 1964)

***Gila bicolor mohavensis* (Snyder)** **Mohave tui chub**
Native range: Mohave River, California (Moyle 1976)
Records in Mexico: Arroyo Santo Tomás, Baja California Norte (Follett 1961, Miller 1968)
Date of release: 1955
Source: Soda Lake, California (Miller 1968)
Purpose: species protection
Status: failed
Impact: unknown

***Gila orcutti* (Eigenmann and Eigenmann)** **arroyo chub**
Native range: Malibu Creek and Los Angeles River south to the San Luis Rey River, California; widely introduced north to the Santa Inez River and Mohave River of the Death Valley drainage, California (Swift 1980)
Records in Mexico: see entry for *Gila bicolor mohavensis*
Date of release: see entry for *Gila bicolor mohavensis*
Source: San Luis Rey River, San Diego County, California (Miller 1968)
Purpose: see entry for *Gila bicolor mohavensis*
Status: failed
Impact: unknown, but is known to undergo mass hybridization with *Gila bicolor mohavensis* (Hubbs and Miller 1943)

***Hypophthalmichthys molitrix* (Valenciennes)** **silver carp**
Native range: Asia
Records in Mexico: Hidalgo (Escalante and Contreras, in press)
Date of release: 1965
Source: unknown, but probably from People's Republic of China
Purpose: unknown
Impact: unknown

***Notemigonus crysoleucas* (Mitchill)** **golden shiner**
Native range: Atlantic slope from the Maritime Provinces south to Florida, west to Texas and north to Saskatchewan (Lee 1980)
Records in Mexico: two border records in California and Arizona (Lee

1980). One specimen collected at Marte R. Gómez Reservoir (Barajas and Contreras, in press)

Date of release: 1973

Source: probably by fishermen

Purpose: release of excess bait

Status: unsuccessful

Impact: probably none

***Notropis lutrensis* (Baird and Girard) red shiner**

Native range: in the Mississippi and Gulf drainages of the United States from South Dakota and Illinois to northern Mexico (Matthews 1980). Several subspecies from the Río Grande to the Río Pánuco, interior drainages of the northern plateau, and the headwaters of the Río Yaqui (Contreras 1975, 1978).

Records in Mexico: lower Río Colorado (Hubbs 1954)

Date of release: on or before 1953 (Hubbs 1954)

Source: an escape from Arizona Fish Farms near Blythe, California (Hubbs 1954)

Purpose: perhaps none; may also have been released as excess bait or on purpose as a forage fish (Hubbs 1954)

Status: established

Impact: most native fishes are gone from the lower Río Colorado, presumably due to many introduced fishes and increased salinity

***Pimephales promelas* (Rafinesque) fathead minnow**

Native range: Chihuahua, Mexico, north to Great Slave Lake, east to Nebraska, and west to Alberta (Lee and Shute 1980)

Records in Mexico: three border localities in California and Arizona (Lee and Shute 1980)

Date of release: unknown

Source: probably by fishermen

Purpose: release of excess bait

Status: established

Impact: unknown

***Pimephales vigilax* (Baird and Girard) bullhead minnow**

Native range: Mississippi River basin from Maryland and South Dakota south to Mexico, Texas, Louisiana, and Mississippi, east to Georgia and Alabama (Lee and Kucas 1980)

Records in Mexico: middle Río Grande (Hubbs et al. 1977)

Date of release: unknown

Source: probably by fishermen

Purpose: release of excess bait

Status: established

Impact: unknown

CATOSTOMIDAE – SUCKERS

***Carpiodes carpio* (Rafinesque) river carpsucker**

Native range: Mississippi basin from Pennsylvania to Montana, south to the Gulf of Mexico drainages from the Mississippi River to Mexico (Lee and Platania 1980).

Records in Mexico: twelve localities reported for the Río Yaqui (Hendrickson et al. 1980)

Date of release: unknown

Source: unknown

Purpose: unknown

Status: established

Impact: unknown

COBITIDAE – LOACHES

***Misgurnus anguillicaudatus* (Cantor) oriental weatherfish**

Native range: eastern Asia (Hensley and Courtenay 1980)

Records in Mexico: canals around Chapingo, Mexico

Date of release: before 1961

Source: Chapingo Fish Hatchery

Purpose: a food fish (aquaculture)

Status: unknown. This fish has not been seen since the hatchery was closed.

Impact: unknown

ICTALURIDAE – BULLHEAD CATFISHES

***Ictalurus furcatus* (Lesueur) blue catfish**

Native range: major rivers of the Mississippi, Missouri, and Ohio basins, and to Mexico and northern Guatemala (Glodek 1980). Southern populations represent the nominal *I. meridionalis.*

Records in Mexico: Río Yaqui at Huasabas (Hendrickson et al. 1980)

Date of release: unknown

Source: unknown

Purpose: probably as a food fish (aquaculture)

Status: established

Impact: unknown

***Ictalurus melas* (Rafinesque) black bullhead**

Native range: southern Ontario, Great Lakes, and the St. Lawrence River to the Gulf of Mexico and northern Mexico, and Montana to the Appalachians (Glodek 1980). We are not aware of any records that indicate this fish is native to Mexico.

Records in Mexico: Río Yaqui near Ciudad Obregón (Branson et al. 1960) and localities in the Río Casas Grandes (Contreras et al. 1976, Hendrickson et al. 1980, Miller and Chernoff 1980). Río Pánuco, Río Lerma-Santiago and Río Balsas (Instituto Politécnico Nacional).

Date of release: unknown

Source: unknown

Purpose: a food fish (aquaculture)

Status: established

Impact: appears associated with replacement of native fishes at Casas Grandes, Chihuahua (Contreras et al. 1976)

***Ictalurus punctatus* (Rafinesque)** **channel catfish**

Native range: central drainages of the United States to southern Canada and northern Mexico (Glodek 1980)

Records in Mexico: Río Colorado in Baja California Norte (Follett 1961), Río Yaqui at Tónichi (Hendrickson et al. 1980) and Presa Oyul, Tamaulipas (Vergara 1976). Stockings reported in Jalisco and Nayarit (Departamento de Pesca and other agencies).

Date of release: 1975–76 for the Río Pánuco (Vergara 1976), unknown for other localities

Source: escapes from aquaculture facilities

Purpose: a food fish (aquaculture)

Status: established

Impact: unknown, but appears to have been associated with replacement of native fishes in the Río Colorado (Miller 1961, 1963)

Taxonomic note: closely related species or subspecies are known south of the Río Grande to at least the Río Pánuco.

CYPRINODONTIDAE – KILLIFISHES

***Fundulus zebrinus* Jordan and Gilbert** **plains killifish**

Native range: southeastern Montana to Missouri and Texas (Pecos River) (Poss and Miller 1983). We concur with Hubbs and Wauer (1973) and Hubbs et al. (1977) that this fish is not native to the Río Grande.

Records in Mexico: three localities in the middle Río Grande (Hubbs et al. 1977)

Date of release: unknown

Source: probably by fishermen

Purpose: release of excess bait

Status: established

Impact: unknown

Taxonomic note: synonymous with *F. kansae* (see Poss and Miller 1983).

POECILIIDAE – LIVEBEARERS

***Gambusia affinis* (Baird and Girard) mosquitofish**

Native range: southern Indiana and Illinois to Florida, northward along the Atlantic slope to New Jersey, south to Veracruz (Lee and Burgess 1980) as far as the Río Cazones.

Records in Mexico: Río Yaqui (Hendrickson et al. 1980), Arroyo San José of Baja California (Miller and Hubbs 1954) and other nearby localities (Follett 1961), and the interior drainage of the Río Santa María, Chihuahua (Miller and Chernoff 1980). We record this fish for Ojo la Rosita, San Antonio, Ojo San Bartolo, Todos Santos, Laguna Atascadero, and San Luis, Río Colorado, Sonora.

Date of release: unknown except since 1931 in Arroyo San José, Baja California (Miller and Hubbs 1954)

Source: unknown

Purpose: mosquito control

Status: established

Impact: associated with replacement of native fishes in the Río Casas Grandes (Contreras et al. 1976)

***Gambusia panuco* Hubbs Pánuco gambusia**

Native range: Río Pánuco (Alvarez 1970)

Records in Mexico: La Media Luna springs, San Luis Potosí (Hubbs and Miller 1977)

Date of release: before 1972

Source: unknown

Purpose: unknown

Status: unknown

Impact: unknown

***Poecilia latipinna* (Lesueur) sailfin molly**

Native range: coastal waters of the Gulf of Mexico and the Atlantic slope from North Carolina to central Veracruz (González de la Rosa 1977)

Records in Mexico: Río Colorado at San Luis, Sonora (González de la Rosa 1977) to the Río Colorado delta (Burgess 1980); upper Río Pánuco (Instituto Politécnico Nacional)

Date of release: unknown

Source: unknown

Purpose: probably release of an ornamental (aquarium) fish

Status: established

Impact: unknown

***Poecilia reticulata* (Peters) guppy**

Native range: northeastern South America and nearby Antilles (Hensley and Courtenay 1980)

Records in Mexico: Parras (Contreras 1969), Monterrey (Contreras 1978); also in several localities of the Río Balsas (Morelos), upper Río Pánuco, around Lago de Chapala, and Laguna Cortés and Todos Santos, Cabo San Lucas, Baja California Sur
Date of release: before 1961 in Morelos; unknown at other localities
Source: release of an ornamental (aquarium) fish
Status: established
Impact: unknown

***Xiphophorus helleri* Heckel green swordtail**
Native range: Gulf of Mexico basins from the Río Nautla to northwestern Belize (Hensley and Courtenay 1980)
Records in Mexico: Río Balsas basin of Michoacán (no specimens; Alvarez and Cortés 1962). Morelos state, Monterrey, Parras, Arroyo Chorro of Coahuila, Cointzio of Michoacán.
Date of release: unknown
Source: unknown
Purpose: release of an ornamental (aquarium) fish
Status: established
Impact: massive hybridization with introduced *X. maculatus* and *X. variatus,* and the endangered endemic *X. couchianus* to Ojo de la Peñita, one of its only three surviving populations

***Xiphophorus maculatus* (Günther) southern platyfish**
Native range: Gulf of Mexico basins from Ciudad Veracruz to northern Belize (Hensley and Courtenay 1980)
Records in Mexico: springs around Monterrey, Parras, Cabo San Lucas (Ojo la Rosita, Todos Santos) and near Morelia (Instituto Politécnico Nacional).
Date of release: unknown
Source: unknown
Purpose: release of an ornamental (aquarium) fish
Status: established
Impact: see account for *X. helleri*

***Xiphophorus variatus* (Meek) variable platyfish**
Native range: southern Tamaulipas, eastern San Luis Potosí and northern Veracruz (Hensley and Courtenay 1980)
Records in Mexico: around Monterrey (Contreras 1978). Also in the Río Colorado at San Luis, Sonora.
Date of release: unknown
Source: unknown
Purpose: release of an ornamental (aquarium) fish
Status: established

Impact: associated with replacement of native fishes around Monterrey (Contreras et al. 1976). Also see account for *X. helleri.*

ATHERINIDAE – SILVERSIDES

***Chirostoma estor* Jordan** **blanco de Pátzcuaro**

Native range: Lagos de Pátzcuaro and Zirahuén (Barbour 1973)

Records in Mexico: introduced to more than 150 reservoirs (Rosas 1976c); also in Veracruz and Puebla (Instituto Politécnico Nacional); introduced in Morelos, Nuevo León, and Tamaulipas (Departamento de Pesca).

Date of release: unknown

Source: official stocking programs

Purpose: a food fish (aquaculture)

Status: established (Puebla, Veracruz, Morelos) to unsuccessful (Nuevo León, Tamaulipas)

Impact: unknown

***Chirostoma grandocule* (Steindachner)** **charal blanco**

Native range: Lago de Pátzcuaro (Barbour 1973)

Records in Mexico: widely stocked (Rosas 1976c); also introduced in Chihuahua, Coahuila, Hidalgo, Mexico, Michoacán, Nuevo León, Puebla, San Luis Potosí, Sinaloa, and Veracruz by the Departamento de Pesca

Date of release: unknown

Source: official stocking programs

Purpose: a food fish (aquaculture)

Status: established in Mexico and Veracruz, unsuccessful in other areas

Impact: unknown

***Chirostoma jordani* Woolman** **charal comun**

Native range: Lago de Chapala and the Valley of Mexico (Meek 1904, Barbour 1973).

Records in Mexico: Camargo, Chihuahua (see *Menidia beryllina;* Minckley 1965); also in Amistad (Coahuila) and Falcon (Tamaulipas) reservoirs (C. Hubbs, personal communication to W. R. Courtenay, Jr., 1975)

Date of release: unknown, but before 1975 in Río Grande reservoirs

Source: not listed as included in official stocking programs. It is probable that common Chapalan silversides were substituted for species from Pátzcuaro for unknown reasons and stocked north of Chapala.

Purpose: probably as a food fish (aquaculture)

Status: established at Camargo, Chihuahua; unsuccessful in Amistad and Falcon reservoirs on the Río Grande

Impact: associated with replacement of native fishes at Camargo, Chihuahua (Contreras et al. 1976); unknown in Amistad and Falcon reservoirs

***Chirostoma labarcea* Meek** **charal de Chapala**

Native range: area of Lago de Chapala (Barbour 1973)

Records in Mexico: Camargo, Chihuahua (as *Chirostoma* sp.; Minckley 1965; later identified by Barbour as *C. labarcae*). Other specimens from around Lerdo, Durango, apparently are of this species.

Date of release: unknown

Source: not listed in official stocking programs but probably released by the federal government

Purpose: perhaps as a food fish (aquaculture) or an accidental introduction in mixed stocks of other introduced fishes

Status: established

Impact: associated with replacement of endemic fishes at Lerdo, Durango (Contreras et al. 1978), if this species is involved

***Chirostoma sphyraena* Boulenger** **blanco de Chapala**

Native range: area of Lago de Chapala (Barbour 1973)

Records in Mexico: Camargo, Chihuahua (Minckley 1965); also in Durango.

Date of release: unknown

Source: part of official stocking programs

Purpose: a food fish (aquaculture)

Status: established

Impact: two specimens tentatively identified as *C. consocium* from Camargo, Chihuahua, may be of this species; if so, they are associated with replacement of native fishes there (Contreras et al. 1976)

***Menidia beryllina* (Cope)** **tidewater silverside**

Native range: coastal streams from Massachusetts to Veracruz (Gilbert and Lee 1980) and the lower Río San Juan in Nuevo León and Tamaulipas (Contreras 1967)

Records in Mexico: Camargo (see under Comments, below) and Delicias, Chihuahua (Minckley 1965); also in Don Martín Reservoir, Coahuila, and Colombia, Nuevo León

Date of release: unknown

Source: probably by fishermen

Purpose: possible release of excess bait

Status: established

Impact: associated with replacement of native fishes at Camargo, Chihuahua (Contreras et al. 1976)

Comments: The specimens from Camargo, Chihuahua, have been reidentified recently as *Chirostoma jordani.* The extensive stocking programs conducted by the federal government using silversides suggests that common Chapalan species (i.e., *C. estor, C. jordani,* and *C. labarcae*) were probably substituted for species from Pátzcuaro for unknown reasons and stocked north of Chapala.

PERCICHTHYIDAE – TEMPERATE BASSES

***Morone chrysops* (Rafinesque) white bass**

Native range: Mississippi and Ohio valleys and Great Lakes to the Río Grande (Burgess 1980). We do not regard this species as native to Mexico.

Records in Mexico: lower and middle Río Grande (Hubbs et al. 1977); also in the Río San Juan to Dr. Coss, Nuevo León.

Date of release: unknown

Source: unknown

Purpose: a food fish (aquaculture) and sport

Status: established

Impact: unknown

***Morone saxatilis* (Walbaum) striped bass**

Native range: St. Lawrence River south to the St. Johns River, Florida, and disjunctly from the Suwannee River, Florida, to Lake Pontchartrain, Louisiana (Burgess 1980)

Records in Mexico: Río Colorado and Río Grande near Del Río (Burgess 1980)

Date of release: unknown

Source: probably by U.S. state agencies

Purpose: a food fish (aquaculture) and sport

Status: established in the Río Grande; possibly established south of Lake Mead in the Río Colorado

Impact: unknown

CENTRARCHIDAE – SUNFISHES

***Ambloplites rupestris* (Rafinesque) rock bass**

Native range: Massachusetts to the Red River drainage of North Dakota, southern Ontario around the Great Lakes to the St. Lawrence River to Quebec, west of the Appalachian divide south to the Tennessee River drainage in northern Alabama and in the eastern and western tributaries of the upper and middle Mississippi drainage (Cashner 1980)

Records in Mexico: Río Piedras Verdes at Colonia, Juárez, and in the Casas Grandes basin of Chihuahua (Miller and Chernoff 1980)

Date of release: unknown

Source: unknown

Purpose: a food fish (aquaculture)

Status: unknown

Impact: unknown

***Lepomis auritus* (Linnaeus) redbreast sunfish**

Native range: Atlantic slope from New Brunswick to central Florida (Lee 1980)

Records in Mexico: areas of Amistad (Coahuila) and Falcon (Tamaulipas) reservoirs (Lee 1980) without mention of specific localities
Date of release: unknown
Source: unknown
Purpose: perhaps as a forage species or food fish (aquaculture)
Status: unknown
Impact: unknown

***Lepomis cyanellus* (Rafinesque) green sunfish**
Native range: east-central North America west of the Appalachians from Ontario to eastern North Dakota, south to Georgia and northeastern Mexico (Lee 1980)
Records in Mexico: Baja California (Follett 1961) and Río Yaqui (Hendrickson et al. 1980)
Date of release: unknown
Source: unknown
Purpose: perhaps as a forage species or food fish (aquaculture)
Status: established
Impact: unknown

***Lepomis gulosus* (Cuvier) warmouth**
Native range: eastern North America from Kansas to Iowa to southern Wisconsin, Minnesota, western Pennsylvania, to Florida and the Río Grande (Lee 1980). We do not accept the Río Grande as part of the native range of this fish.
Records in Mexico: Marte R. Gómez Reservoir, Tamaulipas (Barajas and Contreras, in press), and San José de Gracia in Aguascalientes (Contreras and Contreras, in press); also in Mal Paso Reservoir, Zacatecas
Date of release: unknown
Source: unknown
Purpose: a food fish (aquaculture)
Status: established
Impact: unknown

***Lepomis macrochirus* (Rafinesque) bluegill**
Native range: coastal Virginia to Florida, west to Texas and northern Mexico, western Minnesota to western New York (Lee 1980)
Records in Mexico: Baja California (Follett 1961), lower Río Yaqui (Branson et al. 1960), and twelve other localities in the Río Yaqui (Hendrickson et al. 1980), Valley of Mexico (Alvarez and Navarro 1957), and Aquascalientes (Contreras and Contreras, in press); also in Saúz springs, most reservoirs in Chihuahua, El Zarco in Coahuila, and Las Adjuntas Reservoir in Tamaulipas
Date of release: unknown
Source: unknown

Purpose: a forage species and for food (aquaculture)
Status: established
Impact: associated with replacement of native fishes at Bustillos, Chihuahua, Peña del Aguila, Tunal and Lerdo in Durango (Contreras 1975, 1976; Contreras et al. 1976)

***Lepomis megalotis* (Rafinesque) longear sunfish**
Native range: west of the Appalachians, from southern Quebec to Alabama and western Florida to Río Grande tributaries in northeastern Mexico to Oklahoma, northeastward to southern Ontario (Bauer 1980)
Records in Mexico: Río Yaqui at Guerrero (Hendrickson et al. 1980), Bustillos, Chihuahua (Contreras 1975, Contreras et al. 1976)
Date of release: unknown
Source: unknown
Purpose: a forage and food fish (aquaculture)
Status: established
Impact: associated with replacement of native fishes at Bustillos, Chihuahua (Contreras et al. 1976)

***Lepomis microlophus* (Günther) redear sunfish**
Native range: peninsular Florida, lower Atlantic slope and Gulf slope drainages west to Texas, north to southern Indiana (Lee 1980). May be native to the lower Río Grande, but we doubt it.
Records in Mexico: Río Grande (Hubbs et al. 1977) and Río Yaqui at Presa Novillo (Hendrickson et al. 1980)
Date of release: unknown
Source: unknown
Purpose: probably as a forage species and a food fish (aquaculture)
Status: unknown
Impact: unknown

***Lepomis punctatus* (Valenciennes) spotted sunfish**
Native range: southeastern United States from eastern Texas to and including peninsular Florida, Atlantic slope, north to southeastern North Carolina, and the Mississippi basin north to Illinois (Lee 1980)
Records in Mexico: Falcon Reservoir, Tamaulipas (Lee 1980)
Date of release: unknown
Source: unknown
Purpose: probably as a forage species and food fish (aquaculture)
Status: unknown
Impact: unknown

***Micropterus dolomieui* Lacepède smallmouth bass**
Native range: Minnesota eastward to Quebec, southward to the Tennessee River system in Alabama, and westward to eastern Oklahoma (Lee 1980)

Records in Mexico: areas of Amistad Reservoir (Lee 1980). Rosas (1976c) mentions culture of this species in Mexico without further data.
Date of release: unknown
Source: unknown
Purpose: a sport and food fish (aquaculture)
Status: unknown
Impact: unknown

***Micropterus salmoides* (Lacepède) largemouth bass**
Native range: northeastern Mexico to Florida, much of the Mississippi basin to southern Quebec and Ontario, and the Atlantic slope north to southern or central South Carolina (Lee 1980)
Records in Mexico: Lago de Pátzcuaro (De Buen 1941; Alvarez and Cortés 1962), Baja California (Follett 1961), Valley of Mexico (Alvarez and Navarro 1957), Río Yaqui (Hendrickson et al. 1980), and Aguascalientes (Contreras and Contreras, in press). Also in the Río Colorado at San Luis, Sonora, and Navojoa, Sonora. Numerous localities from the stocking programs of the Departamento de Pesca.
Date of release: unknown
Source: unknown in part; government stocking programs in part; stocks probably from the United States
Purpose: a sport and food fish (aquaculture)
Status: established
Impact: damage to local fisheries at Lago de Pátzcuaro (De Buen 1941), predation and displacement of native fishes and perhaps the replacement of another bass at Cuatro Cíenegas (Contreras 1969), predation on and displacement of endemic fishes at Potosí, Nuevo León (Contreras 1978), and replacement of native fishes at Bustillos, Chihuahua, and Peña del Aguila and Río Tunal, Durango (Contreras 1975, Contreras et al. 1976)

***Pomoxis annularis* Rafinesque white crappie**
Native range: east-central North America from southern Ontario and southwestern New York west of the Appalachians, south to the Gulf coast and west to Texas, South Dakota, and Minnesota (Lee 1980). We do not recognize this fish as native in the Río Grande.
Records in Mexico: Marte R. Gómez Reservoir (Barajas and Contreras, in press), Falcon Reservoir, Tamaulipas (Lee 1980), lower Río Yaqui (Hendrickson et al. 1980), and Baja California Norte (Follett 1961); also in the Río San Juan at Dr. Coss, Nuevo León
Date of release: unknown but probably early 1950s
Source: probably the United States
Purpose: a food fish (aquaculture)
Status: established
Impact: unknown

***Pomoxis nigromaculatus* (Lesueur) black crappie**

Native range: Atlantic slope from Virginia to Florida, along the Gulf coast to central Texas, north to North Dakota and eastern Montana, eastward to the Appalachians (Lee 1980)

Records in Mexico: lower Río Grande (Lee 1980) and lower Río Colorado (Follett 1961, Lee 1980); Falcon Reservoir

Date of release: unknown

Source: United States

Purpose: a food fish (aquaculture)

Status: established

Impact: unknown

CICHLIDAE – CICHLIDS

***Cichlasoma cyanoguttatum* (Baird and Girard) Río Grande cichlid**

Native range: Río Pánuco-Tamiahua system to the Río Grande

Records in Mexico: Cuatro Ciénegas Valley (LaBounty 1974)

Date of release: unknown

Source: perhaps nearby Río Salado

Purpose: unknown

Status: established

Impact: hybridizes with an endemic *Cichlasoma* (LaBounty 1974)

***Cichlasoma "urophthalmus"* (Günther) mojarra criolla**

Native range: Río Usumacinta to Yucatán Peninsula and Isla Mujeres (Alvarez 1970)

Records in Mexico: Río Papaloapan near Temazcal (Delgadillo 1976), but the photograph does not correspond to typical *C. urophthalmus*

Date of release: 1968

Source: wild stocks from Tabasco

Status: unknown

Impact: unknown

Note: Delgadillo (1976) mentions a total of four cichlids including this species and *Petenia splendida* and other aquarium fishes stocked in the Temazcal Reservoir.

***Petenia splendida* Günther tenhuaycaa**

Native range: Río Usumacinta basin (Alvarez 1970)

Records in Mexico: Río Papaloapan at Temazcal Reservoir (Delgadillo 1976)

Date of release: 1968

Source: wild stock

Purpose: a food fish (aquaculture)

Status: unknown
Impact: unknown

In regard to the tilapias below, the official records of the Departamento de Pesca are all based on introductions of tilapia, regardless of species. Therefore, our records may apply to any introduced species of these Old World cichlids. Our accounts are based on published records and letters from reliable sources. The official records show that *T. nilotica, T. mossambica,* and *T. melanopleura* (probably *T. zilli)* were imported originally.

***Tilapia aurea* (Steindachner) blue tilapia**
Native range: Senegal, middle Niger, Chad, lower Nile and Jordan river system (Hensley and Courtenay 1980)
Date of release: unknown
Source: first imported into Mexico in 1964 from Auburn, Alabama (Delgadillo 1976). Introductions in the lower Río Colorado were probably made in the early 1960s (Minckley 1973)
Purpose: in Mexico to establish a food fish; in the United States for aquatic weed control
Status: established
Impact: unknown
Taxonomic note: Specimens from lower Río Colorado (Palo Verde Valley, California, southward) are probably hybrids (W. R. Courtenay, Jr., and W. L. Minckley, personal communications).

***Tilapia zilli* (Gervais) redbelly tilapia**
Native range: north and central Africa to Jordan (Hensley and Courtenay 1980)
Records in Mexico: lower Río Colorado (Hensley and Courtenay 1980)
Date of release: unknown in Mexico. Introductions in the lower Río Colorado could have been made in the early 1960s (Minckley 1973)
Source: first imported into Mexico in 1964 as *Tilapia melanopleura* from Auburn, Alabama (Delgadillo 1976; see also Thys van den Audenaerde 1968)
Purpose: in Mexico, probably to establish a food fish; in the United States for aquatic weed control
Status: established
Impact: unknown

Other records for *Tilapia:* We have observed specimens of tilapias from Marte R. Gómez Reservoir (1974), Falcon Reservoir in Tamaulipas (1975), Peña del Aguila and other localities in the upper Río Mezquital of Durango (1981), the Río Nazas at El Salvador, Durango (1981), the coastal plains of Sinaloa (1978), and La Boca Reservoir near Monterrey (1979). They appear to be established in all localities.

Table 6-2. REPLACEMENT OF NATIVE FISH SPECIES BY EXOTICS IN MEXICO

Locality	Species number[a]		Time span	Other factors
	Natives	Exotics		
Casas Grandes, Chihuahua	4 - 2	1 - 4	1964–75	dams
Laguna Bustillos, Chihuahua	1 - 0	0 - 3	1901–64	
Río Chihuahua, Chihuahua	13 - 2	0 - 1	1908–75	channels, dams, drying
Camargo, Chihuahua	20 - 8	0 - 2	1901–75	dams
Lerdo, Durango	7 - 0	0 - 3 - 0	1903–81	dams
Peña del Aguila, Durango	3 - 1	2 - 5	1963–81	dams
Río Tunal, Durango	7 - 0	0 - 2 - 0	1961–81	dams
Parras, Coahuila	8 - 3	0 - 5	1880–1973	
Monterrey, Nuevo León	12 - 0	0 - 2 - 0	1903–81	channels, drying

[a] The numbers correspond to the time span: the one on the left to the beginning date, the one on the right to the later date. The middle number of instances with three numbers existed at an undeterminate middle date.

Known Impacts of Exotic Fishes in Mexico

There have been few investigations on the impacts of introduced fishes on native fishes and aquatic habitats in Mexico. The Mexican government is understandably concerned with the low protein diet of many Mexicans, particularly country people. The government, therefore, measures success of introduced fishes in economic and social terms and not on real or potential adverse impacts to native biota or habitats.

Contreras (1969, 1975, 1978) and Contreras et al. (1976) found circumstantial evidence that introductions of exotic fishes have had adverse impacts on some native fish populations. Selected data from these studies are shown in table 6-2. We chose to use graphs (figure 6-1) to show population trends following introductions and declines in native fishes in certain localities: Casas Grandes, Chihuahua, and Lerdo and Peña del Aguila, Durango. We caution, however, that introduced fishes may not be the major cause of declines of native fishes at these localities. These are areas of major development, including intense agriculture, and urban and highway construction. In some instances, the entire fish fauna, both native and introduced, has been destroyed.

Known impacts include hybridization with endemics, as was demonstrated by the introductions of *Xiphophorus helleri, X. maculatus,* and *X. variatus* at La Peñita, near Monterrey; the result was massive gene swamping of the endangered endemic *X. couchianus* population. A similar situation occurred

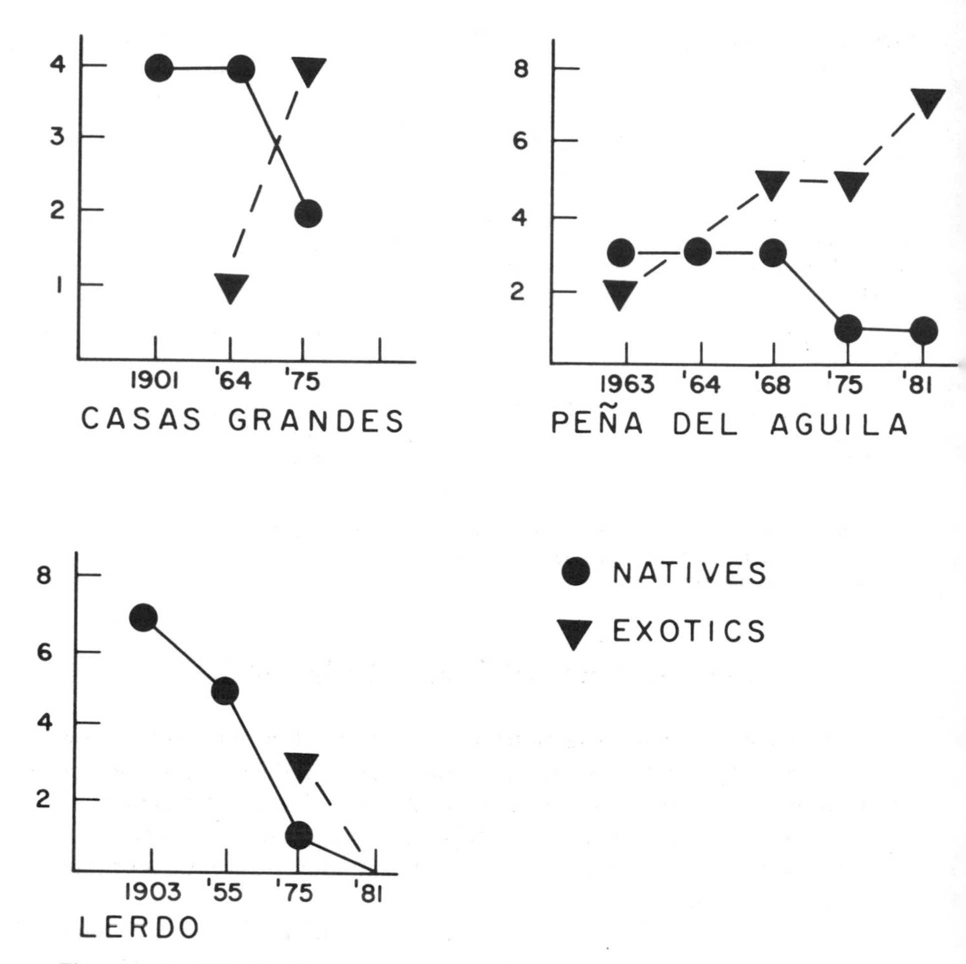

Figure 6–1. Trends of replacement of native fishes by exotics at three localities in northern Mexico.

in Cuatro Ciénegas when the exotic *Cichlasoma cyanoguttatum* hybridized with an endemic species of *Cichlasoma* (LaBounty 1974).

Other introductions have resulted in severe predation on endemic species. bringing these fishes close to extinction. This happened when *Micropterus salmoides* was introduced at El Potosí, Nuevo León; the major impact there was on *Megupsilon aporus* and to a lesser degree on *Cyprinodon alvarezi* (Contreras 1978).

Some impacts are unexpected, such as the extirpation of a newly introduced but as yet unidentified tilapia in the Balsas basin. The native *Cichlasoma istlanum* was host to an innocuous nematode parasite. That parasite attacked the introduced tilapia and has nearly destroyed what was to have been a new fishery (Rosas 1976b). Simultaneously, the tilapia almost completely eliminated the native *Cichlasoma.* Another (perhaps) unexpected event was the change in diet of some introduced tilapias from herbivorous to carnivorous when they became infested by certain parasites. We used *perhaps* parenthetically because tilapias, like most cichlids, will freely take and utilize various kinds of prepared aquarium fish foods designed for omnivores.

Most introductions of exotic fishes in Mexico were made without previous thought of negative impacts or methods of control. Many species were imported from the United States, including fishes that share drainages in both nations. Some records are incomplete or nonexistent; hence, the earliest records of introductions do not include the date of release or the source of the stock. Therefore, the full history of introductions in Mexico will never be known.

There is, however, the important aspect of increasing the availability, production, and consumption of fish protein in a nation with a largely protein-deficient society. One means of accomplishing these goals is through introductions. To biologists concerned with reductions and potential extinctions of native fishes, particularly of endemic species, such introductions may appear incongruous. Nevertheless, when social and economic parameters are considered, introductions can be viewed as justifiable, particularly by governmental agencies.

While we share the concern of governmental agencies in providing protein for the citizens, we emphasize that introducing fishes toward such goals and also protecting native fishes can be accomplished through improved planning. We recommend that a protocol be developed for Mexico by Mexican biologists that can and should serve the needs of its citizens properly while simultaneously protecting and preserving our native biotic communities. An approach for developing such a protocol is provided by Kohler and Stanley in chapter 18.

Conclusion

Mexico has at least 55 exotic fishes, 26 of which are foreign and 29 are transplants within the country. Additional species can be expected in our waters since they are now established near the border. Reproductive populations have been established by 41 species, 1 is not reproducing, 3 have been unsuccessful and the success of another 11 species is unknown and should be verified. Negative impacts on native biota, often complicated by other factors, is manifested by regression of native fishes, sometimes through hybridization with endemics, and some unexpected biological backlashes to introduced species. For most exotics, their impact is viewed officially as positive, economically and socially, as a food source for underdeveloped areas. A protocol for introductions that will meet the needs of Mexican citizens while protecting native fishes is necessary.

Acknowledgments

Many persons made this study possible. We thank the authorities of the Departamento de Pesca (Dirección General de Acuacultura) for permission to examine their stocking records. Dr. José Alvarez del Villar kindly permitted access to the fish collection of the Escuela Nacional de Ciencias Biológicas, Instituto Politécnico Nacional, from which numerous distributional records, primarily in southern Mexico, were obtained. Most northern records are ours and specimens are deposited in the fish collection of the Escuela de Graduados en Ciencias Biológicas, Universidad Autónoma de Nuevo León. Several students completed their theses under the supervision of the senior author and their results provided many new records; their thesis references are provided in the Literature Cited section.

Many persons too numerous to list helped in this effort. We thank Ana María Robledo for typing the manuscript. To all who assisted, we express our profound gratitude.

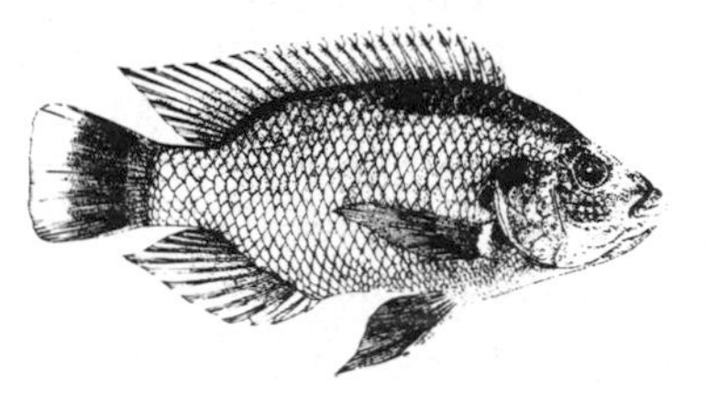

Literature Cited

(Note: Many references included here are from the Atlas of North American freshwater fishes, edited by D. S. Lee, C. R. Gilbert, C. H. Hocutt, R. E. Jenkins, D. E. McAllister, and J. R. Stauffer, Jr., published in 1980 by the North Carolina State Museum, Raleigh. In the interest of saving space, all references from the Atlas will be so indicated without repeating names of editors and the publisher.)

Allen, A. W. 1980. *Cyprinus carpio* Linnaeus, 152. *In:* Atlas of North American freshwater fishes.

Alvarez, J. 1959. Nota preliminar sobre la ictiofauna del estado de San Luis Potosí. Acta Científica Potosina 3(1):71–88.

Alvarez, J. 1970. Peces Méxicanos. Instituto Nacional de Investigaciones Biológico Pesquerías. Boletin 2.

Alvarez, J., and M. T. Cortés. 1962. Ictiología Michoacána. I. Claves y catálogo de las especies conocidas. Anales Escuela Nacional Ciencias Biológicas, México 9(1-4):85-142.

Alvarez, J., and L. Navarro. 1957. Los peces del Valle de México. Secretaría de Marina. Comisión para el Fomento de la Piscicultura Rural, México.

Arredondo, L. 1973. Especies acuáticas de valor alimenticio introducidas en México. Reunión Continental sobre la Ciencia y el Hombre. AAAS/México, D. F. Consejo Nacional Ciencia y Tecnología.

Arredondo, L. 1975. Algunos aspectos sobre la taxonomía de tilapia. Piscis, México 1(2):24-28.

Banarescu, P. 1964. Pisces – Osteichthyes (Pesti Ganoizi si Ososi). Fauna Republici Populare Romine 13.

Barajas, M. L., and S. Contreras. In press. Variación estacional y morfología de los peces de la Presa Marte R. Gómez, noreste de México. IV Congreso Nacional de Zoológica, México.

Barbour, C. D. 1973. The systematics and evolution of the genus *Chirostoma* Swainson (Pisces, Atherinidae). Tulane Studies in Biology and Zoology, 18(3): 97-141.

Bauer, B. H. 1980. *Lepomis megalotis* (Rafinesque), 600. *In:* Atlas of North American freshwater fishes.

Behnke, R. J. 1980. *Salmo gairdneri* Richardson, 106. *In:* Atlas of North American freshwater fishes.

Branson, B. A., C. J. McCoy, and M. E. Sisk. 1960. Notes on the freshwater fishes of Sonora with an addition to the known fauna. Copeia 1960(3):217-220.

Burgess, G. H. 1980. *Dorosoma petenense* (Günther), 70; *Poecilia latipinna* (Lesueur), 549; *Morone chrysops* (Rafinesque), 574; *Morone saxatilis* (Walbaum), 576. *In:* Atlas of North American freshwater fishes.

Cashner, R. C. 1980. *Ambloplites rupestris* (Rafinesque), 581. *In:* Atlas of North American freshwater fishes.

Cházari, E. 1884. Piscicultura de agua dulce. Secretaría de Fomento, México.

Contreras, S. 1967. Lista de peces del Estado de Nuevo León. Cuadernos del Instituto de Investigaciones Científicas, Universidad de Nuevo León, 11:1-12.

Contreras, S. 1969. Perspectives de la ictiofauna en las zonas aridas del norte de México. Memorias del Simposio Internacional sobre el Aumento de Producción de Alimentos Zonas Aridas. International Center for Arid Lands Studies Publications, 3:293-304.

Contreras, S. 1975. Cambios de composición de especies en comunidades de en zonas semiáridas de México. Publicaciones Biológicas del Instituto de Investigaciones Científicas. Universidad Autónoma de Nuevo León, México 1(7):181-194.

Contreras, S. 1976. Impacto ambiental de obras hidráulicas. Informe Técnico, Secretaría de Agricultura y Recursos Hidráulicos, México.

Contreras, S. 1978. Speciation aspects and man-made community composition changes in Chihuahua desert fishes, 405-431. *In:* Transactions of the Symposium on Biological Resources of the Chihuahuan Desert Region, U.S. and Mexico. R. H. Wauer and D. H. Riskind (eds.). National Park Service Transactions and Proceedings, sec. 3.

Contreras, A. J., and S. Contreras. In press. Los peces de Aguascalientes, México. III Congreso Nacional de Zoología, México.

Contreras, S., V. Landa, T. Villegas, and G. Rodriguez. 1976. Peces, piscicultura, presas, polución, planificación pesquera, y monitoreo en México, o la danza de las P. Memorias del Simposio de Pesquerías en Aguas Continentales 1:315–346.

De Buen, F. 1941. El *Micropterus (Huro) salmoides* y los resultado de su aclimatación en el Lago de Patzcuaro. Revista de la Sociedad Méxicana de Historía Natural 2(1):69–78.

Delgadillo-T., M. S. 1976. La Estación de Temascal como factor de desarrollo en la acuacultura de la cuenca del Papaloapan. Memorias del Simposio de Pesquerías en Aguas Continentales 1:55–86.

Escalante, M. A., and S. Contreras. In press. Distribución de algunas especies exóticas (transfaunadas de sus ecosistemas nativos) en la República Méxicana. III Congreso Nacional de Zoología, México. 1979.

FIDEFA. 1975. La piscicultura en el medio rural Méxicano. Fideicomiso para el Desarrollo de la Fauna Acuática, México. 44 pp.

Follett, W. I. 1961. The fresh-water fishes – their origins and affinities. *In:* Symposium on the biogeography of Baja California and adjacent seas. Systematic Zoology 9(3–4):212–232.

Gibbs, R. H. 1957. Cyprinid fishes of the subgenus *Cyprinella* of *Notropis*. II. Variation and subspecies of *Notropis venustus* (Girard). Tulane Studies in Zoology 5(8):175–203.

Gilbert, C. R., and D. S. Lee. *Menidia beryllina* (Cope), 558. *In:* Atlas of North American freshwater fishes.

Glodek, G. S. 1980. *Ictalurus furcatus* (Lesueur), 439; *Ictalurus melas* (Rafinesque), 441; *Ictalurus punctatus* (Rafinesque), 446. *In:* Atlas of North American freshwater fishes.

Gonzáles de la Rosa, C. 1977. Biogeografía del complejo de *Poecilia:* Grupo *Latipinna – velifera petenensis* en México (Poeciliidae – Atheriniformes). Tésis Inédita, Facultad de Ciencias Biológicas, Universidad Autonoma de Nuevo León, México.

Guillory, V. 1980. *Ctenopharyngodon idella* Valenciennes, 151. *In:* Atlas of North American freshwater fishes.

Hendricks, M. L. 1980. *Salvelinus fontinalis* (Mitchill), 114. *In:* Atlas of North American freshwater fishes.

Hendrickson, D. A., W. L. Minckley, R. R. Miller, D. J. Siebert, and P. H. Minckley. 1980. Fishes of the Río Yaqui basin, Mexico and United States. Journal of the Arizona Nevada Academy of Sciences 15(3):65–106.

Hensley, D. A., and W. R. Courtenay, Jr. 1980. *Carassius auratus* (Linnaeus), 147; *Misgurnus anguillicaudatus* (Cantor), 436; *Poecilia reticulata* Peters, 550; *Xiphophorus helleri* Heckel, 554; *Xiphophorus maculatus* (Günther), 555; *Xiphophorus variatus* (Meek), 556; *Tilapia aurea* (Steindachner), 771; *Tilapia mossambica* (Peters), 774; *Tilapia zilli* (Gervais), 775. *In:* Atlas of North American freshwater fishes.

Hubbs, C. L. 1954. Establishment of a forage fish, the red shiner (*Notropis lutrensis*), in the lower Colorado River system. California Fish and Game 40(3):287–294.

Hubbs, C. L., and R. R. Miller. 1943. Mass hybridization between two genera of cyprinid fishes in the Mohave Desert, California. Papers of the Michigan Academy of Science, Arts and Letters 28:343–378.

Hubbs, C. L., and R. R. Miller. 1977. Six distinctive cyprinid fish species referred to *Dionda* inhabiting segments of the Tampico embayment drainage of Mexico. Transactions of the San Diego Society of Natural History 18(17):267–336.

Hubbs, C. 1982. Occurrence of exotic fishes in Texas waters. Pearce-Sellards Series, Texas Memorial Museum 36:1–19.

Hubbs, C., R. R. Miller, R. J. Edwards, K. W. Thompson, E. Marsh, G. P. Garrett, G. L. Powell, D. J. Morris, and R. W. Zerr. 1977. Fishes inhabiting the Río Grande, Texas and Mexico, between El Paso and the Pecos confluence. *In:* Importance, preservation and management of riparian habitat. A symposium. U.S.D.A. Forest Service General Technical Report RM-43:91–97.

Hubbs, C., and R. H. Wauer. 1973. Seasonal changes in the fish fauna of Tornillo Creek, Brewster County, Texas. Southwestern Naturalist 17:375–379.

LaBounty, J. 1974. Materials for the revision of cichlids from northern Mexico and southern Texas, U.S.A. (Perciformes, Cichlidae). Ph.D. dissertation, Arizona State University. 120 pp.

Lee, D. S. 1980. *Notemigonus crysoleucas* (Mitchill), 219; *Lepomis auritus* (Linnaeus), 500; *Lepomis cyanellus* Rafinesque, 591–592; *Lepomis gulosus* (Cuvier), 595; *Lepomis macrochirus* Rafinesque, 597–598; *Lepomis microlophus* (Günther), 601; *Lepomis punctatus* (Valenciennes), 602; *Micropterus dolomieui* Lacepède, 605; *Micropterus salmoides* (Lacepède), 608; *Pomoxis annularis* Rafinesque, 611-614. *In:* Atlas of North American freshwater fishes.

Lee, D. S., and G. H. Burgess. 1980. *Gambusia affinis* (Baird and Girard), 538. *In:* Atlas of North American freshwater fishes.

Lee, D. S., and S. T. Kucas. 1980. *Pimephales vigilax (Baird and Girard), 343. In:* Atlas of North American freshwater fishes.

Lee, D. S., and S. P. Platania. 1980. *Carpiodes carpio* (Rafinesque), 367. *In:* Atlas of North American freshwater fishes.

Lee, D. S., and J. R. Shute. 1980. *Pimephales promelas* Rafinesque, 341. *In:* Atlas of North American freshwater fishes.

Matthews, W. J. 1980. *Notropis lutrensis* (Baird and Girard), 285. *In:* Atlas of North American freshwater fishes.

Meek, S. E. 1904. The freshwater fishes of Mexico north of the Isthmus of Tehuantepec. Field Columbian Museum of Zoology. Ser. 5:1–252.

Miller, R. R. 1961. Man and the changing fish fauna of the American southwest. Papers of the Michigan Academy of Science, Arts and Letters 46(1960):365–404.

Miller, R. R. 1963. Extinct, rare and endangered American freshwater fishes. XVI International Congress of Zoology, 8:4–11.

Miller, R. R. 1968. Records of some native freshwater fishes transplanted into various waters of California, Baja California, and Nevada. California Fish and Game 54(3):170–179.

Miller, R. R. 1978. "Native" trout of the Río Yaqui. Desert Fishes Council Annual Meeting 1978:39.

Miller, R. R., and B. Chernoff. 1980. Status of the endangered Chihuahua chub, *Gila nigrescens,* in New Mexico and Mexico. Desert Fishes Council Proceedings 1979:74–84.

Miller, R. R., and C. L. Hubbs. 1954. An erroneous record of the California killifish, *Fundulus parvipinnis,* from Cabo San Lucas, Baja California. Copeia 1954(3):234–235.

Minckley, W. L. 1965. Records of atherinid fishes at inland localities in Texas and northern Mexico. Great Basin Naturalist 25(3–4):73–76.

Minckley, W. L. 1973. Fishes of Arizona. Arizona Game and Fish Department. Phoenix. 293 pp.

Moyle, P. B. 1976. Inland fishes of California. University of California Press, Berkeley. 405 pp.

Obregon, F. 1960. Cultivo de la carpa seleccionada en México. Secretaría de Agricultura y Ganadería, México. 63 pp.

Poss, S., and R. R. Miller. 1983. The taxonomic status of the plains killifish, *Fundulus zebrinus.* Copeia 1983(1):55–67.

Robins, C. R., R. M. Bailey, C. E. Bond, J. R. Brooker, E. A. Lachner, R. N. Lea, and W. B. Scott. 1980. A list of common and scientific names of fishes from the United States and Canada. Special Publication of the American Fisheries Society 12:1–174.

Romero, H. 1967. Catálogo sistemático de los peces del Alta Lerma con descripción de una nueva especie. Anales de la Escuela Nacional de Ciencies Biológicas, México 14:47–77.

Rosas-M., M. 1976a. Reproducción natural de la carpa herbívora en México, *Ctenopharyngodon idellus* Cyprinidae. Memorias del Simposio sobre Pesquerías en Aguas Continentales 1:1–28.

Rosas-M., M. 1976b. Sobre la existencía de un nematodo parasito de *Tilapia nilotica* (*Goezia* sp. Zeder 188 Goezidae) de la Presa Adolfo Lopez Mateos (Infiernillo, Mich.). Memorias del Simposio sobre Pesquerías en Aguas Continentales 1:1–28.

Rosas-M., M. 1976c. Peces dulceacuícolas que se explotan en México y datos sobre su cultivo. Centro de Estudios Económicos y Sociales del Tercer Mundo 2:1–135.

Smith, M. L., and R. R. Miller. In press. Conservation of desert spring habitats and their endemic fauna in northern Chihuahua. Proceedings of the Desert Fishes Council, 1981.

Swift, C. C. 1980. *Gila orcutti* (Eigenmann and Eigenmann), 169. *In:* Atlas of North American freshwater fishes.

Tellez-R., C. 1976. Alimentación, hábitos alimenticio y su relación entre *Cyprinus carpio* y *Carassius auratus* (L.) en 13 cuerpos de agua de la parte central de México. Fideicomiso para el Desarrollo de la Fauna Acuática. Serie Técnica. 4:1–20.

Thys van den Audernaede, D.F.E. 1968. An annotated bibliography of tilapia. Musée royale de l'Afrique centrale, Documentation zoologique, 14:1–406.

Treviño-Robinson, D. 1959. The ichthyofauna of the lower Rio Grande, Texas and Mexico. Copeia 1959(3):253–256.

Uyeno, T., and R. R. Miller. 1979. Chromosomes of American trouts, subgenus *Parasalmo,* including Mexican species. Abstracts of the 59th annual meeting, American Society of Ichthyologists and Herpetologists, Orono, Maine. 1 p.

Vergara de los Ríos, M. 1976. Engorda de bagre (*Ictalurus punctatus*) en Jaulas. Memorias del Simposio de Pesquerías en Aguas Continentales, México. 1:89–97.

Welcomme, R. L. 1981. Register of international transfers of inland fish species. Food and Agriculture Organization Fisheries Technical Paper 213. 120 pp.

CHAPTER 7

Exotic Fishes in Hawaii and Other Islands of Oceania

J. A. Maciolek

The most recent comprehensive review of the status and consequences of fish introductions to the Hawaiian islands was that by Brock (1960), which listed 48 species. The only broad coverage of exotic fishes elsewhere in Oceania was a short article by Devambez (1964), who accounted for 9 established fishes in seventeen island areas but did not note the many other introductions including unsuccessful ones.

This report is intended primarily as an update of the status of fish introductions in the tropical insular Pacific with a consideration of the effects of successful introductions. Area coverage (figure 7-1) includes archipelagoes and isolated islands of Oceania for which information was found, but excludes New Guinea (and associated islands) and the region south of the Tropic of Capricorn. Reference to obscure literature, agency reports, and personal communications was necessary to ensure comprehensive coverage.

Environmental Basis for Exotic Fish Establishment

Among habitats available for colonization by introduced fishes, marine waters in Oceania predominate and generally overshadow the existence of diverse inland waters, particularly on high islands. Hawaii provides an exam-

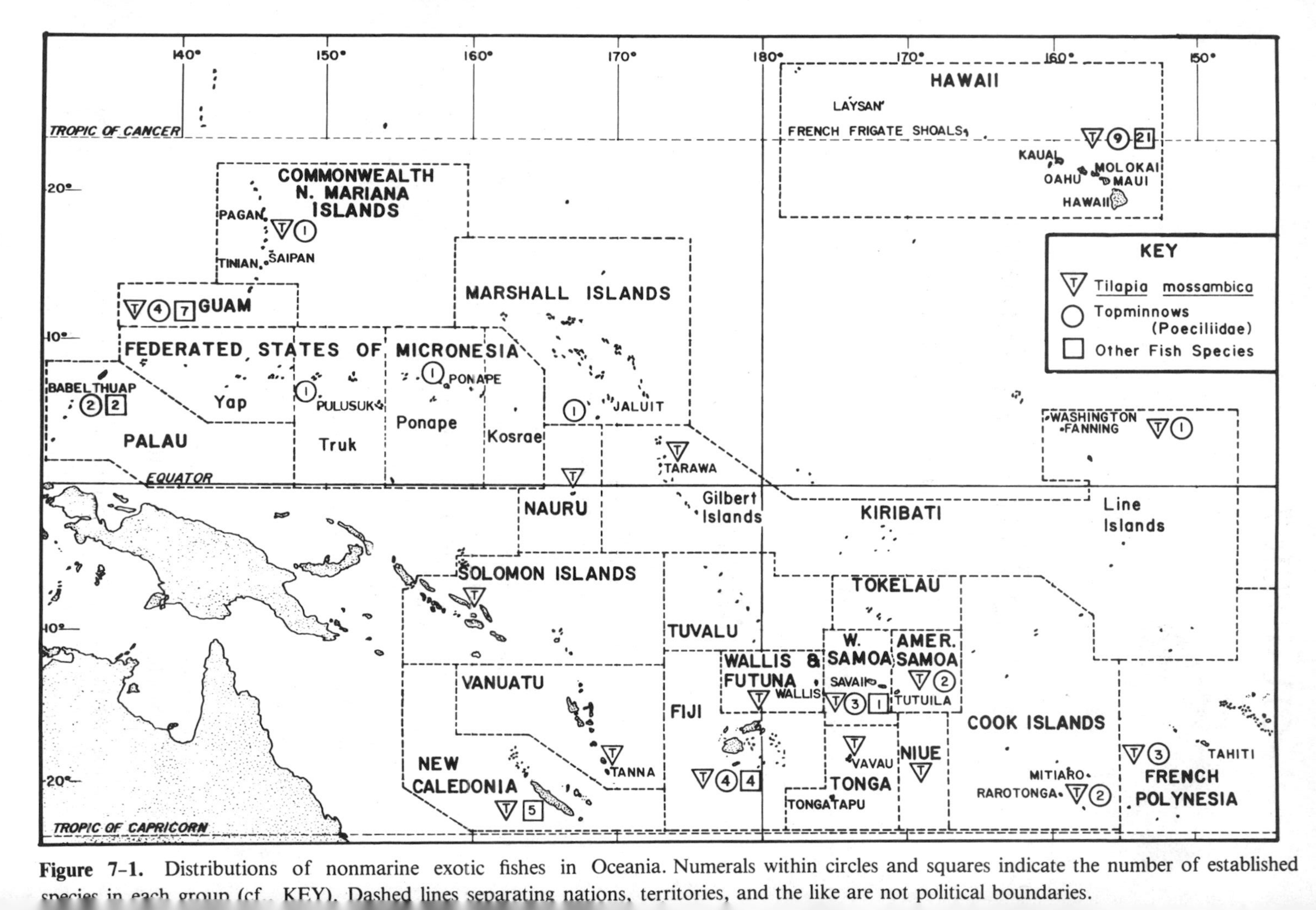

Figure 7–1. Distributions of nonmarine exotic fishes in Oceania. Numerals within circles and squares indicate the number of established species in each group (cf. KEY). Dashed lines separating nations, territories, and the like are not political boundaries.

ple of aquatic habitat diversity. Among the five major islands in the southeastern archipelago there are about 400 reservoirs, 360 perennial streams, and 30 stream-mouth estuaries, as well as numerous irrigation ditches, a few lakes, and many fresh and saline wetlands (Maciolek 1978). Water temperatures favor warmwater species but coldwater fishes are marginally accommodated at higher altitudes. Inland waters on flat islands, however, are less abundant and limited to closed lagoons and small wetlands. Dahl (1980) included the distribution of inland waters in a survey report covering twenty biogeographic provinces in Oceania.

Native fauna also is a factor in the potential success of introductions, especially through trophic interactions. Faunal biomasses of insular inland waters may not differ much from those of comparable continental waters from whence most exotics are derived, but insular waters usually have much lower species diversities. For example, primary and secondary freshwater fishes are absent as natives in Oceania where streams are inhabited by diadromous and itinerant marine fish taxa (Herre 1940, Myers 1953). Passive, often herbivorous gobies are the more common fishes in such streams. Crustaceans, such as atyid shrimps, predominate among invertebrate fauna that can provide excellent forage for aggressive foreign fishes. If we assume that other environmental requirements are met, insular inland waters appear to be satisfactory habitats for many continental species.

Hawaiian Fish Introductions

The story of exotic fishes in Oceania centers in Hawaii where the first and by far the greatest number of foreign species have been introduced, and where the best continuing records of such fishes exist. Also, Hawaii apparently is the only location in Oceania where marine exotics have been established. Table 7-1 contains a listing of all exotic fishes for which information was found.* Included are 12 species not discussed by Brock (1960), 8 of which were released after his paper was published. Locations (islands) of significance are shown in figure 7-1.

In the following text fishes are referred to by common name as given by Robins et al.(1980) where applicable. Reference to scientific name is given by a number in parentheses indicating the listing in column 1 of table 7-1. The exact identity of some species of Poeciliidae, especially members of the *Poecilia sphenops* complex (Schultz and Miller 1971), is questionable, and the occurrence of *P. butleri* in Oceania is possible. In the absence of recent authoritative identifications, poeciliid species are reported here as given in reference sources. Also in table 7-1 are numbers of specimens imported,

* There have been unconfirmed reports of other established fishes such as the hybrid red tilapia and *Tilapia rendalli* (R. Yamada, personal communication).

Table 7–1. HAWAIIAN FISH INTRODUCTIONS WITH INFORMATION ON IMPORTATION AND ESTABLISHMENT SUCCESS. QUESTION MARKS INDICATE THAT INFORMATION IS INDEFINITE OR CONFLICTING IN REFERENCE SOURCES

	Importation and release				Establishment			
			Releases					
Scientific name[a]	Stock source[b]	Number imported	Year[c]	Island[d]	Success[e]	Island[d]	Major habitat	Information[f] source
Clupeidae								
1 *Dorosoma petenense*	CA	› 3,200	1958 +	H,K,Ma,O	+ +	K,O	reservoir	6, 15, 28, 48, 54, 55, 59, 63, 94, 105
2 *Sardinella marquesensis*-m	Marquesas	143,800	1955 +	O	+ + +	All	marine	15, 45, 53, 54, 63, 81, 84, 92, 94, 105, 108, 123
Engraulidae								
3 *Anchoa compressa*-m	CA	6,000	1932	O	–			13, 15, 63, 94, 105
Salmonidae								
4 *Oncorhynchus tshawytscha*-d	CA	› 66,000	1876 +	K	–			13, 15, 60, 63, 85, 86
5 *Salmo gairdneri*	CA,OR,WA	› 10^6	1920 +	All	+	K	stream	6, 13, 15, 45, 46, 50, 56, 63, 67, 69, 85, 92
6 *Salmo trutta*	United States	23,590	1935 +	K	–			13, 15, 63, 67, 71, 85, 86, 92
7 *Salvelinus fontinalis*	CA	› 1,012	1876 +	H,K	–			13, 15, 60, 63, 67, 70, 85, 86
Plecoglossidae								
8 *Plecoglossus altivelis*	Japan	250,000	1925 +	H,K,O	–			13, 15, 63
Cyprinidae								
9 *Barbus semifasciolatus*	?	?	1940	O	+	O	reservoir	13, 15, 63
10 *Carassius auratus*	China	?	‹ 1900	All?	+ + +	All?	reservoir	6, 13, 15, 26, 46, 55, 60, 63, 67, 79
11 *Cyprinus carpio*	Asia?	?	‹ 1900	All?	+ + +	All?	reservoir	13, 15, 26, 46, 55, 60, 63, 67, 75, 76, 111
12 *Ctenopharyngodon idella*	Taiwan	2,000	1968?	H	–			67
13 *Misgurnus anguillicaudatus*	Asia?	?	‹ 1900	K,Ma,O	+ + +	K,Ma,O	stream	13, 15, 18, 55, 63, 67, 73, 76, 79, 87, 109, 111, 112

Ictaluridae								
14 *Ictalurus nebulosus*	CA	?	1893+	H,O	–			13, 15, 60, 63, 79
15 *Ictalurus punctatus*	CA,KS,OK,TX	> 200,000	1953+	H,K,Ma,O	+	O	reservoir	15, 28, 45, 46, 50, 55, 63, 67, 73
Clariidae								
16 *Clarias fuscus*	China	?	< 1900	All?	+ + + +	H,K,Ma, O	stream	13, 15, 26, 46, 51, 53, 60, 63, 66, 73, 74, 76, 79, 87, 109, 111, 112
Oryziatidae								
17 *Oryzias latipes*	Asia?	?	1922	O	–			13, 15, 63
Cyprinodontidae								
18 *Aplocheilus lineatus*	?	?	> 1960	O?	–			131
19 *Fundulus grandis*	TX	< 420	1905	All	–			13, 15, 63, 115
20 *Nothobranchius guentheri*	?	?	1967?	O?	– ?			63, 131
Poeciliidae								
21 *Gambusia affinis*	TX	< 420	1905+	All	+ + +	All?	various	2, 9, 13, 15, 18, 28, 48, 55, 63, 64, 65, 72, 79, 105, 109, 111, 112, 115, 131
22 *Poecilia latipinna*	TX	< 420	1905+	All	+ + +	All	estuary	13, 15, 63, 67, 72, 76, 105, 111, 115, 121, 131
23 *Poecilia mexicana*	?	?	> 1960?	O	+ + +	O	stream	84, 105, 110, 111, 112
24 *Poecilia reticulata*	?	?	1922+	All?	+ + + +	All	stream	6, 13, 15, 55, 63, 66, 77, 79, 87, 109, 111, 112, 131
25 *Poecilia sphenops*	?	?	< 1950?	O	+	O	estuary	7, 13, 105
26 *Poecilia vittata*	?	?	< 1950?	O	+	O	stream?	13, 15, 63, 64, 105, 111, 131
27 *Xiphophorus helleri*	?	?	1922+	All?	+ + + +	All?	stream	13, 15, 18, 55, 63, 66, 76, 77, 87, 109, 111, 112, 131
28 *Xiphophorus maculatus*	?	?	1922+	H,Ma,O	+ +	H,Ma,O	reservoir	13, 15, 55, 63, 79, 87, 109, 111, 112, 131
29 *Xiphophorus variatus*	?	?	> 1960	O?	+ ?	O?	pond	131
Channidae								
30 *Ophicephalus striatus*	China	?	< 1900	O	+ +	O	reservoir	13, 15, 25, 26, 46, 55, 60, 63, 67, 79, 111, 121

Table 7–1. HAWAIIAN FISH INTRODUCTIONS WITH INFORMATION ON IMPORTATION AND ESTABLISHMENT SUCCESS. QUESTION MARKS INDICATE THAT INFORMATION IS INDEFINITE OR CONFLICTING IN REFERENCE SOURCES *(continued)*

	Importation and release				Establishment			
			Releases					
Scientific name[a]	Stock source[b]	Number imported	Year[c]	Island[d]	Success[e]	Island[d]	Major habitat	Information[f] source
Synbranchidae								
31 *Monopterus albus*	Asia?	?	‹ 1900	O	+ +	O	pond	13, 15, 51, 63, 76
Percichthyidae								
32 *Morone saxatilis*-d	CA	?	1920? +	K	–			13, 15, 63, 94
Serranidae								
33 *Cephalopholis guttatus*-m	Moorea	2,285	1956 +	H,O	+ + +	All	marine	15, 45, 63, 67, 89, 90, 92, 94, 106
34 *Cephalopholis urodelus*-m	South Pacific	1,808	1958	O	–			15, 45, 63, 89, 90, 92, 94
35 *Epinephelus fasciatus*-m	Marquesas	51	1958	O	–			15, 63, 89, 90, 92, 94
36 *Epinephelus hexagonatus*-m	Moorea	› 978	1956	Ma,O	–			15, 63, 89, 90, 92, 94
37 *Epinephelus merra*-m	Moorea	1,631	1956 +	K,O	–			15, 63, 89, 90, 92, 94
38 *Epinephelus spiniger*-m	Marquesas	22	1958	O	–			15, 63, 89, 90, 92, 94
Kuhliidae								
39 *Kuhlia rupestris*-d	Guam	213	1957 +	O	–			6, 13, 15, 63, 94
Centrarchidae								
40 *Lepomis macrochirus*	CA	14	1946 +	All	+ + +	All	reservoir	6, 13, 15, 27, 28, 46, 55, 63, 67, 73, 74, 76, 85, 92, 109, 111
41 *Micropterus dolomieui*	CA	1,274	1953 +	H,K,O	+ + +	H,K,O	stream	6, 15, 27, 45, 55, 63, 76, 92, 97, 111
42 *Micropterus salmoides*	CA	› 181	1897 +	All	+ + +	H,K,O	reservoir	6, 13, 15, 26, 27, 28, 46, 49, 55, 56, 60, 63, 67, 73, 85, 97
Lutjanidae								
43 *Lutjanus fulvus*-m	South Pacific	› 262	1955? +	O	+ +	All	marine	15, 45, 63, 67, 89, 90, 91, 92, 94, 106, 107, 108
44 *Lutjanus gibbus*-m	South Pacific	177	1958 +	O	–			15, 63, 89, 90, 92, 94
45 *Lutjanus guttatus*-m	Mexico	3,400	1960 +	Mo,O	–			15, 45, 63, 89, 94
46 *Lutjanus kasmira*-m	South Pacific	› 3,163	1955 +	O	+ + + +	All	marine	15, 45, 63, 67, 89, 90, 91, 92, 94, 106, 107, 108

Lethrinidae								
47 *Lethrinus* sp.-m	Marquesas	32	1956 +	O	–			63, 89, 90, 92, 94
Mullidae								
48 *Upeneus vittatus*-m	Marquesas	?	1955	O	– ?			93, 94
Cichlidae								
49 *Astronotus ocellatus*	CA	64	1952 +	K,O	+	K,O	reservoir	6, 13, 15, 28, 46, 53, 63, 67, 89
50 *Cichla ocellaris*	NY	2[g]	1961 +	K,O	+ +	K,O	reservoir	6, 28, 45, 49, 63, 67, 93, 121
51 *Cichlasoma meeki*	?	?	1940	O	+	O	reservoir	13, 15, 63
52 *Cichlasoma* sp.	?	?	> 1960	O	+ ?	O	stream	111
53 *Tilapia macrochir*	Congo	52	1958 +	Ma,O	+	Ma,O	reservoir	6, 15, 46, 47, 55, 63, 80, 121
54 *Tilapia melanopleura*	Congo	50	1957	K,Ma,O	+ +	K,Ma,O	reservoir	6, 15, 45, 55, 63, 80, 92
55 *Tilapia melanotheron*	?	?	> 1970	O	–	O	reservoir	47, 105, 121
56 *Tilapia mossambica*	Singapore	14[g]	1951 +	All	+ + + +	All	various	6, 13, 14, 15, 18, 29, 47, 53, 55, 63, 64, 72, 75, 76, 77, 87, 92, 105, 111, 112, 121, 131
57 *Tilapia zilli*	West Indies	19	1957 +	H,Ma,O	+ +	H,Ma,O	reservoir	6, 15, 45, 46, 55, 63, 92
Mugilidae								
58 *Chelon engeli*-m	Marquesas	?	1955	O	+ + +	All?	estuary	93, 94, 108
Osphronemidae								
59 *Osphronemus goramy*	Philippines	> 9	1950?	O	–			13, 15, 63
60 *Trichogaster leeri*	Asia?	?	1940	O	–			13, 15, 63

[a] Phyletic family arrangement follows Greenwood et al. (1966) except for separate listing of Percichthyidae (Robins et al. 1980). Lower case letter following species name refers to marine (m) or diadromous (d) fishes; all other species are inland-water fishes. Other nomenclature used in literature sources: 2, *Harengula vittata*; 5, *Salmo irideus*; 14, *Ameiurus nebulosus*; 15, *Ictalurus lacustris*; 16, *Clarias batrachus* and *C. magur*; 17, *Aplocheilus latipes*; 22, *Mollienesia latipinna*; 24, *Lebistes reticulatus*; 25, *Mollienesia sphenops*; 26, *Limia vittata*; 28, *Platypoecilus maculatus*; 31, *Fluta alba*; 32, *Roccus saxatilis*; 33, *Cephalopholis argus*; 43, *Lutjanus vaigensis*; 55, *Tilapia macrocephala* and 56, *Sarotherodon mossambicus*.

[b] Stock source is located where fish was obtained; U.S. state sources are abbreviated by two-letter postal codes.

[c] Year given is that of first (or only) release; plus sign after year indicates more than one release.

[d] Islands include the five largest of the Hawaiian Archipelago: H = Hawaii, K = Kauai, Ma = Maui, Mo = Molokai, and O = Oahu.

[e] Minus signs (–) indicate species not established; plus signs (+) increase in number with increasing establishment success (based on numbers of populations and abundance within populations).

[f] Numbers refer to entries in Literature Cited section.

[g] Number refers to reproducing survivors from lot of fish imported.

which is the figure most frequently found in the records and usually represents the number released. These fishes fall into three general ecological groups: marine, diadromous (broadly defined as migrating between marine and stream environments), and nonmarine (including both freshwater and euryhaline inland-water species).

DIADROMOUS AND MARINE FISHES (19 SPECIES)

None of the four introduced diadromous fishes (table 7–2) was successful. Chinook salmon (4) were imported first in 1876. Subsequently, 66,000 fry were reported (Brock 1960) to have been released on Kauai in 1925 and 1927. However, Brock (1952) listed annual stockings of this salmon in two of the larger Kauai rivers during the years 1925 through 1929. Ayu (8), a Japanese fish related to salmonids, was imported in 1925. Six lots totaling 250,000 fish were planted on three islands but the massive stocking failed. In either 1920 or 1922, an unspecified number of striped bass (32) were stocked in seven large Kauai streams. Nato (39), the fourth diadromous species, occurs as adults and juveniles in the source Guamanian streams. The 170 nato released into a small Oahu stream in 1958 did not survive.

Randall (1960) and Randall and Kanayama (1973) discussed the background for and history of all marine fish introductions. Hawaii has a distinctive inshore fish community that lacks many fishes (especially among lutjanids and serranids) that are common elsewhere in Oceania. Nineteen species representing 7 families (table 7–2) have been introduced, 5 of which became established. The only introduction attempted before 1955, the deep body anchovy (3), was released in 1932 and failed to establish. Along with the introduction of the "Marquesan" sardine (2) in 1955, two fishes were inadvertently released: a goatfish (48), that was not observed in the following two decades; and a mullet (58) that is well established. Seven other releases of the sardine were made, and in 1958, establishment was evident. Snappers and sea basses comprised the 11 other marine fishes released between 1955 and 1962. Only 3 of these introductions were successful: roi (33), toau (43), and taape (46). All of the 5 established marine species are distributed throughout the southeastern part of the archipelago, and 2 of them recently were reported (Parrish et al. 1980) from the leeward islands: taape abundant to Laysan and toau at French Frigate Shoals (cf. figure 7–1).

Several new and uncommon inshore marine fishes were recently reported from Hawaii by Randall (1980). Such new finds raise the question of whether they might be introduced species, natural nonreproducing strays from elsewhere, or reproducing indigenous species that have gone unnoticed because of rarity and other factors that tend to obscure them. One such fish is a grouper (*Chromileptes altivelis*) known from one specimen and one observation. Randall speculated that individuals may have been released aquarium stock. Another is the goatfish (48) noted above, now known from two speci-

Table 7–2. HAWAIIAN MARINE AND DIADROMOUS (d) FISH INTRODUCTIONS AND ESTABLISHMENTS ARRANGED BY FAMILY

Family	Number of species		Relative abundance	Principal habitat
	Introduced	Established		
Clupeidae	1	1	+ + +	marine
Engraulidae	1	0		
Kuhliidae (d)	1	0		
Lethrinidae	1	0		
Lutjanidae	4	2	+ + +	marine
Mugilidae	1	1	+ + +	estuary
Mullidae	1	0?		
Percichthyidae (d)	1	0		
Plecoglossidae (d)	1	0		
Salmonidae (d)	1	0		
Serranidae	6	1	+ +	marine

mens collected in Kaneohe Bay, Oahu, one in 1977 and the other in 1978. Randall proposed that they were progeny of the 1955 release. However, the appearance of the species after a lapse of twenty-two years suggests that other explanations of its presence are equally plausible. In the absence of stronger evidence to the contrary, the goatfish is not treated here as an established exotic.

NONMARINE FISHES (41 SPECIES)

Information regarding the introduction and establishment of 41 nonmarine species, representing 14 families, is summarized in table 7–3. Because of their diversity and long history, these fishes are discussed here by time period of initial introduction. Three such periods are identified: (1) early introductions (mostly undocumented); (2) species introduced between 1920 and 1950 (mostly casual introductions); and (3) fishes introduced from 1950 to the present (mostly with planned effort).

Introductions before 1906 (12 species).

No record exists of the importation of 6 fishes that were established in Hawaii by 1901: goldfish (10), common carp (11), oriental weatherfish (13), Chinese catfish (16), snakehead (30), and ricefield eel (31). Presumably they were brought to Hawaii by Asian immigrants. All species are found in standing waters and 2 of them (catfish, weatherfish) are abundant in many streams.

Among 6 intentional introductions made during this period, 3 succeeded and 3 failed. Brook trout (7), planted at least three times (1876, 1894, 1896), did not survive. Brock (1952, 1960) noted only one unsuccessful release of the brown bullhead (14) in 1893 on Hawaii Island. However, Mainland (1939) collected it from several Oahu habitats including four streams. It has not been

Table 7-3. NONMARINE FISHES INTRODUCED INTO HAWAII ARRANGED BY FAMILY

Family	Number of species		Relative abundance	Principal habitat
	Introduced	Established		
Centrarchidae	3	3	+ + +	stream
Channidae	1	1	+	stream
Cichlidae	9	9	+ +	all
Clariidae	1	1	+ + +	stream
Clupeidae	1	1	+ +	reservoir
Cobitidae	1	1	+ + +	stream
Cyprinidae	4	3	+ +	reservoir
Cyprinodontidae	3	0		
Ictaluridae	2	1	+	reservoir
Oryziatidae	1	0		
Osphronemidae	2	0		
Poeciliidae	9	9	+ + +	all
Salmonidae	3	1	+	stream
Synbranchidae	1	1	+	pond

reported since, and Mainland's record suggests that the bullhead had been planted variously on Oahu in the 1930s. Largemouth bass (42) were imported from California in 1897, 1908, and 1911, and progeny of successful reproduction have been widely distributed into reservoirs. In 1905, gulf killifish (19), mosquitofish (21), and sailfin molly (22) were imported from Texas as a mixed lot of about 450 individuals. They were reared in special ponds on Oahu and thence planted on all major islands. The killifish did not survive but the other 2 species are among the more abundant and widely distributed exotics in Hawaii.

Introductions from 1920 to 1950 (12 species).

Rainbow trout (5) were introduced first in 1920 by the plantings of eggs and fry in five Kauai streams. During the next two decades, well over 200,000 fry and eggs were stocked on the major islands (Brock 1952):

Island	Minimum number habitats stocked	Total number releases
Hawaii	10	11
Kauai	13	35
Maui	11	13
Molokai	2	4
Oahu	8	17

Despite this repetitive and widespread stocking, reproducing populations developed only in a few streams at Kokee, an elevated plateau on Kauai where releases are continued annually to supplement irregular reproduction

Figure 7–2. Male rainbow trout caught in Haipuaena Stream (elevation 1,340 m), E. Maui Island, in May 1967, a few years after fry were stocked. This trout, test-planted on the five largest islands, grows well in Hawaiian streams but is known to reproduce only in those at Kokee, Kauai. Pocket knife indicates relative size.

in naturalized stocks. After 1950, at least 14 additional releases of rainbow trout were made on islands other than Kauai; some of these survived for several years but did not reproduce (figure 7–2). Needham and Welch (1953) attributed such lack of reproduction to water temperatures not being sufficiently low (i.e., below 15 C) for a long enough period to permit gonad maturation. Brown trout (6) reacted similarly to the Hawaiian environs; nine plantings were made in Kauai streams in 1935 and the species flourished for a few years but was last found in 1949.

Nine fish species representing aquarium stock (source undetermined) were first introduced on Oahu, most of them either in 1922 or in 1940. The 1922 releases included medaka (17) which failed, and 3 successful introductions: guppy (24), green swordtail (27) and southern platyfish (28). These topminnows have been transplanted to other islands and are among the more abundant Hawaiian exotics. Of the 3 species introduced in 1940, the pearl gourami (60) failed but the half-banded barb (9) and the firemouth (51) succeeded marginally (tenuous populations are reported only from Oahu). Two other "aquarium" topminnows, liberty molly (25) and Cuban limia (26), were established by 1950 but circumstances of their introductions are unknown.

The bluegill (40) was imported from California in 1946. Progeny of the initial lot have been distributed into reservoirs on all major islands.

Introductions from 1950 – present (17 species).

Four introductions that occurred during the past twenty-two years were unsuccessful: a gourami (59) released on Oahu in 1950; 2 cyprinodonts (18, 20) imported by the state Department of Health;* and the grass carp (12), released on the island of Hawaii in about 1968. The grass carp was cultured in Hawaii even before this introduction and may have been released elsewhere.

Eight cichlid fishes were among 13 successful introductions during this period. Six of the cichlids, intentional introductions imported in small numbers, were bred in captivity before release to the wild: oscar (49), tucunare (50), and 4 tilapias (53, 54, 56, 57). These fishes occur as reservoir populations on several islands. One of them, the Mozambique tilapia (56), also has large populations in many streams, estuaries, and low wetlands. The 2 other cichlids (52, 55) were cursory introductions now restricted to single habitats on Oahu.

Two fishes first introduced from California in 1953, channel catfish (15) and smallmouth bass (41), are now widely distributed. Both species inhabit reservoirs and the bass also lives in several streams. Threadfin shad (1) was also imported from California (in 1958) and now inhabits reservoirs on two islands. Introductions of the shortfin molly (23) and the variable platyfish (29) were not noted by Brock (1960) and apparently occurred more recently. The former species is abundant in at least nineteen Oahu streams.

SUMMARY OF HAWAIIAN INTRODUCTIONS

If all of the 60 fishes introduced into Hawaii are examined at different taxonomic levels for establishment success, little difference is apparent:

	Number of species		
Taxon	Introduced	Established	Percent success
Family	23	14	61
Genus	39	23	59
Species	60	36	60

However, differences in success by primary habitat are striking:

	Number		
Species by habitat	Introduced	Established	Percent success
Diadromous	4	0	0
Marine	15	5	33
Nonmarine	41	31	76

* Another cyprinodont, *Cynolebias bellottii,* was reported as an introduction by Kanayama (1968) but it apparently was never stocked in the wild (P. Nakagawa, personal communication).

Hawaiian inland waters have been highly vulnerable to invasion by exotics, three-fourths of the known introductions having become established. It is noteworthy that so many nonmarine exotics have succeeded in an area as small as Hawaii; it is even more remarkable that Oahu alone has 30 of the 31 species (35 of 36, if marine species are included). Only the rainbow trout does not occur there.

Some distributions and abundances of established species given in table 7-1 are from older reports (e.g., Broch 1960) that cannot reflect recent changes, especially those resulting from later introductions. Unfortunately, exotic fishes have not been monitored directly, and information on their status has resulted largely from surveys done for other purposes. Some extensive assessments of stream fish communities were conducted by the Hawaii Cooperative Fishery Research Unit during 1967–79. Those data, published or not, were used in the evaluations of distributions and abundances of many species listed in table 7-1. A segment of such data concerning fifty-eight streams in which one or more populations of topminnows were found during surveys from 1975 to 1978 provides an extended appraisal of five stream-dwelling members of this successful exotic family:

	Number of populations where species was			
Topminnow species	Numerous	Common	Occasional	Total populations
Mosquitofish (21)	1	8	4	13
Shortfin molly (23)	19	1	1	21
Guppy (24)	27	15	3	45
Cuban limia (26)	2	6	1	9
Green swordtail (27)	20	9	6	35

As a group, topminnows were abundant in 69 of the populations, common in 39, and scarce in 15. Two of the species, mosquitofish and Cuban limia, were common but rarely found in large numbers and the other 3 were abundant in most of the streams where they were found. The data suggest that the recently introduced shortfin molly is becoming a dominant fish in the streams it has colonized (thus far only on Oahu).

Fishes Introduced Elsewhere in Oceania

Twenty Pacific locations beyond Hawaii, identified as having exotic fishes, are shown in figure 7-1. Among these locations, at least 24 introduced species (table 7-4) have established populations and several unsuccessful introductions have been noted. All of the established fishes are nonmarine. Prominent among them are the Mozambique tilapia and a few species of topminnows

Table 7-4. FISHES INTRODUCED ON ISLANDS OF OCEANIA OTHER THAN HAWAII (SEE FIGURE 7-1)

Politico-geographic location (Island of special concern)	Scientific name[a]	Information source[b]
A. North Pacific		
Commonwealth of the Northern Mariana Islands (Pagan, Saipan, Tinian)	*Gambusia affinis*	10, 65
	Tilapia mossambica	2, 4, 10, 19, 24, 58
Federated States of Micronesia (Ponape, Pulusuk)	*Gambusia affinis*	65, 127, 132
Guam	[*Arius* sp.]	1
	Astronotus ocellatus	10, 104
	Betta brederi	10, 43, 104
	Cichla ocellaris	10, 21, 41
	Clarias batrachus	1, 8, 10, 31, 43, 51, 62, 95, 104
	Cyprinus carpio	10, 95
	Gambusia affinis	8, 10, 17, 43, 95, 104, 114
	[*Ictalurus punctatus*]	40
	[*Micropterus dolomieui*]	24, 37, 38, 97
	Micropterus salmoides	24, 38, 39, 40, 97
	[*Ophicephalus striatus*]	1
	Poecilia reticulata	10, 16, 43, 95, 104, 114
	Poecilia latipinna	10, 43, 104
	Tilapia mossambica	8, 10, 16, 19, 24, 42, 43, 61, 95, 96, 104, 114, 116
	Tilapia zilli	10, 24, 36, 43, 95, 104, 118
	Xiphophorus helleri	10, 43, 104
Kiribati		
1. Line Islands (Fanning, Washington)	*Gambusia affinis*	44
	Tilapia mossambica	44, 68
2. Gilbert Islands	*Tilapia mossambica*	24
Marshall Islands (Jaluit)	*Gambusia affinis*	35
Palau (Babelthuap)	*Misgurnus anguillicaudatus*	11, 12
	Poecilia reticulata	11, 12, 30, 58
	Puntius sealei	11, 12, 30
	Xiphophorus maculatus	11, 12, 30
B. Southwest Pacific		
Fiji (Viti Levu)	[*Ctenopharyngodon idella*]	102, 130
	Gambusia affinis	65, 102, 130
	[*Macquaria colonorum*?]	102, 122
	[*Madigania unicolor*?]	102
	[*Mesopristes argenteus*?]	102
	Micropterus dolomieui?	24, 130
	Ophicephalus striatus?	24, 119, 130
	Poecilia reticulata	102, 130
	Poecilia mexicana[c]	102
	Puntius gonionatus	102, 126, 130
	Tilapia mossambica	4, 19, 24, 102, 116, 126, 130
	Tilapia nilotica?	126, 130
	Xiphophorus helleri	102
Nauru	*Tilapia mossambica*	24
New Caledonia	[*Cyprinus carpio*?]	4

Table 7-4. FISHES INTRODUCED ON ISLANDS OF OCEANIA OTHER THAN HAWAII (SEE FIGURE 7-1) *(continued)*

Politico-geographic location (Island of special concern)	Scientific name[a]	Information source[b]
	Micropterus salmoides	5, 23, 24
	Ophicephalus striatus	24
	Osphronemus goramy	4, 24, 117
	Tilapia mossambica	4, 115, 117, 119
	Tilapia zilli	24, 119
	Trichogaster pectoralis	4, 24, 117
Solomon Islands	*Tilapia mossambica*	4
Vanuatu (Tanna) [= New Hebrides]	*Tilapia mossambica*	4, 19, 24
C. Southeast Pacific		
American Samoa (Aunuu, Tutuila, Swains)	*Gambusia affinis*?	65
	Poecilia mexicana[c]	113, 137
	[*Poecilia vittata*]	137
	Tilapia mossambica	4, 24, 119, 137
Cook Islands (Rarotonga, Mitiaro)	*Gambusia affinis*	65
	Poecilia reticulata	134
	Tilapia mossambica	19, 25, 128
French Polynesia (Tahiti)	[*Cyprinus carpio*?]	20, 99, 100, 101
	Gambusia affinis	65, 99, 100, 101, 124
	[*Ictalurus punctatus*?]	20, 99, 100, 101
	[*Micropterus salmoides*?]	20, 99, 100, 101
	Poecilia mexicana[c]	124
	Poecilia reticulata	33, 100, 124
	[*Salmo gairdneri*]	20, 98, 101
	Tilapia mossambica	24, 124
Niue	*Tilapia mossambica*	24
Tonga (Tongatapu, Vavau)	*Tilapia mossambica*	24, 78
Wallis and Futuna (Wallis)	*Tilapia mossambica*	24
Western Samoa (Savaii)	*Carassius auratus*	133
	Gambusia affinis?	65
	Poecilia mexicana[c]	133
	Poecilia reticulata?	32
	Tilapia mossambica	4, 24, 120, 133

[a] Species in brackets [ ] failed to establish; question marks indicate vague or conflicting information in reference sources.
[b] Numbers refer to entries in Literature Cited section.
[c] Two Tahitian poeciliid specimens, tentatively identified by Dr. R. R. Miller (Museum of Zoology, University of Michigan) as *Poecilia butleri,* are provisionally included here with *P. mexicana* (see introductory remarks, section on Hawaiian Fish Introductions).

that are discussed separately here for the entire region. Other fishes are considered in context with regional subdivisions.

MOZAMBIQUE TILAPIA AND TOPMINNOWS (POECILIIDAE)

The Mozambique tilapia is noted (figure 7-1) as being established in sixteen locations. Following Hawaii's lead, Fiji imported the species in 1954.

The following year it was introduced into Guam, the Northern Marianas (Saipan), and New Caledonia. Thereafter, it was introduced into various other developing countries in Oceania, apparently with assistance from the South Pacific Commission (Van Pel 1955 et seq., Anonymous 1958). It was introduced into American Samoa (Aunuu Island) in 1957 by transfer from Western Samoa. The distribution of the Mozambique tilapia given here probably is incomplete; the species may occur in nations not noted, and very likely exists on islands other than those reported.

Initial introductions of Mozambique tilapia into the localities reported here except for that at Fanning Atoll are assumed to be deliberate attempts to establish the species. At Fanning Atoll small tilapia apparently were released in 1958 from a research vessel and established atypical populations described by Lobel (1980). Fanning's island ring contains shallow embayments around the inner margin that are euhaline to hyperhaline except after heavy rainfall. In 1978, Lobel observed populations of tilapia in five of these "estuaries" and noted their excursions into the lagoon. It is well known that the hardy Mozambique tilapia is broadly euryhaline and readily inhabits mixohaline estuaries. Its ability to spawn in seawater in an artificial habitat was noted by Brock (1954) and it is believed to migrate along continental coastlines (Robins et al. 1980). However, invasion of an open tropical lagoon and mingling with typical reef fishes possibly signal a new dimension in the ecological evolution of this versatile tilapia.

Six species of topminnows are known from at least twelve locations in Oceania (figure 7–1, table 7–4). One of them, the mosquitofish (*Gambusia affinis*), is widely distributed (nine or more areas) and was one of the earlier intentional introductions of exotic fishes. Mosquitofish and the guppy (*Poecilia reticulata,* five areas) were introduced on Tahiti more than fifty years ago (Rougier 1926a et seq., Fowler 1932b). Krumholz (1948) noted the occurrence of the mosquitofish in several Pacific island areas largely as a result of stocking by the U.S. Army during World War II. Its distribution in Oceania may even be wider than that of the Mozambique tilapia. Because of their long-term establishment and unobtrusive nature, the mosquitofish and guppy often are considered native fishes by residents of islands where they occur. Other topminnows were later introductions and have more restricted distributions: sailfin molly (*Poecilia latipinna*) and green swordtail (*Xiphophorus helleri*) on Guam; southern platyfish (*X. maculatus*) on Babelthuap (Palau); and shortfin molly (*P. mexicana*) in Fiji, Tahiti, and the Samoan Archipelago.

OTHER EXOTIC SPECIES: NORTH PACIFIC

North of the equator, other introduced fishes have been reported only on the islands of Babelthuap (Palau) and Guam. Intrusion of exotics has been minor in Palau where only the oriental weatherfish (*Misgurnus anguillicaudatus*) and a barb (*Puntius sealei*) are reported to be established; both probably were introduced before 1945. Guam, however, is second only to Hawaii in the numbers of introductions and establishments (table 7–4).

Six additional Guamanian introductions succeeded. The walking catfish (*Clarias batrachus*), one of three fishes imported from Manila in 1910, was the earliest naturalized species. Other established species, with the possible exception of *Cyprinus carpio* (no data), are relatively recent introductions. They include three cichlids (*Astronotus ocellatus, Cichla ocellaris,* and *Tilapia zilli*) and one anabantid (*Betta brederi*). Several other species not listed in table 7–4 have been stocked in breeding and rearing ponds, and a governmental committee has approved 30 species of Southeast Asian fishes for reservoir stocking (Guam 1979). Although none of the approved fishes has been released thus far, a small reservoir was stocked with Taiwanese red tilapia (*T. mossambica* x *T. hornorum*), locally called "cherry snapper." Illegal poisoning killed this stock before reproduction was evident (A. Hosmer, personal communication). Notable failures of species after large releases into a reservoir included the channel catfish and centrarchid basses.

OTHER EXOTIC SPECIES: SOUTH PACIFIC

In Fiji and New Caledonia, several fishes (table 7–4) were introduced since 1953, mostly for aquaculture. Successful Fijian introductions include a barb (*Puntius gonionatus*) and perhaps Nile tilapia (*Tilapia nilotica*). There are questionable reports of the establishment of smallmouth bass (*Micropterus dolomieui*), snakehead (*Ophicephalus striatus*), and three Australian fishes (*Macquaria colonorum, Madigania unicolor,* and *Mesopristes argenteus*). Grass carp (*Ctenopharyngodon idella*), accidentally released from a culture facility (S. Cavuilati, personal communication), survived but apparently failed to reproduce (T. Lichatowich, personal communication). Exotics established in New Caledonia are the largemouth bass (*Micropterus salmoides*), snakehead, giant gourami (*Osphronemus goramy*), redbelly tilapia (*Tilapia zilli*), and sepat Siam (*Trichogaster pectoralis*). The kissing gourami (*Helostoma temmincki*) is said to be established in the South Pacific (Van Pel 1958) but the location is unspecified.

Four fishes (common carp, channel catfish, largemouth bass, and rainbow trout), imported to Tahiti from San Francisco fifty-five years ago, failed to survive (Rougier 1926a et seq.); indigenous anguillid eels were blamed for the failure of bass and catfish. The only other fish reported to be established in the subregion is the goldfish (*Carassius auratus*) which occurs in a Western Samoan lake as a result of a planting early in the century (A. Philipp, personal communication).

Some Consequences of Exotic Fish Introductions

Impacts of immigrant species may be beneficial, negligible, adverse, or even a combination of these for a given species under some circumstances. Beneficial and negligible effects generally relate to the purpose for a species introduction, such as for food, forage, recreation, or biological control. Adverse effects center on changes in natural ecosystems induced by exotics, particularly on native species that may be direct (competition and predation) or indirect (e.g., introduction and transmission of disease or parasites). Benefits generally are self-evident, as are some adverse impacts, but even these are difficult to quantify. Direct studies of impacts have been few and the appraisals that follow are, therefore, largely qualitative and subjective.

Apart from probable control of water weeds and mosquitoes locally by introduced fishes and minor fisheries for tilapia in Fiji and Guam, there is little information on effects of exotics in most of Oceania beyond Hawaii. Some remarks I received indicated the tendency of Mozambique tilapia to become a nuisance in the wild and noted its poor acceptance as a food fish. The Republic of Kiribati is said to consider the eradication of tilapia as a problem requiring major effort by its Department of Natural Resources (L. Taylor, personal communication). In the Cook Islands, where tilapia was introduced as a food and mosquito-control fish, it has developed large populations that are fished infrequently by the islanders for food and barracuda bait (J. Dashwood, personal communication). The prevalent public attitude toward the Mozambique tilapia is exemplified by a statement (Guam 1979) made in context with the marketing of red tilapia: "hybrids were marketed under the name 'Cherry Snappers' . . . to avoid the local prejudice against tilapia as a food fish. (Locally, the term 'tilapia' refers to *S. mossambicus* which is considered an undesirable fish by most residents.)" The following discussion of the consequences of introduced fishes refers essentially to the situation in Hawaii.

MARINE FISHES

The Marquesan sardine was introduced into Hawaii as potential bait for skipjack tuna to supplement inadequate stocks of native bait fish (Murphy 1960). Despite a wide distribution that coincides in habitat with the much-desired native bait fish (nehu), the sardine is not caught in sufficient numbers to make it significant to the fishery (Hida and Morris 1963, Shomura 1977). Ecologically, it probably provides forage for larger marine fishes but it also may be suppressing the abundance of nehu by competition. The Marquesan sardine was implicated recently in a human fatality attributed to "clupeid poisoning" (Tabata 1981b).

Two of three marine species introduced for fishery enhancement, toau and roi, do not occur in large numbers and do not enter the commercial catch. Evidently they are minor catches of noncommercial inshore fishermen. Taape, the third species, occurs in great abundance at depths from 10 m to more than 100 m and has become a commercially significant species (landings have increased steadily from 500 kg in 1970 to 47,000 kg in 1980) attendant with much controversy (Tabata 1979, 1981a, 1981b). The taape is relatively small (< 20 cm) and has gained only minor consumer acceptance. Consequently, there are marketing difficulties and its wholesale value (ca. $1.00/kg in 1979) is less than half that of most co-occurring commercial species. Many persons feel that it is displacing the more valuable fishes. Diet studies indicate that taape feeds primarily on small fishes and certain crustaceans (Oda and Parrish 1981). Thus it possibly competes with and may prey upon more desirable species of fishes.

Randall and Kanayama (1972) first noted the establishment and population expansion of the inadvertently introduced Marquesan mullet, a small (< 15 cm) species. They suggested that it was making inroads on the valuable native mullet.* Their premise is substantiated by collections I made on Kauai in typical striped mullet nursery areas (estuaries) during the period 1975–78. In those collections, the exotic species comprised about three-fourths of the mullets captured by gill net. A few years earlier, it was found to comprise 44 percent of the mullet catch in an Oahu estuary (Maciolek and Timbol 1981). The Marquesan mullet is not a commercial species but a minor food fish and bait fish locally.

NONMARINE FISHES

Effects of selected species.

Many of the introduced inland-water species provide recreational fishing on the five major islands. However, compared with those of other states, this fishery is not large. Licenses, required for anglers of age nine or older, totaled 7,118 (less than 1 percent of the state's population) in fiscal year 1980 (Hawaii 1981b). A month-long season for rainbow trout attracts about 500 fishermen trips and yields an average (four years, 1977–80) of 570 fish weighing 136 kg (Hawaii 1977 et seq.). Warmwater fishing centers on reservoirs on the islands of Hawaii, Kauai, Maui, and Oahu. Oahu's Wahiawa Reservoir (135 ha), a premier site, yields an annual average (four years, 1977–80) of 5,374 kg for about 50,000 hours of angling effort (Hawaii 1977 et seq.). The catch consists of more than 11 species: 65 percent tilapias (3 species, mainly *T. mossambica*); 12 percent bluegill, 8 percent tucunare, 5 percent largemouth bass; 4 percent common carp; and less than 2 percent each of channel catfish,

* In 1979, the wholesale price of local (fresh) striped mullet (*Mugil cephalus*) was nearly $5.00/kg.

goldfish, Chinese catfish, snakehead, oscar, and smallmouth bass. Native aquatic fauna generally does not inhabit Hawaiian reservoirs.

An interesting sidelight resulting from the stocking of centrarchid fishes on Kauai has been the appearance of an intergeneric hybrid, *Lepomis macrochirus* x *Micropterus salmoides,* first noted in 1963. According to Dr. R. M. Bailey (personal communication) who identified the hybrid, its occurrence under natural conditions appears to be unique. The hybrid is superficially intermediate between the parent species (figure 7–3). It persists in at least two reservoirs and may be fertile (S. Shima, personal communication), possibly leading to the evolution of a new "species."

Several exotic species function as forage (including bait fish). Threadfin shad, originally introduced as a tuna bait fish (Hida and Thomson 1962), is valuable as a reservoir forage fish (Kanayama 1968). Tilapias and topminnows, forage fishes in reservoirs and occasionally the bait of inshore marine anglers, have been tested extensively in tuna bait fish programs (Shomura 1977). Continuing efforts have been made since 1957 to mass-culture such species to ensure a reliable bait fish supply. Naturally produced native nehu, however, remain the principal and most desired bait fish. Another application tested was the use of Mozambique tilapia and two topminnows (mosquitofish and guppy) as bioassay organisms. The sensitivity of those exotics to organic toxicants was compared with that of two native fishes by Nunogawa et al. (1970). As one might suspect, the native fishes were more sensitive to all toxicants tested, thus indicating their greater suitability as bioassay animals.

The mosquitofish and sailfin molly were the first fishes introduced into Hawaii as control agents (Van Dine 1907). These and other topminnows have been useful in controlling mosquito larvae in certain artificial habitats such as urban ponds and ditches, water tanks, effluent ponds, and reservoirs (P. Nakagawa, personal communication). Mosquitofish and probably other topminnows may be significant predators despite their small size (Myers 1965). Predation by topminnows on carp fry was noted by Rougier (1928), and an incidence of mosquitofish devouring tilapia fry was observed in Wahiawa Reservoir (Hawaii 1980c). Such evidence indicates that topminnows may also be serious predators on native fauna in natural aquatic environments.

Certain tilapias (*T. melanopleura, T. mossambica, T. zilli*) were introduced in Hawaii partly or mostly for aquatic weed control (Brock 1960). They have been effective in that role in many developed environments such as reservoirs and ditches. The ubiquitous Mozambique tilapia has become dominant in many of its habitats, particularly in coastal wetlands where it has developed nuisance populations (Maciolek 1971). Amid throngs of stunted tilapia in the shallow saline ponds on Oahu's Mokapu Peninsula, I observed nuptially colored males no longer than 5 cm constructing nests. Water birds such as the black-crowned night heron probably feed on the tilapia, but bird predation is generally insufficient to control its population in most wetlands.

Figure 7–3. Intergeneric centrarchid hybrid, *Lepomis macrochirus* x *Micropterus salmoides,* appeared spontaneously in at least two reservoirs on Kauai Island, Hawaii, where it has persisted since about 1963. Local fisherman refer to it as "blue-bass." (Photo courtesy of the Waikiki Aquarium, Honolulu.)

In a survey on Hawaii Island, Maciolek and Brock (1974) found tilapia and topminnows in 33 of 318 coastal pools sampled. An endemic hypogeal shrimp was found in 187 of the pools, but in only 5 pools did it co-occur with the exotic fishes and then in very low numbers. Like topminnows, the Mozambique tilapia would likely be a significant predator on native aquatic fauna, especially in streams and estuaries.

The smallmouth bass, which provides a minor sport fishery on two islands, appears to be a serious piscine predator on native stream fauna. In the midreach of the Wailua River on Kauai, a stream now dominated by smallmouth bass, Timbol (1977) collected no native fishes at ten sampling stations. Apparently, this bass had displaced earlier populations of largemouth bass and bluegill that had already suppressed indigenous stream fauna. On Oahu Island, no native fishes or crustaceans were found at locations on Maunawili and Nuuanu streams where smallmouth bass was collected in 1975 and 1976. A few years earlier (1967–71) I used the same collecting sites to demonstrate native fishes and crustaceans to students; at that time, the smallmouth bass was not evident.

Overview of exotics in Hawaiian streams.

As a group, nonmarine exotic fishes have had the most extensive adverse effects in streams. One study (Norton et al. 1978) showed that exotics predominated among fish communities in three Oahu streams, and that channel modification strongly favored exotics over native fishes and invertebrates. In the absence of definitive studies, the relative importance of exotics as competitors (for food, shelter, etc.) or as predators is unknown. Especially because prominent native fauna (fishes and crustaceans) are diadromous, entering streams as postlarvae about 2 cm long or less, I suspect that predation is more significant.

Twenty of the 31 established exotic fishes have been collected from streams, but not all of them exist in abundances sufficient to have significant impacts. The other 11 species, most of them reservoir or pond fishes, may sometimes occur in streams if only accidentally. On the basis of preferred habitat, food habits, and numbers of individuals within stream populations, and on the assumption that the abundance of an introduced species is the primary determinant of its effects on the native aquatic community, established Hawaiian exotics can be grouped as follows:

1. Fishes with negligible stream impact

Dorosoma petenense
Barbus semifasciolatus
Ictalurus punctatus
Poecilia vittata
Ophicephalus striatus
Monopterus albus
Lepomis macrochirus
Micropterus salmoides
Astronotus ocellatus
Cichla ocellaris
Cichlasoma meeki
Cichlasoma sp.
Tilapia macrochir
Tilapia melanopleura
Tilapia melanotheron
Tilapia zilli
Xiphophorus variatus

2. Fishes with intermediate stream impact

Salmo gairdneri
Carassius auratus
Cyprinus carpio
Misgurnus anguillicaudatus
Poecilia latipinna
Poecilia sphenops
Xiphophorus maculatus

3. Fishes with high stream impact

Clarias fuscus
Gambusia affinis
Poecilia mexicana
Poecilia reticulata
Xiphophorus helleri
Micropterus dolomieui
Tilapia mossambica

Table 7–5. HAWAIIAN NONMARINE EXOTIC FISHES: TIME OF INTRODUCTION AND ESTABLISHMENT SUCCESS. PREVALENT SPECIES ARE THOSE IN TABLE 7–1 WITH 3 OR 4 PLUS SIGNS (COLUMN 6).

	Period of introduction		
Factor	Before 1906	1920 to 1950	1950 to present
Species introduced			
Number	12	12	17
Species established			
Number	9	9	13
% introduced spp.	75	75	76
Species prevalent			
Number	7	3	3
% introduced spp.	58	25	18
% established spp.	78	33	23

SPECIES MOBILITY

Some established exotics have spread much faster than others, the mobility in developing new populations depending partly on intrinsic factors (size, fecundity, environmental tolerances, etc.) and partly on human assistance. The Mozambique tilapia was well distributed throughout the major islands within ten or fifteen years after its introduction and new populations continue to appear (e.g., Maciolek and Timbol 1981). It is probably the most prevalent exotic in Hawaii as judged by both numbers of populations and abundances within populations. At the other extreme, a few established species such as the firemouth and half-banded barb show no mobility and may not spread beyond their present habitats.

Although it is possible to estimate the relative mobility of a species, it is difficult to predict how much time would be needed for the species to "saturate" an island after initial establishment, and impossible to say when (or if) it may appear on other islands. If we consider all habitable islands and all introduced nonmarine exotics in historical perspective, time to attain prevalence can be demonstrated roughly on a species-group basis. Table 7–5 indicates that, whereas fishes introduced before 1906 had about the same establishment success as those introduced after 1949, group prevalence of the early importations is about three times that of the more recent establishments (or introductions). This comparison suggests that many decades may be needed for at least some species to express fully their prevalence potential. Thus, the ultimate distributions and abundances (and therefore the eventual impacts) of recent Hawaiian introductions may not be manifest until sometime during the next century.

As noted earlier, the chief effect of exotics is on natural environments that they invade, thereby displacing native species. Attrition of native aquatic fauna also results from other environmental disturbances and modifications which often favor exotic fishes. All are related to human population density and socioeconomic development, which are most evident on Oahu Island where nearly all established exotics exist today, where four-fifths of the state's population lives, and where there is a severe depletion of native freshwater fauna. Hubbell's (1968) observation applies well to the Oahu situation: "Destruction of native island biotas by man and his introductions is proceeding rapidly . . . and on many islands is already almost complete." Much needed now are studies on the ecology of exotic fishes and their populations distributions, especially on islands not yet as severely invaded as Oahu.

Acknowledgments

Many individuals and agencies supplied information used in the preparation of this chapter. Besides those cited, comments and suggestions were received from R. E. Brock and J. D. Parrish at the University of Hawaii, L. G. Eldredge at the University of Guam, and J. I. Ford at the Honolulu offices of the U.S. Fish and Wildlife Service. A. S. Timbol, Kauai Community College, provided unpublished data from stream surveys. Henry Sakuda, director, Hawaii Division of Aquatic Resources, supplied agency data and review comments.

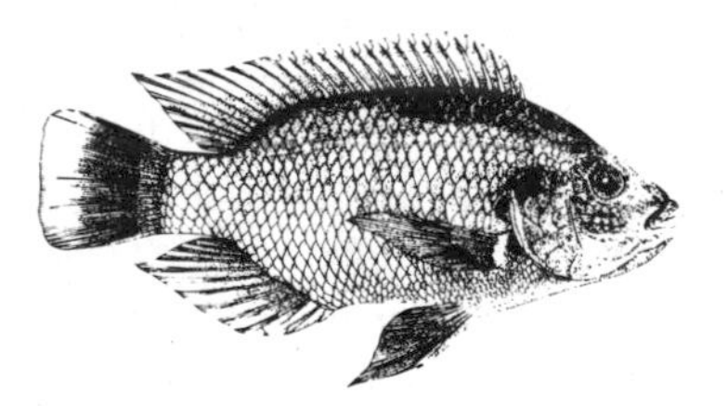

Literature Cited

(Note: The citations are numbered to facilitate use of references in the tables.)

1. Anonymous. 1910. Guam News Letter. Agana, Guam, 23 May 1910, 2:1.
2. Anonymous. 1924. The use of fish for mosquito control. International Health Board, Rockefeller Foundation, New York.
3. Anonymous. 1955. Saipan's Sablan. The New Yorker, 3 September 1955:21–22.
4. Anonymous. 1958. New introduction of edible pondfish from Philippines. South Pacific Commission Quarterly Bulletin 8(2):19.

5. Anonymous. 1963. Black bass success in New Caledonia. South Pacific Bulletin 13(4):54.
6. Anonymous. 1965. Checklist of introduced fresh water fish stockings (to November, 1965). Hawaii Division of Fish and Game, 11/24/65. Mimeographed.
7. Baldwin, W. J. 1974. Raising mollies for skipjack bait may eliminate use of frail nehu. National Fisherman 54(9):7-c.
8. Belk, D., M. J. Merton, and J. E. Shafer. 1971. Agana Springs nature reserve. Guam Science Teachers Association, Agana, Guam.
9. Bell, F. T., and E. Higgins. 1939. A plan for the development of Hawaiian fisheries. United States Bureau of Fisheries, Investigation Report 42:1–25.
10. Best, B., and C. Davidson. 1981. Inventory and atlas of the inland aquatic ecosystems of the Marianas Archipelago. University of Guam, Marine Laboratory Technical Report 75.
11. Bright, G. 1979. The inland waters of Palau, Caroline Islands. Report, Office of the Chief Conservationist, Trust Territory of the Pacific Islands, Koror, Palau.
12. Bright, G., and J. June. 1981. The freshwater fishes of Palau, Caroline Islands. Micronesica 17:107–111.
13. Brock, V. E. 1952. A history of the introduction of certain aquatic animals to Hawaii. Report, Board of Agriculture and Forestry, Territory of Hawaii, 1952: 114–123.
14. Brock, V. E. 1954. A note on the spawning of *Tilapia mossambica* in sea water. Copeia 1954:72.
15. Brock, V. E. 1960. The introduction of aquatic animals into Hawaiian waters. Internationale Revue der gesamten Hydrobiologie und Hydrographie, 45:463–480.
16. Brock, V. E., and M. Takata. 1956. A limnological resurvey of Fena Reservoir, Guam, Marianas Islands. Territory of Hawaii, Board of Commissioners of Agriculture, Honolulu, Hawaii.
17. Brock, V. E., and Y. Yamaguchi. 1955. A limnological survey of Fena River Reservoir, Guam, Marianas Islands. Territory of Hawaii, Board of Commissioners of Agriculture and Forestry, Honolulu, Hawaii.
18. Broshears, R. E., and J. D. Parrish. 1980. Aquatic habitat and aquatic food sources for endangered waterbirds at Hanalei National Wildlife Refuge. Hawaii Cooperative Fishery Research Unit, Technical Report 80-3, University of Hawaii, Honolulu, Hawaii.
19. Chimits, P. 1957. The tilapias and their culture. Food and Agriculture Organization, Fisheries Bulletin 10:1–24.
20. Comité, S.E.O. 1927. Pisciculture. Bulletin de la Société des Etudes Océaniennes 20:283–284.
21. Courtright, A. M. 1970. Fresh water fishing adds new sport to Guam scene. Pacific Daily News (Agana, Guam), 16 April 1970:16.
22. Dahl, A. H. 1980. Regional ecosystems survey of the south Pacific area. South Pacific Commission Technical Paper no. 179.
23. Devambez, L. C. 1960. American game fish for New Caledonia. South Pacific Bulletin 10(4):25 and 38.
24. Devambez, L. C. 1964. Tilapia in the south Pacific. South Pacific Bulletin 14(4): 27, 28, and 52.
25. Devick, B. 1978a. The devilfish. Hawaii Fishing News (Honolulu, Hawaii), August 1978.

26. Devick, B. 1978b. Common fish diseases. Hawaii Fishing News (Honolulu, Hawaii), October 1978.
27. Devick, B. 1978c. Hawaii's bass population wormy: Not yet. Hawaii Fishing News (Honolulu, Hawaii), November 1978.
28. Devick, B. 1980. The importance of fish population balance. Hawaii Fishing News (Honolulu, Hawaii), May 1980.
29. Elliot, E. 1955. Its mouth is its nursery. Natural History Magazine 64:330–331.
30. Fehlmann, H. A. 1960. Ecological distribution of fishes in a stream drainage in the Palau Islands. Ph.D. dissertation, Stanford University, Palo Alto, California. 172 pp.
31. Fowler, H. W. 1925. Fishes of Guam, Hawaii, Samoa, and Tahiti. Bernice P. Bishop Museum, Bulletin 22.
32. Fowler, H. W. 1932a. Fishes obtained at Samoa in 1929. Bernice P. Bishop Museum, Occasional Papers 9(18):1–16.
33. Fowler, H. W. 1932b. Fresh-water fishes from the Marquesas and Society Islands. Bernice P. Bishop Museum, Occasional Papers 9(25):4–11.
34. Greenwood, P. H., D. E. Rosen, S. H. Weitzman, and G. S. Myers. 1966. Phyletic studies of teleostean fishes, with a provisional classification of living forms. Bulletin of the American Museum of Natural History 131:341–455.
35. Gressitt, J. L. 1961. Terrestrial fauna, 69–73. *In:* D. I. Blumenthal (ed.). A report on typhoon effects upon Jaluit Atoll. Atoll Research Bulletin no. 75.
36. Guam. 1957. Annual report to the governor. Guam Division of Fish and Wildlife.
37. Guam. 1963. Annual report to the governor. Guam Division of Fish and Wildlife.
38. Guam. 1964. Annual report to the governor. Guam Division of Fish and Wildlife.
39. Guam. 1966. Annual report to the governor. Guam Division of Fish and Wildlife.
40. Guam. 1967. Annual report to the governor. Guam Division of Fish and Wildlife.
41. Guam. 1968. Annual report to the governor. Guam Division of Fish and Wildlife.
42. Guam. 1978. Job progress report, federal aid to fish and wildlife restoration. Project no. FW-2R-15. Guam Aquatic and Wildlife Resources Division, Department of Agriculture.
43. Guam. 1979. Annual report. Guam Aquatic and Wildlife Resources Division, Department of Agriculture.
44. Guinther, E. B. 1971. Ecologic observations on an estuarine environment at Fanning Atoll. Pacific Science 25:249–259.
45. Hawaii. 1962. Annual report. Hawaii Division of Fish and Game, Department of Agriculture and Forestry.
46. Hawaii. 1977 et seq. Creel census and fishermen checking station operation. Job Progress Reports, Hawaii Division of Fish and Game, Job 1, Study IV: 1977 — Project no. F-14-R-1; 1978 — Project no. F-14-R-2; 1979 — Project no. F-14-R-3; 1980a — Project no. F-14-R-4.
47. Hawaii. 1980b. Disturbances and fluctuations in the Wahiawa Reservoir ecosystem. Job Progress Report, Hawaii Division of Fish and Game, Project no. F-14-R-4, Job no. 4 (Study I).
48. Hawaii. 1980c. Studies on fish population balance, methods of harvest, and rate of harvest. Job Progress Report, Hawaii Division of Fish and Game, Project no. F-14-R-3, Job no. 3 (Study I).
49. Hawaii. 1981a. Studies on the relationship between tucunare and largemouth bass. Job Progress Report, Hawaii Division of Fish and Game, Project no. F-14-R-4, Job no. 1 (Study I).

50. Hawaii. 1981b. Report to the governor. Annual Report, Hawaii Department of Land and Natural Resources, Honolulu, Hawaii.
51. Herald, E. S. 1961. Living fishes of the world. Doubleday and Company, New York.
52. Herre, W.C.T. 1940. Distribution of fresh-water fishes in the Indo-Pacific. Scientific Monthly 51:165–168.
53. Hida, T. S., J. S. Harada, and J. H. King. 1962. Rearing tilapia for tuna bait. U.S. Fish and Wildlife Service, Fishery Bulletin 62:1–20.
54. Hida, T. S., and R. A. Morris. 1963. Preliminary report on the Marquesan sardine, *Harengula vittata,* in Hawaii. Pacific Science 17:431–437.
55. Hida, T. S., and D. A. Thomson. 1962. Introduction of the threadfin shad to Hawaii. Progressive Fish-Culturist 24:159–163.
56. Hosaka, E. Y. 1944. Sport fishing in Hawaii. Bond's, Honolulu, Hawaii.
57. Hubbell, T. R. 1968. The biology of islands. Proceedings of the National Academy of Sciences 60:22–32.
58. Ikebe, K. 1939. Records of the introduction of top minnows into the south sea islands. South Sea Fishery News 3(9):7–10. In Japanese: Translation no. 10, Southwest Fisheries Center, National Marine Fisheries Service, Honolulu, Hawaii.
59. Iversen, R. T. 1971. Use of threadfin shad, *Dorosoma petenense,* as live bait during experimental pole-and-line fishing for skipjack tuna, *Katsuwonus pelamis,* in Hawaii. National Oceanic and Atmospheric Administration, Technical Report NMFS, SSRF-641.
60. Jordan, D. S., and B. W. Evermann. 1905. Aquatic resources of the Hawaiian Islands. Part 1: The shore fishes. Bulletin of the United States Fish Commission 23:1–574.
61. Kami, H. T., N. Drahos, R. J. Lujan, and J. J. Jeffrey. 1974. Biological study of the Geus River Basin. University of Guam, Marine Laboratory Technical Report no. 16.
62. Kami, H. T., I. I. Ikehara, and F. P. DeLeon. 1968. Check-list of Guam fishes. Micronesica 4:95–131.
63. Kanayama, R. K. 1968. Hawaii's aquatic animal introductions. Proceedings of the Annual Conference of Western Association of State Game and Fish Commissioners 47:123–131.
64. King, J. E., and P. T. Wilson. 1957. Studies on tilapia as skipjack bait. U.S. Fish and Wildlife Service, Special Scientific Report—Fisheries no. 225.
65. Krumholz, L. A. 1948. Reproduction in the western mosquitofish, *Gambusia affinis affinis* (Baird and Baird), and its use in mosquito control. Ecological Monographs 18:1–43.
66. Kubota, W. T. 1972. The biology of an introduced prawn, *Macrobrachium lar* (Fabricius) in Kahana Stream. M.S. thesis, University of Hawaii, Honolulu, Hawaii.
67. Lachner, E. A., C. R. Robins, and W. R. Courtenay, Jr. 1970. Exotic fishes and other aquatic organisms introduced into North America. Smithsonian Contributions in Zoology 59:1–29.
68. Lobel, P. S. 1980. Invasion by the Mozambique tilapia (*Sarotherodon mossambicus;* Pisces; Cichlidae) of a Pacific atoll marine ecosystem. Micronesica 16:349–355.
69. MacCrimmon, H. R. 1971. World distribution of rainbow trout (*Salmo gairdneri*). Journal of the Fisheries Research Board of Canada 28:663–704.

70. MacCrimmon, H. R., and J. S. Campbell. 1969. World distribution of brook trout, *Salvelinus fontinalis*. Journal of the Fisheries Research Board of Canada 26:1699–1725.
71. MacCrimmon, H. R., and T. L. Marshall. 1968. World distribution of brown trout, *Salmo trutta*. Journal of the Fisheries Research Board of Canada 25:2527–2548.
72. Maciolek, J. A. 1971. Aquatic ecosystems of Kealia floodplain and Maalaea Bay, Maui. Hawaii Institute of Marine Biology, Technical Report 27.
73. Maciolek, J. A. 1978. Insular aquatic ecosystems: Hawaii, 103–120. *In:* Classification, inventory, and analysis of fish and wildlife habitat, proceedings of a national symposium. U.S. Fish and Wildlife Service FWS/OBS-78/76.
74. Maciolek, J. A. 1982. Lakes and lake-like waters of the Hawaiian Archipelago. Bernice P. Bishop Museum, Occasional Papers 25(1):1–14.
75. Maciolek, J. A., and R. E. Brock. 1974. Aquatic survey of the Kona Coast ponds, Hawaii Island. University of Hawaii Sea Grant Advisory Report, AR-74-04.
76. Maciolek, J. A., and A. S. Timbol. 1980. Electroshocking in tropical insular streams. Progressive Fish-Culturist 42:57–58.
77. Maciolek, J. A., and A. S. Timbol. 1981. Environmental features and macrofauna of Kahana Estuary, Oahu. Bulletin of Marine Science 31:712–722.
78. Maciolek, J. A., and R. Yamada. 1981. Vai Lahi and other lakes of Tonga. International Association of Theoretical and Applied Limnology, Proceedings 21:693–698.
79. Mainland, G. B. 1939. Gobioidea and fresh water fish on the island of Oahu. M.S. thesis, University of Hawaii, Honolulu, Hawaii.
80. Malecha, S. R. 1968. Studies on the serum protein polymorphisms in some populations of introduced fresh water fishes. M.S. thesis, University of Hawaii, Honolulu, Hawaii.
81. Murphy, G. I. 1960. Introduction of the Marquesan sardine, *Harengula vittata* (Cuvier and Valenciennes), into Hawaiian waters. Pacific Science 14:185–187.
82. Myers, G. S. 1953. Paleogeographical significance of fresh-water fish distribution in the Pacific. Proceedings of the 7th Pacific Science Congress, 4:38–48.
83. Myers, G. S. 1965. *Gambusia,* the fish destroyer. Tropical Fish Hobbyist 13(5): 31–32, 53–54.
84. Nakamura, E., and R. Wilson. 1970. The biology of the Marquesan sardine, *Sardinella marquesensis*. Pacific Science 24:359–376.
85. Needham, P. R. 1949. A fisheries survey of the streams of Kauai and Maui with special reference to rainbow trout (*Salmo gairdnerii*). Territory of Hawaii, Division of Fish and Game, Special Bulletin no. 1.
86. Needham, P. R., and J. P. Welch. 1953. Rainbow trout (*Salmo gairdneri* Richardson) in the Hawaiian Islands. Journal of Wildlife Management 17:233–255.
87. Norton, S. E., A. S. Timbol, and J. D. Parrish. 1978. Stream channel modification in Hawaii. Part B: Effect of channelization on the distribution and abundance of fauna in selected streams. U.S. Fish and Wildlife Service FWS/OBS-78/17.
88. Nunogawa, J. H., N. C. Burbank, R.H.F. Young, and L. S. Lau. 1970. Relative toxicities of selected chemicals to several species of tropical fishes. University of Hawaii, Water Resources Research Center, Technical Report no. 40.
89. Oda, D. K., and J. D. Parrish. 1981. Ecology of commercial snappers and groupers introduced to Hawaiian reefs. Proceedings 4th International Symposium on Coral Reefs, Manila, 1981.

90. Parrish, J. D., and D. K. Oda. 1980. Ecology of some exotic marine fishes introduced to Hawaii. Proceedings of the Association of Island Marine Laboratories of the Caribbean, 15:25 (Abstract).
91. Parrish, J. D., L. Taylor, M. DeCrosta, S. Feldcamp, L. Sanderson, and C. Sorden. 1980. Trophic studies of shallow-water fish communities in the northwestern Hawaiian Islands, 175–188. *In:* R. W. Grigg and R. T. Pfund (eds.). Proceedings of the Symposium on Status of Resource Investigations in the Northwestern Hawaiian Islands. University of Hawaii, Sea Grant Miscellaneous Report, MR-80-04.
92. Randall, J. E. 1960. New fishes for Hawaii. Sea Frontiers 6:33–43.
93. Randall, J. E. 1980. New records of fishes from the Hawaiian Islands. Pacific Science 34:211–232.
94. Randall, J. E., and R. K. Kanayama. 1972. Hawaiian fish immigrants. Sea Frontiers 18:144–153.
95. Randall, R. H., and R. T. Tsuda. 1974. Field ecological survey of the Agana-Chaot River Basin. University of Guam, Marine Laboratory Technical Report 12.
96. Raulerson, L., M. Chernin, and P. Moore. 1978. Biological study of the potential Ugum dam site. Report prepared for the U.S. Army Corps of Engineers, Fort Shafter, Hawaii. Contract no. DACW 84-78-C-0003.
97. Robbins, W. H., and H. R. MacCrimmon. 1974. The black bass in America and overseas. Biomanagement Research Enterprises, Sault Sainte Marie, Ontario, Canada. 196 pp.
98. Robins, R. C., R. M. Bailey, C. E. Bond, J. R. Brooker, E. A. Lachner, R. N. Lea, and W. B. Scott. 1980. A list of common and scientific names of fishes from the United States and Canada, 4th ed. American Fisheries Society, Special Publication no. 12.
99. Rougier, E. 1926a. Essai de pisciculture. Bulletin de la Société des Etudes Océaniennes 11:16–20.
100. Rougier, E. 1926b. Pisciculture. Bulletin de la Société des Etudes Océaniennes 12:40–41.
101. Rougier, E. 1928. Pisciculture. Bulletin de la Société des Etudes Océaniennes 26:107–109.
102. Ryan, P. A. 1980. A checklist of the brackish and freshwater fish of Fiji. South Pacific Journal of Natural Science 1:58–73.
103. Schultz, R. J., and R. R. Miller. 1971. Species of the *Poecilia sphenops* complex (Pisces: Poeciliidae) in Mexico. Copeia 1971:282–290.
104. Shepard, J. W., and R. F. Myers. 1981. A preliminary checklist of the fishes of Guam and the southern Mariana Islands, 60–88. *In:* A working list of marine organisms from Guam. University of Guam, Marine Laboratory Technical Report no. 70.
105. Shomura, R. S. (ed.). 1977. Collection of tuna baitfish papers. National Oceanic and Atmospheric Administration Technical Report, NMFS Circular 408.
106. Tabata, R. 1979. Taape—What needs to be done? Hawaii Fishing News (Honolulu, Hawaii), October 1979.
107. Tabata, R. 1981a. Taape in Hawaii, new fish on the block. University of Hawaii, Sea Grant College Marine Advisory Program, UNIHI-SEAGRANT-AB-81-03.
108. Tabata, R. (ed.). 1981b. Transcripts of "Taape: What needs to be done?" workshop. University of Hawaii, Sea Grant College Program, Working Paper no. 46.
109. Timbol, A. S. 1977. A report on the aquatic survey of stream macrofauna for

the hydroelectric power study for Hawaii. Prepared for United States Army Corps of Engineers, Honolulu, Hawaii.

110. Timbol, A. S. 1979. Freshwater macrofauna and habitats in the Hawaiian Archipelago and U.S. Oceania, 4–1 to 4–23. *In:* J. E. Byrne (ed.). Literature review and synthesis of information on Pacific island ecosystems. U.S. Fish and Wildlife Service, FWS/OBS-79/35.

111. Timbol, A. S., and J. A. Maciolek. 1978. Stream channel modification in Hawaii. Part A: Statewide inventory of streams, habitat factors, and associated biota. U.S. Fish and Wildlife Service, FWS/OBS-78/16.

112. Timbol, A. S., A. J. Sutter, and J. D. Parrish. 1980. Distribution, relative abundance, and stream environment of *Lentipes concolor* (Gill 1860), and associated fauna in Hawaiian streams. University of Hawaii, Water Resources Research Center Cooperative Report no. 5.

113. U.S. Army Corps of Engineers. 1981. American Samoa stream inventory. Part A: Tutuila. American Samoa Water Resources Study, U.S. Army Corps of Engineers, Honolulu, Hawaii.

114. U.S. Fish and Wildlife Service. 1980. Fish and Wildlife Coordination Act, Section 2(b) Report, U.S. Fish and Wildlife Service, Department of the Interior, Appendix K-I. *In:* Ugum River Interim Report and Environmental Impact Statement. U.S. Army Engineer District, Honolulu, Fort Shafter, Hawaii.

115. Van Dine, D. L. 1907. The introduction of top-minnows into the Hawaiian Islands. Hawaii Agricultural Experiment Station, Honolulu, Press Bulletin no. 20.

116. Van Pel, H. 1955. Pond culture of tilapia. South Pacific Commission Quarterly Bulletin 5(2):30–31.

117. Van Pel, H. 1956. Introduction of edible pond fish from the Philippines. South Pacific Commission Quarterly Bulletin 6(1):17–18.

118. Van Pel, H. 1958. Fresh water fish for the Pacific. South Pacific Commission Quarterly Bulletin 8(4):48–49, 52.

119. Van Pel, H. 1959. Fisheries in American Samoa, Fiji, and New Caledonia. South Pacific Commission Quarterly Bulletin 9(3):26–27.

120. Van Pel, H. 1961. S.P.C. fisheries investigations in Western Samoa. South Pacific Bulletin 11(1):20–22.

121. Welcomme, R. L. 1981. Register of international transfers of island fish species. Food and Agricultural Organization, Fisheries Technical Paper no. 213.

122. Whitley, G. P. 1927. A check-list of fishes recorded from Fijian waters. Journal of the Pan-Pacific Research Institution 2(1):3–8.

123. Williams, V. R. 1980. Growth and reproduction of the Marquesan sardine (*Sardinella marquesensis*) in Hawaii. M.S. thesis, University of Hawaii, Honolulu, Hawaii.

Personal Communications Cited

124. Bagnis, R., Chief, Medical Oceanographic Unit, Institut de Recherches Médicales "Louis Malardé," B.P. 30, Papeete, Tahiti.

125. Bailey, R. M., Curator of Fishes, Museum of Zoology, University of Michigan, Ann Arbor, Michigan 48109.

126. Cavuilati, S. T., Ministry of Agriculture and Fisheries, P.O. Box 358, Suva, Fiji.

127. Croft, R. A., State Fisheries Officer, Marine Resources Division, Federated States of Micronesia, P.O. Box B, Kolonia, Ponape, ECI 96941.
128. Dashwood, J., Chief Fisheries Officer, Ministry of Agriculture and Fisheries, P.O. Box 96, Rarotonga, Cook Islands.
129. Hosmer, A., Fisheries Biologist, Aquatic and Wildlife Resources Division, Department of Agriculture, Box 23367, GMF, Guam 96921.
130. Lichatowich, T. J., 670 Allan Avenue, Hubbard, Oregon 97032.
131. Nakagawa, P. Y., Chief, Vector Control Branch, Hawaii Department of Health, P.O. Box 3378, Honolulu, HI 96801.
132. Nelson, S. G., Associate Professor of Marine Sciences, University of Guam, UOG Station, Mangilao, GU 96913.
133. Philipp, A. L., Chief Fisheries Officer, Economic Development Department, Government of Western Samoa, Apia, Western Samoa.
134. Ryan, P. A., University of the South Pacific, P.O. Box 1168, Suva, Fiji.
135. Shima, S., Aquatic Biologist, Hawaii Division of Fish and Game, Department of Land and Natural Resources, 1151 Punchbowl Street, Honolulu, Hawaii 96813.
136. Taylor, L. R., Director, Waikiki Aquarium, 2777 Kalakaua Avenue, Honolulu, Hawaii 96815.
137. Wass, R. C., Fisheries Biologist, Office of Marine Resources, Government of American Samoa, Pago Pago, AS 96799.
138. Yamada, R., Department of Genetics, Biomedical Science Building, 1960 E-W Road, University of Hawaii, Honolulu, Hawaii 96822.

Addendum

While this manuscript was in press, I became aware of three additional references worthy of note. W. A. Bryan. [1915. Natural history of Hawaii. Hawaiian Gazette Co., Ltd., Honolulu.] discussed introduced Hawaiian freshwater fishes, indicating that the brown bullhead was stocked on Oahu Island early in the century. In context with the mention of that species in the foregoing text, the bullhead either persisted on Oahu by natural reproduction for a few decades, or was reintroduced subsequently. M. Laird. [1956. Studies of mosquitoes and freshwater ecology of the South Pacific. Bulletin Royal Society of New Zealand 6:1–288.] found the guppy (*Poecilia reticulata*) at two island groups, New Caledonia and Vanuatu, not previously noted (table 7–4). Finally, V. R. Williams and T. A. Clarke. [1984. Reproduction, growth, and other aspects of the biology of the gold spot herring, *Herklotsichthys quadrimaculatus* (Clupeidae), a recent introduction to Hawaii. Fishery Bulletin 81(3):587–597.] described the establishment of yet another exotic marine species in Hawaiian waters.

CHAPTER 8

Exotic Fishes in Puerto Rico

Donald S. Erdman

The tropical island of Puerto Rico lies between latitude 17°55′ and 18°31′ N and longitudes 65°37′ and 67°17′ W. Average annual rainfall varies from 100 cm along the semiarid south coast to nearly 500 cm in a mountainous rain forest (El Yunque, 1065 m). The highest recorded temperature is 39.4 C (San Lorenzo in 1911) and the lowest 4.4 C (Aibonito in 1911). Winter water temperatures at the government fish hatchery (Maricao) rarely fall below 18 C. There are some 1,200 streams (ríos) and rivulets (quebradas) in Puerto Rico.

Introductions of exotic fishes, made primarily to create freshwater fisheries, began in early 1900s following the construction of four reservoirs. The purpose of this chapter is to trace the history of introductions of exotic fishes in Puerto Rico and to summarize their status.

Native Fishes Inhabiting Inland Waters

There are no native primary freshwater fishes in Puerto Rico (Hildebrand 1935, Myers 1938, Erdman 1972). Three secondary freshwater species (*Poecilia vivipara, Fundulus fonticola,* and *Rivulus marmoratus*), interpreted by some authors as native, have been recorded from the island. However, Myers (1938), for distributional reasons, strongly suspected that *Poecilia vivipara* was introduced, and Miller (1955) showed the record of *Fundulus fonticola* to be erroneous. *Poecilia vivipara* is found predominantly in mangrove lagoons and rarely occurs in streams or rivulets any distance above sea level. Although *Rivulus marmoratus* in Puerto Rico also may be an introduction (Erdman 1972), it is possible that this is the only native secondary freshwater fish on the island (Rosen 1973). It lives in mangrove lagoons and breeds in brackish and fresh water. If *Poecilia vivipara* and *Rivulus marmoratus* are exotic to Puerto Rico, their source or times of introduction are unknown. Of the remaining peripheral freshwater species (Darlington 1957), by far the majority are primarily marine species that occur sporadically in fresh water. Exceptions to this are peripheral species that spend a greater and more predictable part of their life cycles in fresh waters. The degree to which these fishes have adapted to a freshwater existence is variable. In Puerto Rico some species that are extreme in this regard are the catadromous species: the American eel (*Anguilla rostrata*), probably the mountain mullet (*Agonostomus monticola*), the hognose mullet (*Joturus pichardi*), and gobies such as *Gobiomorus* and *Awaous*. *Sicydium plumieri* goby adults ("olivos") live and breed upstream. The larval stage ("seti") is passed at sea, after which postlarvae enter streams (Erdman 1961), metamorphose into juveniles, and migrate upstream.

More than 80 fish species have been found in Puerto Rican estuaries, but few of these ascend rivers beyond 10 m in altitude above mean sea level, and only three (*Agonostomus monticola, Sicydium plumieri,* and *Gobiomorus dormitator*) ascend above 80 m. Table 8-1 is an example of a survey of native fishes in a Puerto Rican river. Comprehensive treatments of native Puerto Rican freshwater fishes include those of Evermann and Marsh (1900), Nichols (1929), Hildebrand (1934, 1935), Erdman (1972), and Corujo-Flores (1980).

Between 1910 and 1914, four large dams were constructed to create reservoirs for sugar cane irrigation along the southern coast. Four reservoirs, Comerío, Guayabal, Patillas near the south coast, and Carite in the mountains, were formed. Only three native fishes succeeded in entering one or more of the lowland reservoirs, but none was able to establish reproducing populations within the reservoirs; these are the mountain mullet, the bigmouth sleeper (*Gobiomorus dormitator*), and the river goby (*Awaous tajasica*). No native fishes have been found in Carite Reservoir.

At present there are more than 28,000 surface hectares in twenty-two major reservoirs of Puerto Rico. These reservoirs are unsuitable habitat for native river fishes. Therefore, the planned introduction of suitable food and game fishes was believed warranted. This policy has also been practiced in Hawaii (Morita 1970) where problems with freshwater fisheries management are similar.

Introductions of Exotic Fishes

The first intentional releases were made by the U.S. Bureau of Fisheries into Carite and Comerío reservoirs in 1915 and 1916 (Johnson 1915, O'Malley 1916, Robbins and MacCrimmon 1974). The brown bullhead (*Icta-*

Table 8–1. NATIVE FISHES KNOWN TO OCCUR IN THE AÑASCO RIVER, WESTERN COAST OF PUERTO RICO

Species	English name	Spanish name	Lower 0-10m	Middle 10-80m	Upper 80m +
Elops saurus	ladyfish	macabí	+	–	–
Megalops atlanticus	tarpon	sábalo	+	–	–
Anguilla rostrata	American eel	anguila	+	+	–
Anchoa hepsetus	striped anchovy	bocúa	+	–	–
Strongylura timucu	timucu	agujón	+	–	–
Oostethus brachyurus	opossum pipefish		+	–	–
Centropomus ensiferus	swordspine snook	robalo	+	–	–
Centropomus parallelus	fat snook	robalo	+	+	–
Centropomus pectinatus	tarpon snook	robalo	+	–	–
Centropomus undecimalis	snook	robalo	+	–	–
Caranx latus	horse-eye jack	jurel ojón	+	+	–
Oligoplites saurus	leatherjacket	cueriduro	+	–	–
Trachinotus falcatus	permit	pámpano	+	–	–
Lutjanus apodus	schoolmaster	pargo rubio	+	–	–
Lutjanus griseus	gray snapper	pargo prieto	+	–	–
Lutjanus jocu	dog snapper	pargo colorado	+	–	–
Eucinostomus melanopterus	flagfin mojarra	muniama	+	–	–
Pomadasys crocro	burro grunt	viejo	+	+	–
Agonostomus monticola	mountain mullet	dajao	+	+	+
Joturus pichardi	hognose mullet	morón	+	+	–
Mugil curema	white mullet	jarea	+	+	–
Mugil liza	liza	lisa	+	+	–
Polydactylus virginicus	barbu	barbudo	+	–	–
Dormitator maculatus	fat sleeper	mapiro	+	–	–
Eleotris pisonis	spinycheek sleeper	morón	+	+	–
Gobiomorus dormitator	bigmouth sleeper	guavina	+	+	–
Awaous tajasica	river goby	saga	+	+	–
Evorthodus lyricus	lyre goby		+	–	–
Gobionellus boleosoma	darter goby		+	–	–
Gobionellus oceanicus	highfin goby		+	–	–
Sicydium plumieri	tri-tri	olivo	+	+	+
Citharichthys spilopterus	bay whiff		+	–	–
Trinectes inscriptus	scrawled sole		+	–	–
Sphoeroides spengleri	bandtail puffer	tamboril	+	–	–
Sphoeroides testudineus	checkered puffer	tamboril	+	–	–

lurus nebulosus), bluegill (*Lepomis macrochirus*), and largemouth bass (*Micropterus salmoides*) were released; only the first two species became established from those early stockings.

Hildebrand (1934) made the first extensive survey of fishes of the rivers and reservoirs for the specific purpose of making recommendations for the development of Puerto Rico's freshwater fisheries. He recommended introducing rainbow trout (*Salmo gairdneri*) but subsequent releases failed in becoming established. A Division of Ornithology and Fish Culture was established in the Puerto Rico Department of Agriculture soon after Dr. Hildebrand's survey.

Until 1973, when authority was transferred to the new Department of Natural Resources (Schulte 1974, Rivera-González 1979), the Department of Agriculture had responsibilities for managing freshwater sport fisheries (including making introductions) and the fish hatchery. From 1954 through 1972, funding for sport fishery research and development was augmented through the Dingell-Johnson Federal Aid in Fish Restoration Program. During that period, sport and forage fishes were successfully introduced and tilapias were imported for aquatic weed control and experimental aquaculture.

A more detailed history of exotic fishes introduced into Puerto Rico, arranged in phylogenetic sequence, follows and is summarized in table 8-2.

CLUPEIDAE – HERRINGS

***Dorosoma petenense* (Günther) threadfin shad**

On 30 May 1963, 40 adult threadfin shad in spawning condition were stocked in Guajataca reservoir. They were flown from Atlanta, Georgia, to Ramey Field. By February 1964 shad fry and juveniles were found in the reservoir. Since then, this species has been stocked successfully in Cidra, Carite, Garzas, and La Plata reservoirs. Most threadfin shad average about 50 mm in length but are recorded to 184 mm. Introduced as a forage fish for largemouth bass, the bass had higher condition factors following the introduction.

SALMONIDAE – TROUTS

***Salmo gairdneri* Richardson rainbow trout**

Following the recommendations of Hildebrand (1934), 10,000 fingerlings and 10,000 eggs of rainbow trout were imported in 1934. The eggs were hatched at a field station at El Yunque and fingerlings were raised to 75 to 150 mm in length. In 1938 a trout hatchery was constructed at Maricao and trout were cultured until 1942. In 1940, 6,775 trout were stocked into inland waters from Maricao; 5,262 were stocked in fiscal year 1942. There has been no evidence of natural reproduction of rainbow trout in Puerto Rico.

***Salmo trutta* Linnaeus brown trout**

Bonnet (1941), Iñigo (1949), and MacCrimmon and Marshall (1968) record the importation of brown trout into Puerto Rico around 1934. Introductions were made in Río Espiritu Santo at the El Yunque Forest. These and subsequent releases failed to establish, and stocking ceased after 1942 (MacCrimmon and Marshall 1968).

CYPRINIDAE – CARPS AND MINNOWS

Barbus conchonius (Hamilton-Buchanan) rosy barb

The rosy barb, apparently released from aquaria, was introduced in Río Arroyata, a tributary of the Río La Plata, between Cidra and Naranjito near route 172, before 1971. On 18 March 1971, I saw large schools, composed of hundreds of individuals, swimming upstream. Included were several bright red males in nuptial coloring. On 30 March 1971, I returned with the Maricao hatchery staff to collect individuals for culture purposes. By spring 1972 they spawned at Maricao. Since then they have been stocked in Loiza reservoir where they became established. Adults reach about 50 mm in length in Puerto Rican waters.

***Carassius auratus* (Linnaeus) goldfish**

Nichols (1929) reported an abundant population of goldfish at Isabela and the stocking of a small pond in the hills of Guayama. Hildebrand (1934) collected goldfish in Guayabal and Guajataca reservoirs where a few adults may still occur.

Although goldfish have spawned readily in concrete tanks with running water at Maricao, this fish does not appear to maintain populations in the wild in Puerto Rico.

***Ctenopharyngodon idella* (Valenciennes) grass carp**

Puerto Rico received its first shipment (200 fingerlings) of grass carp from Lonoke, Arkansas, on 30 September 1972. They were stocked in golf course ponds at the Dorado Beach Hotel to control heavy growths of southern naiad (*Najas guadaloupensis*). Some grew at a rate of 0.45 kg per month during the first year. Weights of 11 grass carp, seined from the ponds on 10 April 1973, ranged from 4.5 to 5.3 kg and the southern naiad was gone from the ponds.

On 12 October 1974, the Carib King Shrimp Company imported 100 grass carp fingerlings to help control southern naiad in their ponds at Cabo Rojo. Although the fish grew rapidly, too few were stocked to achieve weed control.

***Cyprinus carpio* Linnaeus common carp**

In 1953 several common carp (Israeli mirror carp variety) arrived at the Maricao Fish Hatchery from the Dominican Republic. They lived and grew well, but were eliminated after two years because they were not considered useful for the fisheries program.

***Hypophthalmichthys molitrix* (Valenciennes) silver carp**

Two silver carp apparently entered Puerto Rico as fingerlings mixed with the grass carp received from Lonoke, Arkansas, in September 1972. Each specimen removed from the Dorado Beach Hotel golf course ponds in April 1973 weighed 1.0 kg.

***Pimephales promelas* Rafinesque fathead minnow**

On 24 June 1957, 150 fathead minnows, about 50 mm in length, were received by air from the U.S. Fish and Wildlife Service hatchery at Welaka, Florida. From 1957 until September 1975, when floods from hurricane Eloise destroyed the stocks, culture of fathead minnows was conducted at Maricao. At least four reservoirs and several farm ponds were stocked to provide forage for largemouth bass. The fathead minnow failed to establish, perhaps because of its extreme vulnerability to such predators as largemouth bass.

ICTALURIDAE – BULLHEAD CATFISHES

***Ictalurus catus* (Linnaeus) white catfish**

How white catfish got to Puerto Rico is a mystery, but they may have accompanied a shipment of channel catfish from Baltimore, Maryland. This fish is established in Guajataca, Caonillas, Dos Bocas, Guayabal, Garzas reservoirs (Erdman 1972) and in Matrullas reservoir (Ortiz-Carrasquillo 1980, 1981). The largest specimen I have seen weighed 5 kg and was caught at Dos Bocas reservoir on 26 August 1958.

***Ictalurus nebulosus* (Lesueur) brown bullhead**

Some biologists recognize two subspecies: the brown bullhead (*I. n. nebulosus)* from the northern United States and the marbled bullhead (*I. n. marmoratus*) from the southeastern United States. While these may only be color or geographical variants (Smith 1979) and are so treated herein, both are present in Puerto Rico.

The U.S. Bureau of Fisheries introduced the brown variety into Carite reservoir in 1915 and 1916. This variety remains limited to reservoirs in the east-central and southeastern part of the island in Carite, Comerío, Melanía, and Patillas reservoirs and the central Río La Plata; it is probable that this fish occurs in the La Plata reservoir to the north and downstream from there.

Hildebrand (1935) reported the introduction of the black bullhead (*Ictalurus melas*) in 1914, an apparent misidentification of the brown bullhead released in 1915.

The marbled variety was introduced in 1946 in western parts of the island. It occurs in Caonillas, Dos Bocas, Guajataca, Loco, and Luchetti (Yauco) reservoirs, Lajas Valley drainage canals, Río Hondo of the Guanajibo drainage and Caño Tiburones (Barceloneta), and probably elsewhere.

The marbled variety averages consistently larger than the brown variety in Puerto Rico (0.45 to 1.03 kg).

***Ictalurus punctatus* (Rafinesque) channel catfish**

Channel catfish were first introduced by the U.S. Bureau of Fisheries in 1938, and later to Cidra, Dos Bocas, and Loiza reservoirs where they became established and provided a useful fishery. Nevertheless, this species has a tendency to overpopulate in these reservoirs.

CYPRINODONTIDAE – KILLIFISHES

***Fundulus diaphanus* (Lesueur) banded killifish**

One banded killifish, about 51 mm in length, was discovered in a shipment of other fishes from the U.S. Fish and Wildlife Service hatchery in Welaka, Florida, in June 1957 to the Maricao Fish Hatchery. It grew to 153 mm and was not seen after 1958. I mention this species because it became so well adapted and because of its bait fish potential. It may be expected to be introduced and become established in the future.

POECILIIDAE – LIVEBEARERS

***Gambusia affinis* (Baird and Girard) mosquitofish**
Hildebrand (1934) recorded mosquitofish as introduced for mosquito control about 1923. He found it in Patillas reservoir and it also occurs in Melanía reservoir near Guayama. This species has not spread much since 1935, although it is often confused with *Poecilia vivipara* which may account for some inconsistencies in distributional records.

***Poecilia reticulata* (Peters) guppy**
Erdman (1947) first reported guppies from Puerto Rico as abundant in Adjuntas, Aibonito, Comerío, and Cayey. This species now occurs throughout the island. Oliver-González (1946) reported guppy predation on schistosome cercariae. This species is now found in streams from sea level to 700 m, and Ortiz-Carrasquillo (1981) records guppies from Río Matrullas to 736 m.

***Xiphophorus helleri* Heckel green swordtail**
Erdman (1947) reported this fish from Quebrada Honda, just above its junction with Río La Plata near Aibonito. Since then, I have found this species in the Arecibo drainage near Utuado.

***Xiphophorus maculatus* (Günther) southern platyfish**
Erdman (1947) reported the southern platyfish from Quebrada Honda at its junction with Río La Plata at La Plata, 11 km northwest of Aibonito. Since then, I have found this fish in the Loiza drainage near Loiza reservoir and at Río Abajo Forest Station north of Utuado.

PERCICHTHYIDAE – TEMPERATE BASSES

***Morone chrysops* (Rafinesque) white bass**
On 17 April 1972, 78 adult white bass were received by air from Atlanta, Georgia. Only 12 arrived alive and were in weakened condition. These survivors were introduced in Loiza reservoir above Carraizo dam that evening but apparently died. A possible future introduction of fingerlings of this species may result in establishment.

CENTRARCHIDAE – SUNFISHES

***Ambloplites rupestris* (Rafinesque) rock bass**
See entry for *Lepomis gulosus*.

***Lepomis auritus* (Linnaeus) redbreast sunfish**
The redbreast sunfish was received in a shipment of fishes from the U.S. Fish and Wildlife Service hatchery in Welaka, Florida, in June 1957. The stock was caught in the wild in the upper St. Johns River and trip survival was nearly 100 percent. The species adapted immediately to hatchery ponds at

Maricao and spawned before year's end. It spawns readily at the hatchery at any time of year, has proven to be the easiest sunfish to culture, and survives well when transported by truck to reservoirs. Ortiz-Carrasquillo (1980) reported successful stockings of this fish in Coamo, Dos Bocas, Garzas, Guajataca, Guineo, Loco, and Matrullas reservoirs. It was also stocked in Río Jayuya where it spawned. It occurs in the upstream portion of Río Manatí. It is established in the Río Maricao near the Fish Hatchery and farther downstream in the Río Guanajibo drainage.

***Lepomis gulosus* (Cuvier) warmouth**

According to U.S. Bureau of Fisheries stocking records in O'Malley (1916) and as published in the "Informe del Comisionado de Agricultura, 1934–1935," 1,200 rock bass, *Ambloplites rupestris,* were included in the shipment of bass, bluegills, and bullheads to Puerto Rico. To date no *A. rupestris* have been found. On August 16, 1971, however, at Carite reservoir, Fernando Villa, superintendent of the Maricao Fish Hatchery and staff, seined one 121 mm TL warmouth along with largemouth bass, bluegills, and bullheads. The warmouth was caught under "poma rosa" bushes and branches along shore with many bluegills. Since warmouth are sometimes confused with rock bass, it is possible that this fish was a survivor of a breeding population descended from the 1916 stocking.

***Lepomis macrochirus* Rafinesque bluegill**

The bluegill is probably the most common species of *Lepomis* in Puerto Rico since its first introduction into Carite and Comerío reservoirs in 1915. It now occurs island-wide in reservoirs, farm ponds, and a few rivers such as the La Plata at Aibonito to Comerío where currents are not swift. In 1957, this species was abundant in Cartagena Lagoon but has since been replaced by introduced *Tilapia mossambica.*

***Lepomis microlophus* (Günther) redear sunfish**

The redear sunfish has been a valuable introduction to the fresh waters of Puerto Rico. It has proven sport and food qualities but also feeds on snails including the pulmonate *Biomphalaria glabrata,* common in farm ponds and an intermediate host for the blood fluke, *Schistosoma mansoni.*

The redear sunfish was introduced in Loiza reservoir from stocks obtained from the U.S. Fish and Wildlife Service hatchery at Welaka, Florida. It has been established in Garzas and Guajataca reservoirs since before 1948, but stocking records, dates, and sources have not been located.

Redear sunfish, stocked with largemouth bass, have been found to be the best sportfish combination in Puerto Rico farm ponds.

***Micropterus coosae* Hubbs and Bailey redeye bass**

This fish was introduced successfully in 1958 and has been cultured at the Maricao Fish Hatchery since then. It is more suited to rivers than the largemouth bass and was introduced to improve sportfishing in rivers. There is a small wild population established in the Río Maricao near the hatchery. In 1963, redeye bass and redbreast sunfish were successfully stocked in Río Jayuya, upstream from the town of Jayuya. On 1 August 1964, I observed fingerling redeye bass and redbreast sunfish in El Cantil, a rock-sided pool upstream from Jayuya. Redeye bass have been reported from Caonillas reservoir, downstream from Jayuya.

***Micropterus salmoides* (Lacepède) largemouth bass**

Attempts by the U.S. Bureau of Fisheries to successfully introduce the largemouth bass to newly created reservoirs in 1915 and 1916 failed. In 1946, the U.S. Fish and Wildlife Service provided Félix Iñigo, superintendent of the Maricao Fish Hatchery, with 2,170 fingerlings and adults for transfer by ship to Puerto Rico; these bass became the basis of a successful reservoir sport fishery in following years.

The largemouth bass now occurs in all major Puerto Rican reservoirs and has been stocked island-wide in farm ponds. This fish has been the most successful predator in reservoirs. Largemouth bass from Comerío reservoir have become established in Río La Plata upstream as far as Cayey where stream gradients are not very steep. Introductions to the Río Añasco, Cartagena, and Tortuguero lagoons failed, probably because of unfavorable ecological conditions.

Additional references on introductions of largemouth bass are those of Meehean et al. (1948), Iñigo (1949), Soler (1951), Bird (1960), Schulte (1974), and Rivera-González (1979).

***Pomoxis annularis* Rafinesque white crappie**

More than 100 white crappie were air-shipped to Puerto Rico from the U.S. Fish and Wildlife Service hatchery at Welaka, Florida, in June 1957. All died in transit or shortly after arrival, before they could be cultured or released.

CICHLIDAE – CICHLIDS

***Astronotus ocellatus* Agassiz oscar**

The oscar is a popular aquarium fish in Puerto Rico. In March 1971, I examined a 0.3-hectare farm pond at Aibonito that contained oscars, some up to 0.2 kg in weight. I am unaware of any reproducing populations on the island.

***Cichla ocellaris* Schneider peacock bass**

As a courtesy of the government of Colombia, on 27 January 1967, a shipment of 200 peacock bass fingerlings (64–76 mm in length) was received at San Juan by air from the fish culture station at Buga. About 25 percent arrived alive and 30 fish survived the four-hour trip by truck to Maricao. Lengths of 22 survivors on 30 March were about 102 mm and on 3 July ranged from 152 to 203 mm. In May 1968, 17 fish had grown to adults of an average 305 mm in length. The first spawning occurred in late May but fry disappeared after 9 June. The first successful spawning took place in early August.

Although successful spawning populations could be established on occasion, they did not maintain themselves over time in farm ponds and this stocking program was discontinued. Better success was achieved in reservoirs but production did not meet our expectations. Large populations became established in a year or so at Toa Vaca and La Plata reservoirs but were greatly reduced by uncontrolled fishing.

***Tilapia aurea* (Steindachner) blue tilapia**

Blue tilapia were introduced in 1971 from Auburn, Alabama, for experimental aquaculture in ponds as a potential food fish (Pagan-Font 1973).

***Tilapia hornorum* Trewavas Wami tilapia**

The Wami tilapia was introduced in November 1963 from stocks obtained from the Cooperative Fishery Research Unit in Tucson, Arizona. Males of this species were crossed with female Mozambique tilapia (*Tilapia mossambica*) to produce hybrid males. The experiment succeeded at Maricao and all-male hybrids were raised in several island ponds.

To insure that the stocks of Wami tilapia would be kept pure, I stocked some 100 fingerlings in a small (0.04 hectare) pond on Mona Island. Descendants of this stock are still reproducing there. Because of the small volume of water, none of the fish grow longer than 67 mm and they can reproduce at this or even smaller sizes.

***Tilapia mossambica* (Peters) Mozambique tilapia**

The Mozambique tilapia was introduced to island waters on 11 July 1958 as recommended by the secretary of agriculture of Puerto Rico. Stocks came from Auburn, Alabama. The primary purpose of the introduction was control of algae in canals for sugar cane irrigation. This tilapia has the widest distribution on the island.

***Tilapia nilotica* (Hasselquist) Nile tilapia**

The Nile tilapia was introduced from Brazil in 1973 for experimental purposes (production of all-male hybrids) by the University of Puerto Rico.

***Tilapia rendalli* (Boulenger) redbreast tilapia**

On 27 July 1963, 19 juvenile *T. rendalli,* each about 51 mm long, were received from Auburn, Alabama. They were introduced to control certain rooted macrophytes such as southern naiad. This tilapia has also been observed to be an efficient snail predator in Puerto Rico.

In 1966, a serious problem with blue-green algae occurred in Loiza reservoir near Carraizo dam. The algae were being treated at great expense with

Table 8–2. SUMMARY LIST OF FISHES INTRODUCED IN PUERTO RICO

Species	Family	Year	Status	Impact
Dorosoma petenense	Clupeidae	1963	in reservoirs	beneficial
Salmo gairdneri	Salmonidae	1964	failed	
Salmo trutta	Salmonidae	1934	failed	
Barbus conchonius	Cyprinidae	1970?	La Plata and Loiza rivers	beneficial
Carassius auratus	Cyprinidae	1914	low rate of reproduction	no impact
Ctenopharyngodon idella	Cyprinidae	1972	no reproduction	beneficial
Cyprinus carpio	Cyprinidae	1953	no longer present	
Hypophthalmichthys molitrix	Cyprinidae	1972	no reproduction	beneficial
Pimephales promelas	Cyprinidae	1957	low reproduction in the wild	no impact
Ictalurus catus	Ictaluridae	1938	in reservoirs	beneficial
Ictalurus nebulosus	Ictaluridae	1915 and 1946	brown variety in reservoirs and La Plata River; marbled variety in reservoirs and rivers	beneficial
Ictalurus punctatus	Ictaluridae	1938	in six reservoirs	beneficial
Fundulus diaphanus	Cyprinodontidae	1957	one lived one year at Maricao	
Gambusia affinis	Poeciliidae	1923	Patillas and Melania reservoirs	beneficial
Poecilia reticulata	Poeciliidae	1940?	islandwide	beneficial
Xiphophorus helleri	Poeciliidae	1940	La Plata River and Utuado	beneficial
Xiphophorus maculatus	Poeciliidae	1940	La Plata and Loiza reservoirs	beneficial
Morone chrysops	Percichthyidae	1972	stocked in Loiza; failed	
Lepomis auritus	Centrarchidae	1957	certain rivers	beneficial
Lepomis gulosus	Centrarchidae	1915	Carite reservoir	beneficial
Lepomis macrochirus	Centrarchidae	1915	rivers, reservoirs, and ponds	beneficial
Lepomis microlophus	Centrarchidae	1957	rivers, reservoirs, and ponds	beneficial
Micropterus coosae	Centrarchidae	1958	Jayuya and Maricao rivers	beneficial
Micropterus salmoides	Centrarchidae	1946	reservoirs, ponds, and La Plata River	beneficial
Pomoxis annularis	Centrarchidae	1957	failed	
Astronotus ocellatus	Cichlidae	1970?	Aibonito farm pond	beneficial
Cichla ocellaris	Cichlidae	1967	certain reservoirs	beneficial
Tilapia aurea	Cichlidae	1971	experimental ponds	controversial
Tilapia hornorum	Cichlidae	1963	Mona Island pond near airport	controversial
Tilapia mossambica	Cichlidae	1958	islandwide	controversial
Tilapia nilotica	Cichlidae	1974	experimental ponds	controversial
Tilapia rendalli	Cichlidae	1963	less common than *T. mossambica*	controversial

copper sulfate. In April 1966, *T. rendalli* and Mozambique tilapia were released in the reservoir and shortly thereafter the algal problem disappeared.

Periodically, especially during flood periods, a few hundred kg of tilapia are fished for sale in the market at Río Piedras and elsewhere. *T. rendalli,* although less abundant than Mozambique tilapia, is easier to catch by hook and line and, therefore, appears in this fishery.

Impact of Foreign Fishes on Native Fishes

With the possible exception of tilapias, there is little evidence of harmful impacts of exotic fishes on native Puerto Rican fishes. The primary reason is that most of the approximately 70 species found in inland waters are marine spawners. Populations of the few native fishes that reproduce in fresh or brackish waters have declined in numbers mainly because of reduced minimum river flows created by agricultural and industrial development rather than competition with exotic fishes.

While tilapias have become abundant in reservoirs, they are less numerous in rivers and coastal lagoons. Centrarchids and ictalurids have low populations outside of reservoirs and rivers entering reservoirs. Repeated stockings of largemouth bass into the Río Añasco failed despite the fact that this fish is in the La Plata and Loiza rivers at considerable distance above dams.

Marine predators such as ladyfish (*Elops saurus*), tarpon (*Megalops atlanticus*), jacks (Carangidae), and snooks (*Centropomus* spp.) keep the tilapias reduced in numbers in the rivers and coastal lagoons.

Conclusion

The future of freshwater fisheries in Puerto Rico will depend on strict pollution control. Pesticides and heavy metals have been found in concentration in some freshwater fishes. Moreover, agricultural and industrial development must occur in such a manner as to prevent further depletion of river flows.

With respect to reservoirs, one or more predatory fishes that are open water spawners, such as the white bass, snakehead (*Channa striata*), and tigerfish (*Hydrocyon* sp.) are needed to help control tilapia populations. Tilapia numbers also can be reduced by using large cast nets and fish pots. Nest builders such as largemouth bass and peacock bass are at a disadvantage during large reservoir drawdowns in the dry season.

With proper management and protection of water quality, freshwater fishes will continue to be a valuable resource for the people of Puerto Rico.

Acknowledgments

I gratefully acknowledge the support of the Fishery Research Laboratory of CODREMAR (Corporation for the Development and Administration of the Marine, Lacustrine and Fluvial Resources of Puerto Rico), and the Department of Natural Resources of Puerto Rico, of which CODREMAR is a branch agency.

The Fishery Research Laboratory is supported by the government of Puerto Rico with additional funding by the Fisheries Research and Development Act of 1964 (PL 88-309). The Department of Natural Resources administers the Sport Fisheries Research and Restoration Act of 1950 (popularly known as the Dingell-Johnson Act).

I have been associated with these two agencies since their inception, in 1973 for DNR and 1979 for CODREMAR. From 1954 through 1972, I was fishery biologist for the D-J program under the Fish and Wildlife Division of the Department of Agriculture of Puerto Rico, and assistant director of the fish lab from 1974 to 1979.

For review of the manuscript, I am indebted to Félix Iñigo, former chief of the Fish and Wildlife Division and later assistant secretary of agriculture, to Dannie A. Hensley of the Department of Marine Sciences, University of Puerto Rico, and Harvey R. Bullis, Jr., formerly with the National Marine Fisheries Service.

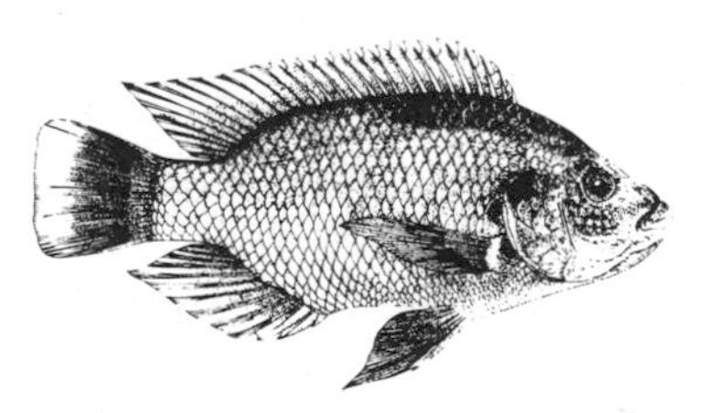

Literature Cited

Bird, E. A. 1960. Fishing off Puerto Rico. A. S. Barnes & Co., New York. 111 pp.

Bonnet, L. C. 1941. Desarrollo y planes para la pesca de agua dulce en Puerto Rico. Revista de Agricultura, Industría y Comercío, Departamento de Agricultura Puerto Rico 4(3):339–346.

Corujo-Flores, I. N. 1980. A study of fish populations in the Espiritu Santo River estuary. M.S. thesis, University of Puerto Rico. 87 pp.

Darlington, P. J. 1957. Zoogeography: The geographical distribution of animals. John Wiley and Sons, Inc., New York. 675 pp.

Erdman, D. S. 1947. Aquarium fishes in Puerto Rico. Revista de Agricultura 28(1): 90–91.

Erdman, D. S. 1961. Notes on the biology of the gobiid fish *Sicydium plumieri* in Puerto Rico. Bulletin of Marine Science Gulf & Caribbean 11(3):448–456.

Erdman, D. S. 1972. Inland game fishes of Puerto Rico. Departamento de Agricultura de Puerto Rico, San Juan 4(2):1–96.

Evermann, B. W., and M. C. Marsh. 1900. The fishes of Porto Rico. Bulletin of the U.S. Fish Commission 1900:49–350.

Hildebrand, S. F. 1934. An investigation of the fishes and fish cultural possibilities of the fresh waters of Puerto Rico, with recommendations. Typewritten report to Commissioner of Agriculture and Commerce, Puerto Rico. 38 pp.

Hildebrand, S. F. 1935. An annotated list of fishes of the freshwaters of Puerto Rico. Copeia 49–56.

Iñigo, F. 1949. Introduction of North American fishes into Puerto Rico. Revista de Agricultura de Puerto Rico 40(1):107.

Johnson, R. S. 1915. The distribution of fish and fish eggs during the fiscal year 1915. Report of the U.S. Bureau of Fisheries 1915. 138 pp.

MacCrimmon, H. R., and T. L. Marshall. 1968. World distribution of brown trout, *Salmo trutta.* Journal of the Fisheries Research Board of Canada 25(12):2527–2548.

Meehean, O. L., E. J. Douglass, and L. M. Duncan. 1948. Propagation and distribution of food fishes for the calendar years 1944–1948. U.S. Fish and Wildlife Service Statistical Digest 24. 82 pp.

Miller, R. R. 1955. An annotated list of the American cyprinodontid fishes of the genus *Fundulus,* with the description of *Fundulus persimilis* from Yucatán. Occasional Papers of the Museum of Zoology, University of Michigan 568:125.

Morita, C. M. 1970. Freshwater fishing in Hawaii. Hawaii Division of Fish & Game, Honolulu. 20 pp.

Myers, G. S. 1938. Fresh-water fishes and West Indian zoogeography. Annual Report of the Smithsonian Institution 1937:339–364.

Nichols, J. T. 1929. The fishes of Porto Rico and the Virgin Islands. Scientific Survey of Porto Rico and the Virgin Islands. New York Academy of Sciences 10(2):161–295.

Oliver-Gonzalez, J. 1946. The possible role of the guppy, *Lebistes reticulatus,* on the biological control of *Schistosoma mansoni.* Science 104:605.

O'Malley, H. 1916. The distribution of fish and fish eggs during the fiscal year 1916. Report of the U.S. Bureau of Fisheries 1916. 111 pp.

Ortiz-Carrasquillo, W. 1980. Resumen historico de la introducción de los peces de agua dulce en los lagos artificiales de Puerto Rico desde 1915 hasta 1975. Science-Ciencia 7(3):95–107.

Ortiz-Carrasquillo, W. 1981. Notas sobre los crustaceos y peces del Rio Matrullas, Orocovis, P.R. Science-Ciencia 8(1):9–13.

Pagan-Font, F. A. 1973. Potential for cage culture of the cichlid fish *Tilapia aurea.* Tenth meeting, Association of Island Marine Laboratories of the Caribbean 1973. 52 pp.

Rivera-González, J. E. 1979. Estudio de las poblaciónes pisicolas en los lagos Loiza y Guajataca. Tercer Simposio del Departamento de Recursos Naturales, 1976:1–10.
Robbins, W. H., and H. R. MacCrimmon. 1974. The black bass in America and overseas. Biomanagement Enterprises, Sault Sainte Marie, Ontario, Canada. 196 pp.
Rosen, D. E. 1973. Suborder Cyprinoidei. *In:* Fishes of the western North Atlantic. Part 6: Memoirs of the Sears Foundation for Marine Research 1:229–262.
Schulte, T. S. 1974. An initial analysis of the fishery populations of five Puerto Rican reservoirs. First Symposium, Department of Natural Resources of Puerto Rico, San Juan.
Smith, P. W. 1979. The fishes of Illinois. University of Illinois Press, Urbana, Illinois. 314 pp.
Soler, P. J. 1951. A bibliography on the fishes and fisheries of Puerto Rico. Proceedings of the Gulf & Caribbean Fisheries Institute 3:143–149.

CHAPTER 9

Introductions of Exotic Fishes in Australia

Roland J. McKay

Australia, largely because of its position, size, and topography, is, with the exception of Antarctica, the driest of the continents; about 93 percent of the land mass receives less than 510 mm and 37 percent receives less than 250 mm of rainfall per year. Evaporation in inland Australia is exceedingly high and, therefore, permanent bodies of standing water – lakes, lagoons, and ponds – are rare; much of this area is characterized by ephemeral rivers draining into salt lakes, and vast areas of uncoordinated drainage. The area of exceedingly low rainfall occupies a major portion of South Australia, Western Australia, and the Northern Territory; the only areas receiving fairly reliable rainfall are the southwestern portion of Western Australia and the southern area of South Australia during the winter, and the northern part of the continent during monsoonal summer rains. The eastern coastal part of Australia is more reliably watered but is subject to long and severe droughts and widespread flooding. The only river system of any magnitude, the Murray River with a total catchment of about 1 million sq km, has a low and unreliable runoff. Rivers such as the Darling and Lachlan may dry up to a chain of isolated waterholes during drought periods while smaller rivers may be reduced to ribbons of sand for hundreds of kilometers. The sparseness of the freshwater system and its unreliability has resulted in the development of a depauperate freshwater fish fauna. Australia has some 145 species of fishes that complete their life cycle in fresh water and a large number of marine and estuarine species that commonly enter fresh water for prolonged periods.

Many of the major groups of freshwater fishes found on other continents are absent from Australia, New Zealand, and Papua New Guinea. There are no native cyprinids, mormyrids, channids, cyprinodonts, anabantids, mastacembelids, cichlids, and salmonids, and apart from plotosids and ariids derived from marine ancestors, no siluriform fishes. The absence of the dominant groups of primary freshwater fishes from Australia is not easily explained. The only "primary" freshwater fishes in Australia are the lungfish (*Neoceratodus forsteri*), the saratogas (*Scleropages leichardti* and *S. jardini*), the enigmatic *Lepidogalaxias salamandroides,* and possibly the Galaxiidae and Retropinnidae. All are regarded as "primitive" fishes. The lungfishes were widely distributed throughout Pangaea during the Devonian period and persisted as the Lepidosirenidae in Africa and South America and the Ceratodidae in Australia. *Neoceratodus* had marine relatives in *Paraceratodus* and *Ceratodus* and, therefore, can be regarded as telolimnic rather than primary or archaeolimnic (Patterson 1972), despite its long history in fresh water in Australia. *Scleropages* is an osteoglossid genus with an ancient lineage, but the Osteoglossomorpha may also be a teleolimnic group and, thus, have probably achieved their distribution by means of seaways (Patterson 1972). The small southern Western Australian *Lepidogalaxias* is an archaeolimnic relict with a restricted range. Its closest relatives are possibly the Northern Hemisphere esocoids (Rosen 1974), and therefore the family Lepidogalaxiidae, now represented by one species, has survived in Australia since the breakup of Pangaea some 180 million years ago. Rosen (1974) has drawn attention to the possibility that the southern galaxiids and retropinnids are Gondwanian and may, therefore, be at least 90 million years old. The older Australian freshwater groups have ancient origins and have persisted with surprisingly little modification.

In addition to their conservatism, many are adapted to a harsh environment and irregular droughts. The lungfish is a hardy fish capable of surviving low oxygen tension, especially at night when it is most active in areas of aquatic vegetation in rivers and isolated waterholes; their diet is omnivorous, and worms, crustacea, fish, tadpoles, frogs, and large quantities of aquatic vegetation are eaten. *Lepidogalaxias* can survive almost dry conditions (McDowall 1978), and some members of the Galaxiidae are drought tolerant (*Galaxiella, Galaxias, Neochanna*) and may aestivate (McDowall 1970, 1978, 1980). The larvae and juveniles of some galaxiid species have been found up to 700 km from land (McDowall et al. 1975), and the retropinnids and aplochitonids have larvae that go to sea. The saratoga *Scleropages jardini* may survive salinities of up to about 14 ppt (Roberts 1978) and may, therefore, survive drought conditions by moving into more estuarine waters.

The antiquity of the older elements of the freshwater fish fauna indicate that before the breakup of Pangaea the freshwater fish fauna was more or less cosmopolitan, but may have been divided into three zones—the Boreal Salmoniform Zone, the Ostariophysan Zone, and the Austral Salmoniform Zone—as Rosen (1974) has suggested. It appears that the ostariophysan fishes were confined to West Gondwana, the characins to Africa–South

America, the siluroids to Africa–South America and later to Asia, the cyprinoids to Asia and later to Africa after its separation. The earliest ostariophysans were probably in tropical Pangaea but did not have a wide distribution in Gondwana as they are absent from Australasia and Madagascar. A barrier to the movement of freshwater fishes from West Gondwana to East Gondwana, either temperature or a physical one such as a transantarctic mountain range, probably existed. When the Australian plate moved away from Antarctica, some 50 million years ago, the older elements of the Australian freshwater fish fauna, including *Percalates* and *Maccullochella,* were present (Hills 1934). With the exception of some uplifting in southeast Australia during the Eocene, the continent had a low well-watered landscape with a fairly stable mild and humid climate. Seasonal fluctuations were minimal and rainforest grew over vast areas (Gill 1975).

Widespread climatic changes commenced in early Pleistocene times and major faunal extinctions occurred. At about 44,000 years BP both global and local evidence suggests a substantial change in temperature and hydrological regimes. In semiarid southern Australia before 44,000 BP lakes were dry, dunes were stable, while the rivers of the Murray-Darling system provided the arterial waterways as they do today. Then followed a period of increased floods with water to the dry inland and the formation of permanent lakes. In the Willandra system, Murray cod and golden perch entered the lakes. By 32,000 BP the lakes reached their maximum stands and by about 26,000 BP commenced to disappear or become ephemeral. By about 22,000 BP, an accelerated trend toward aridity resulted in the drying of lakes and by 20,000 BP increased evaporation and low precipitation resulted in a phase of extreme aridity. By 15,000 BP, a slight amelioration in climate occurred and by 13,000 BP with some minor fluctuations the landscape and the hydrologic systems had assumed its main physical characters that have persisted until the present day. Of the various Quaternary glacial episodes, that from which we have emerged within the last 20,000 years was perhaps the most severe (Bowler 1976).

The period of extreme aridity, coupled with periods of marine transgressions during the Eocene and Miocene, and areas of southern glaciation in Tasmania and Victoria no doubt resulted in the widespread extinction of the freshwater fish fauna and the substantial reduction in range of the surviving species. With the retreat of the glaciers and increased precipitation, a reinvasion of the lakes and river systems occurred. The Tasmanian lakes system was occupied by *Galaxias* spp. while the mainland was colonized by marine species such as the clupeoids, ariids, plotosids, belonids, melanotaenids, atherinids, centropomids, teraponids, kuhliids, apogonids, toxotids, mugilids, a large number of gobiids, and a few soles. It appears that this invasion of the fresh waters of Australia and the adaptive radiation of recent colonizers is still in progress today. Many of the Australian freshwater fishes, although fairly well adapted to the harsh environment and pronounced seasonal fluctuations, show little evidence of specialization, and lack of diversity of species characteristic of many other continents.

Thus, the Australian freshwater fish fauna can be considered "permissive." Courtenay et al. (1974) maintain that "permissive" habitats (those not fully occupied with regard to trophic levels and species niche concepts) and weakened environments (those altered naturally or by man), are particularly susceptible to establishment of nonindigenous exotic species. The paucity of Australian freshwater fish species may, therefore, increase the vulnerability of our native freshwater fishes to domination, reduction, and perhaps complete eradication by some of the exotic fishes that have evolved in direct competition or association with the highly diversified fish communities of other continents (McKay 1977).

Freshwater Fishes Introduced into Australia

Approximately 22 species of exotic freshwater fishes have been successfully introduced into the Australian environment (table 9–1) and 20 exotic fishes introduced to Papua New Guinea (West and Glucksman 1976).

Introductions to Australia can be separated into three main phases: the earliest was the period of the acclimatization societies from 1862 to 1896, the second the liberation of *Gambusia* during the Second World War, and the latest the spread of European carp and the release of aquarium fish species. Weatherley and Lake (1967) have given a comprehensive account of freshwater fishes introduced to southern Australia, and the native and exotic freshwater fishes of Australia have been reviewed by Lake (1971, 1978) and those from southeastern Australia by McDowall (1980).

The first species to be introduced into Australian waters was the European perch (*Perca fluviatilis*) to Tasmania in 1862. Another shipment from England arrived in Victoria in 1868, and in 1888 some of these fish were sent to New South Wales. Perch are now widespread in sluggish weedy waters in the southeast on both sides of the Great Dividing Range, in central and southern New South Wales, Victoria, South Australia, and Tasmania. The distribution of European perch in Australia has been fully treated in a zoogeographic study by Weatherley (1963) and Weatherley and Lake (1967).

In 1864, brown trout (*Salmo trutta*) were imported as fertilized ova after four unsuccessful attempts. Eggs from this stock were distributed to mainland Australia and New Zealand (Roughley 1951). An attempt to introduce Atlantic salmon (*Salmo salar*) to Tasmania failed in 1866 and was tried again unsuccessfully in 1884. In 1876, European carp (*Cyprinus carpio*) were liberated in Victoria and in 1888 in Queensland. The history of the European carp in Victoria has been outlined by Wharton (1971, 1979) and Shearer and Mulley (1978). Three strains of *C. carpio* are present in Australia, the "Pros-

Table 9–1. EXOTIC FISHES SUCCESSFULLY INTRODUCED INTO AUSTRALIAN FRESHWATERS

Atlantic salmon	*Salmo salar,* N.S.W.
Brown trout	*Salmo trutta,* W.A., S.A., Vic., Tas., N.S.W.
Rainbow trout	*Salmo gairdneri,* W.A., S.A., Vic., Tas., N.S.W. (restocked)
Brook trout	*Salvelinus fontinalis,* Tas., N.S.W.
Quinnat salmon	*Onchorynchus tshawytscha,* Vic. (restocked)
Carp	*Cyprinus carpio,* W.A., S.A., Vic., N.S.W., Qd.
Goldfish	*Carassius auratus,* W.A., S.A., Vic., Tas., N.S.W., Qd.
Crucian carp	*Carassius carassius,* W.A., N.S.W., Qd.?
Tench	*Tinca tinca,* S.A., N.S.W., Tas.
Roach	*Rutilus rutilus,* Vic.
Mosquitofish	*Gambusia affinis affinis,* all states except Tasmania
Mosquitofish	*Gambusia affinis holbrooki,* W.A.
Mosquitofish	*Gambusia dominicensis,* N.T.
Guppy	*Poecilia reticulata,* Qd.
Sailfin molly	*Poecilia latipinna,* Qd.
Swordtail	*Xiphophorus helleri,* Qd.
Platy	*Xiphophorus maculatus,* Qd.
European perch	*Perca fluviatilis,* W.A., S.A., N.S.W., Vic., Tas.
Mozambique tilapia	*Tilapia mossambica,* Qd.
Spotted tilapia	*Tilapia mariae,* Vic.
Convict cichlid	*Cichlasoma nigrofasciatum,* Vic.
Jack Dempsey	*Cichlasoma octofasciatum,* Vic.?

pect" strain introduced into Prospect Reservoir prior to 1900, a "Singapore" strain introduced into the Murrumbidgee Irrigation Area, and a "River" strain that was cultured by a German fish farmer at Boolarra and subsequently became established in Gippsland waters and elsewhere. The last strain was almost certainly illegally imported from Germany. In 1963, the Victorian government enacted legislation that gave power for government officers to enter private property and kill European carp. An extensive program was mounted to eradicate the fish from all private and public waters. Although this was thought to be successful, it was found later that carp had survived in two areas at least. One of these was the Yallourn Storage Dam in Gippsland and the other was Lake Hawthorn, a body of water in northwestern Victoria that received drainage water and overflows to the Murray River. From these two points European carp have now spread throughout the Murray-Darling river system to Queensland and through most of the waters in Gippsland. The rate at which this strain has extended its range in Victoria in the last fifteen years has been rapid, and contrasts with the sedentary behavior of the Prospect and Singapore strains. A commercial fishery on European carp has been established in Victoria.

Tench (*Tinca tinca*) were first introduced into Tasmania and by 1882 were firmly established. In 1876, they were widely distributed in Victorian streams and in 1886 were liberated in New South Wales where they bred successfully only to the west of the Great Dividing Range in the waters of the Murray-

Darling system. The distribution and ecology of Tench in Tasmanian waters are described by Weatherley (1959, 1962) and Weatherley and Lake (1967).

The first rainbow trout (*Salmo gairdneri*) were introduced to New South Wales from New Zealand in 1894; eggs from these were distributed to Tasmania, Victoria, and Western Australia (Roughley 1951). Brook trout (*Salvelinus fontinalis*), Atlantic salmon, and quinnat salmon (*Oncorhynchus tshawytscha*) are present in Australia, but apart from *S. fontinalis* in Clarence Lagoon, Tasmania, did not form self-reproducing populations.

The mosquitofish (*Gambusia affinis*) was liberated in the Botanic Gardens, Sydney, in 1925. From this population some were sent to Brisbane (Hamlyn-Harris 1929) but apparently not released. The first liberations in Queensland were by military personnel in 1941. The fish were sent to Brisbane from Sydney, held in breeding ponds at Enoggera, Brisbane, and later liberated in the Brisbane area. Early in 1943, a supply of *Gambusia* was brought from Fingal, New South Wales, and released in the south Brisbane area. The species multiplied rapidly and in September 1943 they were released in Cairns. Other releases were made in the Northern Territory, South Australia, Victoria, and Western Australia. Local shire councils and state health departments continued to distribute the fish; it is now widespread in mainland Australia, but absent from Tasmania. *Gambusia dominicensis* is present in streams and waterholes near Alice Springs, and is probably the result of release of aquarium fish (Lake 1978).

Crucian carp (*Carassius carassius*) and goldfish (*Carassius auratus*) were introduced into Australia about 1876 (Lake 1959). The goldfish is widespread in New South Wales, Victoria, and South Australia and present in Queensland, Western Australia, and Tasmania, sometimes locally abundant. The roach (*Rutilus rutilus*) is known from the Yarra River in Victoria and apparently occasionally in the Murray (McDowall 1980).

Swordtails (*Xiphophorus helleri*) were known to be present in Kedron Brook, for some fifteen years, and in Enoggera and Ithaca creeks, Brisbane, for at least twelve years. In 1977, a grant from the Australian National Parks and Wildlife Service, Canberra, enabled the Queensland Museum to survey the creeks of the Moreton Bay and Hervey Bay regions. An additional grant from the Australian National Parks and Wildlife Service extended the survey northward to Cairns (McKay 1978). Of thirty-six separate drainage systems in the Brisbane region, 94 percent contained *Gambusia* and 47 percent contained swordtails. The guppy (*Poecilia reticulata*) was found in five drainage systems and the platy (*Xiphophorus maculatus*) in three. The rosy barb (*Barbus conchonius*) was found in one suburban creek. In the Hervey Bay area north of Brisbane, the platy and the sailfin molly (*Poecilia latipinna*) were collected from a drain below an aquarium fish farm. In the area from Maryborough to Cairns, the swordtail was present at Gladstone and the platy at Babinda. Guppies were abundant at Cairns, Innisfail, Ingham, Mackay, Rockhampton, Gladstone, and at Ban Ban springs near Gayndah. In 1977, the guppy was abundant in Ithaca, Toowong, and Seven Hills Creek, but by

1981 had been reduced to low numbers in Seven Hills Creek only. The rosy barb was abundant in Seven Hills Creek in 1976 when large open areas of creek were available, but as the terrestrial grasses, mainly para grass (*Brachiaria mutica*), invaded the free water the population declined and by 1981 had disappeared. Goldfish have been recorded from Kedron Brook, Sandy Creek, Enoggera Reservoir, and in a number of western Queensland rivers, but were not found to be abundant during this survey.

The original stocks of the swordtail, guppy, platy, molly, rosy barb, and goldfish were imported via the ornamental fish trade. The release of unwanted aquarium fish undoubtedly led to the establishment of the swordtail in Kedron Brook, Enoggera, Ithaca, and other Brisbane creeks. The Seven Hills Creek introduction is suspected to be caused by the escape of juvenile fish via a drain from ornamental ponds located near the headwaters of this creek. Schoolchildren were responsible for some translocations of the swordtail, but a number of quite separate introductions have occurred. The Brisbane flood of January 1974 led to the widespread flooding of suburban creeks and a number of outdoor ponds and indoor aquarium fish were liberated. Ornamental ponds in Brisbane and other Queensland cities are well suited to the maintenance of hardy species such as the livebearers, barbs, and anabantid fish that reproduce freely and are not stressed by the high summer temperatures. Unfortunately, most outdoor ponds are not fitted with an outlet screen and discharge into storm water drains during periods of heavy summer rainfall. The frequent use of botanic gardens and civic parks to dispose of unwanted aquarium fish may be preferable to dumping unwanted fish into natural waters, but the practice often establishes a population of exotic fishes that may be harvested and dispersed by schoolchildren. One aquarium dealer in Gympie informed me that on rare occasions local farmers had purchased guppies and swordtails to control mosquito larvae in farm dams.

In 1978, specimens of the Niger cichlid or spotted tilapia (*Tilapia mariae*) and the convict cichlid (*Cichlasoma nigrofasciatum*) were taken in the cooling pondage of the Hazlewood power station near Morwell, Victoria. A survey in 1979 showed that both species were self-maintaining and were present in a creek below the cooling pondage. A Jack Dempsey (*Cichlasoma octofasciatum*) was also taken but this species may not have become established (Cadwallader et al. 1980). *Tilapia* species were available from some aquarium retailers in Brisbane (McKay 1977) and, although a prohibited genus, were maintained by aquarists in outdoor ponds. In April 1979, one Mozambique tilapia (*Tilapia mossambica*) was taken by line in Leslie Harrison Dam, Tingalpa, Brisbane. A survey of this dam in April 1980 confirmed the presence of a self-maintaining population. By July 1980 adult *T. mossambica* had escaped from the dam via the spillway into Tingalpa Creek. A report of tilapia in urban drains at Townsville by the Queensland Fisheries Service was investigated in June 1980, and *T. mossambica* was found to be abundant in the Botanical Gardens at Anderson Park where they had been discarded some two years previously; from this site they had been flushed into the urban drainage system and into Ross Creek via the tidal gates. A recent report confirms their presence in the Ross River and indicates their presence in the Burdekin River, Queensland. In April 1981, specimens of *T. mossambica* were captured in the North Pine Dam, Lake Samsonvale, Brisbane, where they have become established.

Tilapia (possibly *T. mossambica*) are established in the Gascoyne River, Western Australia (N. Morrissy, personal communication), and possibly in the Chapman River, Western Australia.

Effects of Introduced Species on Native Freshwater Fishes

> With few exceptions, information on the effects of introduced fish on the native fish fauna is anecdotal and fragmentary, not only because so little is known about native fish, but also because the effects of introduced fish have been overshadowed by the effects of the great physical changes that have taken place in native fish habitats. There is much overseas evidence of the adverse effects of introduced fish on native fish, e.g., Miller (1961), Regier (1968), Zaret and Paine (1973), Zaret (1974) and Moyle (1976) and *a priori* it must be assumed that the introduction of any new species will have repercussions as far as the native fauna is concerned. (Cadwallader 1978:219)

Weatherley and Lake (1967) and Lake (1971) contended that man-made changes to the environment have been the principal factors causing a decline in the Australian freshwater fish fauna. The dramatic effects of stream flow regulation by weirs and dams in changing the regimes of rivers, and preventing the upstream migration of fish, are obvious and far-reaching. The clearing of the catchment area for agriculture resulting in increased siltation, water temperatures, and salinity has had a detrimental effect on native fishes as has the clearing of snags and logs, straightening of stream courses, and other river "improvement" or flood mitigation measures. Usually the effects of such physical alterations, including water pollution, are difficult to isolate from those of biological pollution.

Jackson (1981) has summarized the evidence relating to the impact of trout on the native fishes of southeastern Australia. He concluded that, although information on the relationships between trout and native species is scarce, clearly the mountain galaxias *G. olidus* is particularly susceptible to displacement by trout and that galaxiids appear to be susceptible to displacement by predatory game fish throughout their range as previous authors had demonstrated (McDowall 1968, 1978; Tilzey 1976; Cadwallader 1978).

The European carp (*Cyprinus carpio*) is now a threat to Australian fresh waters. A study to assess the biological and environmental impact of carp on

Victorian waters has commenced. It appears that in shallow waters carp may increase the turbidity of the water and disrupt plants, especially in warm slow-flowing waters, but in rivers and large deep lakes the effects of carp may not be so pronounced (Cadwallader 1978).

European perch are widespread throughout the major rivers to the west of the Great Dividing Range and in southwestern Western Australia. They are most abundant in still or sluggish weedy areas. From a study of the freshwater fish catch of the Kerang lakes between 1919 and 1949, Cadwallader (1978) has shown that when there was a large population of European perch the Murray native fish population was low and did not increase until the European perch population declined. European perch are voracious fish that prey on small fishes and freshwater crayfish (*Cherax* spp.). In Western Australia they were suspected of causing a decline in the populations of small *Galaxias nigrostriatus* (McKay 1977).

The mosquitofish is widespread throughout mainland Australia. Originally introduced to control mosquitoes, the species may be the most harmful of our introduced fishes. Myers (1965) concluded that *Gambusia* is a dangerous fish to introduce and is little or no better as a mosquito destroyer than many other species. He noted that almost everywhere *Gambusia* has been introduced it has gradually eliminated most or all of the smaller native species of fish and has also taken a heavy toll of the young of important game or food fishes. In many Australian streams near centers of human habitation, *Gambusia* has become the dominant fish, sometimes to the complete exclusion of native fishes. Ahuja (1964) states that *Gambusia* will survive temperatures ranging from 6 to 35 C, a dissolved oxygen content as low as 0.18 ppm, and given time to adapt may survive salinities more than twice that of seawater. The ability of *Gambusia* to withstand extremes of temperature well outside those experienced in the natural environment is noteworthy.

Cherry et al. (1976) observed *G. affinis* in a drainage system where the water temperature exceeded 44 C, and Otto (1973) found that although they tolerate temperatures of up to 42 C for brief periods, the critical thermal maxima of a hot-adapted population was 35 C and the lower lethal temperature of the cold-adapted population was 0.5 C. Otto (1973) discussed selection for increased tolerance of heat and cold. Cold-adapted populations of *Gambusia affinis* occur in southern Australia. *Gambusia* are exceptionally hardy fishes (Hildebrand 1918) and can tolerate extremely low oxygen tensions. In a study of an anaerobic spring in Florida, Odum and Caldwell (1955) found that *G. affinis holbrooki* and sailfin mollies survived oxygen levels of less than 0.3 ppm by gulping air at the surface and would survive in cages held below the surface at 1.3 ppm oxygen. Lewis (1970) has observed that *G. affinis* and *Poecilia latipinna* survive in oxygen-depleted waters by virtue of their upturned mouth and their habit of skimming below the surface; this morphological adaptation allows these fishes to utilize surface oxygen and feed at the atmosphere-water interface for prolonged periods. Their presence in oxygen-deficient drains in southern Queensland attests to their ability to tolerate almost anaerobic conditions. Their tolerance to pesticides and pollutants is well documented (Culley and Ferguson 1969), and this permits the establishment of resistant strains in polluted drains. According to Maglio and Rosen (1969), *Gambusia* are secondary consumers that feed primarily on zooplankton and aquatic insect larvae. The eggs and larvae of native freshwater fishes are also consumed, and large fishes may have their fins nipped or, in some instances, entirely removed (McKay 1977). *Gambusia* may alter the ecosystem by reducing zooplankton populations (Hurlbert et al. 1972), and may have a marked influence on Entomostraca and aquatic insects of small lakes (Stephanides 1964). Their aggressive behavior, especially when in dense populations, may contribute to the decline of other freshwater fishes (Minckley and Deacon 1968, field observations).

In coastal Queensland streams containing large numbers of *Gambusia, Poecilia,* and *Xiphophorus,* the native freshwater fishes, particularly the surface-feeding or mosquito-eating *Melanotaenia, Pseudomugil, Craterocephalus,* and *Retropinna,* are usually rare or absent. In Enoggera Creek above the reservoir where introduced species are not present, 8 species of native fish are abundant. The dominant species is *Melanotaenia fluviatilis.* Immediately below the reservoir where *Gambusia* and *X. helleri* are present, 5 native species occur: the gudgeon (*Hypseleotris galii*) is dominant and no *Melanotaenia* or *Retropinna* is present. Further downstream where *Gambusia* and *X. helleri* are abundant, *H. galii* is rare and the dominant native species is the freshwater eel (*Anguilla reinhardtii*) (Arthington et al. 1981). In 1977, *Gambusia* and *X. maculatus* were absent from College's Crossing on the Brisbane River and the dominant fish present was the native blue-eye (*Pseudomugil signifer*) which comprised 73 percent of the catch. By 1981, both exotic fishes had invaded this area and the blue-eye had become much less abundant (1.5 percent).

It appears that the introduction of one poeciliid fish has an appreciable effect on the small surface inhabiting native fishes. This effect is markedly increased when two or more livebearers are present, and in such cases almost all the surface-feeding indigenous species become rare or disappear. A study is now in progress to investigate the ecology and interactions of exotic and endemic freshwater fishes in southeastern Queensland streams (Arthington et al. 1981). Despite warnings of the danger of translocating this fish, many health officers and local councils continue to transplant *Gambusia* into any body of water containing or suspected of containing mosquito larvae. *Gambusia dominicensis* has colonized waterholes in Central Australia, where it threatens the existence of a rare desert goby, *Chlamydogobius eremius* (Connell 1974).

Gambusia is remarkably successful in many Australian freshwater habitats, particularly those lacking predatory fishes or those altered directly or indirectly by man, because of a combination of adaptations, all of which enhance its survivability and competitiveness. It is a ovoviviparous fish that matures early and produces multiple broods of young during an extended period of the year.

> Some idea of the reproductive capacity of the species may be gained by estimating the number of young that would be added to the pond during a single beginning with 10 adult females and assuming no mortality, and ample food and space. With an average brood size of 50 young, an average maturation rate of young-of-the-year of 20 days, and an average brood interval of 25 days for all mature fish over the six-month breeding season, the population, in October, would be expected to total 4,802,500 individuals. (Maglio and Rosen 1969:27)

The maturation rate varies according to temperature and in tropical areas females may reproduce throughout the year. The growth of a population from a single gravid female capable of giving birth to multiple broods of young, after initial fertilization by the male, may be spectacular and is largely correlated with net and gross primary productivity (Goodyear et al. 1972). In southern Queensland, *Gambusia* is largely restricted to shallow drains and suburban streams where water flow, except for periods of heavy rainfall, is minimal. Fast-flowing streams and rivers are not suited to *Gambusia* and, when present, these generally inhabit the quieter backwaters away from the main flow of the current. Reddy and Pandian (1974) found that the predatory efficiency of *Gambusia affinis* was greatly decreased with an increase in current speed.

It is too early to assess the impact of tilapias on Australian freshwater fishes. In the Brisbane reservoirs large numbers of the native spangled perch (*Leiopotherapon unicolor*) occur and it is possible that the presence of this voracious and highly aggressive species may retard the population growth of tilapia species. As the reservoir matures and the population of the spangled perch declines, the population growth of *T. mossambica* may be an explosive one. West and Glucksman (1976) have documented the introductions of tilapia to Papua New Guinea, and state that, although introduction and natural dispersal have been extensive, there is insufficient information to assess their effect on the native fish fauna; subjective observations suggest that to date there have been no ill effects. Berra et al. (1975:323) recorded 6 exotic species from the Laloki River, New Guinea, and remarked: "It is sad commentary on man's propensity for juggling freshwater fish faunas that the species collected at the most localities was an introduced one, *Tilapia mossambica.* This species also ranked second in total numbers of individuals, and most of these were juveniles indicating that reproduction is taking place." Courtenay et al. (1974) have reviewed the problems associated with *Tilapia* and other fishes introduced to Florida waters.

The Aquarium Industry and Possible Introductions of Ornamental Fishes

Exotic freshwater fishes are imported into Australia in considerable numbers to support the thriving aquarium trade and hobby. Importations of goldfish and other coldwater fishes that entered the country occurred before the development of rapid air services and modern methods of fish transportation, but their number and species diversity were restricted. By 1963–64, 277,863 live aquarium fish were imported, and it became obvious that a system of classification was necessary to prevent some undesirable fish species from entering the country. Very little was known about the biology of the large variety of exotic fishes that were available to the hobby. Indeed, some exotic fishes became known to the trade before they were named scientifically, and even today new species are being found. Some of the popular aquarium fishes are poorly known taxonomically and accurate identification is, therefore, difficult. The first classification of approved and prohibited aquarium fishes was compiled by the Advisory Committee on the Importation of Live Aquarium Fish, a group of ichthyologists and trade representatives functioning as a subcommittee of the Commonwealth/State Fisheries Conference. This classification entitled "Alphabetical List of Exotic and Indigenous Aquarium Fishes Classified into Categories 'A' and 'D'" was published by the Fisheries Division, Department of Primary Industry, Canberra, in November 1963, with addendum sheets. The list contained some 700 species approved for importation and over 100 prohibited species. All unlisted species were not "approved" and, therefore, were prohibited.

The rapid growth of the industry resulted in a spectacular increase in the number of fishes imported and within ten years of the compilation of the Alphabetical List, over 8 million exotic freshwater fishes were imported (table 9–2). With this volume of imported fishes the inspectors at the port of entry found the task of identification of approved and prohibited fishes overwhelming, especially when juvenile fishes or mixed consignments were encountered. The lack of adequate facilities at the airport prevented all but a cursory examination of the bags of fish for unfamiliar fish species, aquatic plants, and invertebrate animals. It was not surprising, therefore, that in recent years a number of prohibited fishes gained entry. The growing concern of fisheries biologists to the possibility of the introduction of exotic fish species, parasites, and disease organisms was outlined by McKay (1977) who largely reflected the view of the Advisory Committee on Endangered Species and Import and Export of Live Fish, which reports to the Standing Committee of Fisheries. Feral populations of *Xiphophorus helleri, X. maculatus, Poecilia reticulata, Tilapia mariae, Tilapia mossambica, Cichlasoma nigrofasciatum,* and instances of the introduction of *Misgurnus anguillicaudatus, Geophagus*

Table 9-2. NUMBER AND VALUE OF LIVE FISHES IMPORTED INTO AUSTRALIA 1963 TO 1980

Year	Number	Value $A
1963-64	227,863	21,156
1964-65	935,500	–
1965-66	1,055,293	73,000
1966-67	1,429,103	103,000
1967-68	1,970,300	133,000
1968-69	1,524,289	184,000
1969-70	3,241,086	240,000
1970-71	4,544,832	332,000
1971-72	6,027,387	460,341
1972-73	6,200,863	494,093
1973-74	8,315,336	687,987
1974-75	10,856,062	1,029,218
1975-76	12,148,252	1,186,505
1976-77	11,203,877	1,349,553
1977-78	11,546,208	1,432,097
1978-79	8,193,977	1,221,247
1979-80	9,718,313	1,346,555

$A = Australian dollars

brasiliensis, Cichlasoma octofasciatum, and *Barbus conchonius* increased this concern. The tilapia introductions were from illegal importations or stocks retained since 1963 by aquarists, as after this date they were prohibited species.

An analysis of over one million aquarium fishes imported into Melbourne in 1978-79 showed that 20 species accounted for about 80 percent of live fish imports and about 80 species accounted for over 96 percent; the remainder was made up of about 90 species. The goldfish accounted for some 24 to 30 percent of live fish imports. It appeared that a reduction in the number of live aquarium fishes approved for importation would not seriously harm the trade, although the specialist hobbyist or those seeking the unusual or bizarre would be disadvantaged.

The Advisory Committee then formulated a set of criteria to apply to aquarium fishes imported into Australia, and applied these criteria to lists of aquarium fishes submitted by the trade some years previously. A list, termed the "List of 100," containing about 200 or so species, was proposed by the Advisory Committee but met strong opposition by the Australian Federation of Aquarium Fish Importers and Traders (AFAFIT) and some aquarium societies. The list was rejected on the grounds of being out of date, too restrictive, and of an arbitrary nature. A list of prohibited fishes would, of course, be too difficult to police as it would be lengthy and require inspectors to be familiar with a large number of fishes not regularly imported. The submission from AFAFIT and several aquarium societies of a comprehensive list of fishes was then used as a foundation to which the Advisory Committee applied its criteria. In consultation with the trade it compiled a list of genera and species for continued importation (table 9–3). If this list is adopted, the Advisory Committee will review it annually in consultation with the trade.

Fisheries authorities recognize that no exotic freshwater fish can be considered environmentally safe. A complete ban on the importation of all nonindigenous fishes would be unacceptable to the estimated 900,000 hobbyists, and destroy the Australian aquarium trade reported to have a retail turnover of $200 million annually.

The acceptance of a list of fishes approved for importation does not resolve the problem of a considerable number of potentially harmful fishes now maintained by aquarists throughout Australia (*Astronotus, Cichlasoma,* etc.). A permit to keep some "nonapproved" fishes may be possible under state legislation; this is being considered by some states and should allow the specialist hobbyist and clubs to preserve their interests as well as import fish from time to time under special permit. Sale of "nonapproved" fishes would be prohibited.

State legislation is also required to control the breeding, marketing, and distribution of aquarium fish within Australia because of the high potential risk to health and survival of native species.

It is quite impossible to predict the consequences of liberating aquarium fish. Some species have a demonstrated ability to form feral populations in unlikely habitats, and others fail to survive despite repeated attempts to introduce them. With the wide range of aquatic habitats in Australia, particularly the northern tropical and subtropical areas, it is difficult to assess the degree of risk associated with an aquarium fish species. The Advisory Committee employed the following criteria in their rejection of aquarium fishes proposed for importation to Australia. Some examples are given.

1. *Have a proven capacity for causing environmental damage by rapid reproduction, method of feeding, or having a wide capacity for survival or dispersal.*

A number of fishes fall into this category, particularly those that have been widely introduced throughout the world to the detriment of native freshwater fishes. The commonly imported swordtail *Xiphophorus helleri,* platys *X. maculatus, X. variatus,* guppy *Poecilia reticulata,* and molly *P. latipinna* have been liberated in coastal Queensland and are now well established, and are, therefore, included in the list of "approved" fishes despite their capacity to harm native freshwater fishes. A ban on their importation might lead, because they are easily bred, to backyard farming in areas subject to cyclonic rainfall. Some fishes of the family Cichlidae are also widespread in Florida (*Cichlasoma, Tilapia,* and *Astronotus*) and some have formed feral populations in Australia (*Cichlasoma, Tilapia mariae,* and two introductions of *Tilapia mossambica* in Queensland). The Florida introductions were the result of fish farming activities. The European carp, widespread in eastern Australia, is included here. Any fish species that becomes feral in other coun-

Table 9–3. LIST OF SPECIES AND GENERA OF AQUARIUM FISHES PROPOSED FOR CONTINUED IMPORTATION TO AUSTRALIA – 1981

Genus	Species	Common name
Abramites	microcephalus	headstander
Acanthophthalmus	all species	kuhli loaches
Achirus	fasciatus	freshwater sole
Aequidens	curviceps	curviceps
	maronii	keyhole
	pulcher	blue acara
Alestes	longipinnis	African tetra
"Anoptichthys"	jordani	blind cave fish
Anostomus	all species	headstanders
Aphyocharax	rubripinnis	blood fins tetra
Aphyosemion	all species	killifish
Apistogramma	all species	dwarf cichlid
Aplocheilus	all species	panchax
Astronotus	ocellaris	oscar
Balanteocheilus	melanopterus	silver shark
Barbodes	everetti	clown barb
	fasciatus	striped barb
	hexazona	tiger barb
	lateristriga	spanner barb
	pentazona	banded barb
Betta	all species	fighting fish
Botia	macracantha	clown loach
	sidthimunki	dwarf loach
Brachydanio	albolineatus	pearl danio
	kerri	Kerr's danio
	frankei	leopard danio
	nigrofasciatus	spotted danio
	rerio	zebra danio
Brachygobius	all species	bumblebee fish
Brochis	coeruleus	blue catfish
Capoeta (= Barbus)	arulius	longfin barb
	oligolepis	checker barb
	partipentazona	tiger barb
	semifasciocatus	golden barb
	tetrazona	tiger barb
	titteya	cherry barb
Carassius	auratus	goldfish
Carnegiella	all species	hatchet fish
Chanda	ranga	glass perchlet
Cheirodon	axelrodi	cardinal tetra
Chilodus	punctatus	spotted headstander
Colisa	chuna	honey dwarf gourami
	fasciata	"giant" dwarf gourami
	labiosa	thick-lipped gourami
	lalia	dwarf gourami
Corydoras	all species	armoured cats
Danio	malabaricus	giant danio
Dermogenys	pusillus	half beak
Dianema	urostriata	stripe tailed catfish
Epalzeorhyncus	kallopterus	flying fox
	siamensis	Siamese flying fox
Esomus	danrica	flying barbs
	goddardi	flying barbs
	malayensis	flying barbs

Table 9–3. LIST OF SPECIES AND GENERA OF AQUARIUM FISHES PROPOSED FOR CONTINUED IMPORTATION TO AUSTRALIA – 1981 *(continued)*

Genus	Species	Common name
Farlowella	acus	twig catfish
Gasteropelecus	all species	hatchet fish
Gastromyzon	myersi	dwarf stone sucker
Geophagus	jurupari	earth-eating cichlid
Gnathonemus	macrolepidotus	elephant nose
	petersi	elephant nose
Gymnocorymbus	ternetzi	black widow tetra
Gyrinocheilus	aymonieri	sucking catfish
Helostoma	rudolfi	pink kissing gourami
	temmincki	green kissing gourami
Hemigrammus	all species	tetras
Homaloptera	orthogoniata	lizard fish
Hyphessobrycon	all species	tetras
Julidochromis	all species	dwarf cichlids
Kryptopterus	bicirrus	glass catfish
	macrocephalus	poormans glass catfish
Labeo	bicolor	redtail shark
	erythrurus	red fin shark
	variegatus	variegated shark
	frenatus	rainbow shark
Laubuca	laubuca	Indian hatchet fish
Leporinus	arcus	lipstick leporinus
	fasciatus	banded leporinus
	frederici	Frederici's leporinus
	maculatus	spotted leporinus
	melanopleura	spot-tailed leporinus
	multifasciatus	multi-banded leporinus
	striatus	striped leporinus
Loricaria	filamentosa	whiptail catfish
Macrognathus	aculeatus	spiny eel
Macropodus	opercularis	paradise fish (males only)
Megalamphodus	megalopterus	black phantom tetra
	sweglesi	red phantom tetra
Melanotaenia (Nematocentrus)	all species	rainbow fish
Micralestes (Phemacogrammus)	interruptus	Congo tetra
Misgurnus	anguillicaudatus	weather loach
Moenkhausia	all species	tetra
Monodactylus	argenteus	mono
	sebae	African mono
Morulius	chrysophekadion	black shark
Nanacara	anomala	golden dwarf acara or dwarf golden-eye cichlid
	taenia	dwarf lattice cichlid
Nannostomus	all species	pencil fish
Nematobrycon	all species	emperor tetras
Oryzias	latipes	medaka
	javanicus	java medaka
Osteocheilus	vittatus	bony-lipped barb
	hasselti	bony-lipped barb
Otocinclus	arnoldi	sucker catfish
Oxygaster	oxygateroides	glass barb
Pantodon	bucholzi	butterfly fish
Papiliochromic (Apistogramma)	ramirezii	ram

Table 9–3. LIST OF SPECIES AND GENERA OF AQUARIUM FISHES PROPOSED FOR CONTINUED IMPORTATION TO AUSTRALIA – 1981 *(continued)*

Genus	Species	Common name
Paracheirodon (Hyphessobrycon)	innesi	neon tetra
Pelmatochromis	subocellatus	kribensis
	pulcher	kribensis
	taeniatus	kribensis
Petitiella	georgiae	false rummy nose
Pimelodus	ornatus	pictus cat
Poecilia	reticulata	guppy
	latipinna	sailfin mollie
	velifera	sailfin mollie
	sphenops	black mollie
Poecilobrycon	all species	pencil fish
Prionobrama	filigera	glass bloodfish
Pristella	riddlei	pristella
Pterophyllum	all species	angelfish
Puntius (= Barbus)	bimaculatus	two spot barb
	conchonius	rosy barb
	"asoka"	asoka barb
	filamentosus	black spot barb
	cumingi	Cummings barb
	nigrofasciatus	ruby barb
	stoliczkai	Stoliczka's barb
	vittatus	Kooli barb
Rasbora	argyrotaenia	silver rasbora
	boraptensis	redtail rasbora
	caudimaculata	redtail rasbora
	dorsiocellata	emerald eye rasbora
	dusonensis	yellow tail rasbora
	einthoveni	blue line rasbora
	elegans	two-spot rasbora
	hengelsii	harlequin rasbora
	heteromorpha	harlequin rasbora
	kalochroma	clown rasbora
	leptosoma	copper-striped rasbora
	maculata	dwarf-spotted rasbora
	pauciperforata	red line rasbora
	sarawakensis	Sarawak rasbora
	steineri	gold line rasbora
	taeniata	blue line rasbora
	trilineata	black scissortail
	vaterifloris	flame rasbora
Rhodeus	amarus	bitterling
	sericeus	bitterling
Sphaerichthys	osphronemoides	chocolate gourami
Symphysodon	all species	discus
Synodontis	nigriventris	upsidedown cat
Tanichthys	albonubes	white cloud
Telmantherina	ladigesi	Celebes rainbow
Thayeria	all species	hockeystick tetra
Thoracocharax	all species	hatchet fish
Toxotes	jaculator	archer fish
Trichogaster	leeri	pearl gourami
	microlepis	moonbeam gourami
	trichopterus	blue, opaline, golden gourami

Table 9-3. LIST OF SPECIES AND GENERA OF AQUARIUM FISHES PROPOSED FOR CONTINUED IMPORTATION TO AUSTRALIA – 1981 *(continued)*

Genus	Species	Common name
Trichopsis (Ctenops)	pumilus	gourami
	vittatus	croaking gourami
Trinectes	maculatus	freshwater flounder
Tropheus	duboisii	dwarf Tanganyikan cichlid
	moorei	dwarf Tanganyikan cichlid
Xiphophorus	helleri	swordtail
	maculatus	southern platy
	variatus	variegated platy

tries should be carefully considered before being approved for importation to Australia.

2. *Are voracious or aggressive.*

Examples are, of course, piranhas (*Serrasalmus* spp.), African tigerfish (*Hydrocynus*), and others. The oscar (*Astronotus ocellatus*) is of particular concern as it is kept in Australian aquaria and has demonstrated its capacity to form feral populations. A permit to keep oscars may be necessary, and their sale prohibited.

3. *Are harmful to man, livestock, or native animals because of their capability of inflicting wounds by the possession of strong, serrated, or venomous spines that lock into position* (most freshwater catfishes), *in having electric organs* (electric eels and catfish) *or poisonous flesh* (*Tetraodon, Sphaeroides,* etc.).

While such fishes are relatively harmless when kept under aquarium conditions, introductions, given sufficient time, become almost inevitable. The successful introduction of the walking catfish (*Clarias batrachus*) in Florida is now regarded as a disaster. The walking catfish was on sale recently in Australia. Other catfish with venomous spines could be harmful to bathers, fishermen, and aquatic birds if they became a pest, despite the fact that Australia has its own native freshwater catfish. The families of catfish that may possess accessory breathing organs (Clariidae, Loricariidae, Callichthyidae, etc.) have been largely excluded (e.g., *Hoplosternum, Hypostomus-Plecostomus* introduced to the United States).

4. *Are proven carriers of any economically important disease or parasite not already recorded from Australia.*

Very little is known of the diseases and parasites associated with aquarium fish. No importation is restricted because of this criterion.

5. *May be readily confused with noxious fish species (e.g., Metynnis* and piranha*)*.

Small piranha could be illegally imported in bags containing large numbers of the silver dollar, *Metynnis* spp., especially with the lack of adequate facilities at most ports of entry. A proposed system of quarantine now under review by the Commonwealth Department of Health may alter this situation in the future.

6. *Grow to a large size suitable for food or sport fish stocking.*

Some fishes, although employed as aquarium fishes, may grow to large size and were, therefore, excluded (some *Labeo* spp.; some catfish, e.g., *Auchenoglanis, Pangassius;* some cichlid fishes, e.g., *Haplochromis* spp., *Lamprologus* spp., and the giant gouramy, *Osphronemus*).

A few fish species were removed from the list supplied by the trade and aquarium societies as species or genera that were difficult to determine at the airport. A few species of catfish, although having strong spines and accessory breathing organs, were allowed entry because of their long trouble-free history with the aquarium hobby throughout the world (*Corydoras*) or their spectacular color pattern (*Pimelodus ornatus,* the "pictus cat"). Other species were retained on the list subject to an early review, e.g., *Astronotus Dianema, Geophagus, Misgurnus, Pimelodus,* and *Synodontis.*

Some of the fish genera approved in 1963 were removed from the list because of more recent information concerning their biology, behavior, or capacity to become feral in other countries. They include: *Acanthodoras, Anabas, Astyanax, Badis, Belonesox, Belontia, Callichthys, Channa, Chelonodon, Cichlasoma, Cnesterodon, Cobitis, Crenicara, Datnioides, Etroplus, Fundulus, Jordanella, Luciocephalus, Mastacembelus, Micropoecilia, Monocirrhus, Mystus, Nandus, Phallichthys, Phalloceros, Phalloptychus,* most *Pimelodella, Plecostomus,* some *Poecilia, Polycentropsis, Polycentrus, Quintana, Scatophagus, Selenotoca,* some *Synodontis, Tetraodon, Xenocara,* and *Xenomystus.* A few species not listed in the Alphabetic List of 1963 were approved.

It is hoped that this list will provide a more secure foundation for the continuation of the aquarium hobby and industry while protecting the Australian freshwater environment as much as possible. It is essential to begin a public education system to acquaint fish fanciers with the grave dangers of introducing exotic fishes discarded from aquarium tanks.

The Australian Department of Health is considering a system of quarantine for live fish imports. If adopted, it would require importers to hold fish under supervision in licensed premises and would greatly facilitate identification of fish species, diseases, and parasites.

Introduction of Sport or Food Fishes

The Australian Fisheries Council controls the introduction of sport or food fishes. It has recently granted approval to Queensland Fisheries Service to import Nile perch (*Lates nilotica*) for research purposes only. Nile perch has been proposed as a possible introduction to Queensland reservoirs to provide recreational fishing in an area now devoid of the native barramundi (*Lates calcarifer*), a species that, unlike the Nile perch, will not breed in fresh water. The study will compare the advantages and disadvantages of *Lates nilotica* with *Lates calcarifer.*

Literature Cited

Ahuja, S. K. 1964. Salinity tolerance of *Gambusia affinis*. Indian Journal of Experimental Biology 2:9–11.

Arthington, A. H., R. J. McKay, and D. Milton. 1981. Consultancy on ecology and interactions of exotic and endemic freshwater fishes in southeastern Queensland streams. Report 1. Australian National Parks and Wildlife Service, Canberra. 96 pp.

Berra, T. M., R. Moore, and L. F. Reynolds. 1975. The freshwater fishes of the Laloki River system of New Guinea. Copeia 1975:316–325.

Bowler, J. M. 1976. Recent developments in reconstructing late Quaternary environments in Australia, 55–57. *In:* R. L. Kirk and A. G. Thorne (eds.). The origins of the Australians. Canberra, I.A.A.S.

Cadwallader, P. L. 1978. Some causes of the decline in range and abundance of the native fish in the Murray-Darling river system. Proceedings of the Royal Society of Victoria 90:211–224.

Cadwallader, P. L., G. N. Backhouse, and R. Fallu. 1980. Occurrence of exotic tropical fish in the cooling pondage of a power station in temperate south-eastern Australia. Australian Journal Marine and Freshwater Research 31:541–546.

Cherry, D. S., R. K. Guthrie, J. H. Roger, Jr., J. Cairns, Jr., and K. L. Dickson. 1976. Responses of mosquitofish (*Gambusia affinis*) to ash effluent and thermal stress. Transactions of the American Fisheries Society 105:686–694.

Connell, D. W. 1974. Water pollution. Causes and effects in Australia. University of Queensland Press. 132 pp.

Courtenay, W. R., Jr., H. F. Sahlman, W. W. Miley II, and D. J. Herrema. 1974. Exotic fishes in fresh and brackish waters of Florida. Biological Conservation 6: 292–302.

Culley, D. D., and D. E. Ferguson. 1969. Patterns of insecticide resistance in the mosquitofish *Gambusia affinis.* Journal of the Fisheries Research Board of Canada 26:2395–2401.

Gill, E. D. 1975. Evolution of Australia's unique flora and fauna in relation to the Plate Tectonics Theory. Proceedings of the Royal Society of Victoria 87:215–234.

Goodyear, C. P., C. E. Boyd, and R. J. Beyers. 1972. Relationships between primary productivity and mosquitofish (*Gambusia affinis*) production in large microcosms. Limnology and Oceanography 17:445–450.

Hamlyn-Harris, R. 1929. The relative value of larval destructors and the part they play in mosquito control in Queensland. Proceedings of the Royal Society of Queensland 41:23–38.

Hildebrand, S. F. 1918. Notes on the life history of the minnows *Gambusia affinis* and *Cyprinodon variegatus.* Report of the U.S. Commissioner of Fisheries 1917. Appendix 6.

Hills, E. S. 1934. Tertiary freshwater fishes from southern Queensland. Memoirs of the Queensland Museum 10:157–174.

Hurlbert, S. H., J. Zedler, and D. Fairbanks. 1972. Ecosystem alteration by mosquitofish (*Gambusia affinis*) predation. Science 175:639–641.

Jackson, P. D. 1981. Trout introduced into southeastern Australia: Their interaction with native fishes. Victorian Naturalist 98:18–24.

Lake, J. S. 1959. The freshwater fishes of New South Wales. New South Wales State Fisheries Research Bulletin 5. 19 pp.

Lake, J. S. 1971. Freshwater fishes and rivers of Australia. Nelson, Melbourne.

Lake, J. S. 1978. Australian freshwater fishes. An illustrated field guide. Nelson, Melbourne.

Lewis, W. M., Jr. 1970. Morphological adaptations of cyprinodontoids for inhabiting oxygen deficient waters. Copeia 1970:319–326.

Maglio, V. J., and D. W. Rosen. 1969. Changing preferences for substrate color by reproductively active mosquitofish *Gambusia affinis* (Baird and Girard) (Poeciliidae, Atheriniformes). American Museum Novitates 2397:1–39.

McDowall, R. M. 1968. Interactions of the native and alien faunas of New Zealand and the problem of fish introductions. Transactions of the American Fisheries Society 97:1–11.

McDowall, R. M. 1970. The galaxiid fishes of New Zealand. Bulletin of the Museum of Comparative Zoology 139:341–342.

McDowall, R. M. 1978. A new genus and species of galaxiid fish from Australia (Salmoniformes: Galaxiidae). Journal of the Royal Society of New Zealand 8:115–124.

McDowall, R. M. (ed.). 1980. Freshwater fishes of south-eastern Australia. Reed, Sydney. 208 pp.

McDowall, R. M., D. A. Robertson, and R. Saito. 1975. Occurrence of galaxiid larvae and juveniles in the sea. New Zealand Journal of Marine and Freshwater Research 9:1–9.

McKay, R. J. 1977. The Australian aquarium fish industry and the possibility of the introduction of exotic fish species and diseases. Fisheries Division, Department of Primary Industry, Fisheries Paper 25, Canberra, 36 pp.

McKay, R. J. 1978. The exotic freshwater fishes of Queensland. Report to Australian National Parks and Wildlife Service, Canberra. 144 pp.

Miller, R. R. 1961. Man and the changing fish fauna of the American southwest. Papers of the Michigan Academy of Science, Arts and Letters 46:365–404.

Minckley, W. L., and J. E. Deacon. 1968. Southwestern fishes and the enigma of "endangered species." Science 159:1424–1432.

Moyle, P. B. 1976. Fish introductions in California: History and impact on native fishes. Biological Conservation 9:101–118.

Myers, G. S. 1965. *Gambusia,* the fish destroyer. Tropical Fish Hobbyist 13:31–32, 53–54.

Odum, H. T., and D. K. Caldwell. 1955. Fish respiration in the natural oxygen gradient of an anaerobic spring in Florida. Copeia 1955:104–106.

Otto, R. G. 1973. Temperature tolerance of the mosquitofish *Gambusia affinis* (Baird and Girard). Journal of Fish Biology 5:575–585.

Patterson, C. 1972. The distribution of Mesozoic freshwater fishes. Proceedings of the International Congress of Zoology 17:1–22.

Reddy, S. R., and T. J. Pandian. 1974. Effect of running water on the predatory efficiency of the larvivorous fish *Gambusia affinis.* Oecologia 16:253–256.

Regier, H. A. 1968. The potential misuse of exotic fish as introductions, 92–111. *In:* Introductions of Exotic Species. Ontario Department of Lands and Forests, Research Report 82.

Roberts, T. R. 1978. An ichthyological survey of the Fly River in Papua New Guinea with descriptions of new species. Smithsonian Contributions to Zoology 281:1–72.

Rosen, D. E. 1974. Phylogeny and zoogeography of salmoniform fishes and relationships of *Lepidogalaxias salamandroides.* Bulletin of the American Museum of Natural History 153:269–325.

Roughley, T. C. 1951. Fish and Fisheries of Australia. Angus and Robertson, Sydney. 343 pp.

Shearer, K. D., and J. C. Mulley. 1978. The introduction and distribution of the carp, *Cyprinus carpio* Linnaeus in Australia. Australian Journal of Marine and Freshwater Research 29:551–563.

Stephanides, T. 1964. The influence of the antimosquitofish *Gambusia affinis,* on the natural fauna of a Corfu lakelet. Proceedings of the Hellenic Hydrobiological Institute 9:3–6.

Tilzey, R.D.J. 1976. Observations on the interactions between indigenous Galaxiidae and introduced Salmonidae in the Lake Eucumbene catchment, New South Wales. Australian Journal of Marine and Freshwater Research 27:551–564.

Weatherley, A. H. 1959. Some features of the biology of the tench, *Tinca tinca* (Linnaeus) in Tasmania. Journal of Animal Ecology 28:73–87.

Weatherley, A. H. 1962. Notes on distribution, taxonomy and behaviour of tench, *Tinca tinca* (Linnaeus) in Tasmania. Annals and Magazine of Natural History (13), 4(48):713–719.

Weatherley, A. H. 1963. Zoogeography of *Perca fluviatilis* Linnaeus and *Perca flavescens* Mitchell with special reference to the effects of high temperature. Proceedings of the Zoological Society of London 141:557–576.

Weatherley, A. H., and J. S. Lake. 1967. Introduced fish species in Australian island waters, 217–239. *In:* A. N. Weatherley (ed.). Australian inland waters and their fauna – eleven studies. Australian National University, Canberra.

West, G. J., and J. Glucksman. 1976. Introduction and distribution of exotic fish in Papua New Guinea. Papua of the New Guinea Agricultural Journal 27:19–48.

Wharton, J.C.F. 1971. European carp in Victoria. Fur, Feathers and Fin 130:3–11.

Wharton, J.C.F. 1979. Impact of exotic animals, especially European carp *Cyprinus carpio,* on native fauna. Victorian Fisheries and Wildlife Papers 20:1–13.

Zaret, T. M. 1974. The ecology of introductions—a case-study from a Central American lake. Environmental Conservation 1:308–309.

Zaret, T. M., and R. T. Paine. 1973. Species introduction in a tropical lake. Science (New York) 182:449–455.

CHAPTER 10

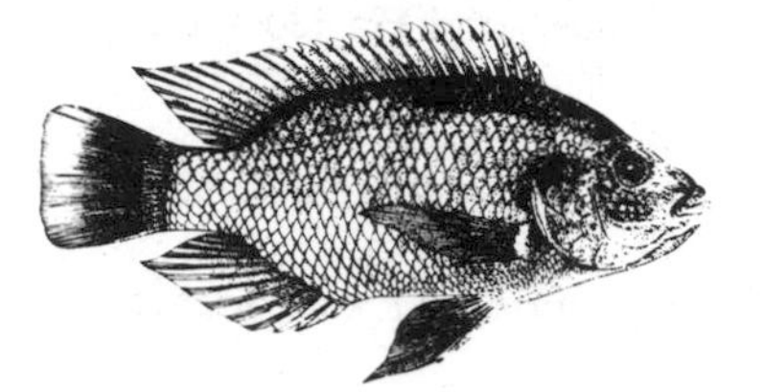

Exotic Fishes: The New Zealand Experience

R. M. McDowall

New Zealand, isolated far from other land masses in the cool temperate southwestern Pacific, has a sparse freshwater fish fauna. There are just 27 species in 7 families – a southern lamprey (family Geotriidae), a couple of eels (Anguillidae), a couple of southern smelts and a southern grayling (Retropinnidae and Prototroctidae), a moderate number of galaxiids (Galaxiidae), a mugiloidid (Mugiloididae – the only freshwater representative of the family), 6 eleotrids (Eleotridae), and a flounder (Pleuronectidae) (McDowall 1978). These few species (most of them small, all of them inconspicuous) were all that populated the numerous lakes and rivers when European man arrived in the mid-nineteenth century. Reasons for the sparseness of the fauna are not certain but it is probably in part because of New Zealand's geographical isolation and in part because of its unstable tectonic history, there being relatively little emergent land in the Oligocene (Fleming 1979, McDowall and Whitaker 1975). As settlers experienced nostalgia for the countryside and life style they had left behind, they sought to engage in activities and have experiences similar to those of their homeland. This is clearly expressed by the use of European common names for many of the endemic fish: "trout," "grayling," "minnow," "smelt," and others were all found here.

Partly as a result of this nostalgia and partly because of the need for more diverse sources of food, European man, in the mid-1800s, began to introduce and establish a wide variety of animals and plants (Thomson 1922). In this regard, New Zealand is no different from other countries, but the sparseness of the fauna, the great fertility and productivity of the land, and its moist temperate climate meant that a great many of the introduced species thrived, many of them becoming serious pests (McDowall 1968a). The history of these introductions and the trouble they have caused have been documented many times (Thomson 1922, Wodzicki 1950, Laycock 1966). What seems often forgotten is that some of the species were successful in forming the basis of important recreational, agricultural, and industrial resources.

Early Introductions – 1860s to early 1900s

Because of the sparseness of the freshwater fish fauna, additions to this fauna were natural. Introductions began as early as 1864 and continued steadily for more than forty years. The process has been described by several authors (Thomson 1922; Stokell 1955; McDowall 1978, 1979). Since these introductions began, more than 30 species have been introduced, with varying success. In 1864 the North Canterbury Acclimatization Society was the first of many such societies set up to encourage the introduction and establishment of exotic fauna and flora. The initial effort involved Atlantic salmon (*Salmo salar*), brown trout (*Salmo trutta*), alpine char (*Salvelinus alpinus*), some type of carp, probably the common goldfish (*Carassius auratus*), maybe also the European carp (*Cyprinus carpio*), a Japanese minnow (possibly *Pseudorasbora parva*), gudgeon (*Gobio gobio*), barbel (*Barbus barbus*), bleak (*Alburnus alburnus*), tench (*Tinca tinca*), rudd (*Scardinius erythrophthalmus*), dace (*Leuciscus leuciscus*), roach (*Rutilus rutilus*), minnow (*Phoxinus phoxinus*), and perch (*Perca fluviatilis*) – 15 species in all and not a single fish or egg arrived in New Zealand alive. There followed a successful consignment: brown trout, tench, and carp were introduced in 1867, and Atlantic salmon, additional brown trout, tench, European perch, and the Australian bass, *Macquaria novemaculeata,* arrived in 1868 (Thomson 1922).

From these beginnings developed a massive effort for the establishment of exotic species in New Zealand's largely uninhabited lakes and rivers. Apart from these early efforts, attention was concentrated on salmonids, primarily brown trout and Atlantic salmon but also including chinook salmon (*Oncorhynchus tshawytscha*), brook char (*Salvelinus fontinalis*), and, later in the 1800s, rainbow trout (*Salmo gairdneri*) (Scott et al. 1978).

Species that did not require access to the sea, but were able to establish wholly freshwater populations, often became established rapidly. Thus, brown trout succeeded from the first consignment (1867, from Tasmania, originating in the United Kingdom); so did tench, European perch, brook char, and others, brought in at that time or in the few years that followed. Although these species did succeed, continued importations were made. Brown trout stocks, including sea-run, riverine, and lake stocks, were imported from England, Scotland, Germany, and Italy. Several importations of brook char arrived from the United States, one of these via the United Kingdom, and several of rainbow trout (1883) from California (Scott et al. 1978).

A seemingly inexplicable introduction was the North American brown bullhead, brought in 1877, released, and forgotten until a small boy caught one and took fright at its "ferocious looking head" (Hunter 1915).

It was much more difficult to establish species that required spending part of their life cycle in the sea. Massive efforts to establish Atlantic salmon were made between 1864 and 1910, with at least 24 different introductions, involving nearly 5 million ova that produced more than 2.5 million fry for release. Apart from a landlocked stock established in the Lakes Te Anau-Manapouri system in southern New Zealand, the venture was a failure (Thomson 1922, Stokell 1959). There is no compelling evidence that a single sea-run salmon has ever returned to New Zealand's rivers.

The same is true of early attempts to establish chinook salmon. Four consignments arrived in the late 1870s but again no reliable evidence of success exists.

The intent of these early introductions under the auspices of the proliferating acclimatization societies (more than twenty of them were set up before 1890) was to establish recreational fisheries. However, in the early years of the twentieth century the New Zealand government became involved. It saw a potential for the establishment of commercial salmon fisheries in New Zealand comparable in character to the salmon fisheries of the western coast of the United States and Canada.

Beginning in 1901 and continuing for about five years, chinook salmon (*Oncorhynchus tshawytscha*) and whitefish (*Coregonus albus*) ova were imported into New Zealand from North America on a substantial scale. While the early introduction of chinooks in the 1870s and 1880s had failed, the later efforts succeeded largely because of the intensive effort expended. In the early 1900s, New Zealand developed the first successful transplants of Pacific salmon. The whitefish liberations, however, failed.

Also introduced during this period, in 1901–2, with just one small consignment in each instance were North American lake trout (*Salvelinus namaycush*) and sockeye salmon (*Oncorhynchus nerka*). Lake trout were released in a shallow subalpine lake where they have survived (Stokell 1951). The sockeye were released into a large east coast braided river in the South Island and, although there is no evidence of a sea-run, a landlocked stock has persisted in an upland lake.

Period of Consolidation

By about the early 1900s, the major effort at introduction of freshwater fish species had been expended, although intermittent subsequent introductions of brown trout and Atlantic salmon occurred. There followed a period of consolidation during which various established species, but primarily rainbow trout and brown trout, were released in localities throughout New Zealand.

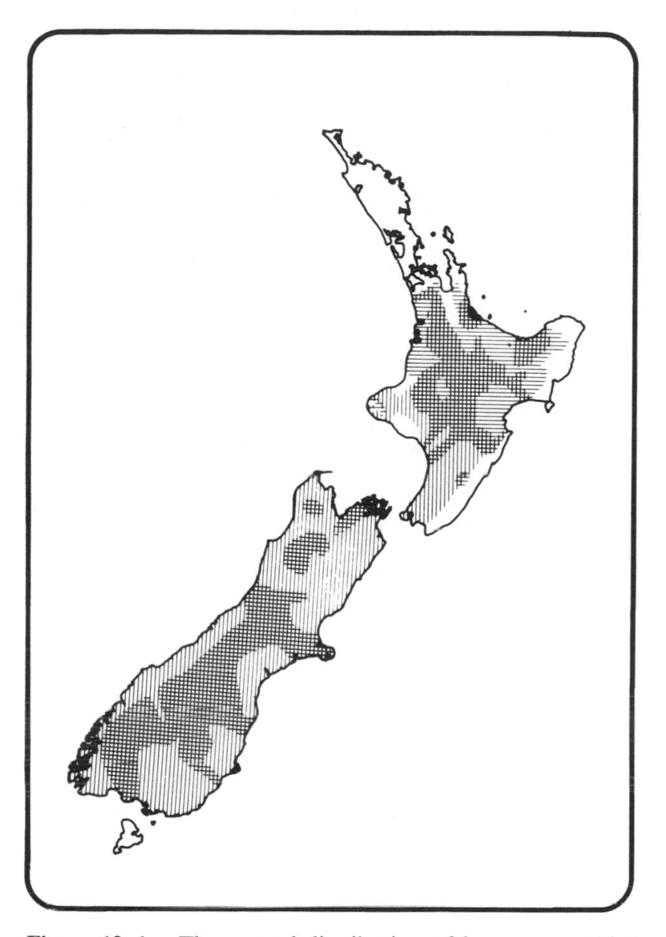

Figure 10–1. The general distribution of brown trout (*Salmo trutta*) (vertical lines) and rainbow trout (*Salmo gairdnerii*) (horizontal lines) in New Zealand. Populations that are artificially maintained in some Northland lakes are not included (after Waugh, 1973).

As seems typical of trout introductions into new environments, there was an initial flush of success: the species occupied hitherto untenanted niches and exploited virgin food resources without significant competition or predation. Brown trout were successful, especially in the river fisheries, while rainbows excelled in the lakes. In the meantime, steady but not spectacular runs of chinook salmon had developed. Acclimatization societies pursued a policy of widespread and massive releases of various salmonids, but largely in spite of these continuing release policies, by the 1940s the three principal salmonid species – rainbow trout, brown trout, and chinook salmon – had probably established a distribution suited to their habitat tolerances. Existing brown trout populations include both sea-run and wholly freshwater stocks. Rainbows have failed to establish sea-run populations, and chinooks, while

Figure 10–2. Distribution of salmons in New Zealand: Atlantic salmon (*Salmo salar*), sockeye salmon (*Oncorhynchus nerka*) and Quinnat salmon (*Oncorhynchus tshawytscha*).

primarily sea-run, have established a series of voluntarily landlocked populations in high elevation lakes.

Brook char became established in localized pockets in headwater streams, primarily in inland South Island localities, while lake trout, Atlantic and sockeye salmon persisted in single river-lake systems, with only Atlantic salmon making much of a contribution to the fisheries.

Of the other species involved in early introductions, European perch created minor local interest among anglers, but failed to achieve wide popularity. Tench, catfish (brown bullhead), and goldfish made no contribution. Few people knew of the existence of either tench or catfish in New

Zealand in the 1930s and 1940s. Other species listed as being among the early importations – dace, barbel, rudd, and others – failed to establish.

The only significant addition to the fauna before about 1950 was the arrival of the mosquitofish (*Gambusia affinis*). This species was introduced from North America via Australia in 1930, ostensibly to control mosquitoes. The introduction and the reasons for it were never properly documented, nor were the results ever studied or reported, and the whole process can only be regarded as virtually frivolous. Mosquitofish became established in warmer North Island waters, primarily in margins of swamps.

By the 1930s the early spectacular success of brown and rainbow trout was beginning to fade away. Lakes, particularly in the central North Island, were becoming overpopulated with poor conditioned fish, and revised management practices were adopted to increase harvest rates and improve food resources that had become seriously depleted. These practices included capture of large numbers of trout for public sale. Following this, the populations largely settled down, an equilibrium developed among environment, food resources, salmonid populations, and exploitation. All that was required to sustain good fisheries was periodic adjustment of the rates of exploitation. Some environmental deterioration was beginning to be noticed. Acclimatization societies persisted in their policies of massive releases of young trout, but scientific evidence now available indicated that this was largely an unproductive activity.

Modern Era in Introductions

In the early 1950s several things changed. Anglers became dissatisfied with the unavailability of game fish in warmer northern waters. Under pressure from anglers, government fisheries staff investigated possible candidates for introduction into warm, shallow lakes, and in the late 1950s recommended that largemouth bass (*Micropterus salmoides*) be introduced. Opposition came primarily from those who were concerned at the potential impact of bass on trout populations. In 1968, the proposal was rejected (McDowall 1968b). In spite of rumors that bass had been privately introduced and released (Anonymous 1966), bass did not come to New Zealand.

Moves were made by various groups to bring in fast-growing rainbow trout and coho salmon (*Oncorhynchus kisutch*), the latter because its wider temperature tolerances, it was thought, might allow it to establish in North Island waters that lacked salmon (Beam 1970). Neither species was introduced.

Although introductions of established species had virtually ceased by the early 1900s, the 1950–60s saw two interesting arrivals. A stock of sea trout was brought to New Zealand for experiments aimed at determining the origin and status of New Zealand's sea-run stocks of this species. A consignment of Atlantic salmon of Baltic origin was procured and released; it was hoped that these fish, accustomed to local migrations in the Baltic, would retain such habits in New Zealand waters and, therefore, be more likely to return from the sea (McDowall 1978). The experiment failed.

Growing awareness in New Zealand of the serious impact of introduced animals and plants (both as pest species and in relation to their impact on the native biota) led to an increasingly careful scrutiny of proposed introductions and a tightening of rules and regulations related to them. Interest in the establishment of additional fish species continued, nevertheless, partly, no doubt, owing to the flush of European immigration into New Zealand in the 1950s following World War II. Species like barbel, dace, and rudd were again nominated, but government attitudes were consistently negative. Frustrations developed, and eventually a consignment of rudd was brought in illegally, apparently along with or misidentified as goldfish, which were legal imports until the early 1970s. The importer bred from these rudd stocks and made clandestine liberations in small lakes and dams in a widespread area of northern New Zealand (McDowall 1978). As a result the importation of all coldwater fish into New Zealand was banned. The same individual was responsible for widespread release of tench derived from populations present since the 1860s.

Eventually the brood stock of rudd was destroyed but it was too late. Rudd were already well established and widespread in northern New Zealand, as were tench (Cadwallader 1977). The problems generated by these illicit releases resulted in the passage of Noxious Fish Regulations in 1980. These regulations specify that possession of certain named fish species, including rudd, was illegal and authorized the confiscation of such fish and prosecution of the person possessing them.

During the 1960s and 1970s other exotic species were also expanding their ranges. For decades, catfish (brown bullhead) had remained little known and were present in only two waters. In recent years they have become both more abundant and widespread in catchments and have spread more widely. Mosquitofish have done the same. Further spread of these species seems possible only by man.

Several other developments have affected the status of exotic species. Aquarium escapees have been discovered in thermal waters in the central North Island. A population of guppies (*Poecilia reticulata*) was once present in a tributary of Lake Taupo, but was exterminated. Another is present in a tributary of the Waikato and has been there for many years. Sailfin mollies (*Poecilia latipinna*) were released into thermal waters at the southern end of Lake Taupo. Populations of both species are confined to limited areas by adjacent cold waters.

Recent Experiments with Herbivorous Carps

Eutrophication of lakes has become a serious problem in a relatively small number of New Zealand lakes over the past twenty to thirty years, resulting in excessive beds of aquatic macrophytes around shallow lake margins and phytoplankton blooms in open lake basins. Drainage ditches have for many years had problems with aquatic macrophytes, and considerable effort and expense are required to keep them clear, mostly by mechanical means or herbicidal sprays. To seek ways of combating the symptoms of eutrophication, New Zealand introduced brood stocks of Chinese grass carp or white amur (*Ctenopharyngodon idella*) and silver carp (*Hypophthalmichthys molitrix*). Research staff have spent about ten years developing culture techniques for the production of stock and evaluating the effects of releases of these species on a wide range of environmental factors. They have examined their effectiveness in controlling problem plant growths, the effects on water quality, interactions with other indigenous and imported fish species, and impact on wildfowl habitats (Edwards and Hine 1974, Edwards 1974, Edwards and Moore 1975, Mitchell 1977, 1980).

These studies continue and, although the number and diversity of experimental catchments and waters in use are rising slowly, no decision has yet been made to approve release of fish to either farmers (for use in weed-infested drainage ditches) or fisheries/water managers (for use in lakes). Studies are being carried out to investigate the use of sterile hybrid stocks.

Early opposition to the use of grass carp came from fishing clubs and some fishery managers. Some of this persists, but has been partly replaced by opposition from managers of wildfowl populations who are concerned that grass carp may destroy marginal vegetation in lakes occupied by ducks and swans. Ill-informed critics have confused grass carp with European carp (*Cyprinus carpio*) (Anonymous 1981). The future uses of silver carp and grass carp remain uncertain. At present the most likely use for grass carp is the control of weed growth in farm drainage ditches.

The mid-1970s saw a developing interest in the ocean ranching of chinook salmon (Waugh 1981). This activity has not yet had any major impact on the distribution or abundance of the species although the latter will occur as the numbers of salmon released rise. Some expansion of the range of the species in New Zealand is possible as already two ocean-ranching ventures, operating outside the "natural range" of chinook salmon in New Zealand, seek to generate runs in new river systems. Success to date has not been notable.

Another management strategy under study seeks the establishment of sea-run rainbow trout. Attempts are being made in several areas, but to date the results are not encouraging. Rainbow trout in low elevation coastal rivers tend to migrate to sea and never return.

Legislation, Regulations, and Policy

There is currently prohibition on the private importation of "coldwater" fishes into New Zealand that would be regarded as likely to survive in the wild at ambient water temperatures. There are no such controls on tropical species which may be imported by licensed persons with approved quarantine facilities. Although these controls cover importations of fish into New Zealand, they do not deal with possession. The recently passed Noxious Fish Regulations now cover the prohibition or possession of nominated species. The controls on coldwater species mean that virtually only government agencies may import such species.

As discussed above, from time to time, there have been proposals to introduce additional freshwater game fishes, such as largemouth bass and coho salmon. In New Zealand the Freshwater Fisheries Advisory Council, made up of a membership of acclimatization societies and government officials, is responsible to the Minister of Fisheries for advice on policy and administration of freshwater fisheries. This includes advice on proposals to introduce fish. The council established a subcommittee to examine proposals for fish introductions and to establish guidelines to govern them. These guidelines are designed to ensure that introductions are made with proper safeguards and consideration, that the most suitable species are considered, and that any introduction is environmentally safe.

The guidelines include the following:

> Having identified a water-type which would appear suitable to support a game fish, it would be necessary to select a specific water body suitable for introductions, with detailed studies of the fauna, flora, chemical, and physical characteristics. After this there would need to be selection of fishes appropriate for the chosen waters, a suitable quarantine period for imported fishes, and introduction of trial stocks of such fishes for controlled studies in artificial ponds. Observations on behaviour, growth, food, parasites, and diseases would follow, and once these were completed, release of disease-free fish into an isolated lake could occur, with a continuing study of the lake environment, the native biota, and the introduced fish. Particular attention would be paid to interaction of the native and introduced faunas. Subsequently, stocks would be released into a water with a trout population to examine any interactions with trout. If the candidate for release proved suitable, it would then be necessary to establish rearing-hatchery facilities and to develop a technology to build up stocks of fish for release, thereby avoiding introduction of further ova or fish. Important factors that need to be considered include the ability of the fish to exist in areas beyond those contemplated for release, effects on existing fisheries, and effects on the native fauna.

The government is now applying these guidelines to the consideration of grass carp and silver carp as biological controls for lake weeds and plankton blooms. At present there are no "live" proposals to establish additional species.

The virtual absence of important salmonid diseases in New Zealand makes the likelihood of introducing additional salmonids slight. Whirling disease was first identified in New Zealand in 1971 and recurred in 1979. Otherwise New Zealand can claim freedom from significant salmonid diseases.

Status and Utility of Existing Exotic Fauna

The chief exotic species in New Zealand are salmonids of recreational value. Their distribution and status may be summarized as follows (McDowall 1978, 1979):

Brown trout: Widespread, in most river systems from Auckland south, distribution stable, probably the most important recreational freshwater fish in New Zealand.

Rainbow trout: Also widespread but more dominant in inland lakes, present in some larger river systems mostly in headwaters; some attempts underway to establish sea-run populations; important as a recreational species in lakes.

Chinook salmon: Present and abundant along most of the east coast of the South Island, limited sea-run populations in a restricted area of the west coast, landlocked populations in inland lakes in both areas. Some slight range extension possible as a result of ocean-ranching activities; a popular angling species in South Island rivers.

Sockeye salmon: Present only in lakes in the headwaters of the Waitaki River system, status fragile and presently subject to major changes because of planned modifications of river flow patterns in relation to hydroelectricity construction; unexploited.

Atlantic salmon: Present only in Lakes Fergus and Gunn in the upper Eglinton Valley, status threatened; some minor exploitation.

Brook char: Widespread but local, mostly in small headwater streams where other salmonids are absent; mostly small fish of no interest to anglers but a population of catchable fish in a few small lakes, some maintained by hatchery releases; some exploitation.

Lake trout: Present only in Lake Pearson in subalpine Canterbury, status threatened, exploited only incidentally by anglers seeking other species; largely unrecognized.

Catfish: Distribution not clear, but primarily Waikato River in North Island and Lake Mahinapua in the South Island; abundant where present and regarded as a nuisance; unexploited.

Goldfish: Widespread and locally present in many small lakes; unexploited.

Tench: Widespread in waters in and north of the Waikato River, plus populations in a small lake in the southern North Island and a few eastern South Island rivers; largely unexploited.

Rudd: Widespread and locally abundant in lakes and slowly flowing rivers in and north of the Waikato; largely unexploited; designated a noxious species.

European carp: Localized populations of "Koi" hybrids in ponds and occasional farm dams, actual occurrence not documented; unexploited; designated a noxious species.

Grass carp: Present in captivity in government facilities and experimental releases under controlled conditions in a few northern lakes and drains; unexploited.

Silver carp: Present only in government facilities and in Lake Orakai, as a study/control area; unexploited.

Mosquitofish: Widespread and locally abundant in Northland waters; actual range not documented; of no value for exploitation.

Sailfin molly: Present only in thermal waters, near Tokaanu at the southern margin of Lake Taupo; possibly exploited by aquarists.

Guppy: Present in thermal streams flowing into the Waikato River; unexploited.

European perch: Widespread and abundant in many areas but distribution erratic, mostly in small lakes and sluggish streams; locally exploited by anglers.

Several of the species listed above are best considered as "threatened exotic species." Both anglers and fisheries managers share reservations on importation of new species or stocks, and consequently regard it important that existing stocks be maintained to keep future uses and options open.

Sockeye populations appear threatened by dam construction, but as the dams are recent it is a little too soon to reach any conclusion on their effects. However, an immediate objective must be the development of strategies to be put into effect should the 1982 sockeye run into the principal spawning streams fail (as the 1981 run largely did).

Atlantic salmon were once abundant and a sought-after game fish in the Te Anau-Manapouri system. Good catches were taken in the 1920s and 1930s and fish up to 6 kg were reported. A massive decline in Atlantic salmon abundance seems to have followed heavy exploitation of ova for releases elsewhere and the release of rainbow trout into these lakes in the 1930s. Today, few, if any, Atlantic salmon are caught in Te Anau and Manapouri, but a relict population persists in lakes higher in the system. This population is augmented by the progeny of a brood stock established in a hatchery. While it is likely that this stock is based on a restricted gene pool, efforts at reestablishment seem to be making progress.

Perhaps the most precarious species is the lake trout. Lake Pearson is a shallow basin lake and the survival of lake trout there for seventy-five years is a surprise. They are small and usually in poor condition, reaching a kilogram or a little more. The current strategy with lake trout is to establish and maintain a hatchery brood stock and at present the hatchery fish are surviving. The status of the wild lake population is hard to assess but is regarded as endangered.

Management practices for exotic species, apart from the regulation of rates of exploitation and the protection of habitat values, are minimal. It has long been believed that good habitat will support adequate trout or salmon populations and that in general releases of hatchery fish are either unnecessary or unproductive. Release of hatchery-reared browns and rainbows are made but they cannot be regarded as a prominent aspect of New Zealand's salmonid management. Some interest exists in capture of stunted wild fish from overpopulated streams and their release in lakes with inadequate spawning.

There has, in addition, been some interest in increasing the populations of brook char, and recent releases have been made into a variety of lakes with, as yet, undetermined success.

Impact of Exotic Fishes

Little is known about the relationship between the exotic and indigenous fish faunas. However, a number of inferences can be made. It probably *seems* that exotic fishes have invaded the New Zealand freshwater environment without significant harmful effects. This is possibly an illusion deriving from the fact that very little is known about the former and present status of so many of the native species.

Only one native species is extinct, the so-called grayling *Prototroctes oxyrhynchus*. The fish went into rapid decline as early as the 1870s and has not been reliably reported since 1923 (McDowall 1976). Trout could have had something to do with this, but the early decline seems likely to have preceded the widespread distribution of trout.

In addition, it is evident that stocks of other species are in a state of decline, in large measure because of widespread and often intense deterioration in the aquatic environment available to fish in New Zealand fresh waters. There is little direct and explicit evidence but it seems that in only small measure is decline in the native fauna related to the presence of introduced species. Some native species appear to cope with the presence of trout. The eels, eleotrids, *Retropinna* species (smelt), and others give little indication of harmful effects. While lake populations of *Galaxias brevipinnis* (koaro) are in no danger of extinction there is no doubt, from early descriptions, that the abundance of this species in the important lakes of the central North Island is greatly reduced. Best (1929) and Phillipps (1940) described how this species, once caught in large quantities, can no longer be taken. In fact, it seems likely that the early decline in the condition of trout in Lake Taupo was related to their overexploitation of the population of *G. brevipinnis.*

In lowland streams, the large galaxiid *Galaxias argenteus* has essentially disappeared from habitats occupied by brown trout while the small, higher elevation species *Galaxias divergens* seems incompatible with trout in small trout nursery streams in the foothills. *Galaxias divergens* populations tend to retreat into the hills and be found in localities above barriers to the upstream migration of the spawning trout and their progeny. Fish (1966) reported that when rainbow trout were introduced into a small coastal dune lake the native freshwater crayfish (*Paranephrops planifrons*), the crab (*Halicarcinus lacustris*), and a pelagic galaxiid, *Galaxias gracilis,* disappeared.

There are few conclusive data that trout are detrimental to stocks of native fish on a widespread basis, but it seems clear that in some localities and for some species extinction or marked stock reductions of native species have been associated with the presence of introduced trout. New Zealand's freshwater fish fauna evolved in almost an absence of piscivorous predators, facilitating successful invasion of New Zealand's fresh waters by piscivores, evident in the early successes of trout in New Zealand.

It seems that, for their survival, New Zealand's indigenous fauna needs some protection from the introduced salmonids and that the current criteria being applied to the introduction and establishment of any additional species should be carefully applied.

Literature Cited

Anonymous. 1966. Largemouth rumour. Hunting and Fishing in New Zealand 2(5):38.

Anonymous. 1981. Rather weed than carp. New Zealand Fishing News 3(5):11.

Beam, M. 1970. Background data relating to proposed introduction of the coho salmon (*Oncorhynchus kisutch*). Appendix D to Chairman's Report, Freshwater Fisheries Advisory Council no. 26.

Best, E. 1929. Fishing methods and devices of the Maori. Bulletin of the Dominion Museum, Wellington 12:1–230.

Cadwallader, P. L. 1977. Introduction of rudd *Scardinius erythrophthalmus* into New Zealand. Part 1. Review of the ecology of rudd and the implication of its introduction into New Zealand. Fisheries Technical Report, New Zealand, Ministry of Agriculture and Fisheries 147:1–18.

Edwards, D. J. 1974. Weed preference and growth of young grass carp in New Zealand. New Zealand Journal of Marine and Freshwater Research 8:341–350.

Edwards, D. J., and P. M. Hine. 1974. Introduction, preliminary handling, and diseases of grass carp in New Zealand. New Zealand Journal of Marine and Freshwater Research 8(3):441–454.

Edwards, D. J., and E. Moore. 1975. Control of water weeds by grass carp in a drainage ditch in New Zealand. New Zealand Journal of Marine and Freshwater Research 9:283–292.

Fish, G. R. 1966. An artificially maintained trout population in a Northland lake. New Zealand Journal of Science 9:200–210.

Fleming, C. A. 1979. The geological history of New Zealand and its life. Auckland University Press, Auckland. 141 pp.

Hunter, J. S. 1915. Introduced game in New Zealand. California Fish and Game 1(2):41–44.

Laycock, G. 1966. The alien animals—the story of imported wildlife. Natural History Press, Garden City, New York. 240 pp.

McDowall, R. M. 1968a. Interactions of the native and alien faunas of New Zealand and the problem of fish introductions. Transactions of the American Fisheries Society 97(1):1–11.

McDowall, R. M. 1968b. The proposed introduction of the largemouth black bass *Micropterus salmoides* (Lacepède) into New Zealand. New Zealand Journal of Marine and Freshwater Research 2(2):149–161.

McDowall, R. M. 1976. Fishes of the family Prototroctidae (Salmoniformes). Australian Journal of Marine and Freshwater Research 27:641–659.

McDowall, R. M. 1978. New Zealand freshwater fishes—a guide and natural history. Heinemann Educational Books, Auckland. 230 pp.

McDowall, R. M. 1979. Exotic fishes of New Zealand—dangers of illegal releases. New Zealand Ministry of Agriculture and Fisheries, Fisheries Research Division Information Leaflet 9:1–17.

McDowall, R. M., and A. H. Whitaker. 1975. The freshwater fishes, 227–299. *In:* G. Kuschel (ed.). Biogeography and ecology in New Zealand. Monographiae Biologicae 27:1–689.

Mitchell, C. P. 1977. The use of grass carp for submerged weed control, 145–148. Proceedings of the 30th New Zealand Weed and Pest Control Conference.

Mitchell, C. P. 1980. Control of water weeds by grass carp in two small lakes. New Zealand Journal of Marine and Freshwater Research 14(4):381–390.

Phillipps, W. J. 1940. Fishes of New Zealand. Avery, New Plymouth. 87 pp.

Scott, D., I. Hewitson, and J. C. Fraser. 1978. The origin of rainbow trout *Salmo gairdneri* Richardson in New Zealand. California Fish and Game 64(3):210–215.

Stokell, G. 1951. The American lake char *Cristivomer namaycush*. Transactions of the Royal Society of New Zealand 79(2):213–217.

Stokell, G. 1955. Freshwater fishes of New Zealand. Simpson and Williams, Christchurch. 145 pp.

Stokell, G. 1959. Structural characters of Te Anau salmon. Transactions of the Royal Society of New Zealand 87(3 & 4):255–263.

Thomson, G. M. 1922. The naturalisation of animals and plants in New Zealand. Cambridge University Press, Cambridge. 607 pp.

Waugh, G. D. 1981. Salmon in New Zealand, 277–303. *In:* J. Thorpe (ed.). Salmon Ranching. Academic Press, London. 441 pp.

Wodzicki, K. A. 1950. Introduced mammals of New Zealand. New Zealand Department of Scientific and Industrial Research Bulletin 98:1–250.

CHAPTER 11

Bacteria, Parasites, and Viruses of Aquarium Fish and Their Shipping Waters

Emmett B. Shotts, Jr., and
John B. Gratzek

The potential spread of bacteria, parasites, and viruses through the importation and sale of aquarium species has been of interest to federal and state agencies responsible for animal health for many years. This interest reached a peak in 1973 when it was reported that 300,000 of the estimated 2 million annual cases of human salmonellosis in this country were associated with pet turtles (Wells et al. 1973), primarily *Chrysemys scripta*. This species was subsequently removed from the market by federal regulation.

Canadian scientists showed in 1974 that bacteria of potential public health importance were associated with aquarium fish and their transport water (Trust and Bartlett 1974). The Pet Industry Joint Advisory Council of the United States (PIJAC), noting these findings and realizing the mounting interest, funded a research program at the College of Veterinary Medicine, University of Georgia, to ascertain whether a similar problem existed in the U.S. aquarium fish industry. The program was designed to determine: (1) if a problem existed, (2) the magnitude of the problem, and (3) the possible introduction through importation of exotic tropical species of potential health hazards that might not be currently present in this country.

The worldwide economic value of the pet fish industry has been estimated at approximately $1,800 million retail and about $600 million wholesale annually. Each year in the United States between 75 and 350 million fish are

Table 11-1. SOUTHEAST ASIAN FISH EXAMINED DURING PIJAC STUDIES AT COLLEGE OF VETERINARY MEDICINE, UNIVERSITY OF GEORGIA

Families	Genera	Species
Cyprinidae	Rasbora	heteromorpha
Poeciliidae	Poecilia	reticulata
Cichlidae	Pterophyllum	scalare
Scatophagidae	Apistogramma	ramivezi
Cobitidae	Astronotus	ocellatus
Centropomidae	Scatophagus	argus
Gobidae	Acanthopthlamus	kuhlii
Characidae	Barbus	everetri
Tetraodontidae	Chanda	tetrazona
Hemiramphidae	Brachygobius	titteya
Siluridae	Paracheirodon	ranga
Mastacembelidae	Hyphesodorycon	xanthozona
Callichthyidae	Tetraodon	innesi
Anabatidae	Dermogenys	herbertaxelrodi
Gyrinocheilidae	Labeo	fluviatus
Bothidae	Kryptopterus	pusilius
	Morulius	bicolor
	Xiphophorus	bicirrhis
	Corydoras	chrysophekadion
	Cichlasoma	hymenophysa
	Helostoma	helleri
	Gyrinocheilus	severum
	Hemigrammus	temmincki
	Metynnis	aymonieri
		rubrifrons
		macracanthus
		pentazona hexazona
		flammeus
		innesi
		maculatus
		schreitmuelleri

sold to maintain a pet fish population that averages 20 fish in each of 16.3 million households. This approximates 326 million pet fish in the United States at a value of between $115 and $344 million.

The purpose of this chapter is to report and discuss the findings of a three-year research program funded by PIJAC. Our comments are directed at the potential ecological impact which the introduction of these fish and/or their transport water might incur upon the health of humans, domestic animals, or native game and fish species.

Source of Fish

Previous market surveys have shown that 60–70 percent of the ornamental fish imported into the United States originate from Southeast Asia, approximately 25 percent from South America, and the remainder from domestic

Table 11-2. SOUTH AMERICAN FISH EXAMINED DURING PIJAC STUDIES AT COLLEGE OF VETERINARY MEDICINE, UNIVERSITY OF GEORGIA

Families	Genera	Species
Callichthyidae	Corydoras	punctatus
Characidae	Megalamphodus	metae
Loricariidae	Paracheirodon	sweglesi
Pimelodidae	Plecostomus	axelrodi
Cichlidae	Pimelodella	punctatus
Hemiodontidae	Apistogramma	pictus
Gasteropelecidae	Copeina	ramirezi
Anostomidae	Nannostomus	arnoldi
Gymnotidae	Carnegiella	harrisoni
	Loricaria	discus
	Hemigrammus	parva
	Chilodus	angelicus
	Acanthodoras	innesi
	Hyphessobrycon	rhodostomus
	Gasteropelecus	punctatus
	Cichlasoma	spinosissimus
	Eigenmannia	rubrostigma
	Metynnis	levis
	Anostomus	testivum
	Mimagoniates	agassizi
	Pristella	virescens
		schreitmuelleri
		anostomus
		peruvianus
		microlepis
		elegans
		severum
		riddlei
		striatata

production (Ford 1981). It is estimated that Florida supplies approximately 80 percent of the domestic needs of the United States (Anonymous 1979).

Because these three sources represented the bulk of ornamental fish distributed in the United States, our work centered on studies directed at these sources. In the case of imported species, the general geographic area was further divided. Fish from Southeast Asia were obtained from Singapore, Hong Kong, Bangkok, and Taiwan. The fish in this phase of the study represented 16 families, 24 genera, and 31 species (table 11-1).

The second phase of the study involved imported ornamental species from South America and represented shipments from Colombia, Brazil, Peru, and Guyana. The fish examined from this geographical area represented 9 families, 21 genera, and 29 species (table 11-2).

Studies of the domestic supply of ornamental fish were confined to suppliers located in central Florida. The fish examined here represented 5 families, 14 genera, and 24 species (table 11-3).

Replicate shipments from these predetermined sources within each of the major geographical supply areas were obtained. Each shipment contained

Table 11-3. CENTRAL FLORIDA FISH EXAMINED DURING PIJAC STUDIES AT COLLEGE OF VETERINARY MEDICINE, UNIVERSITY OF GEORGIA

Families	Genera	Species
Cyprinidae	Barbus	titteya
Characidae	Hyphessobrycon	pulchripinnis
Poecillidae	Megalamphodus	tetrazona
Anabatidae	Brachydanio	megalopterus
Belontiidae	Aphyocharax	rerio
	Poecilia	rubripinnis
	Xiphophorus	flammeus
	Helostoma	conchonius
	Puntius	helleri
	Cichlasoma	variatus
	Trichogaster	callistus
	Astyanax	temmincki
	Colisa	oligolepis
	Pterophyllum	schuberti
		meeki
		trichopterus
		serpae
		mexicanus
		nigrofasciatus
		lalia
		albolineatus
		maculatus
		scalare
		reticulata

from three to eight lots (bags) of approximately 50 fish each. The foreign shipments were sent by air from point of origin to a wholesaler who reoxygenated the shipping water and forwarded the shipment via air express to the University of Georgia. Transshipping point for Southeast Asia was Chicago or Los Angeles, and for South America, Miami. All shipments arrived at the laboratory within twenty-four hours of receipt by the wholesaler who transshipped.

Domestic shipments were made directly from the farm to the laboratory, arriving within twenty-four hours. Upon arrival, the fish and shipping water were systematically processed to determine the presence of bacteria, parasites, and viruses following the schema depicted in figure 11-1.

Methods and Materials

BACTERIA

Commercially available bacteriological media were made using the manufacturer's recommendations where possible. Several media not commercially available were made according to directions found in the cited

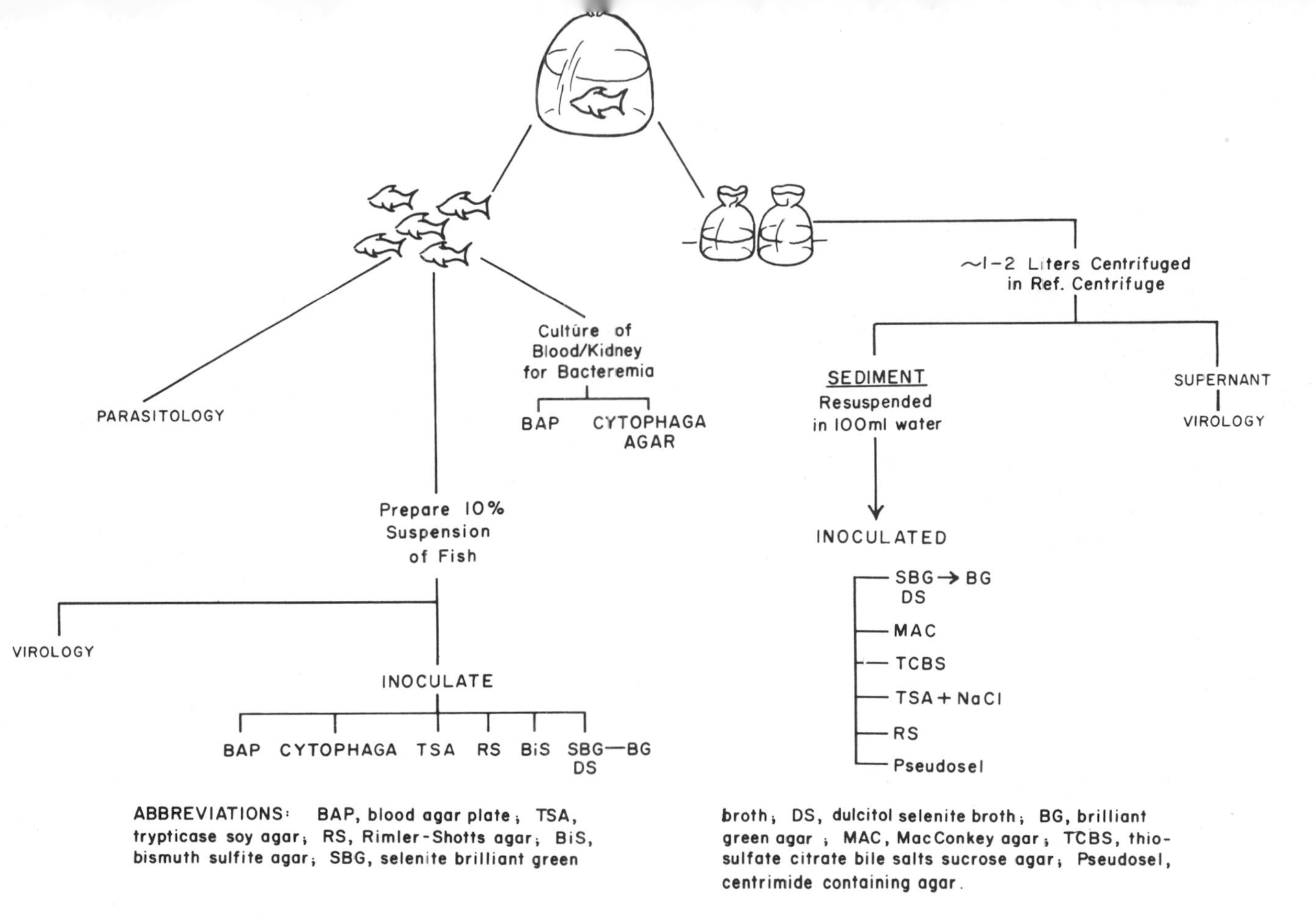

Figure 11-1. Procedural schematic employed for the study of imported aquarium fishes.

literature. All cultures were made in duplicate and incubated at 22 C and 35 C, respectively. Colonies representative of the flora cultured were transferred to trypticase soy agar (TSA) and identified (Lenette et al. 1980, Koneman et al. 1979, Buchanan and Gibbons 1974). Colonies representing all morphological types were selected for identification studies.

After opening an individual bag of fish, blood from five randomly selected fish from each lot was cultured. This was done by killing the fish, sterilizing the exterior surface by dipping in ethyl alcohol, and igniting. The fish were then dissected, exposing the kidney for culture. In smaller fish (< 3 cm) a mixture of kidney and blood was cultured. Initial cultures were streaked on Cytophaga agar (CA) (Anacker and Ordal 1959) and blood agar (BA) plates.

Additional fish were killed and their outer surfaces sterilized as described above. These fish were homogenized whole (without evisceration) in a blender with pH 7.2 sterile phosphate buffered saline (PBS) to obtain a 10 percent wt/vol of macerated fish suspension. This tissue slurry was divided, an aliquot referred for virus isolation studies, and the remainder used to inoculate a series of bacteriological media selected to insure isolation of a wide variety of microorganisms. Selective media were utilized when possible to enhance the probability of specific organisms being recovered even though in low numbers. The media used for this portion of the studies were: BA, utility medium with sensitivity for fastidious organisms; CA, for flexibacterial isolations; TSA, a second utility medium; Rimler-Shotts agar (RS) (Shotts and Rimler 1973) for recovery of *Aeromonas* and other enteric bacteria; bismuth sulfide agar (BiS) specific for *Salmonella typhi.* Tubes of both selenite brilliant green (SBG) and dulcitol-selenite (DS) (Raj 1969) broths were made in double concentration and an equal amount of slurry was added to each. Subsequently, both media were incubated for 48 h at 41.5 C, then streaked to brilliant green agar (BGA) at 24 h and 48 h for detection of salmonellae.

The bacteriological flora associated with the shipping water from each shipment was also examined. At arrival, and concurrent with this portion of the protocol, a pooled water supply was collected that represented water from each lot in the shipment and approximated a volume of 1,500 ml. The water sample was centrifuged at 2,488 xg for twenty minutes at 4 C. Supernatant from this centrifugation was kept for virological isolations. The sediment was collected and resuspended in approximately 100 ml of the supernate to serve as inoculum for a second series of bacteriological media. These media were MacConkey agar (MAC), selective for Gram negative bacteria; thiosulfate-citrate-bile salts-sucrose agar (TCBS), selective for *Vibrio;* trypticase soy agar with 1.5 percent NaCl added (TSA-NaCl), for growth of organisms requiring increased salt concentration; RS, selective for *Aeromonas* and other enteric organisms; Centrimide agar (PA), selective for *Pseudomonas;* and BiS, selective for salmonellae, particularly *Salmonella typhi.* Further examination for salmonellae was done by inoculation of tubes of DS and SBC in 2x concentration with equal amounts of the suspended sediment. These tubes were incubated for 48 h at 41.5 C, then streaked to BGA at 24 h and 48 h.

PARASITES

At the time the individual lots of fish were opened, five fish from each bag were examined for the presence of parasites utilizing standard methods of dissection and examination (Reichenbach-Klinke 1973, Brown and Gratzek 1980, Gratzek 1980).

VIRUSES

Two previously described types of sample collections (slurries and water) were filtered through four layers of cheesecloth and centrifuged at 1,000 xg for fifteen minutes to remove particulate material. The supernatant was retained for further preparation. Both the slurry preparations and the shipping waters were centrifuged at 4 C at 5,000 xg, then filtered through a 0.45 micrometer pore size membrane filter utilizing positive pressure.

These prepared inocula were placed on the following cell cultures to attempt detection of viruses: rainbow trout gonad (RTG-2); fathead minnow (FHM), brown bullhead (BB); VERO African green monkey kidney; pig kidney; bovine kidney; feline kidney; and rabbit kidney. Microculture techniques, medias, and cell culture procedures have been previously described in detail (Gratzek et al. 1973, Rovozzo and Burke 1973). All samples were passed three times at five-day intervals. Samples were considered negative if no cytopathic changes were noted at the completion of the third passage.

ANTIBIOTIC STUDIES

Antibiotic sensitivity studies were done on bacterial isolates identified as belonging to the *Aeromonas hydrophila* complex. These isolates represented all geographical locations and were obtained from both fish and water. Antibiotic sensitivities were determined with the Kirby-Baur disc procedure (Shotts et al. 1976a). Antibiotics used in these studies were ampicillin, sulfa, streptomycin, tetracyline, furans, kanamycin, chloramphenicol, naladixic acid, and penicillin.

Isolates found to be resistant to either tetracycline or to sulfa were studied further to determine whether the resistance noted was a transferable resistance (R-factor) sometimes called plasmid mediated resistance. These studies involved the transfer of the plasmid to another organism and subsequent identification of the magnitude of the resistance associated with the plasmid (Shotts et al. 1976a, 1976b). Studies were confined to sulfa and tetracycline since these are the only drugs currently approved for use in fish.

Results

During the course of these studies 202 lots of fish were examined: 80 from Florida, 55 from South America, and 67 from Southeast Asia. The usual lot of fish contained about 50 specimens. These results then represent data assembled from approximately 10,100 fish obtained from 32 sources.

BACTERIA

Bacteremia

Bacteremia was defined as the recovery of microorganisms from the blood or kidney of one or more of the group of five randomly selected fish cultured from each lot. During this study 71 of 202 lots of fish (36 percent) were bacteremic upon arrival at the laboratory. Those from Southeast Asia had the highest incident (68 percent) whereas fish from South America and Florida had an incidence of 30 percent and 10 percent, respectively.

The organisms most frequently isolated from bacteremic fish were *Aeromonas hydrophila* (50 percent), *Pseudomonas* sp. (36 percent), and *Staphylococcus* sp. (16 percent). From a regional view, *Aeromonas hydrophila* (100 percent) from Florida; *Pseudomonas* sp. (46 percent) from South America and 63 percent from Southeast Asia should be mentioned. Unique organisms more closely associated with geographical regions include *Vibrio* sp. (13 percent) from Florida, *Penicillum* sp. (23 percent) from South America, and *Proteus* sp. (20 percent) and *Flavobacterium* sp. (13 percent) from Southeast Asia (table 11-4).

Slurry

The bacteria recovered from this phase of our study reflect those that were inside the fish (i.e., the flora of the intestine or tissue) since before preparation of the tissue slurry the surface of each fish was sterilized. Bacteria representing 15 genera were recovered from the fish tissue slurries, with *Aeromonas hydrophila* (79 percent) and *Pseudomonas* sp. (64 percent) recovered most frequently. On a regional basis several potential biological markers were observed. These were a high frequency of *Pleisomonas shigelloides* from Florida, an absence of *Pseudomonas putrefaciens* from South America and a high frequency of *Proteus* sp. from Southeast Asia. Of similar importance was the recovery of low levels of *Salmonella arizona* from all geographical areas (table 11-5). *Salmonella arizona* is perhaps the only organism noted of primary potential public health importance.

Shipping Water

Bacteria recovered from the shipping waters were neither unexpected nor particularly remarkable. Cultures of centrifuged sediments were studied from a total of 32 different sources: 11 from Florida, 9 from South America, and

Table 11-4. MICROORGANISMS ASSOCIATED WITH BACTEREMIA OF ORNAMENTAL FISH FROM DIFFERENT GEOGRAPHICAL AREAS

Organism	Source			Total
	Florida	South America	Southeast Asia	
Pseudomonas sp.	0[a]	46	63	36
Aeromonas hydrophila	100	30	20	50
Staphylococcus sp.	13	8	26	16
Proteus sp.	0	0	20	7
Citrobacter sp.	0	8	11	6
Flavobacterium sp.	0	0	13	4
Enterobacter sp.	0	15	6	7
Bacillus sp.	0	0	11	2
Plesiomonas shigelloides	0	0	4	1
Pseudomonas putrefaciens	0	0	2	0.7
Escherichia coli	0	0	4	1
Penicillum sp.	0	23	0	8
Alpha streptococcus	13	0	0	4
Lots Examined	80	55	67	202
Positive	8	17	46	71
Percentage Positive	10	30	68	36

[a] Percentage occurrence in positive lots

Table 11-5. RELATIVE FREQUENCY OF BACTERIA ISOLATED FROM FISH TISSUE SLURRIES OF ORNAMENTAL FISH FROM DIFFERENT GEOGRAPHICAL AREAS

Bacteria	Source			Total (202)
	Florida (80)[a]	South America (55)	Southeast Asia (67)	
Aeromonas hydrophila	96[b]	87	54	79
Pseudomonas sp.	55	78	58	64
Citrobacter sp.	34	53	28	38
Proteus sp.	5	33	57	32
Enterobacter sp.	33	49	25	36
Staphylococcus sp.	3	49	27	26
Plesiomonas shigelloides	61	2	8	24
Escherichia coli	18	16	27	20
Bacillus sp.	0	29	15	15
Flavobacterium sp.	10	16	16	14
Pseudomonas putrefaciens	26	0	12	13
Klebsiella sp.	8	13	0	7
Salmonella arizona	1	4	2	2
Edwardisella tarda	5	2	0	2
Clostridium perfringens	0	0	2	0.7

[a] Number of lots of fish examined
[b] Percentage of lots in which organism occurred

Table 11-6. BACTERIA ISOLATED FROM THE SHIPPING WATERS OF ORNAMENTAL FISH OBTAINED FROM DIFFERENT GEOGRAPHICAL AREAS

Bacteria	Source			Total (32)
	Florida (11)[a]	South America (9)	Southeast Asia (12)	
Pseudomonas sp.	100[b]	100	92	97
Aeromonas hydrophila	100	78	67	82
Citrobacter sp.	91	44	58	64
Enterobacter sp.	55	67	58	60
Proteus sp.	9	44	75	43
Escherichia coli	36	33	42	37
Klebsiella sp.	9	67	17	31
Plesiomonas shigelloides	73	0	8	27
Flavobacterium sp.	45	22	8	25
Pseudomonas putrefaciens	36	0	8	15
Staphylococcus sp.	9	22	8	13
Corynebacterium sp.	0	33	0	11
Salmonella arizona	27	0	8	12
Pseudomonas aeruginosa	9	11	0	7
Vibrio sp.	55	0	0	18

[a] Number of shipping water sources studied
[b] Percentage sources in which organisms occurred

12 from Southeast Asia. The bacteria recovered represented 13 genera: most frequently encountered were *Pseudomonas* sp. (97 percent), *Aeromonas hydrophila* (82 percent), *Citrobacter* sp. (64 percent), and *Enterobacter* sp. (60 percent) (table 11-6). The occurrence of certain genera from only one geographic area was observed. The presence of *Corynebacterium* sp. from South America and *Vibrio* sp. from Florida was unique. The higher frequencies of *Plesiomonas shigelloides* (73 percent), *Pseudomonas putrifaciens* (36 percent), and *Flavobacterium* sp. (45 percent) from Florida, *Klebsiella* sp. (67 percent) from South America, and *Proteus* sp. (25 percent) from Southeast Asia is also noted. The occurrence of *Salmonella arizona* from Florida (27 percent) and Southeast Asia (8 percent) tends to indicate the presence of this organism in these areas. A similar situation can be noted with *Pseudomonas aeruginosa* in Florida (9 percent) and South America (11 percent). Both of these organisms have minor public health importance.

Antibiotic Studies

A total of 207 of 1,384 isolates of *Aeromonas hydrophila* (87 percent) from ornamental fish were found to be resistant to one or more of the antibiotics assayed. By area, the number of resistant isolates varied from 100 per-

Table 11-7. ANTIBIOTIC SENSITIVITIES OF *AEROMONAS HYDROPHILA* ISOLATED FROM ORNAMENTAL FISH AND THEIR SHIPPING WATER OBTAINED FROM DIFFERENT GEOGRAPHICAL AREAS

	Florida (562)[a]	South America (407)	Southeast Asia (215)	Shipping Water (200)[b]	Total (1384)	
Isolates not resistant	0	19	105	53	177 (13%)	
Isolates resistant[c]	562	388	110	147	1207 (87%)	
Antibiotics						
ampicillin	87[d]	89	91	94	77.5	88.8[e]
sulfa	62	49	84	91	56.6	65
streptomycin	23	12	35	38	19.4	22.2
tetracycline	30	32	43	40	29	33.2
nitrofurans	7.2	6	13.6	2.7	6	6.9
kanamycin	2	3	ND	ND	2.6	ND
chloramphenicol	3.7	5	19	6.8	5.2	6
naladixic acid	3	4.6	0	0	2.6	3
penicillin	ND[f]	ND	69	61.9	40	ND
Percent R-factor to tetracycline + sulfa	1.4	1.3	25.5	11.6		

[a] Total isolates from each source
[b] Randomly selected from each area
[c] Resistant to 1 or more antibiotics
[d] Percentage of isolates resistant to drug
[e] Percentage of total resistant isolates resistant to drug
[f] Not done

cent from Florida to 95 percent from South America to 51 percent from Southeast Asia. This represents an inverse relationship to the presence of transferable plasmid resistance noted. A lower percentage of resistant isolates was observed from Southeast Asia (51 percent) but a high percentage of these had R-factors (25.5 percent) (table 11-7). Similarly, Florida (100 percent resistant) and South America (95 percent resistant) had R-factor percentages of 1.4 percent and 1.3 percent, respectively. Individual antibiotics of very little use are those of the beta-lactam family such as ampicillin (88.8 percent resistant). There seemed to be parallel resistance between sulfa and tetracycline with the resistance to sulfa being approximately 2x that of tetracycline. Common antimicrobics used in fish in the past decade appeared to be of equivocal value as noted by sulfa (56.6 percent), tetracycline (29 percent), and streptomycin (19 percent). Others not commonly used appear to be more applicable for use against *A. hydrophila* such as nitrofurans (6.9 percent), chloramphenicol (6 percent), or naladixic acid (3 percent). The occurrence of R-factors in *A. hydrophila* from shipping water at a frequency of 11.6 percent is of some concern.

Table 11-8. SUMMARY OF INFECTED GROUPS[a] OF TROPICAL FISH FROM THREE GEOGRAPHICAL POINTS OF ORIGIN WITH ANY TYPE OF PARASITE

Groups	Southeast Asia[b]	Florida	South America
Percent positive	61	67	98
Percent negative	39	33	2

[a] A group is defined as those fish contained in one shipping bag. Five to 7 fish were examined per bag.
[b] The number of groups examined from each area was: Southeast Asia, 77; Florida, 87; and South America, 57.

PARASITES

The results indicated that the majority of each fish population examined had some type of parasite. The data presented in table 11-8 indicate that 61, 67, and 98 percent of the groups of fish (bags) from Southeast Asia, Florida, and South America, respectively, were infested. The classes of parasites found are presented in table 11-9 while table 11-10 lists the genera of protozoan infestations. Not listed are the trypanosomes (*Crytobia*) that were found in the blood of *Loricaria filamentosa.*

VIRUSES

Samples from 202 lots of ornamental fish and from 32 water sources were examined for the presence of viruses. A single viral isolate was obtained from a tissue slurry made of Kuhli loaches (*Acanthophthalmus* sp.) and from their shipping water. This sample originated in Southeast Asia. The virus, isolated on rabbit kidney, was grouped as a herpesvirus using size, morphology, DNA content, ether susceptibility and lack of hemagglutinating ability. The isolate did not crossreact with channel catfish herpesvirus and it was not pathogenic to channel catfish.

There is no serological reaction with antiserums to infectious bovine rhinotracheitis virus or pseudorabies virus, but the herpesvirus partially crossneutralized with equine rhinopneumonitis virus.

Conclusion

The bacteria most frequently encountered throughout these studies were *Pseudomonas* sp. and *Aeromonas hydrophila.* Those *Pseudomonas* which were placed in this group were those not readily identifiable to species. In all cases, these isolates did not grow at 41.5 C and failed to produce pyocyanin

Table 11-9. SUMMARY OF THE PERCENT INFESTATION OF TROPICAL FISH FROM THREE GEOGRAPHICAL AREAS WITH CLASSES OF PARASITES

	Southeast Asia	Florida	South America
External protozoans	7.8	55.2	19.3
Sporozoans	1.3	2.3	17.5
Trypanasomes	0	0	5.3
Hexamita	3.1	2.3	0
Unidentified protozoans	0	0	5.3
Monogenetic trematodes	23.4	23.0	54.4
Digenetic trematodes:			
Adult	0	0	0
Metacercariae	0	19.5	43.9
Nematodes:			
Encysted	6.3	1.1	56.1
Intralumenal	12.5	5.7	19.3
Cestodes:			
Encysted	0	2.3	24.6
Intralumenal	0	2.3	1.8
Oligochaetes	0	0	1.8
Acanthocephalans:			
Encysted	3.1	0	3.5
Intralumenal	3.1	0	5.3
Copepods	0	1.2	3.5

Table 11-10. GENERA OF PROTOZOA IDENTIFIED IN GROUPS OF FISH FROM THREE GEOGRAPHICAL AREAS

	Southeast Asia			Florida			South America		
	Gill	Skin	Intestine	Gill	Skin	Intestine	Gill	Skin	Intestine
Parasite									
Ichthyophthirius	3.9	1.3	–	2.3	1.2	–	1.8	3.5	–
Trichodina	1.3	–	–	24.1	28.7	–	1.8	1.8	–
Trichodonella	–	–	–	9.2	2.3	–	–	–	–
Tripartiella	–	–	–	1.2	–	–	–	–	–
Costia	–	–	–	8.0	6.9	–	–	–	–
Chilodonella	–	1.3	–	1.2	1.2	–	7.0	–	–
Epistylus	–	–	–	–	1.2	–	–	3.5	–
Glossatella	–	–	–	1.2	2.3	–	–	–	–
Scyphidia	–	–	–	2.3	2.3	–	–	–	–
Oodinium	–	1.3	–	1.2	1.2	–	–	–	–
Tetrahymena	–	–	–	–	1.2	–	–	1.8	–
Hexamita	–	–	3.1	–	–	2.3	–	–	–
Sporozoans	–	1.3	–	2.3	–	–	15.8	1.8	–
Unidentified									5.3

or grow in acetamide broth. Most could be placed in the *Ps. putida*-like group. Another genus, *Citrobacter,* was noted rather frequently. While these isolates were not identified to species, most resembled *Citrobacter freundii.* They were ONPG positive nonlactose fermenters which created the false impression of the presence of salmonellae as has been previously reported (Shotts et al. 1976c, Trust and Bartlett 1974).

The batteries of bacteriological media used provided a comprehensive isolation schema. The predominant flora recovered were gram negative rods; in a few instances gram positive bacteria and fungi were noted. Preliminary attempts to isolate mycoplasmas and mycobacteria from these fish failed; however, mycobacteria were observed in some instances but lost during attempts to recover them in pure culture (Shotts et al. 1976c).

The incidence of bacteremia in shipments of fish seemed to be directly related to the distance they were shipped. The compounding effects of CO_2 buildup in the shipping water and inherent stresses placed upon the fish no doubt entered into the probability of the fish contracting bacteremia.

The presence of *Escherichia coli* and other coliform organisms, particularly from the Southeast Asian shipments, suggests water contamination with fecal material of animal or human origin. The least evidence of such contamination was observed from the Florida shipments. While *Edwardisella tarda* was observed in fish on occasion, these instances were probably associated with the predator birds and reptiles that frequent fish-rearing ponds. The presence of *Plesiomonas shigelloides, Pseudomonas putrefaciens,* and *Flavobacterium* sp. was probably a reflection of the high organic content of the rearing ponds.

The overall similarities of bacteria recovered, particularly from the slurries and water, tend to indicate that this array of bacteria can be expected to occur commonly in ornamental fish. These genera reflect those isolated by Collins (1970) in studies of healthy fish from an oligotrophic lake in England, suggesting that this distribution of organisms in fish may be worldwide in scope. Variations in genera with geographical areas can be accounted for by pond fertilization and varying practices used in these areas for intensive fish culture.

The areas that were sampled for these studies vary greatly in fish culture management practices. In Southeast Asia intense culture procedures are employed predicated upon maximum yield from space and money involved. In Florida, the culture is more of a selective management approach where fish are for the most part pond-raised and normal growth patterns are augmented only when problems arise. Contrasting with both of these is the wild capture "survival of the fittest" that is used in South America to provide ornamental fishes for the world market.

These variations in fish origin may be further delineated by the antibiotic studies done on the *Aeromonas hydrophila* populations obtained from these areas. The Florida and South American strains of *A. hydrophila* have R-factor rates of 1.4 percent and 1.3 percent, respectively, whereas strains from Southeast Asia approximate a rate of 26 percent. This contrast regarding transferable resistance indicates large volume use of antibiotic therapy on a prophylactic basis rather than for treatment of disease conditions.

Although more strains from Florida and South America are resistant to one or more antibiotics than are those from Southeast Asia, most are single antibiotic resistance of a chromosomal nature. This is in sharp contrast to observations in the Southeast Asian strains where multiple antibiotic resistance of a R-factor (transferable plasmid) nature is the dominant feature. High resistance indices are to be expected in gram negative rods when beta lactam drugs such as penicillin or ampicillin are considered since beta lactamase is present to some extent normally in all gram negative bacteria. A high rate of chromosomal resistance to sulfa and streptomycin could be expected because of selection of resistant strains over the past four decades of use. The resistance of sulfa is further complicated by its association with tetracycline resistance on one of the more commonly transferred R-factors noted in *A. hydrophila* (Shotts et al. 1976a, 1976b). Other factors affecting the use of antibiotics in fish culture are reflected by the ability of the fish to absorb the antibiotic and the amount of that antibiotic that may reach the site of infection (Nusbaum and Shotts 1981a, 1981b).

The initial purpose of the parasitological study was to determine whether there were any heretofore undiscovered and perhaps undescribed parasites that could pose a threat to native species. The results suggest that infestations occur frequently with common parasites or at least with parasites that are common in cultivated food fish as well as with cultivated ornamental species. The unidentified group of parasites found in the intestines of wild caught discus fish from South America probably are members of the genus *Protopalina*. This ciliate has been associated with discus fish by Schubert who claims that it is pathogenic (Schubert 1977).

In general the results of this parasitological study cannot be quantitatively compared because of the way in which samples were collected. Fish from Southeast Asia were sent to our laboratory from transfer points. We anticipated that such fish were subjected to treatments by the fish grower and possibly before shipment. Fish from South America were also subjected to a series of transfers which may have involved treatments. Conroy et al. (1981) point out that there is a 50 percent loss in fish between the collection point and the time of arrival at the aquarium of the exporter. We would assume that many of these mortalities are attributable to a variety of causes including parasitic infestations as well as bacterial problems resulting from the stress of shipping and handling. Collaborators from Florida were asked not to treat fish before shipment. This may explain the greater variety of external protozoans seen (table 11-10) when compared to fish from Southeast Asia or South America. Examination of tropical fish from Florida, which were prophylactically treated before shipment, revealed that infestation with external parasites was minimal.

Approximately 98 percent of the fish in bags from South America were infested. Examination of table 11-9 indicates that monogenetic trematodes, metacercariae of digenetic trematodes, plerocercoid stages of cestodes, and nematode infestations comprised the majority of parasitism. Many of the fish from South America were infested with multiple types of parasites; it was not uncommon to examine fish having intermediate forms of cestodes, digenetic trematodes, and nematodes. From our examination of South American fish, we have concluded that the main cause of stunted or unthrifty fish appears to be heavy infestations with such intermediate forms. Furthermore, the incidence of this type of parasitism appears to be seasonal. Fish collected during the months of January and February appear to be most heavily infested. Apparently, the seasonality of the disease is associated with the gradual evaporation of water from collecting ponds, coupled with the presence of fish-eating aquatic birds that harbor the adult stages of the cestodes and trematodes. It is well recognized by South American collectors that fish collected from ponds toward the end of the dry season are inferior; fish collected from small rivers with a dense canopy of trees appear to be less infested, probably because of the reluctance of aquatic birds to hunt in such areas. We assume that aquatic birds are efficient transmitters of parasites throughout their natural flyways.

Conroy et al. (1981), in their study of parasites of northern South American fish, found a few more species of parasites than our study. We did not observe leeches, *Lernea,* or the isopod *Livoneca* – parasites that are commonly observed in tropical fish within the United States. In agreement with the results of Conroy et al. (1981), we did find a variety of sporozoans and *Hexamita* – infestations that can severely debilitate tropical fish.

The results of the parasitological study suggest that, while there are "exotic" aquarium fishes, there do not appear to be "exotic" aquarium-fish parasites.

In summary, ornamental fish are a unique group of pets in that direct contact between owner and pet is minimal to nonexistent. Common hygiene practices should minimize the exchange of the existing bacterial flora between the pet and owner. It would be presumptuous from the findings presented in this study to assume that the introduction of microorganisms already present in this country via aquarium fishes presents a source of potential health hazard to humans, domestic species, or indigenous wild animals or fish in the United States.

Acknowledgments

The authors are indebted to the Pet Industry Joint Advisory Council (PIJAC) for funding these studies. The individuals who contributed in various ways are many and include: J. L. Blue, A. L. Kleckner, John Brown, Beverly Goven, Thomas Nemitz, Kenneth Nusbaum, Jeannine Gilbert,

Deborah Talkington, Lynn Weldon, Linda Goble, Sharon Chitwood, Betty Seckinger, Lucy Campbell, Amy Lohr, Gene Hale, Doug Waltman, and the staff of the media room.

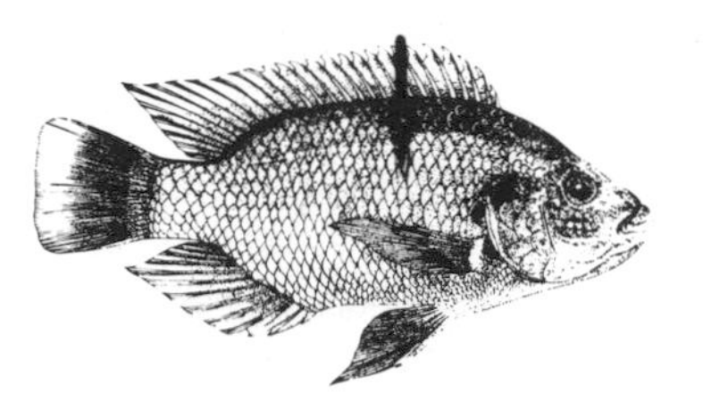

Literature Cited

Anacker, R. L., and E. J. Ordal. 1959. Studies on the myxbacterium *Chrondococcus columnaris* 1. Serological typings. Journal of Bacteriology 78:25–32.

Anonymous. 1979. United Nations study of tropical fish trade. Pet Age, February and April 1979.

Brown, E. E., and J. B. Gratzek. 1980. Fish farming handbook. Avi Publishing Company, Westport, Connecticut. 319 pp.

Buchanan, R. E., and N. E. Gibbons. 1974. Bergey's manual of determinative bacteriology, 8th ed. Williams and Wilkins Company, Baltimore, Maryland. 1,246 pp.

Collins, V. G. 1970. Recent studies of bacterial pathogens of freshwater fish. Water Treatment and Examination 19:3–31.

Conroy, D. A., J. Morales, C. Perdomo, R. A. Ruiz, and J. A. Santicana. 1981. Preliminary observations on ornamental fish diseases in northern South America. Rivista Italiana di Piscicoltura e Ittiopatologia 16:86–98.

Ford, D. M. 1981. The hobby of ornamental fish keeping, 317–322. *In:* D. M. Ford (ed.). Waltham Symposium no. 3, The diseases of ornamental fishes. Pedigree Petfoods Animal Studies Centre, Waltham-on-the-Wolds, Leicestershire, England.

Gratzek, J. B. 1980. An overview of the diseases of ornamental fishes, 25–39. Proceedings of the 4th Kal Kan Symposium.

Gratzek, J. B., W. H. McGlamary, D. L. Dawe, and T. Scott. 1973. Microcultures of brown bullhead (*Ictalurus nebulosus*) cells: Their use in quantitation of channel catfish (*Ictalurus punctatus*) virus and antibody. Journal of the Fisheries Research Board of Canada 30:1641–1645.

Koneman, E. W., S. O. Allen, V. R. Dowell, and H. M. Sommers. 1979. Color atlas and textbook of diagnostic microbiology. J. B. Lippincott Co., Philadelphia. 495 pp.

Lenette, E. H., A. Balows, W. J. Hausler, and J. P. Truant (eds.). 1980. Manual of clinical microbiology, 3d ed. American Society of Microbiology, Washington, D.C. 1,044 pp.

Nusbaum, K. E., and E. B. Shotts. 1981a. Action of selected antibiotics on four common bacteria associated with diseases of fish. Journal of Fish Disease 4:397–404.

Nusbaum, K. E., and E. B. Shotts. 1981b. Absorption of selected antimicrobic drugs from water by channel catfish, *Ictalurus punctatus*. Canadian Journal of Fisheries and Aquatic Science 38:993–996.

Raj, H. D. 1969. Solutions to some problems in food bacteriology. Laboratory Practice 18:157–160.

Reichenbach-Klinke, H. H. 1973. Reichenbach-Klinke's fish pathology. T.H.F. Publications, Neptune, New Jersey. 512 pp.

Rovozzo, G. C., and C. N. Burke. 1973. A manual of basic virological techniques. Prentice-Hall, Englewood Cliffs, New Jersey. 287 pp.

Schubert, G. 1977. Krukheiten der Fische, 654–715. *In:* Kosmos-Handbuch Aquarienkunde. Kosmos-Verlag, Stuttgart, West Germany.

Shotts, E. B., and R. M. Rimler, 1973. Medium for the isolation of *Aeromonas hydrophila*. Applied Microbiology 26:550–553.

Shotts, E. B., V. L. Vanderwork, and W. J. Long. 1976a. Incidence of R factors associated with *Aeromonas hydrophila* complex isolated from aquarium fish, 493–502. *In:* L. A. Page (ed.). Wildlife diseases. Plenum Publishing Co., New York.

Shotts, E. B., V. L. Vanderwork, and L. M. Campbell. 1976b. Occurrence of R factors associated with *Aeromonas hydrophila* isolates from fish and water. Journal of the Fisheries Research Board of Canada 33:736–740.

Shotts, E. B., A. L. Kleckner, J. B. Gratzek, and J. L. Blue. 1976c. Bacterial flora of aquarium fishes and their shipping waters imported from Southeast Asia. Journal of the Fisheries Research Board of Canada 33:732–735.

Trust, T. J., and K. H. Bartlett. 1974. Occurrence of potential pathogens in water containing ornamental fishes. Applied Microbiology 28:35–40.

Wells, J. G., G. McC. Clark, and G. K. Morris. 1973. Evaluation of methods for isolating *Salmonella* and *Arizona* organisms from pet turtles. Applied Microbiology 27:8–10.

CHAPTER 12

Some Parasites of Exotic Fishes

Glenn L. Hoffman and Gottfried Schubert

It is a well-established custom to move desirable species of animals to new locations—e.g., European horses, chickens to North America, American rainbow trout to Europe, South America, Africa, Australia, New Zealand, and so on. Overall, the advantages of such transfers have probably outweighed the disadvantages.

In fisheries, the development of the trade in goldfish includes the most spectacular example of exoticism. Apparently goldfish (mutants of wild black goldfish) appeared in China in about 970 and they had become common by 1500. They were shipped to Japan in 1500, to Portugal in 1600, to Russia in about 1650, to Europe in 1700, and to North America in about 1920 (probably earlier) according to Wallach (1971) and Moznov (1979). Unfortunately some diseases and pests have accompanied such migrations. Therefore, it is recommended that potential exotics be examined carefully by a professional fish parasitologist in an attempt to avoid the transfer of pathogens. It may sometimes be feasible to transfer infected, nondiseased stock to a place where those parasites already exist but are under control.

A few attempts to list the internationally transferred fish parasites have been made (Malevitskaya 1958a, b; Reichenbach-Klinke 1961, Kulakovskaya and Krotas 1961, Hoffman 1970, Bauer and Strelkov 1972, Volovik et al. 1974, Bauer and Hoffman 1976, Gratzek et al. 1976, 1978; Hoffman 1981a) but these records are probably incomplete. The dangers involved in potential fish imports to Australia were discussed by Wharton (1965). The parasites and diseases of ornamental fishes of northern South America were listed and discussed by Conroy et al. (1981); some of these fishes are exported.

In this chapter we discuss some known dangerous or potentially dangerous parasites of exotic fishes, but do not include all of the many parasites reported from exotic fishes, particularly from ornamental fishes, that have not been implicated in disease or in an established new location. We also include some discussion of control and worldwide sources of information on parasites of fishes considered exotic.

Some Known Dangerous or Potentially Dangerous Parasites of Exotic Fishes

We list (in general phylogenetic sequence), and discuss briefly, the parasites of exotic fishes that now seem to be the most important or that represent good examples of the transfer of exotic parasites. A complete listing of all native parasites of fish considered exotic is not possible at this time. Some case records are listed and further information is available from the authors.

Protozoa (listed alphabetically)

Ambiphrya ameiuri (Thompson, Kirkegard, and Jahn 1947): has been reported from the American *Ictalurus punctatus* in the U.S.S.R. (Ivanova 1978). Previously, *A. ameiuri* has been known only from the United States. If numerous on gills this parasite is considered pathogenic.

Balantidium ctenopharyngodonis Chen 1955: in intestine of grass carp, the U.S.S.R. to Hungary (Molnar 1982).

Dermocystidium koi Hoshina and Sahara 1950: in skin of *Cyprinus carpio* (including Koi carp), Japan. We have found it in *C. carpio* from Korea and in Koi carp in the United States (Migaki et al. 1981).

Eimeria sinensis Chen 1956: in intestine of silver and bighead carp, Far East to Hungary (Molnar 1982).

Entamoeba ctenopharyngodonis Chen 1955: intestine of grass carp, U.S.S.R. to Hungary (Molnar 1982).

Heteropolaria colisarum Foissner and Schubert 1977: an epistylid, has been described from exotic anabantid fish in Germany, but the disease did not become established.

Ichthyophthirius multifiliis Fouquet 1876: probably originated in Asia (Hoffman 1970, 1981a) and, because of the lack of host specificity, has been transferred by many fishes throughout the temperate zone. Canella and Canella (1976) mention that Dashu and Lien-Siang (1960) reported that Su-shih, who lived in China during the Sung dynasty, in a work with the translated title, *Interrelations of Organisms,* described a white spot disease of freshwater fish, the causative agent of which was probably Ich, and advised

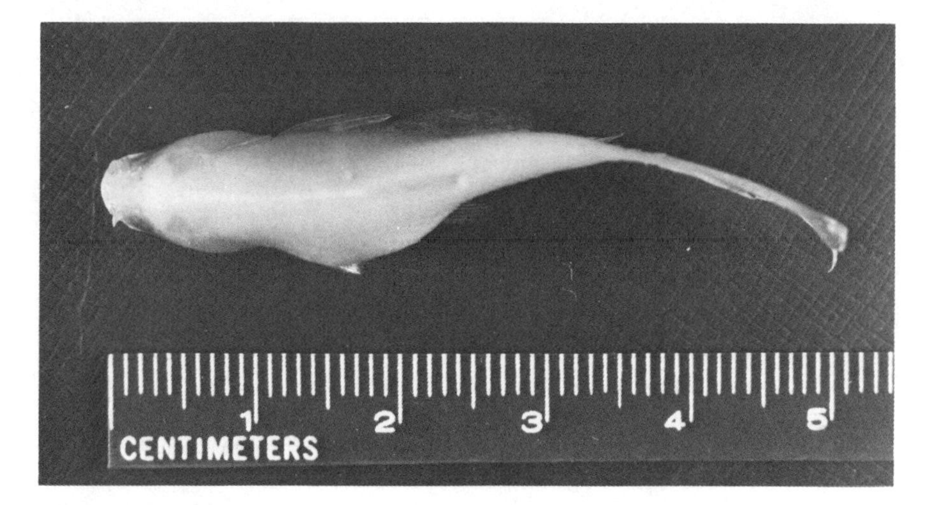

Figure 12–1. Goldfish infected with *Mitraspora cyprini.* Only one kidney is affected, causing unilateral swelling. Usually both kidneys are involved.

people to combat it with leaves of *Acer* (maple). This is probably the most dangerous freshwater fish parasite and its transfer should be denied to places where it is now absent.

Microsporea (various genera and species): one species, *Pleistophora hyphessobryconis,* has been reported often from ornamental fish. Many of these reports are probably in error because the species are difficult to identify. Because it has not been possible to infect adult ornamental fishes experimentally, we assume that they are infected only as fry, particularly in their native habitat.

Mitraspora cyprini Fujita 1912 (*Sphaerospora cyprini*): causes tremendous kidney enlargement (figure 12–1) and death of goldfish in Japan (Ahmed 1973, 1974). Has been reported from carps from Amur River (Bykhovskaya-Pavlovskaya et al. 1962) and goldfish from the United States (Hoffman 1981a, 1981b).

Myxidium giardi (Cepedi 1906): in gills of eel elvers to Hungary (Molnar 1982).

Myxobolus drjagini Akhmerov 1954: head skin of silver and bighead carp, China to Hungary (Molnar 1982).

Myxobolus pavlovskii (Akhmerov 1954): in gills of *Aristichthys nobilis* and *Hypophthalmichthys molitrix,* from Asia to Czechoslovakia and Hungary (Lucký, 1981; Molnar 1982). Species description in Bykhovskaya-Pavlovskaya et al. 1962).

Myxosoma cerebralis (Hofer) (Plehn 1905) Kudo 1933: presumably nonpathogenic in native European salmonids but a serious parasite attacking the cartilage and causing whirling disease in exotic rainbow trout (*Salmo gaird-*

neri) in Europe. It was accidentally imported to the United States in frozen rainbow trout in about 1957. It has been extremely pathogenic to rainbow trout and brook trout (*Salvelinus fontinalis*); however, it is now under control. Since World War II it has also become evident in Italy, Bulgaria, Sweden, Scotland, South Africa, New Zealand, and recently it has been reported from England (Bucke 1982). The ability of the spores to survive freezing has made the remarkable transfers possible. Adequate control measures have greatly reduced its adverse effects (Hoffman 1970, 1976a).

Myxosporidea (various genera and species): There are many known and undescribed myxosporean species in exotic fishes. Usually the whitish cysts are visible and may occur in any organ. Transfer is to be avoided.

Pleistophora hyphessobryconis Schäperclaus 1941: destructive muscle parasite of ornamental fishes including goldfish (Dykova and Lom 1980). Most reports are from Europe but we have also seen it in the United States in *Metynnis* sp. from Brazil (Hoffman Case no. H-80-8) and *Paracheirodon innesi* from South America (Hoffman Case no. S78-209A). However, some of these reports are probably erroneous. The species should be more thoroughly studied.

Protopalina symphysodonis Foissner, Schubert, and Wilbert 1979: an opalinid that has been shipped from Bangkok to Europe and the United States. It is found in the intestine of adult *Symphysodon* in small numbers but is often numerous in fry and apparently kills them.

Sphaerospora carassii Kudo 1919: a gill pathogen of goldfish, common carp (*Cyprinus carpio*) and grass carp (*Ctenopharyngodon idella*) in Europe (Molnar 1979). We have recently found it in goldfish in the United States (Hoffman 1981, Case no. S79-150).

Spironucleus elegans Lavier 1936: a close relative of *Hexamita* spp., found in the intestine of many ornamentals, particularly South American cichlids; often kills fry of *Pterophyllum* and *Symphysodon*. It invades tissues readily when *Capillaria pterophylli,* which damages intestinal mucosa, is present. *Spironucleus elegans* is often seen in Europe and North America and probably worldwide (Molnar 1982).

Thelohanellus dogieli (Akhmerov 1955): causes fin disease of common carp; the Far East to Hungary (Jeney and Molnar, 1981). Species description in Bykhovskaya-Pavlovskaya et al. 1962.

Thelohanellus hovorkai Akhmerov 1960: in carp; from East Asia to Ukraine and later to Hungary via streams (Molnar 1982).

Thelohanellus nikolskii Akhmerov 1955: in carp; from East Asia to the Ukraine; later to Hungary via streams (Molnar 1982).

Trichodina reticulata Hirschmann and Partsch 1955: this parasite of goldfish was described from Europe but probably originated in Asia (Hoffman 1970). It has been found on goldfish in the United States in Arkansas (Hoffman, Case nos. S78-198, S78-221, S79-16A), Alabama, Mississippi,

Louisiana, Georgia, Kentucky, South Carolina (Wellborn 1967), and Pennsylvania (Lom and Hoffman 1964).

Trichodina subtilis Lom 1959: on the gills of goldfish and other fishes in Eurasia (Lom and Haldar 1977) and also found in the United States (Lom and Hoffman 1964, Hoffman 1978).

Trichodinella epizootica (Raabe 1950, Ŝrámek-Huŝek 1953): probably from Asia to Europe on gills of goldfish (Hoffman 1970) and other fishes (Lom 1970, Lom and Haldar 1977, Hoffman 1978) and to the United States (Lom and Hoffman 1964, Hoffman Case nos. S78-221, S79-16A, S79-79). Seemingly this dangerous parasite has become widely distributed.

Trichophrya sinensis Chen 1955: on the gills of silver carp and bighead carp; from China to Hungary (Molnar 1982).

There is a good likelihood that, in addition to the above transplanted protozoa, some important pathogens were transferred before scientific records were kept. Probably the rather extensive transfers of goldfish and common carp took place before the discovery of the protozoan parasites. Possibly the following widespread protozoa were so transferred: *Chilodonella cyprini, C. hexasticha; Cryptobia branchialis, C. carassii (C. agitans); Hexamita* spp.; *Ichtyobodo necatrix;* and *Oodinium pillularis.* Transfer of these parasites should be avoided.

Monogenea (gill and skin flukes)

Many monogeneans (monogenetic trematodes) are known to have been transferred with their hosts (Hoffman 1970, Bauer and Hoffman 1976). They are easily transferred because no intermediate host is necessary. Some of these are destructive to their hosts, but most are specific for one species of fish, or for two or more closely related species. Nevertheless, it is not desirable to transfer them to new locations.

Because of the high host specificity of most Monogenea it is not likely that they will spread to other fish species when they are introduced with their hosts into a new habitat. But there is another danger – they may become more dangerous to their host in the new surroundings. This increased danger was suggested from experiments done by one of us (GS) with ornamental fish. All puffers (Tetradontidae) apparently carry gill flukes in their natural habitat. In freshwater species, the experiments were done with *Tetraodon pelambengensis,* but the findings are known to apply to *T. fluviatilis* as well. Such transfer causes the death of the fish if no precautions are taken and the fish is kept in pure fresh water. The gill flukes (species not determined) propagate rapidly in fresh water and the puffers die within a few months. This effect can be avoided by two means: (1) the fish may be kept in slightly brackish water similar to the natural waters in which puffers live. The number of gill flukes then usually remains low enough for the puffer to survive; or (2) the gill flukes can be eradicated by the administration of medicaments. The fish will then survive for a long time even in fresh water with low conductivity. Probably the defense mechanisms of the puffer against the gill flukes are weakened in pure fresh water, which is not an appropriate habitat, or, less likely, the conditions of the natural habitat are unfavorable for the parasite.

Similar conditions probably prevail for other gill or skin flukes. The responsible factor could be one other than the conductivity response (salinity).

Because of the expansion of the culture of cichlid fishes, mostly *Tilapia* spp., the potential transfer of their parasites has become more important. Several such transfers have already occurred. Six species of *Cichlidogyrus* (gill flukes) have been transferred from Africa to Israel on *Tilapia* spp. (Paperna 1964).

Monogenea of other food fishes have been transferred to many countries where fish culture has been increasing, mostly in the temperate zone—Asia, Europe, and North America. Such transfers were reviewed by Hoffman (1970) and Bauer and Hoffman (1976). *Dactylogyrus extensus* of the gills of common carp has been transferred from Europe to the United States and Israel. Both *D. minutus* and *Gyrodactylus cyprini* of common carp, which are well known in Europe and Central Asia, have been found in the United States. *Pseudacolpenteron pavlovskyi* of Asian common carp has been transferred to Israel and the United States. Three monogeneans of grass carp—*Dactylogyrus lamellatus, D. ctenopharyngodonis,* and *Gyrodactylus ctenopharyngodonis*—have been transferred from Asia to Europe and four of silver carp (*Hypopthalmichthys*)—*Dactylogyrus hypopthalmichthys, D. chenshuchenae, D. skrjabini,* and *D. suchengtai*—have been transferred from Asia to Europe. One "mono," *D. aristichthys* of bighead carp (*Aristichthys nobilis*), has been transferred from Asia to Europe and also, recently, *D. nobilis* (Lucký, 1981). Parasites of American catfishes have been transferred intercontinentally. One common American form, *Cleidodiscus pricei* of the gills of channel catfish (*Ictalurus punctatus*) and from brown bullheads (*I. nebulosus*), has been reported from the U.S.S.R., Hungary, and Poland (Molnar 1963, 1968, 1982; Prost 1973; Musselius and Mirzoeva 1977); in one instance this parasite was discovered in the U.S.S.R. after catfish eggs, presumably free of parasites, had been imported. *Cleidodiscus monticelli* (Cognetti de Marus 1924) (Prost 1973) from the nasal fossae and *Gyrodactylus fairporti* of the gills of American brown bullheads have been found in Poland (Prost 1973). *Gyrodactylus anguillae* Ergens 1960 of European eel (*Anguilla anguilla*) in European waters has been found in Japan, presumably transplanted there by fish transfer (Ogawa and Egusa 1980). *Nitzschia sturionis* causes little damage to the sturgeon *Acipenser stellatus* in the Caspian Sea, but in 1930 it was transferred to the Aral Sea where, in 1936, it caused high mortality of the related native *Acipenser nudiventris* (Bauer and Hoffman 1976).

Monogenea of ornamental fishes have been transferred throughout the world. They usually do not attack other fish, but transfer should be avoided. The parasites of many of the tropical fishes have not been thoroughly studied. Hoffman (1970) listed nine species that are known to have been transferred: *Anacanthorus anacanthorus, Dactylogyrus anchoratus, D. vastator, D. wegeneri, Gyrodactylus bullatarudis, G. elegans, Urocleidoides reticulatus, Urocleidus crescentis,* and *U. orthus.* Since then the following transfers have been reported:

Anacanthorus brevis Mizelle and Kritsky 1969a: from the gills of *Brycon melanopterus,* Brazil to the United States.

Archidiplectanum archidiplectanum Mizelle and Kritsky 1969b: gills of *Gnathonemus petersi,* western Africa to the United States.

Dactylogyrus achmerovi Gussev 1955: gills of carp, East Asia to the Ukraine to Hungary via streams (Molnar 1982).

Dactylogyrus anchoratus, D. baueri, D. formosus: goldfish gills, Japan to the United States (Rogers 1967).

Dactylogyrus ctenopharynogodonis Akhmerov 1952: on gills of grass carp, the U.S.S.R. to Hungary (Molnar 1982).

Dactylogyrus lamellatus Akhmerov 1952: on gills of grass carp, the Far East to Hungary (Molnar 1982).

Dactylogyrus navarroensis Mizelle and Kritsky 1969a: gills of *Hesperoleucas navarroensis,* California, United States.

Dactylogyrus nobilis Long et Yu 1958: on gills of *Aristichthys nobilis,* the Far East to Hungary (Molnar 1982).

Diplozoon tetragonopterini Sterba 1957: *Ctenobrycon spilurus* and *Gymnocorymbus ternetzi,* South America to Germany (Reichenbach-Klinke 1961).

Heteronocleidus gracilis Mizelle and Kritsky 1969b: gills of *Colisa labiosa,* India to California, United States.

Longihaptor longihaptor Mizelle and Kritsky 1969a: gills of *Cichla ocellaris,* Brazil to California, United States.

Trianchoratus acleithrium Price and Berry 1966: gills of *Helostoma rudolfi,* Malaysia to the United States (Mizelle and Kritsky 1969a).

Urocleidoides amazonensis Mizelle and Kritsky 1969a: gills of *Phracto-cephalus hemibiopterus,* Brazil to the United States.

Urocleidoides catus Mizelle and Kritsky 1969a: gills of *Phractocephalus hemibiopterus,* Brazil to the United States.

Urocleidoides megorchis Mizelle and Kritsky 1969a: gills of *Sorubim lima,* South America to the United States.

Urocleidoides robustus Mizelle and Kritsky 1969a: gills of *Symphysodon discus,* Brazil to the United States.

Monogenea of sport fishes have also been transferred intercontinentally. The following records probably represent only a small sample of the total:

Urocleidus dispar (Mueller 1963): pumpkinseed (*Lepomis gibbosus*), gills, the United States to Rumania (Roman 1953), Czechoslovakia (Vojtek

1958), Hungary (Molnar 1963, 1968), throughout Europe (Roman-Chiriac 1960).

Urocleidus furcatus (Mueller 1937): largemouth bass (*Micropterus salmoides*), gills, the United States to Germany (Reichenbach-Klinke 1961), West Germany (Reichenbach-Klinke 1966).

Urocleidus helicis (Mueller 1936): largemouth bass gills, the United States to Italy (Ghittino 1965).

Urocleidus principalis (Mizelle 1936): largemouth bass gills, United States to England (Maitland and Price, 1969).

Urocleidus similis (Mueller 1936): pumpkinseed gills, the United States to Rumania (Roman 1953), Czechoslovakia (Vojtek 1958), Hungary (Molnar 1968), and throughout Europe (Roman-Chiriac 1960).

Trematoda, Digenea

All fish trematodes require a molluscan intermediate host, and consequently the chances of establishing digeneans in foreign countries are decreased. However, where similar faunas exist in different countries or continents, the trematodes that are not strictly host-specific may become established exotically. Manter (1963) believed that the intestinal trematode of salmonids, *Crepidostomum farionis,* was transferred from Europe to North America in trout and became established.

The digenetic trematode, *Amurotrema dombrowskajae,* a diplodiscid, was brought with young grass carp from the rivers of east Asia to several fish farms of European U.S.S.R., central Asia, and Hungary. It was found in the U.S.S.R. during the first few months after shipping but then disappeared. But in the Kapchagay Reservoir (Yli River, Balkhash basin), U.S.S.R., the parasite became established, and was found there often in new generations of grass carp. This trematode has only one intermediate host – a snail. After the cercariae leave the snail they do not penetrate the second intermediate host, but form metacercariae on water plants; grass carp become infected from feeding on such plants. Thus, the life cycle of *A. dombrowskajae* is simpler than that of other freshwater fish trematodes, and this greater simplicity explains the success of its acclimatization in new waters (Bauer and Hoffman 1976, Molnar 1982).

One fish trematode, *Cryptocotyle lingua* (marine blackspot), was transferred from the eastern Atlantic to western Atlantic in infected European snails (*Littorina littorea*) on ships (Sindermann and Farrin 1962). A human pathogen, *Schistosoma mansoni,* was transferred to Hong Kong in 1973–74 in infected South American snails, *Biomphalaria straminea,* on imported water plants for aquaria (Meier-Brook 1975). It is also possible to transfer adult trematodes in their bird hosts. If closely related snails and fish are present such a transfer can be permanent. It is believed that *Bolbophorus confusus,* a Eurasian strigeid trematode, probably came to the United States in a stray pelican, its natural final host. We have recently found it in fathead minnows shipped from South Dakota to Arkansas. Native snails in Montana became infected and the resulting cercariae infected trout (Hoffman 1970). However, a similar "import" of *Scaphanocephalus expansus* of ospreys did not result in an exotic establishment (Hoffman 1953).

We could find no evidence of the exotic establishment of other digeneans in food, ornamental, or sport fishes.

Cestoda

Cestodes require at least one, and often two, intermediate hosts. Although this requirement complicates relocation of parasite species, some have been transferred. An interesting case is that of the pseudophyllaeid cestode *Bothriocephalus opsarichthydis* (*B. acheilognathi, B. gowkengensis*).* This species was described in 1955 from the intestine of young grass carp cultured in the fish farms of south China. Later a low intensity of infection of yearling grass carp in the Amur River was reported (Yukhimenko 1970). In 1954, yearling grass carp were shipped from the Amur River to a Ukrainian fish farm where they died. Two years later Malevitskaya (1958a) reported intense infections of common carp fingerlings at the same fish farm with a previously unknown pseudophyllaeid, which turned out to be *B. opsarichthydis*. Common carps from this farm were used to stock other ponds and a year later more than ten farms of this region were infected with this helminth which caused heavy losses. Thus, *B. opsarichthydis* acquired a new host, the common carp. The common carp can be infected between the ages of two months and three years because it feeds on microcrustaceans, the intermediate host of the parasite, for a longer time than does the grass carp.

In 1958–62, millions of young feral herbivorous fishes captured in China were shipped to different fish farms of the European and central Asian sections of the U.S.S.R. Although hundreds of these fish (100 to 700 mg) were examined, *B. opsarichthydis* was not found. Nevertheless, some of the progeny were infected. A similar epizootic occurred at the Turkemenian Experimental Fish Farm Karametniaz which had been built to culture these fishes. In 1961, eggs were taken from two or three grass carp spawners brought previously from the Amur River and hatched for the first time under artificial conditions. A small pond was stocked with fry and in late autumn there was a high mortality of the fingerlings, and many *B. opsarichthydis* were found. Unfortunately the water supply reservoir had been stocked with a small number of yearling grass carp of Chinese origin. Though the infection rate of the young fish that were shipped was extremely low, it was heavy enough to infect the new generation. This helminth now has been found also in many other fishes inhabiting the great Kara Kum Canal such as wild common carp, various cyprinids, *Silurus glanis,* and American *Gambusia* (Babaev 1965).

---

* Name correction by Dubinina (1982).

Other foci were created in other parts of the U.S.S.R., including some natural bodies of water. During the first decade of the spread of *B. opsarichthydis,* mortalities of grass carp fingerlings were noted. At present, the situation is not so serious because some degree of host-parasite equilibrium has been established. Because *B. opsarichthydis* is a thermophilic species (Musselius 1967, Körting 1974), its numbers are limited in central and northern U.S.S.R.

B. opsarichthydis were also brought to Rumania with young grass carp, and feral foci turned up in the Danube. The parasite was transported to Malaysia (Fernando and Furtado 1963), Hungary (Molnar 1970, 1982) and Yugoslavia (N. Fijan, personal communication). In 1974 it was found in common carp farms of West Germany (Körting 1974) where the incidence and intensity in fingerlings was as high as 80 helminths in a fish 30 cm long; in older fish the intensity was lower. *B. opsarichthydis* was recently transferred from Hong Kong to New Zealand in grass carp fry, along with *Tripartiella* sp., *Ichthyophthirius multifiliis, Dactylogyrus ctenopharyngodonis,* and *Gyrodactylus ctenopharyngodonis.* All the parasites except *I. multifiliis* were eradicated with parasiticides (Edwards and Hine 1974).

The travels of the great Asian tapeworm continued; Hoffman (1976b) reported finding it in cultured bait minnows – golden shiner (*Notemigonus crysoleucas*) and fathead minnows (*Pimephales promelas*) – in North America (figure 12-2). Later (unpublished data) he found it in mosquitofish (*Gambusia affinis*). Presumably it traveled by air in grass carp from Asia. Since then it has been found in mosquitofish in North Carolina and in California. W. Rogers (personal communication, Auburn University) found one *B. opsarichthydis* in channel catfish, but it is rarely found in a nonplankton feeder. Recently one of us (GH) has found *B. opsarichthydis* in an American endangered fish, the Colorado squawfish (*Ptychocheilus lucius*), from a fish hatchery in New Mexico, another new record.

In 1981, one of us (GS) found *Bothriocephalus* in guppies and in *Melanotaenia australis* in West Germany. In both fishes, the identity of the tapeworm could not be fully determined because they were not sexually mature. However, all other data corresponded to *B. opsarichthydis.*

Bothriocephalus claviceps (Goeze 1782) has been brought into Hungary in elvers but has not "spread" to other fishes (Molnar 1982).

About 1950, a high infection rate of common carp with a caryophyllaeid cestode, presumptively identified as *Caryophyllaeus fimbriceps,* was reported from central and northern U.S.S.R. However, Kulakovskaya and Krotas (1961a, b) demonstrated that the cestode was *Khawia sinensis,* previously known in the Amur River and Chinese waters. It was reported from East Germany in 1974. *Khawia sinensis* is more pathogenic than *C. fimbriceps* because it may infect common carp throughout the year rather than only in the spring. Now *K. sinensis* is widespread in fish farms and in natural waters and has displaced *C. fimbriceps* in Ukrainian fish farms. It has recently appeared in Hungary, presumably via streams (Molnar 1982).

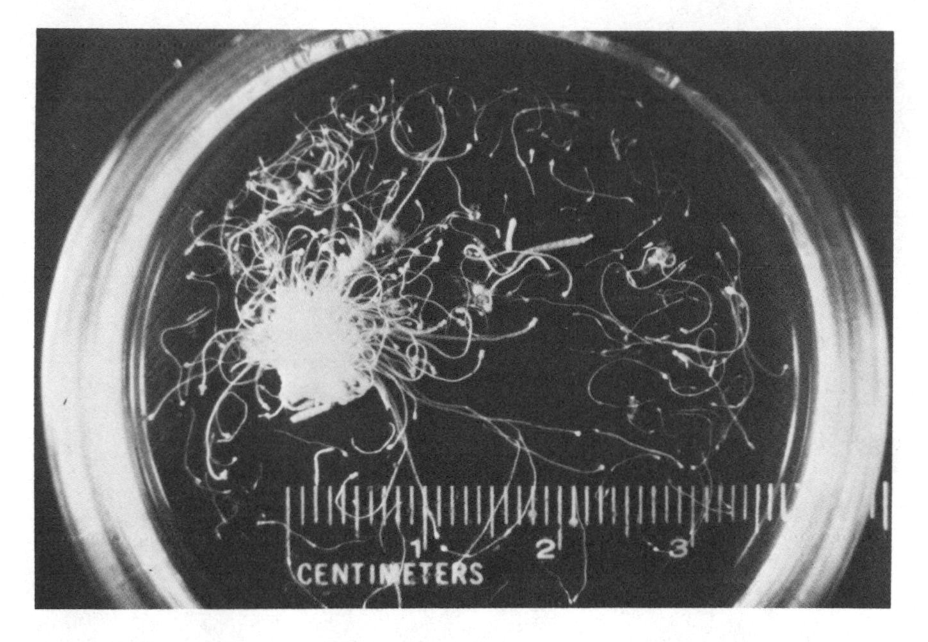

Figure 12–2. *Bothriocephalus acheilognathi* from one small golden shiner.

If one considers largemouth bass as transplanted to the state of Washington, the bass tapeworm, *Proteocephalus ambloplitis,* native to the eastern United States, was established in Washington with its host, and the plerocercoids caused problems in trout (Becker and Brunson 1968). The largemouth bass has been established in about twenty-five foreign countries (Welcomme 1981) but we found no records of the bass tapeworm in those countries.

We found no record of exotic establishment of ornamental fish cestodes.

Nematoda

Some interesting nematode cases involve several species of Philometridae. A new philometrid, *Philometroides lusiana,* was described from common carp cultured in several fish farms in the Latvian Republic (Vismania 1962). Long, red females of this species are located under the scales of the fish. The infection is not lethal, but a high intensity of infection makes the carp undesirable for sale. The first reports indicated that this nematode was only in those fish farms to which wild common carp (*Cyprinus carpio haematopterus*) from the Amur River had been brought. This parasite has not been found in the Amur River, even though parasitologists have searched carefully for it; the incidence there must be extremely low.

Philometra sanguinea (*P. carassii*) is a specific parasite of goldfish, infecting blood vessels of the fins (Vismania and Nikulina 1968). This spectacular parasite was presumably transferred from Japan to North America by fish hobbyists (Hoffman 1970).

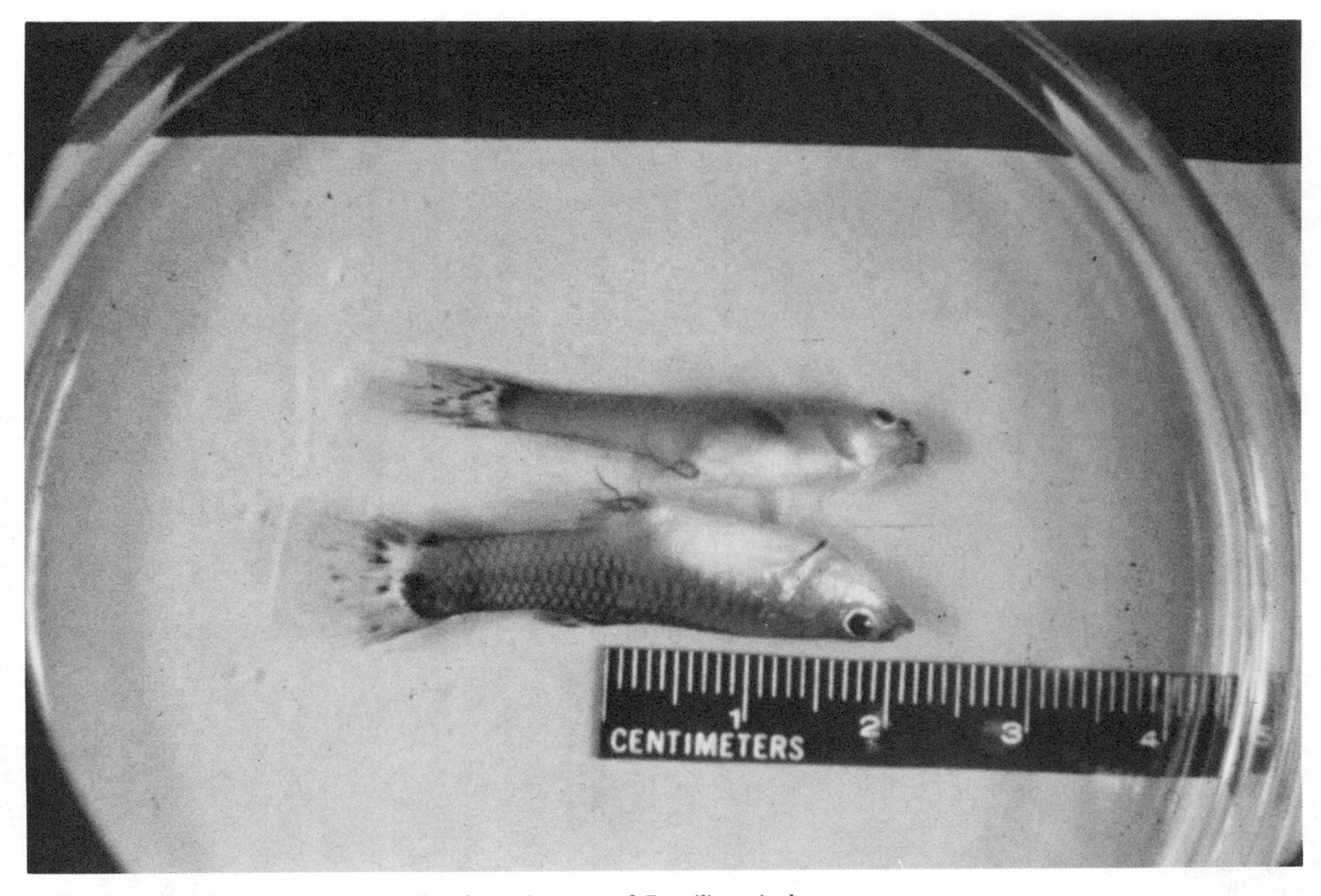

Figure 12–3. *Camallanus cotti* extending from the anus of *Poecilia reticulata.*

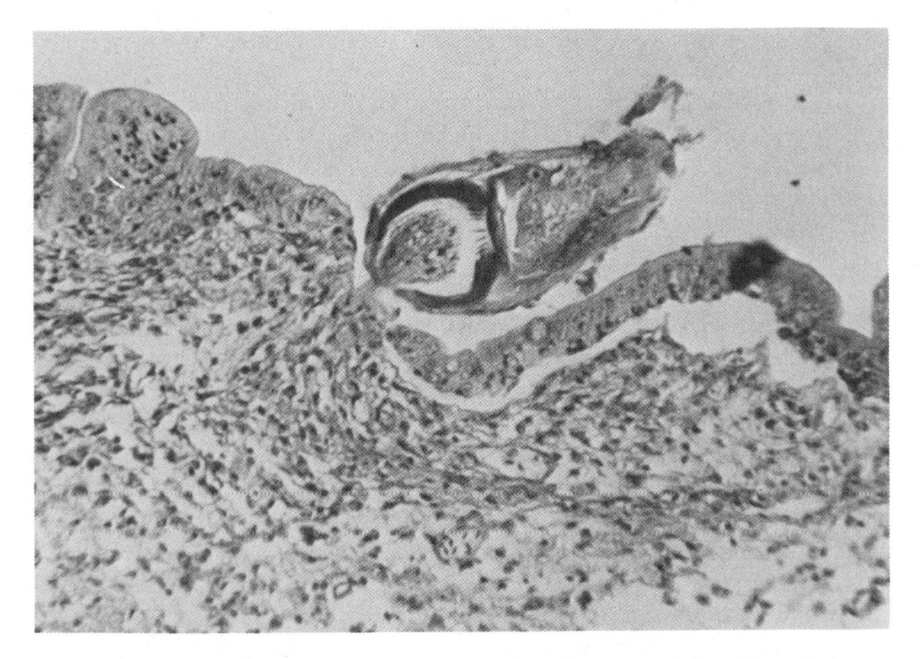

Figure 12–4. *Camallanus cotti* taking a bite of the intestine of *Poecilia reticulata.*

Capillaria pterophylli in *Pterophyllum scalare* of South America was found in Germany (Heinze 1933). Apparently this nematode does not need an intermediate host; consequently there is danger that it could become established in ornamental fish producers' facilities.

Camallanus cotti (figures 12–3 and 12–4), originally described from fish in Japan (Fujita 1927), has become established in ornamental fish culture and has turned up in Malaysia, Europe, the United States, and Australia (Stumpp 1975).

Acanthocephala

Acanthocephala of fish require a crustacean first intermediate host and sometimes also a vertebrate second intermediate host. Perhaps these demands have detracted from their potential exoticism.

Juvenile forms of *Polyacanthorhynchus kenyensis* Schmidt and Canaris 1967 have been found in the liver of largemouth bass and *Tilapia* sp. in Kenya. Both fish are introduced species (Schmidt and Canaris 1967).

Copepoda and Branchiura, Parasitic

Many parasitic copepods are host specific and have caused little trouble because of fish transfers. Some, however, lack host specificity and are major disease agents, particularly in the temperate, subtropical, and tropical zones.

Achtheres micropteri Wright 1882: normally a gill parasite of American centrarchids, particularly *Micropterus* spp., has been found in Africa (Fryer 1968).

Argulus foliaceus Linnaeus: Europe to Ceylon on mirror carp (a form of common carp) and on trout (Kirtisinghe 1964). This branchiuran is potentially dangerous because it is not host specific. It is also the intermediate host for the fish parasite *Molnaria erythrophthalmi* (Nematoda: Skrjabillanidae) and related species in Europe (Rudomentova 1974, Moravec 1978, Tikhomirova, 1980).

Argulus japonicus Thiele 1900 Linnaeus: this branchiuran has been transferred to Africa (Fryer 1960), Ceylon (Kirtisinghe 1964), Israel (Paperna 1964), New Zealand (Hine 1975), and North America (Cressey 1978). Although apparently lacking host specificity, it is usually found on goldfish and common carp. It is also found in Japan and China (Yamaguti 1963).

Lernaea cyprinacea Linnaeus 1758: this devastating copepod has been reported from many species of freshwater fishes, and frog and salamander tadpoles from Africa, Asia, Europe, Israel, Japan, Eurasia, and the United States. In the U.S.S.R. it is known as *Lernaea elegans, L. cyprinacea* being a host-specific parasite of *Carassius carassius;* however, most parasitologists know it as *L. cyprinacea.* It probably originated in Asia and spread with the goldfish trade (Hoffman 1970).

Pseudocaligus sp.: this copepod has been transferred on *Poecilia mexicana* from southern America to Hawaii (M. McGrenra 1978, University of Hawaii at Manoa, personal communication).

Isopoda

Artystone trysibia: imported from Colombia, South America, to the United States in *Corydoras.* Burrows into the fish and lives in the "hole" (Hoffman Case no. H77-1).

Lironeca symmetrica (often reported as *Livoneca*): imported from South America to the United States on ornamental fish. It has become an established parasite and is damaging to many exotic fishes (Herwig 1976).

Control

Because there are often benefits to be derived from fish transfers, but because dangerous parasites may be moved to places where they do not already exist, the parasite must sometimes be controlled. Control can be accomplished by fish inspection, quarantine, and eradication, sometimes aided by fish disease laws. Wharton (1965) discussed this aspect in relation to possible imports to Australia, but there are potential difficulties with species that invade tissues or that have life cycle stages that evade detection. Control of egg-laying monogeneans must be considered because medicaments do not kill the eggs. Therefore, it is necessary to repeat the treatment after the oncomiracidia have hatched but not yet reached maturity. The time depends on species and temperature. Because characteristics are known for very few species, extensive experiments may become necessary. The possibility that there are dormant eggs cannot be denied.

INSPECTION

Samples of fish to be exported should be inspected for the presence of undesirable diseases and parasites. Later in this report we list the laboratories where help or information is likely to be obtainable. To determine the probable absence of a pathogen in a fish population, one must inspect at least 60 fish and find them free of the organism (Ossiander and Wedemeyer 1973). Obviously one cannot inspect (autopsy) 60 fish of a 40-fish shipment, so it is recommended that fish be examined as near the source as possible. Not all parasites should be considered harmful, because many helminths do not multiply in the fish, tanks, or aquaria. Some are larval and cannot complete their development unless the fish are eaten by the vulnerable host.

For parasite inspection, one must examine all organs by naked eye, 10X stereoscope, and wet microscopic squashes of skin, fin, gills, kidney, and intestine (Hoffman 1967, Lucký 1977, McDaniel 1979, Roberts 1978, Schäperclaus 1979, Reichenbach-Klinke 1980). If the fish cannot be examined at the source, they should be inspected by a professional fish parasitologist at the earliest opportunity during their journey to another land.

QUARANTINE

Even if the exotic fish have been thoroughly inspected one should isolate them as completely as possible when they arrive at their destination. If kept in static aquaria, used water should be deposited in municipal sewage that will be adequately treated, in a dry well, or in a disinfecting tank or lagoon. Fish that require running water should be held for quarantine only in facilities possessing an adequate water disinfecting plant. As indicated by Spotte (1979), such a plant involves primary treatment (settling tank), secondary treatment (sand-gravel or similar filtration), and tertiary treatment (ozone, chlorine, or ultraviolet irradiation).

The minimum quarantine time should be the time necessary for certain parasites to complete their life cycles, or time for a sparse parasite population to become numerous enough for easy detection. The exact time has never been determined because of the many variables involved, but one might guess that it should be two weeks or longer if possible because most such life cycles would be completed within that period. One must remember, however, that certain parasites will sometimes not become apparent during short quarantine (Bauer et al. 1977, Schäperclaus 1979). For example, in our own experience, we have had to search the intestine and its contents with a 10X stereoscope to find larval *Bothriocephalus opsarichthydis* in golden shiners. The naked eye does not suffice. Presumably one would have to examine 60 fish of a lot without finding a specific parasite to ensure a 95 percent probability that the lot is free of the specific parasite before the quarantine period is started. Quarantine laws have been reviewed by Evelyn (1982).

ERADICATION

Depending on the intended use of the fish, certain larval helminths could be ignored, particularly if not numerous and if the fish will be kept indoors where there is no chance of the infected fish being eaten by potential final hosts. Treatments for ectoparasites and intestinal helminths must be given repeatedly until inspection confirms that the parasites have been eradicated (Hoffman and Meyer 1974, Wellborn 1979, Mitchell and Hoffman 1981). This eradication may require three or more applications on alternate days. For parasitic copepods, treatment must continue until the adult females die of old age (several weeks) because certain current medicaments kill the larvae but not the adults.

IMPORTATION LAWS

Some nations, provinces, and states have fish disease and importation laws that must be respected. It is beyond the scope of this report to review all such laws; however, U.S. and international laws were reviewed recently (Fryer et al. 1979, Evelyn 1982). These laws are not intended to impede shipments but to try to provide a disease-free product. Also, some of them include enabling legislation providing funds to help in the diagnosis and treatment of fish diseases.

Worldwide Sources of Information on Parasites of Fishes That Might Be Transferred Intercontinentally

Because live fish in international commerce should be inspected for the presence of potentially, or known, dangerous parasites, we present a list of some of the establishments with internationally known fish parasitologists. If they cannot be of assistance they will probably be able to refer the potential shipper to an appropriate person. The U.S.S.R. probably leads the world in fish parasitology, but it was impractical to list the many Soviet institutions where good fish parasitology is being done. A list of American fish disease diagnosticians was published by Hoffman and Mitchell (1982). Information is available on the parasites of fishes of some countries, and it is advisable to obtain such information before arranging importation. These countries are listed below alphabetically with the exception of those located in Africa.

Africa

South Africa: Division of Inland Fisheries, Private Bag 501L, Stellenbosch.

Sudan: Fisheries and Hydrobiological Research Section, P.O. Box 1489, Khartoum.

General: Dr. L. F. Khalil, Commonwealth Institute of Parasitology, St. Albans, Herts., England (has done much research on African fish parasites; his publications are available); Dr. I. Paperna, director, H. Steinitz Marine Biology Laboratory, The Hebrew University, Elat, Israel (has done much research on African fish parasites; his publications are available); Dr. June P. Thurston, Department of Biology, University of Strathclyde, Glasgow, Scotland (has done much research on African fish parasites; her publications are available).

Other Countries

Argentina: Musco Argentino de Ciencias Naturales (Bernardino Rivadabia), Buenos Aires.

Australia: Fish Disease Specialist, Snobs Creek Freshwater Fisheries Research Station and Hatchery, Fisheries and Wildlife Division, Private Bag 20, Alexandra, Victoria, Australia 3714; Fisheries and Wildlife Department, 605 Flinders Street Extension, Melbourne, Australia 3000; Fish Parasitologist, Zoology Department, James Cook University, Townsville, Australia QLD 4811.

Austria: Institute de Ichthyopathology, Veterinary Medicine, Linke Bahngasse 11, A1030, Wien.

Bangladesh: Zoology Department, University of Dacca, Dacca-2, Director of Fisheries, Department of Fisheries, Dacca.

Brazil: Fish Parasitologist, U.F.R.R.J. Km 47 Antiga Rod. Rio-S. Paulo 2 3.460 Seropedica, Rio de Janeiro; Fish Parasitology, Instituto Oswaldo Cruz, Edo Guanavara, Rio de Janeiro.

Burma: FAO/TA Inland Fishery Biologist c/o UNDP, P.O.B. 650, Rangoon.

Canada: Pacific Biological Station, Nanaimo, British Columbia V9R 5K6; Department of Fisheries and Oceans, Resource Branch, P.O.B. 550, Halifax, Nova Scotia B3J 257.

Ceylon: Fish Parasitologist, Department of Zoology, University of Malaya, Kuala Lumpur, Malaysia (work done on fish parasites of Ceylon).

Chile: Department de Pesquerias, Universidad Catolica de Valparaiso, Av. Brazil 2950, Valparaiso.

China: Fish Parasitologist, Institute Hydrobiology, Academia Sinica, Hupeh, Wuchang, People's Republic of China.

Costa Rica: Fisheries Biologist, Projecto de Diversificación Agricola, Apdo. 25, Turrialba; Fish Pathologist, Microbiology, University of Costa Rica, San José.

Cuba: Fish Parasitologist, Centro de Investigaciones Pesquerías, Ave. Lra entre 24 y 26, Miramar, Ciudad Habana.

Czechoslovakia: Institute of Parasitology, Academy of Science, Flemingova Nam. 2, Prague 6, 16632; Department of Poultry, Fish, Bees, and Wildlife, University Veterinary School, Palackeho 1-3, Brno 61242.

Denmark: Fish Pathology, Den Kgl. Veterinaer - og Landbohojskole, Ambulatorisk Klinisk, Bulowsuej 13, 17K 1870 Kobenhavn V.

Finland: Parasitological Institute, Abo Akademi, Porthansgatan 20500, Abo 50; Zoological Station Parasitologie Institute Societas, Scientiarum Fennica, Nylandsgatar 2-Abo.

France: Laboratoire d'Ichtyopathologie, route de Thiverval, 78 Thiverval-Griguan; Laboratoire de Parasitologie, place E. Bataillon, F 34060 Montpellier.

Germany, East: Institute für Binnenfischerei, Muggelseedamm 310, 1162 Berlin-Friedrichshagen; Forschungsstelle für Wirbeltierforschung (im Tierpark Berlin) der Akademie der Wissenschaften der DDR, Am Tierpart 125, 1136 Berlin; Fish Parasitologist, Institute Pathology der Friedrich-Schiller Universität, Zugelmühlenweg 1, Jena 69.

Germany, West: Zool-Parasit Institute de Universität, 37 Kaulbachstrasse, 8 Munich 22; Arbeitsgruppe Biologie der Fische, Universität Hohenheim, Stuttgart; Fischkrankheiten und Fischhaltung, Tierarytliche Hochschule Hannover, Bunteweg 17 (Westfalenhof), Bischofsholer Damm 15, 3000 Hannover 1; Bayerische Lundesanstalt für Wasserforschung Versuchsanlage Wielenbach, 8121 Wielenbach.

Great Britain: Parasitology, National Marine Laboratory, Box 101, Victoria Road, Aberdeen, Scotland AB9; Institute of Agriculture, University of Stirling, Stirling, Scotland FX9 4LA; Commonwealth Institute of Parasitology, 395 A Hatfield Road, St. Albans, Herts, AL4 0XU England; Department of Zoology, University of Liverpool, Box 147, Liverpool, England L69 3BX; National Fish Disease Laboratory, the Nothe, Weymouth, England.

Hungary: Fish Parasitology, Veterinary Medical Research Institute, Academy of Science, H-1581, P.O. Box 18, Budapest; Fish Culture Research Institute, P.O. Box 47, Szarvas H-5541.

India: Central Inland Fisheries Research Institute, 1/644 Sidhnathghat, Buxar (Bihar); Zoology Department, University of Lucknow, Lucknow; Zoology Department, University of Kalyani, Kayani 741 235, West Bengal; Zoology Department, Punjab University, Chandigarh 160014; Zoology Department, Calcutta University, 35 Ballygunge Circular Road, Calcutta 19; Freshwater Aquaculture Research Center, Dhauli, P.O. Kausalyagan, via Bhubaneswar 2, Orissa; Parasitology Department, U.P. College of Veterinary Science and Animal Husbandry, Mathura; Microbiology Department, St. Xaviers College, Bombay; Fish Pathology Unit, Central Indian Fisheries Research Sub-sta., 19 Cantoninent Road, Cuttack, 1 Orissa.

Indonesia: Director, Inland Fisheries Directorate, Djakarta, Indonesia;

School of Biological Sciences, University of Malaya, Kuala Lumpur, Malaysia; Inland Fisheries Research Institute, J1 Sempur No. 1, P.O.B. 51, Bogor, Java; Fisheries Department, University Pertamian Malaysia, Serdang, Selangor, Malaysia; Zoology Department, University of Philippines Diliaa, Quezon City, Philippines; Parasitologist, Tigbauan Research Station, Aquaculture Department, SEAFDEC P.O.B. 256 Iloilo City, Philippines (and Duncan Pierce Laboratory, Philippines).

Iran: Fish Parasitologist, Medical School of Gondi-Shapoor University, Ahwaz.

Iraq: Fish Parasitologist, Biological Research Center, Adhamiya, Baghdad.

Ireland: Fish Parasitologist, Veterinary Research Laboratory, Abbotstown, Castleknock, Dublin.

Italy: Centro per lo Studio Delle Malattie del Pesci, Instituto Zooprofilattico Sperimentale del Premonte e della Ligura, Via Bologna 148, Torino 10154.

Japan: Department of Fisheries, Faculty of Agriculture, University of Tokyo, Yayoii 1-1-1, Bunkyo-ku, Tokyo 113; Laboratory of Fish Pathology, Hokkaido Fish Hatchery, Makanoshima, Toyohira-ku, Sapporo, Hokkaido.

Korea: Laboratory of Fish Diseases, Pusan Fisheries College, Pusan 601-01.

Mexico: Instituto de Biología, Laboratorío de Helmintología, Apdo. Postal 70-153, México (City) 20, D.F.; Laboratory of Helminthología, Universitaria Nacional Autónoma de México, Ciudad Universitaría, México (City) D.F.

Nepal: Director, Fisheries Department, Katmandu; Fisheries Officer, Biratnager Agricultural Station, Tarahara, Sunsari.

New Zealand: Fisheries Research, Box 19062, Ministry of Agriculture, Wellington.

North Vietnam: Parasitologist, Research Station, Freshwater Fisheries, Dihn-Bang.

Norway: Parasitology, Zoological Museum, University of Oslo, Oslo 5; Zoologisk Laboratorium, Universitetet i Bergen Lars Hilleste. 10A N-500 Bergen.

Pakistan: Fish Parasitologist, Zoology Department, University of Karachi, Karachi-32; Fish Parasitologist, Federal B Area, Alnoor Society, Karachi-38; Director, Central Marine Fisheries Department, Karachi.

Peru: Departmento Oceanografía y Pesquería, Universidad Nacional Federico Villareal, Francia 726-Miraflores, Lima; Fish Parasitologist, Univ. Nacional de Trujillo, Dept. de Parasitologia, Trujillo.

Poland: Department of Parasitology, Veterinary Faculty, Agricultural College, Lublin, Akademicka 11.

Portugal: Parasitologist, Inst. Zoologia, Faculdade de Ciencias, 4000 Porto.

Rumania: Parasitologia, Faculty Medicina Veterinaria, Str. Manastur 3 CLU 3400.

Spain: Parasitology Department, Faculty of Veterinary, León; Departmento de Acuicultura, Laboratorío de Ictiopatología, Bioter-Biana, S.A., Emilio Vargas 7, Madrid 27; Associación Espanola Para El Progreso de Las Ciencias, Valverde 24, Madrid 13.

Sweden: Zoologiska Institut, Box 6801, Radmansgaton 70A, Stockholm 113 86.

Switzerland: Lake Research Institute of EAWAG/Swiss Federal Institutes of Technology, CH-8600 Dübendorf.

Taiwan: Parasitologist, Taiwan Fisheries Research Institute, Lu-Kang Branch, Lu-Kang Chang-Hwa 505; Disease Specialist, Fisheries Division, J.C.R.R., Taipei; Fish Parasitologist, Zoology Department, National Taiwan University, Taipei.

Thailand: Chief, Inland Fisheries Division, Department of Fisheries, Ministry of Agriculture, Bangkok.

U.S.S.R.: State Institute of Freshwater Fisheries (GOSNIORKh), Smolonaja 2, Leningrad C-124; Ichthyopathology, All Union Institute of Pond Fisheries (VNIIPRKh), Dmitrowchi Raion, P/O Rybnoe, Moscow 141821; Zoological Institute, Academy of Sciences, Leningrad B-34; Laboratory of Helminthology, Academy of Sciences, Bol. Serpuchovskaya 32 Korp 4, Moscow M-93; Parasitology Laboratory, Institute of Biological Sciences, Sevastopol.

United States: Parasitologist, U.S. Fish and Wildlife Service, Fish Farming Experimental Station, P.O. Box 860, Stuttgart, Arkansas 72160; Fish Pathology Section, F.R.E.D. Division, Alaska Department of Fish and Game, 333 Raspberry Road, Anchorage, Alaska 99502; Fish Parasitologist, Department of Marine Science, Mayaguez, Puerto Rico 00708; Hawaii Institute of Marine Biology, University of Hawaii at Monoa, P.O. Box 1246, Coconut Island, Kaneohe 96744.

Venezuela: Facultad de Ciencias Veterinarías, Universidad Central de Venezuela, Maracay, Venezuela; Instituto Venezolano de Investigaciones Científices, F.V.I.C. Dept. Biofísica y Bioquímica, Apartado 1827, Caracas; Parasitological Laboratory, Dept. Biología, Universidad de Oriente, Apt. P. 105, Cerro Colorado, Cumana.

Yugoslavia: Zavod za riba i pĉele, Veterinarski Fakulete, Sveuĉiliŝta Ŭ Zagrebu, Heinzelova 55, Box 190, Zagreb 41001.

Conclusion

In their natural habitat most fishes are sometimes the hosts of many parasites. It is difficult to document damage done to wild fish, but after the fish are placed in intensive culture or transferred to other countries – sometimes under crowded conditions – some of the parasites are likely to increase in numbers and become pathogenic. Therefore, for the sake of the product if nothing else, the fish should be inspected by a professional parasitologist. Cultured fishes that may be exported are also likely to be parasitized.

To attempt to categorize the parasites of fishes likely to be imported, one must be able to identify the parasites. Because some types are more likely to be dangerous than others, we here attempt to categorize the types. Inasmuch as knowledge about exotic species is gained only by trial and error, many predictions are likely to be erroneous.

HOST-SPECIFIC PARASITES

It is unlikely that host-specific parasites will be of consequence to other fishes in the exotics' new country. However, those that require no intermediate hosts (protozoans other than blood inhabitants, monogeneans, leeches, parasitic crustacea) could continue to flourish on or in the exotic in the new country, particularly if the fish host is cultured intensively. There are, however, control measures for most of these parasites.

Parasites requiring alternative hosts (trematodes, cestodes, nematodes, acanthocephalans) are not so likely to survive because the proper alternative hosts may not be present in the new country.

NON-HOST-SPECIFIC PARASITES

Although many of these non-host-specific parasites have already been transferred to other countries, we should not ignore them. Some are dangerous and should always be excluded from shipment. We list here some of the types with examples of some that have traveled widely.

Ectoparasitic protozoans

These, e.g., *Ichthyophthirius multifiliis, Ichtyobodo necatrix, Chilodonella cyprini, C. hexasticha,* and certain trichodinids are among the most damaging parasites of fishes and should never be transferred. Possibly equally dangerous new species will be found during transfer of exotics. However, because of the large numbers of shipments of exotics that have already been made, it seems less likely that a comparable new problem will appear.

Internal protozoa

Many protozoans are host-specific but some – including *Hexamita, Trypanoplasma, Eimeria,* and a few myxosporeans – have been found in more than one host, but usually the different hosts are closely related. The U.S. exotic *Myxosoma cerebralis* is a good example of this type. It traveled from Europe to the United States in frozen rainbow trout and later parasitized brook trout and salmon.

Adult digenetic trematodes

Few digenetic trematodes have been brought to parasitologists' attention as exotics. The European *Crepidostomum farionis* is one example, however. It was apparently brought to the United States in brown trout and has been of moderate concern in American trout culture.

Metacercarial digenetic trematodes

Like the adults previously mentioned, the metacercariae of trematodes are usually not expected to become exotically established because of the need for certain alternative hosts. *Bolbophorus confusus* of trout in Montana is an exception that apparently came to the United States from Europe in a stray European pelican. Some are human pathogens that should not be shipped internationally.

Cestodes

At least one cestode, the Asian tapeworm, *Bothriocephalus opsarichthydis,* lacks host specificity both as an adult in fish and as a larva in the copepod intermediate host. Thus it has traveled with susceptible fish all over the world. This cestode may be an exception because we have learned of no others that have traveled so well.

Nematodes

With one exception, non–host-specific nematodes have not become exotically important. The exception is *Camallanus cotti* of ornamental fish.

Acanthocephala

We are not aware that any members of this taxon have become important in relation to the transfer of exotic fishes.

Branchiurans and parasitic copepods

We know of two species, *Argulus japonicus* and *Lernaea cyprinacea* (*elegans*?), that lack host specificity and have been established worldwide.

We conclude that many fish parasites are probably of little consequence in world trade. However, because a few have caused great losses worldwide, it behooves us to study them thoroughly before allowing them to be shipped to other countries.

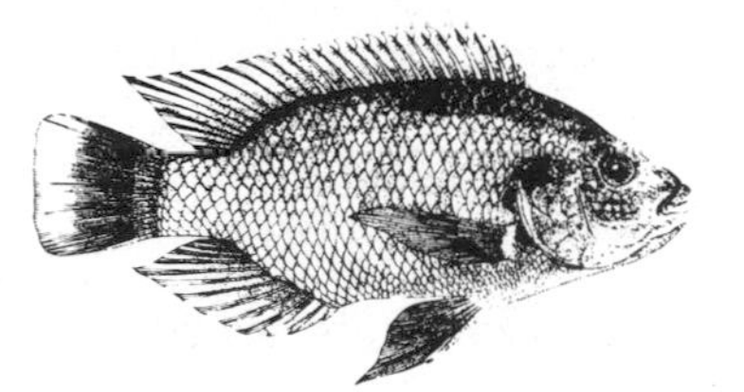

Literature Cited

Ahmed, A.T.A. 1973. Morphology and life history of *Mitraspora cyprini* Fujita, parasitic in the kidney of goldfish. Journal of Medical Science Biology 26(2):87–101.

Ahmed, A.T.A. 1974. Kidney enlargement disease of goldfish in Japan. Japanese Journal of Zoology 17(2):37–65.

Babaev, B. 1965. Distribution of *Bothriocephalus gowkongensis* Yeh, 1955 (Cestoda, Pseudophyllidae) in the water bodies of Kara-Kum Canal. Zoologicheskii Zhurnal 44(9):1407–1408 (in Russian).

Bauer, O. N., and G. L. Hoffman. 1976. Helminth range extension by translocation of fish, 163–178. *In:* L. A. Page (ed.). Wildlife diseases. Plenum Press, New York. 686 pp.

Bauer, O. N., V. A. Musselius, V. M. Nikolayeva, and Yu A. Strelkov. 1977. Fish diseases. Pishchevaya Promyshlennost' Press, Moscow. 431 pp. (English translation, #IP-126, National Marine Fisheries Service, Department of Commerce, Washington, D.C.)

Bauer, O. N., and Yu A. Strelkov. 1972. The effect of acclimatization and transportation on fish parasitc fauna. Izvestia WNIORCh 80:123–131 (in Russian).

Becker, C. D., and W. D. Brunson. 1968. The bass tapeworm: A problem in Northwest trout management. Progressive Fish-Culturist 30:76–83.

Bucke, D. 1982. Whirling disease, the new threat. Fish Farmer (England) 5(2):8–10.

Bykhovskaya-Pavlovskaya, I. E., et al. 1962 (1964). Key to the parasites of freshwater fish of the U.S.S.R. (Opredelitel' Parazitov Presnovoknykh Ryb SSSR, Akademi Nauk SSSR, Zoological Institute Moskva-Leningrad), 1964 English translation TT64-11040, U.S. Department of Commerce, Office of Technical Service, Springfield, Virginia. 919 pp.

Canella, M. F., and Ilda Rocchi-Canella. 1976. Biologie des Ophryoglenina (Cilies Hymenostomes Histophages). Universita Degli Studi di Ferrara, Italy. 504 pp. (English translation U.S. Fish and Wildlife Service.)

Conroy, D. A., J. Morales, C. Pedromo, R. A. Ruiz, J. A. Santicana. 1981. Preliminary observations on ornamental fish diseases in northern South America. Rivista Italiana di Piscicoltura e Ittiopatologia 16(3):86–100.

Cressey, R. F. 1976. The genus *Argulus* (Crustacea: Branchiura) of the United States. Water Pollution Control Research Series 18050 ELDO2/72 (2d printing), Cincinnati, Ohio. 14 pp.

Dashu-Nie and Lee Lien-Siang. 1960. Studies on the morphology and life-cycle of

Ichthyophthirius multifiliis and its control, with a description of a new species. Acta Hydrobiologia Sinica, 2:197–225 (in Chinese with English summary).

Dubinina, M. N. 1982. On the synonomy of species of the genus *Bothriocephalus* (Cestoda, Bothriocephalidae), parasites of Cyprinidae of the U.S.S.R. Parazitologiya 16(1):41–45.

Dykova, E., and J. Lom. 1980. Tissue reaction to microsporidean infections in fish. Journal of Fish Diseases 3:265–283.

Edwards, D. J., and P. M. Hine. 1974. Introductions, preliminary handling, and disease of grass carp in New Zealand. New Zealand Journal of Marine and Freshwater Research 8(3):441–454.

Evelyn, T.P.T. 1982. A partially annotated bibliography on fish quarantine and certification as approaches to disease control in fishes. *In:* Fish Quarantine Workshop Proceedings, International Development Research Centre, Jakarta, Indonesia, December 1982.

Fernando, C. H., and J. Y. Furtado. 1963. Some studies on helminth parasites of freshwater fishes. Proceedings of the Regional Symposium on Scientific Knowledge of Tropical Parasites (University of Singapore) 1:5–9.

Foissner, W., and G. Schubert. 1977. Morphologie der Zooide und Schwarmer von *Heteropolaria colisarum* gen. nov., spec. nov. (Ciliata, Peritrichida), einer symphorionten Epistylidae von *Colisa fasciata* (Anabantoidei, Belontiidae). Acta Protozoologica 16(3/4):231–247.

Foissner, W., G. Schubert, and N. Wilbert. 1979. Morphologie, infraciliatur und silberlinesystem von *Protoopalina symphysodonia* nov. spec. (Protozoa: Opalinata), einer Opanilinidae aus dem Intestinum von *Symphosodon aequifasciata* Pellegrin (Percoidei: Cichlidae). Zoologischer Anzeiger 202(1/2):71–85.

Fouquet, D. 1876. Note sur une espèce d'infusoires parasites des poissons d'eau douce. Archives de Zoologie Expérimentale et Générale 5(2):159–165.

Fryer, G. 1960. Studies on some parasitic crustaceans on African freshwater fishes, with descriptions of a new copepod of the genus *Ergasilus* and a new branchiuran of the genus *Chonopeltis*. Proceedings of the Zoological Society of London 133(4):629–647.

Fryer, G. 1968. The parasitic crustacea of African freshwater fishes; their biology and distribution. Journal of Zoology, London 156:35–43.

Fryer, J. L., and 8 others. 1979. Proceedings from a conference on disease inspection and certification of fish and fish eggs. Oregon State University Sea Grant College Program Publication ORESU-W-79-001.

Fujita, T. 1927. On new species of nematodes from fishes of Lake Biwa. Japan Journal of Zoology 1(5):169–176.

Ghittino, P. 1965. Segnalazion in Italia di una verminosi del persico-trota Nord-Americano d'importazion (*Urocleidus helicis* in *Micropterus salmoides*). Atti della Societa Italiana della Scienze Veterinarie 19:448–450.

Gratzek, J. B., E. B. Shotts, and J. L. Blue. 1976. A survey of parasites, bacteria, and viruses associated with tropical fish imported from Southeast Asia. Aquatic Mammals 4(1):1–5.

Gratzek, J. B., E. B. Shotts, and J. L. Blue. 1978. Ornamental fish: Diseases and problems. Marine Fisheries Review, paper 1299:58–60.

Heinze, K. 1933. Die Gattung *Capillaria* Zeder (1800) als Fischeparasit. Zeitschrift Parasitenkunde 5:393–406.

Henneguy, L. F. 1883. Sur un infusoire flagellé, ectoparasite des poissons. Comptes Rendus de l'Académie des Sciences, Paris 98(10):658–660.

Herkner, H. 1970. Enganzende Beobachtungen zum Thema "Lochenkrankheit." Auraien-und Terrarien-Zeitschrift (DATZ) 5:154–157.

Herwig, N. 1976. Exotic parasite in Florida. Aquarium Industry, October:14.

Hine, P. M. 1975. Fish disease control in New Zealand. FAO Aquaculture Bulletin 7(1-2):21.

Hirschmann, H., and K. Partsch. 1955. Ein einzelliger Fischparasit von überraschender Schönheit. Umschau 55:524–527.

Hoffman, G. L. 1953. *Scaphanocephalus expansus* (Crepl.), a trematode of the osprey in North America. Journal of Parasitology 39(5):568.

Hoffman, G. L. 1967. Parasites of North American freshwater fishes. University of California Press, Berkeley and Los Angeles. 486 pp.

Hoffman, G. L. 1970. Intercontinental and transcontinental dissemination and transfaunation of fish parasites with emphasis on whirling disease (*Myxosoma cerebralis*), 69–81. *In:* S. F. Snieszko (ed.). A symposium on diseases of fishes and shellfishes. Special publication no. 5, American Fisheries Society, Washington, D.C. 526 pp.

Hoffman, G. L. 1976a. Whirling disease of trout. U.S. Fish and Wildlife Service, Fish Disease Leaflet 47. 10 pp.

Hoffman, G. L. 1976b. The Asian tapeworm, *Bothriocephalus gowkongensis,* in the United States, and research needs in fish parasitology. Proceedings 1976 Fish Farming Conference and Annual Convention Catfish Farmers of Texas, Texas A & M University 84–90.

Hoffman, G. L. 1978. Ciliates of freshwater fishes, chap. 8. *In:* J. P. Kreier (ed.). Parasitic protozoa, vol. 2, Academic Press, New York. 730 pp.

Hoffman, G. L. 1981a. Recently imported parasites of baitfishes and relatives, 45–46. *In:* Third Annual Proceedings Catfish Farmers of America Research Workshop, Las Vegas.

Hoffman, G. L. 1981b. Two fish pathogens, *Parvicapsula* sp. and *Mitraspora cyprini* (Myxosporea), new to North America, 184–197. Proceedings International Seminar, Fish, Pathogens and Environment in Polyculture, Fisheries Research Institute, Szarvas, Hungary.

Hoffman, G. L., and F. P. Meyer. 1974. Parasites of freshwater fishes: A review of their control and treatment. T.F.H. Publications, Neptune City, New Jersey. 224 pp.

Hoffman, G. L., and A. J. Mitchell. 1982. Diagnostic services available. Aquaculture Magazine, Buyer's Guide for 1982:22–26.

Hoffman, G. L., and A. J. Mitchell. 1981. Some chemicals that have been used for fish diseases and pests. Fish Farming Experimental Station, U.S. Fish and Wildlife Service, Stuttgart, Arkansas. 8 pp.

Hoshina, T., and Y. Sahara. 1950. A new species of the genus *Dermocystidium, D. koi,* sp. nov., parasitic in *Cyprinus carpio* L. Bulletin of the Japanese Society of Science and Fisheries 15(12):825–829.

Ivanova, N. S. 1978. Schphidiidae Kahl. 1935. Parasites of channel catfish, *Ictalurus punctatus,* 72–74. *In:* V. I. Fedorchenko et al. (eds.). Parasites and diseases of fishes and their prophylaxis. All Union Scientific Institute of River Fish Husbandry 27. Moscow, Pishchevays Promyshelnnost.

Jeney, G., and K. Molnar. 1981. Thelohanellosis of the common carp fry in Hungary. Proceedings Fish, Pathogens and Environment in European Polyculture, Szarvas, Hungary, 205–209.

Kirtisinghe, P. 1964. A review of the parasitic copepods of fish recorded from Ceylon with descriptions of additional forms. Bulletin of the Fisheries Research Station, Ceylon 17(1):45–132.

Körting, W. 1974. Bothriocephalosis of the carp. Veterinary Medical Review no. 2:165–171.

Kudo, R. R. 1919, 1920. Studies on Myxosporidia. A synopsis of genera and species of Myxosporidia. Illinois Biological Monographs 5(3-4):1–265.

Kudo, R. R. 1933. A taxonomic consideration of Myxosporidia. Transactions of the American Microscopical Society 52(3):195–216.

Kulakovskaya, O. P., and B. A. Krotas. 1961a. On *Khawia sinensis* Hsu (Caryophyllaidae, Cestoda) – a parasite transferred from Far East to the fish farms of the U.S.S.R. western districts. Doklody Akademi Nauk SSSR 127(5):1253–1255 (in Russian).

Kulakovskaya, O. P., and B. A. Krotas. 1961b. A parasite introduced into carp hatcheries of the western U.S.S.R. from the Far East. Doklady Akademi Nauk SSSR 137:1253–1255 (AIBS translation, 305–306).

Laird, M. 1961. Quarantines and zoologists. Medical Service Journal 17(8):563–570.

Lavier, G. 1936. Sur quelques flagelles intestinaux de poissons marins. Annales de Parasitologie 14(3):278–289.

Lom, J. 1959. On the systematics of the genus *Trichodinella* Sramek-Husek (= *Brachyspira* Raabe). Acta Parasitologicka Polonica 7:573–590.

Lom, J. 1970. Observations on trichodinid ciliates from freshwater fishes. Archiv für Protistenkunde 112:153–177.

Lom, J., and D. P. Haldar. 1977. Ciliates of the genera *Trichodinella, Tripartiella* and *Paratrichodina* (Peritricha, Mobilina) invading fish gills. Folia Parasitologicka (Praha) 24:193–210.

Lom, J., and G. L. Hoffman. 1964. Geographic distribution of some species of trichodinids (Ciliata: Peritricha) parasitic on fishes. Journal of Parasitology 50(1):30–35.

Lucký, Z. 1977. Methods of the diagnosis of fish disease (Metodicke Navody K. Diagnostice Nemoci Ryb). English translation TT75-58005, U.S. Department Commerce, National Technical Information Service, Springfield, Virginia. 140 pp.

Lucký, Z. 1981. An investigation of invasive diseases of the fry of herbivorous fish and their treatment. Proceedings Fish, Pathogens and Environment in European Polyculture. Szarvas, Hungary, 259–269.

McDaniel, D. (ed.). 1979. Procedures for the detection and identification of certain fish pathogens. Fish Health Section of the American Fisheries Society, Bethesda, Maryland. 118 pp.

Maitland, P. A., and C. E. Price. 1969. *Urocleidus principalis* (Mizelle, 1936), a North American monogenetic trematode new to the British Isles, probably introduced with the largemouth bass *Micropterus salmoides* (Lacepède, 1902). Journal of Fish Biology 1(1):17–18.

Malevitskaya, M. A. 1958a. On the transfer of a parasite of complex development. *Bothriocephalus gowkongensis* Yeh, during acclimatization of Amur fishes. Doklady Akademi Nauk SSSR 123(3):572–575 (in Russian).

Malevitskaya, M. A. 1958b. The introduction of a parasite with a complex life cycle. *Bothriocephalus gowkongensis* Yeh, 1955, during acclimatization of fish from the Amur River. Doklady Akademi Nauk SSSR 123:961–964 (in Russian).

Manter, H. W. 1963. The zoogeographical affinities of trematodes of South American freshwater fishes. Systematic Zoology 12:45–70.

Meier-Brook, C. 1975. A snail intermediate host of *Schistosoma mansoni* introduced to Hong Kong. World Health Organization, WHO/SCHISTO/85:37.

Migaki, G., J. R. Hill, and G. L. Hoffman. 1981. Another exotic (*Dermocystidium koi*). Fish Health Section Newsletter of the American Fisheries Society 9(1):2.

Mitchell, A. J., and G. L. Hoffman. 1981. Preparation of live channel catfish for shipment to states requiring a health permit. Aquaculture Magazine 7(6):28–29.

Mizelle, J. D., and D. C. Kritsky. 1969a. Studies on monogenetic trematodes. XXXIX. Exotic species of Monopisthocotylea with the proposal of *Archidiplectanum* gen. n. and *Longihaptor* gen. North American Midland Naturalist 81(2):370–386.

Mizelle, J. D., and D. C. Kritsky. 1969b. Studies on monogenetic trematodes. XL. New species from marine and freshwater fishes. American Midland Naturalist 82(2):417–428.

Molnar, K. 1963. Mono- and digenetic trematodes from fishes. Allattani kozlemenyek 50:103–106 (in Hungarian).

Molnar, K. 1968. Beitrage zur Kenntnis der Fischparasiten in Ungarn. 3. Weiter Monogeneidenarten aus Fishen. Acta Veterinaria Academiae Scientiarum Hungaricae 18(3):295–311.

Molnar, K. 1970. An attempt to treat fish bothriocephalosis with devermin. Acta Veterinaria Academiae Scientiarum Hungaricae 20(3):325–331.

Molnar, K. 1979. Gill sphaerosporosis in the common carp and grass carp. Acta Veterinaria Academiae Scientiarum Hungaricae 27(1-2):99–113.

Molnar, K. 1982. Parasite range extension by introduction of fish to Hungary. Twelfth meeting of European Inland Fisheries Advisory Commission (EIFAC), Symposium on Stock Enhancement in the Management of Freshwater Fisheries, Budapest, Hungary. Typescript, 14 pp.

Moore, E. (1924) 1925. Fish Diseases. New York State Conservation Commission, 14th Annual Report 83–97.

Moravec, F. 1978. First record of *Molnaria erythrophthalmi* larvae in the intermediate host in Czechoslovakia. Folia Parasitologica (Praha) 25:141–142.

Moroff, T. 1902. *Chilodon cyprini* nov. sp. Zoologica Anzeiger 24:5–8.

Moznov, A. 1979. Cyprinida: A brief history of goldfish breeding in Russia. Tropical Fish Hobbyist 27(2):28–30.

Musselius, V. A. 1967. Parasites and diseases of plant-eating fishes and methods to control them. Izdatel' stoo "Kolos." 81 pp. (in Russian).

Musselius, V. A., and L. M. Mirzoeva. 1977. Parasites and diseases of buffalo and channel catfish. Fabrika Ofsetnoi Pachati no. 2, VNIIPRKH, Dmitrowchi Raion, Moskowskaja Oblast 141821, U.S.S.R.

Ogawa, K., and S. Egusa. 1980. *Gyrodactylus* infestations of cultured eels (*Anguilla japonica* and *A. anguilla*). Fish Pathology (Japan) 15(2):95–100.

Ossiander, F. J., and G. Wedemeyer. 1973. Computer program for sample sizes required to determine incidence in fish populations. Journal of the Fisheries Research Board of Canada 30(9):1383–1384.

Paperna, I. 1964. The metazoan parasite fauna of Israel inland water fishes. Badidgeh (Bulletin Fish Culture of Israel) 16(1/2):1-66.

Prost, M. 1973. Fish Monogenoidea of Poland II. Parasites of *Ictalurus nebulosus* (Le Sueur). Revision of genera *Cleidodiscus* Mueller, 1934 and *Urocleidus* Mueller, 1934. Acta Parasitologica Polonica (Warsaw) 21(22):315-326.

Reichenbach-Klinke, H. (1960) 1961. Die transkontinentale Verschleppung von Parasiten insbesondere der Wassertiere. Sonderdruck Verhandlungen Deutsches Zoologischen Gesellschaft in Bonnam Rhein 24:410-415.

Reichenbach-Klinke, H.-H. 1966. Krankheiten und Schadigungen der Fische. Gustav Fishcher Verlag, Stuttgart, West Germany. 389 pp.

Reichenbach-Klinke, H.-H. 1980. Krankheiten und Schadigungen der Fische. Gustav Fishcher Verlag, Stuttgart (West Germany), New York City. 472 pp.

Roberts, R. J. (ed.). 1978. Fish Pathology. Bailliere Tindall, London, and Lea and Febiger, Philadelphia. 318 pp.

Rogers, W. A. 1967. Studies on *Dactylogyrinae* (Monogenea) with description of 24 new species of *Dactylogyrniae,* 5 new species of *Peleucidhapter,* and the proposal of *Aplodiscus* gen. n., Journal of Parasitology 53(3):501-524.

Roman, E. 1953. Über die Parasitenfauna des Sonnenfisches (*Lepomis gibbosus* L.) bei der Akklimatiation in der Donan. Doklady Akademi Nauk U.S.S.R. 89:765-768.

Roman-Chiriac, Elena. 1960. Clasa Monogenoidea. Fauna republici populare Romine, Platyhelminthes. Bukarest 2(1). 149 pp.

Rudometova, N. K. 1974. On the life cycle of a new parasite of the white amur—*Skrjabillanus amuri*—and the epizootiology of skrjabillanosis. VI. All-Union Conference on Disease and Parasites of Fishes, Moscow, 215-217. (English translation, National Marine Fisheries Service, Washington, D.C.)

Schäperclaus, W. 1941. Eine neue Microsporidienkrankheit beim Neonfisch und seinen Verwandten. Wochenschrift Aquarien Terrareinkunde 38:381-384.

Schäperclaus, W. 1951. Der *Colisa*—Parasit, ein neuer Krankheinserreger bei Aquarienfischen. Die Aquarienund Terrarienzeitschrift 4:169-171.

Schäperclaus, W. 1979. Fischkrankheiten. Akademie-Verlag, Berlin, 2 vols. 1,089 pp.

Schmidt, G. D., and A. G. Canaris. 1967. Acanthocephala from Kenya with descriptions of two new species. Journal of Parasitology 53:634-637.

Sindermann, C. J., and A. E. Farrin. 1962. Ecological studies of *Cryptocotylea lingua* (Tremadoda: Heterophidae) whose larvae cause "pigment spots" of marine fish. Ecology 34:69-75.

Spotte, S. 1979. Fish and invertebrate culture. John Wiley and Sons, New York. 179 pp.

Ŝrámek-Huŝek, R. 1953. Kotazce taxonomie a pathogenicity naŝich ektoparasitickycah brousilek (Urceolariidae). Zoologický a Entomologický Listy (Czech) 2:167-180.

Stumpp, M. 1975. Untersuchengen zur Morphologie und Biologie von *Camallanus cotti* (Fujita, 1927). Zeitschrift Parasitenkunde 46:277-290.

Swezy, O. 1919. The occurrence of Trypanoplasma as an ectoparasite. Transactions of the American Microscopical Society 38(1):20-24.

Thompson, S., D. Kirkegard, and T. L. Jahn. 1947. *Scyphidia ameuiri* n. sp., a peritrichous ciliate from the gills of the bullhead, *Ameiurus melas.* Transactions of the American Microscopical Society 66(4):315-317.

Tikhomirova, V. A. 1980. On nematodes of the family Skrjabillanidae (Nematoda: Camallanata). Parazitologia (Leningrad) 14(3):258–262.

Vismania, K. O. 1962. Carp philometrosis in fish farms of Latvia, Izvestiia Akademi Nauk Latvia SSR 4:93–96 (in Russian).

Vismania, K. O., and V. N. Nikulina. 1968. On the systematics of *Philometra sanguinea* (Rud., 1819) (Nematoda, Dracunculidae) pathogen of crucian carp philometrosis. Parasitologia 2(6):514–518 (in Russian).

Vojtek, J. 1958. *Urocleidus* Mueller, 1934. Novy rod zabrohlistu (Trematoda, Monogenea) pro CSR. Biologia, Casopis Slovenskej Adakemie Vied 13:612–615.

Volovik. 1973. Influence of some gill parasites on fish organisms. Verhandlungen Internationale Vereinigung Limnologie 18(3):1713–1722.

Volovik. S. V., L. M. Mirzoeva, and A. V. Poddubnaya. 1974. On the parasite fauna of a buffalo (Catostomidae, *Ictiobus*) – new objects of pisciculture, acclimatized in the U.S.S.R. All Union Conference on Diseases and Parasites of Fishes, Moscow 6:59–81. (English translation by R. M. Howland, U.S. Fish and Wildlife Service.)

Wallach, J. D. 1971. Management and medical care of goldfish. Journal of the American Veterinary Medical Association 159(5):582–595.

Welcomme, R. L. 1981. Register of international transfers of inland fish species. FAO Fisheries Technical Paper no. 213, FAO Rome, Italy. 120 pp.

Wellborn, T. L., Jr. 1967. Trichodina (Ciliata: Urceolariidae) of freshwater fishes of the southeastern United States. Journal of Protozoology 14(3):399–412.

Wellborn, T. L., Jr. 1979. Control and therapy, 61–85. *In:* J. A. Plumb (ed.). Principal diseases of farm-raised catfish. Southern Cooperative Series 225, Auburn University, Auburn, Alabama. 92 pp.

Wharton, J.C.F. 1965. An assessment of the possibility and significance of introducing fish diseases into Australia. Fisheries and Wildlife Department Publication, Victoria, Australia. 10 pp.

Yamaguti, S. 1963. Parasitic copepods and Branchiura of fishes. Interscience Publishers, New York. 1,104 pp.

Yukhimenko, S. S. 1970. On the occurrence of *Bothriocephalus gowkongensis* Yeh, 1955 (Cestoda, Pseudophyllidae) in the young of Cyprinidae from the Amur River. Parasitologia 4(5):480–483 (in Russian).

CHAPTER 13

Exotic Fishes in Warmwater Aquaculture

William L. Shelton and
R. Oneal Smitherman

The scope of this chapter encompasses the husbandry in the United States of exotic or nonindigenous warmwater finfishes. An exotic fish is defined as one introduced from outside the country. The term *nonnative* is generally applied to introductions from outside a natural range and would include transplants between drainages.

Stickney (1979) defines aquaculture as underwater agriculture or the culture of aquatic organisms for human food, and Huet (1972) distinguishes between "fish culture for restocking" and "fish culture for food." The general principles of aquaculture apply in either realm (Bardach et al. 1972); whether the fish is destined to be stocked for some purpose in water resources management or to be grown to a marketable size, the nursery and early culture methods are similar.

We have limited our review to finfishes even though several imported invertebrates are cultured; for example, the prawn (*Macrobrachium rosenbergii*) and the Pacific oyster (*Crassostrea gigas*) are grown in both insular and continental U.S. waters (Pillay 1972). Coldwater species are not discussed since most recently introduced exotics would be used in warmwater culture. Nevertheless, the introduced brown trout (*Salmo trutta*) is widely cultured and stocked as a sport fish. Other chapters discuss fish introductions pertaining to water resource management, sport fishing, and the aquarium industry. For whatever purpose an exotic fish is used, escape is virtually inevitable; thus, this eventuality should be considered.

The bibliography compiled by Rosenthal (1978) and the summary prepared by Welcomme (1981) provide an overview of the extent of aquatic animal transplantation; the increase of number of exotics in North America was reviewed by Courtenay and Hensley (1980). Considering the potential problems that may result from introductions, why do we persist in this activity? Regier (1968) said that we choose an exotic fish on the basis of what it can do for us and that the desirability of the introduction depends on the benefits compared to any adverse effects. The brown trout and common carp (*Cyprinus carpio*) are good examples to illustrate this enigma.

The rationale for exotic introductions in U.S. aquaculture may involve several considerations. Virtually none of the exotic finfishes were imported to supplement our dietary protein but were considered for water resource management applications such as vegetation control. Currently, aquaculture in the United States is stimulated by economic incentive whereas the principal motivation for future development may be more influenced by nutritional needs (National Academy of Sciences 1978). Increased production can be expected with the present single species culture through genetic, nutritional, and technological improvements, but substantial expansion will require placing more water into production. Availability of land will then become increasingly limiting. An alternative or supplemental approach may be through polyculture. Fertilization and feeding increase the productivity of a pond, but the goal of rational pond management is to utilize the existing ecological niches to produce fish at levels approaching the pond's theoretical carrying capacity (Yashouv 1968, Swingle 1968a).

Polyculture has been little practiced in the United States. This void in our aquaculture may be filled through the use of proven species combinations (e.g., Chinese carps). However, the expediency of adopting known technology should be tempered with additional and more thorough screening of native species for similar functional combinations. While there are some options among our North American fishes, certain trophic niches are not filled, at least from an aquaculture perspective.

Historical Perspectives of the Development of U.S. Aquaculture

Fish culture is centuries old and improvements have evolved through empirical knowledge (Bardach et al. 1972). The chronology of development is concisely outlined by Stickney (1979) and Avault (1980). In the context of man-induced movement of fishes, certain historical milestones have been involved in the culture of exotics. The species cultured in the China of 4,000 years ago was the common carp, which was then introduced into Europe in the Middle Ages and subsequently into most of the rest of the world. Common carp is the most widely cultured fish in the world but is less than appreciated in the United States (Huet 1972).

A second wave of global introductions was stimulated by the culture of Mozambique tilapia, *Tilapia mossambica* (taxonomic nomenclature follows Robins et al. 1980). Several other tilapia species have been cultured and transported in recent years. This group of fishes is probably the next most ubiquitously cultured and is primarily limited by temperature tolerance and its proclivity to reproduce excessively in ponds (Balarin 1979).

A third wave of species translocations involved the Chinese carps during the 1960s, following the development of techniques for induced spawning (Bardach et al. 1972). These fishes have become a significant portion of the polyculture systems in Israel (Halevy 1979, Hepher and Pruginin 1981) and India (Chadhuri et al. 1975, Sinha and Gupta 1975).

Fish culture for stocking game fishes was widely practiced in the United States in the second quarter of the twentieth century, but culture for other purposes was insignificant until the 1950s, at which time it expanded with the culture of trout and buffalofish (*Ictiobus* spp.). The former prospered but the latter dwindled by the early 1960s (Bowen 1970, Swingle 1970). At about that time, the catfish industry rapidly evolved following the development of techniques for inducing spawning (Clemens and Sneed 1957) and pond production (Swingle 1957).

In 1980, a total of 21,100 metric tons of catfish was produced in 22,100 ha of water by about 1,200 producers in ten states. Alabama, Arkansas, and Mississippi have 71 percent of the ponds and 89 percent of the total area under culture. Catfish and other cultured food fish account for only about 2 to 5 percent of the total fish consumed in the United States. Consumers in the United States are not fish eaters by tradition. In 1979, the per capita consumption of fish was 6 kg, only 0.2 kg of which was cultured (USDA 1981).

Traditional marine stocks, once thought unlimited, now are estimated at a maximum harvest of 100–150 million metric tons (mmt) per year (Glude 1977). On a world scale, catch of fish leveled in the 1970s at about 70 mmt per year with an estimated demand in 1980 of about 100 mmt of fish (Hepher 1978). World aquaculture production of finfish in 1975 was about six mmt of which the United States produced only about 0.5 percent. An estimated 10 percent of the world's water-derived protein comes from aquaculture. Because of the sustained yield limit of capture fisheries and the increasing cost of harvesting wild fish, aquaculture will become increasingly important worldwide. Present-day aquaculture represents a step in man's evolution from hunter to herdsman to husbandman (Pillay 1979a).

If aquaculture is to evolve in a parallel fashion to agriculture, then a precedent has been set. Virtually all of our food crops and livestock were derived from outside the United States. Most have undergone genetic selection and represent cultivars rather than truly exotic species (Courtenay and Robins 1975). However, most would not persist without man's continued effort in breeding and few of our domestic animals would survive the rigors beyond the barnyard.

Commercial Producers of Exotic Fishes in the United States

To determine current aquaculture involvement with exotic fishes, we used the 1981 buyer's guide published in *Aquaculture* magazine and requested information on species, production, and marketing from advertisers. Questionnaires were sent to personnel in the main office of each state's Department of Natural Resources, because it was assumed that most states would have a licensing requirement for exotic fishes. In addition, at least one Fisheries Extension specialist was contacted from each state to inquire about producers or dealers of exotic fishes. University, state, and federal researchers known to be working with exotic fishes were contacted about their current activities. The American Fisheries Society (Fish Culture Section) membership list was a useful aid for this endeavor. Finally, follow-up questionnaires were sent to known commercial farmers for specific information. About 250 inquiries were sent, and numerous telephone calls were placed. We received about 50 percent response to the mail requests.

Partial results of the survey (table 13–1) showed that one or more species of the Chinese carps are being cultured by at least thirty commercial dealers in eight states; only six were growing any of them for food. Research was being conducted at about the same number of locations but in a greater number of states, seven of which were oriented toward culture as food fish. The tilapias were similarly distributed but considerably more commercial farmers were growing tilapia for food.

Exotic Species

The following discussion summarizes pertinent aspects of the biology of various exotic fishes that are currently in the United States and that are a part of contemporary aquaculture or that are logical candidates for utilization. The term *exotic* is relative and rationalization often becomes a part of evaluating an introduction. For example, *Tilapia mossambica* is exotic to the continental and insular areas of the United States and has been introduced into both areas (Lee et al. 1980). The threadfin shad (*Dorosoma petenense*) is native to the southeastern United States but has been introduced into Hawaii (Hida and Thomson 1962).

Two fishes exotic to North America are well known but are viewed with contrasting attitudes. The common carp was introduced, complete with halo, but would now receive little praise from most Americans (Hubbs 1968). However, it has as much potential for aquaculture here as in Europe or Asia. In contrast, the brown trout is widely accepted and is annually stocked by at

Table 13–1. ACTIVITIES INVOLVING EXOTIC FISHES THAT HAVE POTENTIAL FOOD VALUE

Activity and fish groups	Locations	States[a]	Food production[b]
Producers/dealers			
Chinese carps	30[c]	8	6
Tilapias	24	12	19
Research[d]			
Chinese carps	30	13	7
Tilapias	18	12	7

[a] Contiguous and insular United States plus Virgin Islands and Puerto Rico
[b] Non-food uses include water resources management (aquatic weed and plankton control)
[c] Twenty locations are in two states (Alabama and Arkansas)
[d] State, university, and federal facilities

least seventeen states (present survey). Both of these species are naturalized and have become an accepted, if not welcome, part of our fauna. Other exotic fishes that have been introduced and can be considered as having potential for U.S. aquaculture are the Chinese carps and some of the cichlids.

CHINESE CARPS

The Chinese carps that have been introduced into the United States include the grass carp (*Ctenopharyngodon idella*), the silver carp (*Hypophthalmichthys molitrix*), and the bighead carp (*Aristichthys nobilis*). The grass carp has caused the greatest controversy, primarily because of the widespread interest in the use of this species for aquatic weed control contrasted with the fear of ecological repercussions. The focal point of the controversy has been the likelihood of natural reproduction, with advocates suggesting that no spawning would occur in stocked lakes, while opponents argued that stocked fish rarely stay where put and that access to riverine habitats might result in successful reproduction. Both contentions were correct; spawning has not been documented in lentic conditions but has occurred in the lotic waters of the United States (Conner et al. 1980).

Grass carp

Grass carp were first introduced into the United States in 1963 in two shipments, one to the U.S. Fish and Wildlife Service, Fish Farming Experimental Station, Stuttgart, Arkansas (Stevenson 1965) from Malaysia, and the other to Auburn University, Alabama, from Taiwan (Sills 1970). The first private shipment was made to Arkansas in 1972 (J. M. Malone, personal communication). Grass carp were first spawned in the United States at the Stuttgart Station and at Auburn in 1966 (Jeffrey 1970). Subsequent spawnings were made annually at the Arkansas Game and Fish Division hatchery at Lonoke, and at Auburn.

Many research organizations have worked with grass carp in the intervening period. Guillory and Gasaway (1978) listed the agencies active in the 1970s. We found that several of these research activities on grass carp have been terminated.

Distributors of grass carp in the United States are primarily in Arkansas and Alabama; a few are in other states where possession is legal. Currently about five fish farmers are known to spawn their own fish; however, most purchase fry and grow them to a stocking size for resale for aquatic weed control. Only two that we contacted were involved with their culture as food fish.

The grass carp is prohibited in most states; this has impeded but not prevented their use in weed control (Guillory and Gasaway 1978). To consider their use in aquaculture, we would need to reexamine the legal constraints, but in the present discussion we emphasize the biological considerations.

The grass carp has been the subject of numerous studies throughout the world. An introduction to the international literature was prepared by Nair (1968). More recently, a number of U.S. studies were collated (Stanley and Lewis 1978); also in 1978 a conference on this species was convened in the United States (Shireman 1979) and another was conducted in Yugoslavia in 1982. The culture of grass carp is discussed in chapters of Huet (1972) and Bardach et al. (1972), and in contributions to a symposium sponsored by the American Fisheries Society, Fish Culture Section (Smitherman et al. 1978). Culture technology is detailed in a manual from the Fisheries Research Institute of Shanghai (1978), by Dah-Shu (1980), and in the more accessible manual by Woynarovich and Horvath (1980).

All of the Chinese carps are river spawners and the reproductive requirements for grass carp were discussed by Stanley et al. (1978) in the context of predicting spawning in the United States. For aquaculture purposes, a dependable supply of fry for stocking is required even though the Chinese polyculture was supported for hundreds of years from wild caught fry. Induced spawning according to southeast Asian tradition is discussed by Lin (1965). More contemporary techniques in China include tank spawning (Tapiador et al. 1977). The Russian approach (Konradt 1968) and Israeli methods (Rothbard 1981) differ little from the Chinese methodology, but the injection schedule used in the United States is quite different (Bailey and Boyd 1971). For overall perspective, reproductive control of fishes was recently reviewed in a paper published from a conference held in Kuwait (Pullin and Kuo 1981).

The nursery phase of culture is important with reference to the rate of growth and survival to fingerling size. General information on pond preparation, fry stocking, and pond management is in Bardach et al. (1972). Traditional fry stocking and subsequent restocking schedules are discussed by Hora and Pillay (1962) and more recently by Pillay (1979b). Stocking during the initial growth period is based on a monoculture since there is an overlap in the diet of fry of the Chinese carps until each species assumes its specialized feeding habits. For example, grass carp do not become herbivorous until they are between 30 to 50 mm total length (Hickling 1966, Desilva and Weerakoon 1981). However, polyculture of young fish was reported to be successful by Murty et al. (1978). An alternative approach is intensification of culture as discussed by Appelbaum and Uland (1979) and Huisman (1981).

Food habits and growth potential are two criteria for selecting certain fish species for aquaculture (Webber and Riordan 1976). The food of grass carp has been examined from several perspectives. Those that have been of main interest in the United States pertain to vegetative feeding such as reported by Stott and Orr (1970), Kilgen and Smitherman (1971), and Opuszyński (1972). However, growth on various prepared diets has been examined to determine the potential for more intensive culture (Fischer 1972a, b; Fischer and Lyakhnovich 1973; Dabrowski 1977; Stroband 1977; Shireman et al. 1978a; Huisman and Valentijn 1981).

Survival during the nursery phase of fry to early fingerling culture is critical. Tamas and Horvath (1976) used the insecticide Dylox to eliminate invertebrate predators and to restructure the plankton community to be more suitable for the diet of young carp. Growth management is a function of stocking density and nutritional input. The traditional approach to growth control is summarized by Hora and Pillay (1962).

The culture practices of most U.S. producers usually involve low density stocking with some supplemental feeding. Fry are stocked at about a one-half million per ha and about 30 percent survival is expected to 4–7 cm fingerlings by early fall. Growth to stocking size is affected by stocking density as discussed by Shelton et al. (1981a). For example, stocking rates of 3,700 or 9,800/ha should produce fingerlings about 15–25 cm or 10–15 cm long, respectively, during the next summer. Fish must be reared to 15–30 cm before stocking to prevent predation, primarily by largemouth bass (Shireman et al. 1978b). Although grass carp readily eat prepared diets, growth is usually slow. At Auburn, we have been investigating more intensive rearing, to a stocking size and also to market size. Fingerlings (7–8 cm) were grown to 23–29 cm at about 8,500/ha in one summer (Mgbenka 1983).

Traditionally, grass carp are not grown for food in a monoculture. They sometimes will be the principal species in polyculture but usually this requires extensive labor to supply vegetation produced from outside the pond. The other species benefit from the recycling of nutrients (Pillay 1979b). More often, grass carp are a subsidiary species and grow rapidly under low density stocking. In polyculture, stock composition is not consistent; different species combinations are used at various locations, making comparison for total productions difficult. Further, examination of growth produced by a species in monoculture may be quite different when compared to its growth in combination with other species. Yashouv (1971) found that production of common carp and silver carp was higher in polyculture than if the species were stocked alone. Theoretically, this relation is a primary goal of polyculture. Total production is increased through simple addition of each component, and it is hoped that there is some mutual benefit between species rather than a reduction in growth due to competition.

In channel catfish monoculture, the maximum safe feeding level has been considered to be about 39 kg/ha/day (Swingle 1968b). If more feed is added, water quality deteriorates adding stress and potential for parasitism and other disease organisms. Consequently, water quality considerations limit greater standing stocks. However, if some of the organic matter is removed, reducing the biological oxygen demand (BOD), feeding rates can be higher. This has been demonstrated by various means of removal of organic wastes with trickling filters, plant uptake, animal biofilters such as clams and finally fish. Various species of fish have been tested as biofilters but the silver carp is the most promising choice because it grazes on phytoplankton.

Production is a function of pond fertility or feed input but higher yields are also associated with increasing stocking densities (Hepher 1978). There is, however, an inverse relation in that higher densities result in slower individual growth (Backiel and LeCren 1967, Shelton et al. 1981a). Table 13–2 summarizes some of the stocking rates and yields (net production) from various areas in the world with differing management inputs. Comparison of these data are confounded by differences in climate, stocking duration, and level of management. The intent of this summary was to present the range in stocking densities, the proportion of species in a polyculture, and an order of magnitude for yield. The daily production was included when it was available or when it could be calculated. Considering the net yield in terms of daily weight gain permitted more equitable comparisons.

Several generalizations are evident from this summary. The management input affects returns (e.g., compare the two systems in Taiwan). Also, the increase in yield that results from higher stocking rates is illustrated by the data from Israel. Israel has a highly technological society and has adopted polyculture and the Chinese carps in its food production scheme.

Grass carp were introduced into Israel in 1965 from Japan. Growth of grass carp in polyculture was quite variable, probably because of different availability of vegetation. At high densities they competed for pelleted food but did not effectively convert food in this form to fish flesh (Tal and Ziv 1978, Hepher and Pruginin 1981). Studies in the United States have indicated similar results, and these data are summarized in table 13–3.

Silver carp

Silver carp were introduced into the United States in 1972 under an agreement of maintenance with the Arkansas Game and Fish Commission (J. M. Malone, personal communication). They were first spawned in Arkansas in 1974. Auburn University received a shipment of young in December 1974 as

Table 13–2. POLYCULTURE YIELDS (KG/HA) AND STOCKING RATES (NO/HA) UNDER DIFFERENT MANAGEMENT INPUTS FROM VARIOUS COUNTRIES

Country	Management[b]	Species – Yield[a] (Stocking rate)					Net yield		Source
		Silver carp	Bighead carp	Grass carp	Common carp	Others[c]	Total	Daily	
China	F, M, O	1360 (900–1050)	1590 (540–580)	2650 (750)[d]	370 (530)	plus 5	8710	24	PRFRI 1980
Taiwan	F, M, O	2700 (3500)	750 (2500)	250 (200)	900 (10000)	2500 (9000)[1]	7300	22	Tang 1970
Taiwan	none	250 (400)	25 (15)	75 (80)	25 (200)	25 (200)[1]	400	1	Tang 1970
Israel	F-h, M, I	1140 (1430–1500)	–	35 (30)	4290 (3500–3580)	1820 (2210–5000)[2]	7200	31	Halevy 1979
Israel	F-h	1250 (2500)	–	340 (750)	4180 (11450)	510 (5000)[2]	6280	50	Moav et al. 1977
Israel	F-h	950 (1250)	–	190 (330)	2890 (3300)[2]	200 (1500)[2]	4240	34	Moav et al. 1977

[a] Yields are means, normal ranges, or typical values and are net (adjusted for initial weights); stocking rates are parenthetic values.
[b] Management – F = fed; F-h = fed high protein diet; M = manure added; O = organic fertilizer; I = inorganic fertilizer.
[c] Total yield only for species listed – 1 = mullet; 2 = tilapia.
[d] Grass carp was the principal species and the fish were fed with terrestrial vegetation.

Table 13-3. NET YIELDS (KG/HA) AND STOCKING RATES (NO/HA) OF EXOTIC FISHES CULTURED IN VARIOUS PARTS OF THE UNITED STATES

State	Species – Yield[a] (Stocking rate)					Net yield		Source
	Silver carp	Bighead carp	Grass carp	Tilapia	Channel catfish	Total	Daily	
AL	220 (250)	–	30 (50)	240 (620)	3490 (7410)	3980	19	Pretto-Malca 1976
AL	1080 (2500)	–	70 (50)	920 (2000)	3420 (7500)	5490	33	Dunseth 1977
AL	680–740 (2470–7410)	–	–	–	4890–5490 (19760–24700)	5630–6170	37–41	Osborn 1977
AL	2480 (2640)	–	–	–	9780 (7600)	12260[b]	16	AU[c] Unpub. 1980
AL	–	430 (120)	–	–	3710 (6750)	4140[b]	8	AU Unpub. 1980
AR	200–1780 (480–1910)	180–620 (120–480)	–	–	400–2970 (4590–7200)	1500–4940	9-17	Henderson 1980
AR	480 (1250)	130 (250)	130 (50)	–	620 (3150)	1370	10	Newton et al. 1978
AR	460 (490)	–	120 (50)	–	770 (3110)	1360	4	Newton 1980
AR	–	620 (490)	120 (50)	–	770 (3110)	1510	4	Newton 1980

[a] All studies included feeding based on only channel catfish biomass; stocking rate are parenthetic values.
[b] Includes program of trapping and restocking over two to three-year period at Auburn University.
[c] Auburn University.

did a number of other institutions (Buck et al. 1978). These fish matured and were spawned at Auburn in 1978, 1981, and 1982. These four year classes have been used in a number of studies over the last eight years at Auburn. At least one other shipment of silver carp, from Yugoslavia, was known to have been received in the continental United States by a private fish farmer.

Silver carp biology is similar to that of grass carp in many respects. Silver carp have been disseminated nearly as widely in the world as grass carp but for use in water quality management as a biofilter for plankton. Information for general culture is similar to that for grass carp; however, no single bibliographic source on silver carp biology is available.

Spawning techniques are comparable to those for grass carp with minor modifications (Yashouv et al. 1970, Makeyeva and Verigin 1971, Henderson 1980). For the nursery phase, the same techniques as described for grass carp should be considered.

The attractiveness of the silver carp in culture is based on its unique filtering mechanism that permits feeding on very small phytoplankters (Wiliamovski 1972, Cremer and Smitherman 1980). A number of food habit studies have focused on the feeding capacity (Panov et al. 1969, Omarov 1970). The BOD has been reduced (Osborn 1977) through removal of phytoplankton (Dunseth 1977, Henderson 1978). Furthermore, the phytoplankton species composition is altered from predominantly bluegreen algae to greens (Lin 1974, Behrends 1977). By altering the composition there has been a reduction in the occurrence of nuisance blooms, subsequent die-offs, and associated odor problems (Henderson 1978, 1980). There also may be a reduction in the incidence of "off-flavors" in fish that are often associated with bluegreen algae (Lovell 1979).

Estimates of filtering capacity based on rates of ingestion vary from 12 to 19 percent of the body weight per day for fish of about 0.75 kg. Silver carp, therefore, benefit from the nutrient supply of a pond but also enhance the general pond water quality through their feeding. Examples of stocking rates and yields from selected studies are summarized in tables 13–2 and 13–3. Because silver carp feed on photoplankton, they can be raised in water receiving a variety of organic enrichment. Silver carp benefit from highly fertile systems whether the origin of the nutrient source is terrestrial animal wastes, heavy feeding, or other species of fishes in polyculture.

In many countries where filter feeders of aquacultural importance were absent, silver carp have been introduced to fill the void. Silver and grass carp were added to polyculture in India in about 1959 (Sinha and Gupta 1975). The term *composite* fish culture is used when referring to the combination of Indian carps and one or more of the Chinese carps. Silver carp were introduced into Israel from Japan in 1966. First-year fingerlings were stocked at about 10,000/ha and reach approximately 100 g, while second year fingerlings were stocked at 800–1,000/ha in polyculture and grew to a marketable size of 1.5–2.0 kg (Tal and Ziv 1978). Israeli biologists cited poor handling characteristics as problems in silver carp culture in that these fish are easily injured and are prone to excessive jumping when disturbed (Hepher and Pruginin 1981).

Bighead carp

Bighead carp were introduced into the United States in 1972 (Cremer and Smitherman 1980) into Arkansas and studied at the State Fish Hatchery at Lonoke. Young were received at Auburn from Arkansas in December 1974 (Pretto-Malca 1976). Workers in other areas also received fish for investigations at about the same time (Buck et al. 1978). The fish received at Auburn matured and were spawned first in 1978 and again in 1981 and 1982. Induced spawning methods are the same as those used for silver carp (Henderson 1980). At least two additional shipments of bighead carp have been imported into the United States by fish farmers; one group came from Israel and the other from Yugoslavia.

The bighead carp is biologically similar to silver carp but is much more easily handled than either the silver carp or grass carp; it jumps less and is easier to seine and feeds primarily on zooplankton (Cremer and Smitherman 1980). It was found to feed readily on a pelleted trout diet while the silver carp would not. In polyculture experiments, bighead carp at 125/ha slightly depressed the production of catfish when compared to monoculture; catfish production was slightly elevated in polyculture with silver carp (Pretto-Malca 1976). Opuszyński (1981) also reported reduced yield of common carp when bighead carp were added. In two pond studies at Auburn, catfish production was good in both the pond with silver carp and the one with bighead carp (table 13-3). The bighead carp, stocked at 120/ha, grew from an average of 0.4 to 3.9 kg in eleven months and the silver carp, stocked at 2,640/ha, grew from 0.3 to 0.8 kg in the same interval.

The bighead carp was introduced into Israel from Germany in 1973 and is considered comparable to silver carp, but has better growth potential and handling characteristics and has more potential as a canned food product (Tal and Ziv 1978). According to Leventer (1981), bighead carp growth is about 10–20 percent better than that of silver carp. Interest in the bighead carp has increased rapidly in the United States since 1979 when information on the production of the triploid hybrid first was obtained; this cross is based on bighead male and female grass carp (Marian and Krasznai 1978). This hybrid cross was made by Russian workers earlier (Andriasheva 1968) but they did not recognize that the offspring were variable in appearance. This hybrid is of prime interest for use in aquatic weed management.

Other hybrids that were made by Russian biologists (Andriasheva 1968) may have aquaculture applications. The Chinese have hybridized silver carp and bighead carp and are now evaluating the hybrid (Tapiador et al. 1977). The Israelis have made these hybrids but without documenting growth rates of a particular cross. During the 1981 spawning season, we produced both reciprocal hybrids between silver and bighead carp. Percent fertilization and hatching success was as high for the hybrids as for their parental types. These fish are being cultured in growth tests to determine their relative value in polyculture. A hybrid that has similar feeding capability to the silver carp, but with better handling temperament, would be highly desirable. Perhaps all four types (parents and hybrids) could be combined for a highly efficient polyculture combination.

TILAPIAS

The literature on tilapias is voluminous. An annotated bibliography prepared by Thys van den Audenaerde (1968) has over 1,600 entries. A booklet published by Balarin (1979) has good coverage for life history of various species with aquaculture potential. Culture information is discussed by Chimits (1955, 1957), Bardach et al. (1972), Huet (1972), and Hepher and Pruginin (1981). Information on more specific topics of aquacultural relationships are contained in the Symposium Proceedings edited by Smitherman et al. (1978) and the Proceedings of the Conference on Tilapia Culture (Pullin and Lowe-McConnell 1982). Distribution of species that are in the United States is summarized by Lee et al. (1980); Courtenay and Hensley (1980) list those that have been introduced but have not become established. With the exceptions of *Tilapia nilotica* and *T. hornorum,* all are listed by Robins et al. (1980); the true *T. nilotica* is mentioned as not being in the United States but this is erroneous. The source of confusion probably originates in the early U.S. literature on the culture of *T. aurea* (e.g., McBay 1962, Avault and Shell 1968). Here the species used was reported as *T. nilotica* which was the identification given when it was brought from Israel in 1957 (Swingle 1960). The species was subsequently recognized as *T. aurea* (Trewavas 1966). The present Auburn stock of *T. nilotica* was imported from Pentecoste, Brazil, in 1974; the origin of this stock was Ivory Coast, West Africa. Also in 1974, *T. hornorum* was brought from the same source and same origin.

Besides the complications associated with the species identity (e.g., *T. nilotica* vs. *T. aurea* and *T. mossambica* vs. *T. hornorum*) the generic designation is unsettled. U.S. taxonomists maintain the genus *Tilapia* for the species under discussion here while Trewavas (1973) has suggested that the mouthbrooders (*T. aurea, T. nilotica, T. hornorum,* and *T. mossambica*) be designated as *Sarotherodon* spp. and that the substrate spawners such as *T. zilli* be retained as *Tilapia*. More recently, Trewavas (1982) has separated the East African maternal mouthbrooders from *Sarotherodon,* designating these as *Oreochromis* spp.

The tilapias of principal interest for aquaculture in the United States and throughout much of the world are *T. nilotica* and *T. aurea* (Schoenen 1982). Both have comparably rapid growth potential and are primarily planktophagic, although food habits vary with size of fish and availability of food (Yashouv and Chervinski 1961, Moriarty and Moriarty 1973, Spataru and Zorn 1978, Sarig and Arieli 1980), and *T. aurea* has a higher minimum temperature tolerance (Lee 1979). *Tilapia mossambica* is cultured in many parts of the world but is less desirable because of its earlier maturity and consequent greater tendency to overpopulate. *Tilapia hornorum* is of interest primarily for its role in hybridization, where the male of this species when mated with several other species produces all-male or nearly all-male broods (Balarin 1979). Hybridization also presents the possibility of enhanced growth because of hybrid vigor, and some behavioral traits of one parent may appear in the hybrid. For example, both *T. nilotica* and *T. aurea* are adept at passing under a seine during harvest whereas *T. hornorum* and its hybrids are much easier to catch (Dunseth 1977, Lovshin 1982).

The early introductions into the United States involved primarily *T. mossambica,* a species considered for use in vegetation control (Kelly 1957), but interest soon turned to culture as a food fish of this and subsequently acquired species (Swingle 1960). In all mouthbrooding species that are commonly cultured, the female carries the eggs and fry. The biology of *T. aurea* is typical of these species and is discussed by McBay (1962).

Many species of the family Cichlidae are well suited for aquaculture. They are tolerant of adverse water quality (Colt et al. 1979) such as low oxygen (Denzer 1968), high temperature (Allanson and Noble 1964, Balarin 1979), and high concentrations of ammonia (Redner and Stickney 1979). Unlike the Chinese carps, tilapias do not have intramuscular bones. However, they lack tolerance for low temperature (Chervinski and Lahav 1976), and this characteristic restricts their use outside warmer areas unless they can be protected during winter. An additional problem is their high reproductive potential because of early maturity, multiple spawns per season, and well-developed parental care.

Overwintering techniques in the United States have been discussed by Avault et al. (1968) and Chervinski and Stickney (1981). From the standpoint of naturalization, the sensitivity to low temperature is a positive aspect and tilapia are unlikely to persist except in warmer parts of the United States or in areas of thermal refuge such as cooling reservoirs for electrogenerating plants. The naturalization and spread in Florida and Texas with the repercussions are discussed in a number of papers (Buntz and Manooch 1968, Noble et al. 1975, Germany and Noble 1978, Hendricks and Noble 1980).

Controlled fry production is discussed for pond systems by Bardach et al. (1972), in plastic pools (Silvera 1978), tanks (Uchida and King 1962, Hida et al. 1962), and in aquaria (St. Amant 1966, Lee 1979). Fry production is affected by brood stock size, stocking density, and sex ratio. Uchida and King (1962) reported monthly averages of about 800 fry/m^2 or about 100 fry/female when *T. mossambica* were stocked at 10 fish/m^2 at a sex ratio of

3:1 (female:male). Calhoun (1981) used lower stocking rates with *T. nilotica* (5 fish/m^2 and a 3:1 sex ratio) and had a monthly average fry production of about 450/m^2 or 125/female.

Control of unwanted reproduction is a major problem in tilapia culture. Guerrero (1982) reviewed various methods of managing reproduction. Each has limitations, and no single approach is appropriate for all conditions. For example, raceway culture (Lauenstein 1978, Ray 1978) and cage culture (Coche 1982) may be appropriate for some areas, and recruitment from reproduction would not be a problem under these conditions; however, grow-out in ponds does require that reproduction be managed. Lovshin (1982) reviewed the potential and limitations for monosexing tilapia by hybridization. Variability in the production of all-male progeny is problematic and, in some species crosses, fry production is low. On the positive side, enhanced growth is indicated and greater seinability is a characteristic of some hybrids. The hybridization approach is being used commercially in Israel but, because of some females in the progeny, the traditional approach of separating the sexes before maturity for final grow-out must be practiced (Hepher and Pruginin 1981). Hormonal sex reversal has potential (Shelton et al. 1978) but it does not consistently yield 100 percent males, and further investigation is needed to resolve the potential conflict with consumption of hormone-treated fish. Predators may be stocked either as a direct means of control or to supplement one of the above approaches. A number of piscivores have been effective: Swingle (1960) used largemouth bass (*Micropterus salmoides*), Dunseth and Bayne (1978) used guapote tigre (*Cichlasoma managuense*), and Fortes (1979) used tarpon (*Megalops* sp.).

Intensification has been used, both as a means of reducing the effect of reproduction and also as a means of increasing production per area. High density tank culture with aeration was discussed by Allison et al. (1979) and Henderson-Arzapalo and Stickney (1981). High density culture with a system for water reuse was described by Rakocy and Allison (1981). Intensification requires more attention to completeness of diets (Davis and Stickney 1978, Winfree and Stickney 1981) and to water quality maintenance. Although production can be high, the expense of water exchange and/or artificial aeration (Allison et al. 1981) must be considered.

Tilapias probably have their greatest potential in enriched tropical or subtropical pond culture. Here they can feed directly on supplementary diet or from autochthonous production. There is a wide variety of dietary approaches to culturing tilapia (e.g., they have been cultured successfully on various wastes). The use of various animal wastes as a source of pond enrichment is practiced throughout the world. Integrated agriculture-aquaculture is an efficient means of producing food. The use of wastes from domestic sources has been traditional in Asian countries but has been tested only on a limited scale in the United States for tilapia (Coleman et al. 1974) and Chinese carps (Henderson 1978). The mode of application varies; diet has been supplemented with byproducts (Bayne et al. 1976, Kohler and Pagan-Font 1978), or animal wastes have either been used to supplement prepared diets (Nerrie 1979), or been used directly in a slurry (Moav et al. 1977), fresh (Collis and Smitherman 1978), dried (Moav et al. 1977), or partially digested (Behrends et al. 1979). In many instances terrestrial animals are cultured in the vicinity of the pond so that their feces fall or are washed directly into the pond (Stickney et al. 1978, Burns and Stickney 1980). In this approach, the manure composition and quantity expected from various animals determines the stocking density of the livestock (Hopkins et al. 1981).

One of the primary concerns of managing ponds that receive manure is high BOD. Delmendo (1980) summarized data on the quantity of manure produced by various domestic animals. The source of animal manure is important and proper management of appropriate amounts is essential, but monitoring DO can avoid catastrophe. Romaire et al. (1978) described a method for anticipating critically low oxygen levels.

Societal customs dictate the emphasis on different domestic animals in various countries. In Israel, for example, cattle and chicken manures are used (Moav et al. 1977, Schroeder and Hepher 1979, Wohlfarth and Schroeder 1979, Schroeder 1980). In the United States, Nerrie (1979) and Burns and Stickney (1980) have used chicken manure in tilapia feeding trials. Swine and duck are commonly integrated with fish production in many countries. Several review articles on integrated agriculture-aquaculture are published in the proceedings of a recent symposium (Pullin and Shehadeh 1980). Data from some of these studies are summarized in tables 13–4 and 13–5. Preliminary tests that have been conducted in the United States are in Smitherman et al. (1978), Buck et al. (1979), and Stickney et al. (1979).

There are impediments to integrated aquaculture, not the least of which are potential disease transmission (Bryan 1974) and aesthetic considerations (Allen and Hepher 1979). In most developing countries this approach is traditional. It is the highly developed societies that will have trouble accepting the change. Israel is a country that is increasingly integrating agriculture and aquaculture.

Another source of waste that is particularly applicable to tilapia culture is waste heat. Sylvester (1975) discusses problems with fish culture in waters warmed by waste heat and Kirk (1972) reviewed tilapia culture under these conditions. A general review on waste heat and aquaculture was the subject of a workshop (Godfriaux et al. 1979).

An additional factor limiting tilapia culture in the United States is the necessity to overwinter fish, either as brood stock or as fingerlings for growout the next year to a marketable size. In portions of the United States where the climate is mild and winterkill is not so much of a threat to the culture, there is concern over escape. The reproductive potential of tilapias is so great

Table 13–4. NET YIELDS (KG/HA) AND STOCKING RATES (NO/HA) FROM INTEGRATED AGRICULTURE-AQUACULTURE POLYCULTURE FROM SEVERAL COUNTRIES USING DIFFERENT LIVESTOCK

Country	Manure Source[a] (no/ha)	Species – Yield (Stocking rate)						Net yield		Source
		Silver carp	Bighead carp	Grass carp	Common carp	Other[b]	Other[b]	Total	Daily	
Taiwan	210 P	760 (1500)	580 (400)	340 (300)	610 (1500)	4560 (30000)[2]	310 (1500)[1]	7170	30	Chen and Li 1980
Taiwan	2200 D	860 (1200)	380 (300)	320 (300)	540 (1000)	2900 (4740)[2]	490 (3000)[2]	5500	23	Chen and Li 1980
Philippines	none	–	–	–	80–180 (1400–2800)	360–370 (8500–17000)[2]	–	450–560	5–6	Cruz and Shehadeh 1980
Philippines	40 P	–	–	–	120–150 (1400–2800)	330–550 (8500–17000)[2]	–	490–680	5–8	Cruz and Shehadeh 1980
Philippines	60 P	–	–	–	200–210 (1400–2800)	540–740 (8500–17000)[2]	–	760–960	8–11	Cruz and Shehadeh 1980
India	50–80 P	2180 (1270)	1220[c] (1700)[3]	1830 (1700)	260 (420)	900 (1700)[4]	740 (1700)[5]	7120	19	Jhingran and Sharma 1980
India	68 D	1350 (950)	790 (630)[3]	740 (635)	100 (1200)	510 (1140)[4]	580 (1770)[5]	4320	11	Jhingran and Sharma 1980

[a] Animal source with equivalent stocking/ha of water – P = pig, D = duck – where wastes directly enter ponds
[b] Other species of fish – 1 = mullet; 2 = tilapia
[c] Composite culture in India; one or more Chinese carp with major Indian carps – 3 = *Catla catla;* 4 = rohu, *Labeo rohita;* 5 = mrigal, *Cirrhinus mrigala*

Table 13–5. NET YIELDS (KG/HA) AND STOCKING RATES (NO/HA) OF EXOTIC FISHES WITH VARIOUS ANIMAL WASTES AS THE NUTRIENT SOURCE

State	Management[a]	Species – Yield (Stocking rate)					Net yield		Source
		Silver carp	Bighead carp	Grass carp	Common carp	Other[b]	Total	Daily	
AL	C	–	–	–	–	2100 (9250)[1]	2100	14	Nerrie 1979
AL	Ch	–	–	–	–	1650 (10000)[1]	1650	16	Collis and Smitherman 1978
AR	S	2550 (12760)	170 (250)	–	–	–	2730	6	Henderson 1978
IL	39–66 P	1800–2500 (4940)	400 (420)	70–90 (130)	–	20–30 (130)[2]	2970–3820	18–20	Buck et al. 1979
IL	67–85 P	2450–3250 (2720–7960)	200–470 (270–280)	90–140 (70)	620–970 (880–910)	20–30 (70)[2]	3530–4580	20–25	Buck et al. 1979
TX	50 P	–	–	–	–	600–890 (5000)[1]	600–890	5–8	Stickney et al. 1978
TX	125 P	–	–	–	–	1040 (10000)[1]	1040	7	Stickney et al. 1978
TX	250 P	–	–	–	–	640 (10000)[1]	640[c]	5	Stickney et al. 1978

[a] Nutrient input – C = cattle wastes; Ch = chicken wastes; S = domestic sewage; P = pig wastes (no. pigs/ha of water)
[b] Other species – 1 = tilapia; 2 = channel catfish
[c] Forty-nine percent mortality

that it presents a problem in culture as well as in potential naturalization; therefore, reproduction in culture situations must be managed. Several applications are workable; however, all require further refinement to approach practical commercial levels for the United States.

Outlook for Exotic Culture in the United States

Under present conditions of the U.S. economy, consumers accept high price crops (trout and catfish) that require high cost, high protein diets. Exotic fish culture has the greatest potential through integration with current production. Polyculture studies at Auburn (Smith 1971, Pretto-Malca 1976, Dunseth 1977) with catfish indicated that catfish yield was reduced in combined culture with tilapia or bighead carp at stocking rates greater than 2,000/ha or 125/ha, respectively, but catfish yield increased in combined culture with silver carp at 250–2,500/ha. Higher yields of silver carp were related to richer environments and higher stocking rates. Some competition can be tolerated; for example, with each 0.4 kg reduction in catfish production, an additional 2 kg of tilapia was produced, which yielded a net increase in profit (Dunseth and Smitherman 1977). Culture of exotics in a monoculture is not presently a feasible replacement for catfish; however, if integrated in suitable combinations, supplemental production and income appear probable.

One obstacle to polyculture in the United States involves the contemporary management of fish farms. Most catfish producers in Mississippi and Arkansas manage large ponds (average 8–12 ha) that are stocked and multiple-harvested by seining at various times during the growing season. Exotic fishes or any other multiple crops integrated into this system create problems in harvesting and sorting. Tilapias are not effectively seined and some of the Chinese carps are infamous for their jumping characteristics. Silver carp are viewed with particular disdain by seining crews. Modified harvesting techniques would need to be developed to remove them efficiently, especially before they grow too large and become hazardous to crewmen. In harvest experiments at Auburn, catfish were fed, then seined. One pond (0.5 ha) contained silver carp and the other (8.9 ha) had bighead carp; over 18 metric tons of catfish were effectively harvested but not a single Chinese carp was captured until draining (unpublished data, table 13–3).

If the harvesting problems are satisfactorily resolved, complications with processing must be addressed. Catfish processors would need to modify operations to accommodate multiple species and to include specialty handling (e.g., deboning, canning) either in a separate facility or operation. If one takes only these practical problems into consideration, it may be difficult to ignore polyculture as an option since 500–1,000 kg/ha of supplemental crop, depending on the species, can be realized with little or no increase in cost of production.

MARKETING

Fish is not a major dietary component in the United States; however, the outlook for increased consumption is good. The U.S. market for food of aquatic origin has been primarily for high cost items such as shrimp, catfish, trout, and salmon. As energy limitations and food costs increase, this scenario may change. The higher nutrient recovery associated with polyculture and the proven systems of particular species combinations may be modifiable to fit U.S. conditions, and the Chinese carps should be considered since they are already in this country.

Efficiency and high production will be of little practicality if there is no market. Crawford and McCoy (1977) examined budgeting for selected aquacultural activities and included tilapia culture as one option. Local marketing studies through retail food chains in Alabama indicated the potential for particular exotic fishes. The bighead carp was well received, and when taste-tested in comparison with catfish, was preferred (McCoy and Hopkins 1977). Consumer acceptance of tilapias was excellent but only fair for silver carp (Dunseth and Smitherman 1977). There was a high repeat market for tilapia, but few indicated subsequent purchase of silver carp (Crawford et al. 1978). Markets have been established in other parts of the country for tilapia (Lauenstein 1978, Ray 1978), but in the Southeast there is competition from wild-harvested fish in Florida that are wholesaled at live weight prices of $0.40–0.42/kg.

Recent awareness of ethnic markets on the East and West coasts for Chinese carps, particularly grass carp and bighead carp, has increased growers' interest. Silver carp have the highest productivity potential and the least marketability. In view of the problems with marketing them as fresh fish, they have been tested as a canned product (Woodruff 1978) and as a deboned product (Lovell et al. 1978). When canned silver carp was taste-tested and market-tested with canned tuna, about one-third of those polled preferred the silver carp. There is potential for these products, particularly as a replacement for the present canned fish and in fast food outlets. The supply of wild caught fish is subject to multiple vagaries and resulting cost instability, whereas an aquacultural product would be more stable and predictable.

CONTROL OF UNWANTED REPRODUCTION

Importation of an exotic for whatever purpose presents the opportunity for naturalization. Grass carp is an excellent example of failure of legislation to prevent the spread of an illegal species (Guillory and Gasaway 1978). Political boundaries are not a part of the biological world. Many advocates of the grass carp emphasized that spawning would not occur; however, their use in lentic waters has not prevented access to lotic waters which provide suitable spawning areas. Stanley (1976a) reviewed instances of natural reproduction of grass carp outside their native range. Stanley et al. (1978) discussed reproductive requirements and predicted that spawning would occur in the United States. Natural spawning has now been reported in the lower Mississippi River (Conner et al. 1980), but recruitment has not been demonstrated and survival will probably be low because of predation. Anticipated adverse effects vary but any negative impact is not likely to be a result of diet overlap with any of our native fishes.

Silver and bighead carps will logically follow a similar pattern of dispersion and probable naturalization since the reproductive biology of these species is similar to that of grass carp. Increased interest in these species is anticipated. The hybridization of bighead carp males and grass carp females produces some individuals that are triploid and reportedly sterile, and they eat vegetation (Marian and Krasznai 1978).

Silver carp are presently only sparingly considered, but their role as a biofilter in culture ponds and waste treatment ponds may result in a broader based interest than for either of the other imported Chinese carps. Diet overlap of silver and bighead carps is more likely if one considers the native species that feed at the same trophic levels (phytoplankton and zooplankton filter feeding), such as the shads (*Dorosoma* spp.), buffaloes (*Ictiobus* spp.), and paddlefish (*Polyodon spathula*).

Control of unwanted reproduction has been approached in several ways. Most attempts have been practically oriented and involve tilapia culture. Their high reproductive capacity has complicated culture in lentic conditions. Predation was tested as a means of controlling young fish (Swingle 1960, Dunseth and Bayne 1978). Attempts to prevent successful reproduction have also included high density stocking (Allison et al. 1979), cage culture where deposited eggs fall through the mesh (Pagan-Font 1975), or various means of monosexing. The most direct approach to monosexing has been by culling females based on the sexually dimorphic papillae (Bardach et al. 1972). Hybrids of some species are usually all males (Lovshin 1982). This approach is being used and investigations continue in Israel (Wohlfarth and Hulata 1981, Hulata et al. 1981). Monosexing through steroid treatment was reviewed by Guerrero (1982) and the applicability to aquaculture was discussed by Shelton et al. (1978). The practicability of this approach for aquaculture will depend on suitability to mass treatment and whether hormone treatment at a small size and months before harvest will be considered a potential human health hazard (Shelton et al. 1981b). Androgens have been used to enhance growth of salmonids and only minute levels were present after brief withdrawal periods (Donaldson et al. 1979). The measured amounts included methyltestosterone and metabolites; thus, actual androgen levels were only small fractions of the indicated levels (Fagerlund and Dye 1979).

With the exception of hybridization, all of the above approaches involved individual treatment of the fish grown. Monosexing through hybridization is based on the genetics of sex determination, a factor poorly understood for fishes (Gold 1979). However, in an attempt to explain the results of certain tilapia crosses that produce male offspring, Chen (1969) proposed a mechanism based on sex chromosomes. Pruginin et al. (1975) and Jalabert et al. (1974) reported inconsistencies with this interpretation. Avtalion and Hammerman (1978) and Hammerman and Avtalion (1979) proposed an alternative hypothesis to explain the results of tilapia hybridization.

On the premise that sex determination is monogenic or involves sex chromosomes, the practical application to tilapia in aquaculture was proposed by Jensen and Shelton (1979). This approach would use hormone treatment to produce brood stock, and the fish for grow-out would be their presumed monosex offering. Small-scale testing with *T. aurea* was encouraging (Hopkins 1979), but a preliminary test of large-scale production with *T. nilotica* was not compatible with the hypothesis (Calhoun 1981). The functional nature of sex chromosomes was not disproved, only that other loci may be involved as discussed by Yamamoto (1969) and Gold (1979).

Genetic manipulation may have promise in aquaculture either through proposed enhancement of growth (Moav 1979) or in reproductive control. Valenti (1975) induced polyploidy in *T. aurea* but there may be greater application for this technique in the Chinese carps. Considerable interest in the United States was generated by the report of a triploid hybrid between female grass carp and male bighead carp (Marian and Krasznai 1978). Hybrid sterility has not been confirmed, and problems with production include low yields of triploids and probable simultaneous production of diploids. Current status was reviewed in eleven reports presented at the 111th annual American Fisheries Society meeting (AFS 1981); publication of a number of these papers is planned in the Transactions of the American Fisheries Society. Various other approaches to sterilization have been attempted (Al-Daham 1970, Dadzie 1975), but these have not generally succeeded.

Stanley (1979) reviewed several approaches to control sex of fishes with specific reference to grass carp. Monosexing of grass carp has been achieved through gynogenesis (Stanley and Sneed 1973, Stanley et al. 1975, Stanley 1976b), and these fish were used in a large-scale weed control test (Theriot and Decell 1978). The low productivity by gynogenesis indicates that this approach is not practical unless yields can be improved as suggested by Nagy et al. (1978). The uncertainty in the mechanism of sex determination as mentioned with reference to tilapia may be less cause for concern with the grass carp. Stanley (1976c) reported that grass carp females are homogamous for sex determination, and substantiation of this finding was reported by Stanley and Thomas (1978) and Shelton and Jensen (1979).

Breeding for monosex offspring in grass carp would permit commercial production as well as wider acceptance (Sutton 1977), but confirmation of the stability of the presumed sex determining mechanism is needed. Sex reversal attempts by oral treatment failed (Stanley and Thomas 1978). Stanley (1979) speculated that treatment should have included treatment at an earlier age by exposure in aquaria, since success of hormone sex reversal depends on efficient exposure throughout the period of gonadal differentiation. Initial results with an androgen implant were encouraging (Jensen et al. 1978). Subsequently, Jensen (1979) described the size and age relationship of gonadal differentiation. Through the application of this information, androgen sex reversal of gynogenetic grass carp has now been successful (Shelton and Jensen 1979). Preliminary breeding tests were conducted in 1980, 1981, and 1982, producing between 85,000 and 250,000 presumed all-female offspring. These fish are now being grown for sexing. To date, all of 600 from the 1980-year class that have been examined were female (Shelton et al. 1982). This sex reversal/breeding approach is also being tested in common carp (Nagy et al. 1981) and in rainbow trout (Johnstone et al. 1979).

An approach to sterilization in which the full reproductive capacity of the grass carp could be used would be the most practical approach. Refstie et al. (1977) suggested such a procedure in which normal diploid fish are crossed with tetraploids. Refstie (1981) reported the successful induction of tetraploidy in rainbow trout by using cytochalasine B. Breeding of the tetraploids will test the theory and its potential. Stanley (1979) also suggested using this approach to produce sterile grass carp.

The combination of monosex production through breeding with polyploid techniques might permit the development of populations of female tetraploids. If these were stocked into an existing population of grass carp, any breeding that involved these females should produce triploid offspring. Whether this could reduce recruitment of diploids in the population is speculative.

The management of reproduction in silver and bighead carp will be similar to any approach found effective for grass carp; current studies are underway at Auburn on these species. The opportunity to apply experience gained on reproductive control of grass carp should be apparent. Equally obvious is the need to waste no time in using this knowledge on species for which distribution is still somewhat limited.

Alternate Species

One consideration that has not been addressed in the present discussion is the option of using native species to supplement or enhance production. There are many species to consider but in a practical cultural sense most can be disregarded.

Attempts have been made to use a variety of native fishes in aquaculture (Fielding 1968). Native species that have been grown with catfish include suckers (Kilgen 1974), mullet (Shireman 1974), and paddlefish (Tuten and

Avault 1981). We have not included the common carp in our discussion of exotics primarily because it has become a ubiquitous part of our fauna. Nevertheless, with its status of naturalization, carp might be considered as a supplemental species in some aquacultural operations. Considerable quantities of carp are marketed as food in some parts of the United States; there are several producers in California, although most fish that are sold are wild-caught.

Buffalofishes might be a profitable combination since they are filter-feeders on zooplankton (Williamson and Smitherman 1976) and might fill a niche similar to that of the bighead carp (table 13–6). Some competition has been indicated when cultured with catfish (Pretto-Malca 1976) and marketability would also have to be considered. In 1979, 4,900 metric tons of buffalofish were marketed from Arkansas alone; however, only about 680 metric tons were pond-raised (Freeze and Henderson, 1981a) and the rest were wild-caught fish (Freeze and Henderson 1981b).

Another zooplankton filter-feeder (Rosen and Hales 1981) which probably has the greatest potential now is the paddlefish. This fish has had variable commercial value in this century (Carlson and Bonislawsky 1981), but exploitation of wild populations is rapidly escalating. The main impetus for the renewed interest is the present market value for their roe as a substitute for sturgeon caviar. There is a high probability that much of the currently harvested paddlefish roe is being illegally marketed as imported caviar.

Paddlefish grow rapidly in ponds (Swingle 1965), but little information is presently available on production at practical stocking levels with catfish (table 13–6). The major deterrent to their use has been an inadequate supply of fry. Personnel of the Missouri Department of Conservation pioneered development of induced spawning techniques, but their main objective has been to restock natural waters. At Auburn, both aspects are now being studied. Slight modifications to induced spawning techniques are being investigated and three year-classes of fish have been produced and stocked in yield studies. Paddlefish culture could easily integrate with catfish systems. For example, fingerlings grown in catfish fingerling ponds during one year can be stocked in catfish grow-out ponds in the second year to produce fish that should be of a marketable size (3–5 kg). Dress-out is lower than for catfish (Mensinger and Brown 1969), but they could be handled by the catfish processor with little or no modification in operation. Further, anticipating the capability to sex second-year fish, a third or fourth growing season for females might produce a caviar crop. Gonadal development of pond-reared fish that are river-spawners is well known (e.g., Chinese carps); preliminary observations on pond-reared paddlefish at Auburn indicate successful maturation. The current price of \$55–\$65/kg for rough processed roe should provide ample incentive.

One other species for consideration is the Sacramento blackfish (*Orthodon microlepidotus*). It has been mentioned as having aquacultural potential (Colt

Table 13-6. NET YIELDS (KG/HA) AND STOCKING RATES (NO/HA) OF NATIVE FISHES IN SIMPLE POLYCULTURE

State	Species – Yield[a] (Stocking rate)				Net yield		Source
	Buffalo[b]	Paddle-fish	Channel catfish	Tilapia	Total	Daily	
AL	–	45 (10)	–	2450 (5070)	2500	12	Swingle 1965
AL	–	100 (70)	3500 (7410)	–	3600	15	Semmens,[c] personal communication
AL	–	140 (250)	–	1760–5120 (14830–22240)	1900–5260	15–41	Stone 1981
LA	170–340 (250)	–	2130–2440 (3950–4940)	–	2300–2690	14–16	Perry and Avault 1976
LA	300 (180)	110 (50)	3200 (6430)	other[d]	3610	12	Tuten and Avault 1981

[a] Principal species (channel catfish or tilapia) fed based on standing crop
[b] Species of buffalo include *Ictiobus cyprinellus, I. niger,* or hybrid
[c] K. Semmens, graduate student, Auburn University
[d] Crayfish production not included

et al. 1979) and it is native to the Sacramento–San Joaquin river system in California (Lee et al. 1980). It is abundant in warm, shallow backwater areas, is tolerant of low dissolved oxygen, and is a filter-feeder on phytoplankton and zooplankton (Moyle 1976). The blackfish apparently has fair growth potential and has an established demand in the oriental fish markets of San Francisco where approximately 68 metric tons are sold annually. Breeding is apparently not dependent on specific conditions, and spawning would probably occur in any part of the country where culture is attempted; therefore, appropriate precaution would be indicated.

Conclusion

The present distribution of exotic fishes in North America, excluding common carp, brown trout, and escaped aquarium species, is primarily a result of movement for purposes of water resource management and not for culture as food fishes. The Chinese carps and several of the tilapias now in the United States have considerable potential in aquaculture, either directly as a food fish or as a functional component of a polyculture.

Development of the aquaculture potential is now limited by legal restrictions that have impeded but have not prevented the use of these exotics. The rationale behind legal constraints is ecologically based and centers on the fear of escape, unwanted natural reproduction, and consequent establishment in open North American waters. To solve this conflict, increased efforts directed toward preventing escape and developing efficient and practical means of reproductive control are necessary to fully utilize these species either in aquaculture or for management.

If these fishes are integrated into aquaculture in the United States, technological developments on harvest and processor handling will be required. Their market potential could result in increased profit to fish farmers, and most of the species considered convert wastes into usable protein.

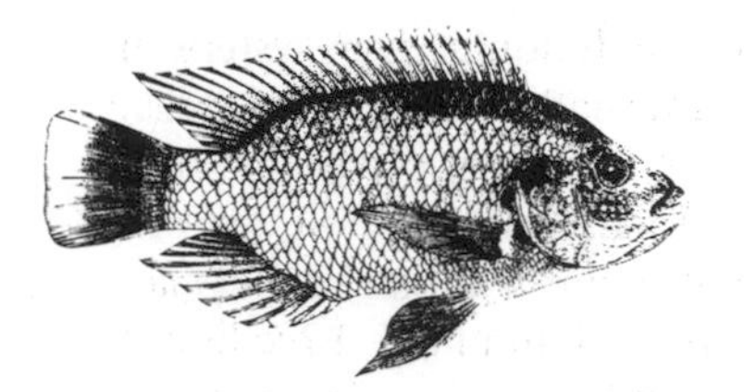

Literature Cited

AFS (American Fisheries Society). 1981. Tentative program, 111th annual meeting. Fisheries 6(3):31–37.

Al-Daham, N. K. 1970. The use of chemosterilants, sex hormones, radiation, and hybridization for controlling reproduction in *Tilapia* species. Ph.D. dissertation, Auburn University, Auburn, Alabama. 160 pp.

Allanson, B. R., and R. G. Noble. 1964. The tolerance of *Tilapia mossambica* (Peters) to high temperature. Transactions of the American Fisheries Society 93:323–332.

Allen, G. H., and B. Hepher. 1979. Wastes and use of recirculating water in aquaculture, 478–487. *In:* T.V.R. Pillay and W. A. Dill (eds.). Advances in aquaculture. Fishing News Books, Surrey, England.

Allison, R., J. E. Rakocy, and D. D. Moss. 1981. A comparison of two closed recirculating systems for culture of tilapia, 979–1019. *In:* J. T. Mannassah and E. J. Briskey (eds.). Advances in food producing systems for arid and semiarid lands. Part B. Academic Press, New York.

Allison, R., R. O. Smitherman, and J. Cabrero. 1979. Effects of high density culture and form of feed on reproduction and yield of *Tilapia aurea,* 168–170. *In:* T.V.R. Pillay and W. A. Dill (eds.). Advances in aquaculture. Fishing News Books, Surrey, England.

Andriasheva, M. A. 1968. Some results obtained by hybridization of cyprinids, 205–214. *In:* T.V.R. Pillay (ed.). Proceedings of the world symposium on warm-water pond fish culture. FAO Fisheries Reports 5(44).

Appelbaum, S., and B. Uland. 1979. Intensive rearing of grass carp larvae, *Ctenopharyngodon idella* (Valenciennes 1884), under controlled conditions. Aquaculture 17:175–179.

Avault, J. W., Jr. 1980. Aquaculture, 379–411. *In:* R. T. Lackey and L. A. Nielsen (eds.). Fisheries Management. Halstead Press, New York.

Avault, J. W., Jr., and E. W. Shell. 1968. Preliminary studies with the hybrid tilapia, *Tilapia nilotica* x *Tilapia mossambica,* 237–242. *In:* T.V.R. Pillay (ed.). Proceedings of the world symposium on warm-water pond fish culture. FAO Fisheries Reports 4(44).

Avault, J. W., Jr., E. W. Shell, and R. O. Smitherman. 1968. Procedures for overwintering tilapia, 343–345. *In:* T.V.R. Pillay (ed.). Proceedings of the world symposium on warm-water pond fish culture. FAO Fisheries Reports 4(44).

Avtalion, R. R., and I. S. Hammerman. 1978. Sex determination in *Sarotherodon*

(Tilapia). I. Introduction to a theory of autosomal influence. Bamidgeh 30:110–115.

Backiel, T., and E. D. LeCren. 1967. Some density relationships for fish population parameters, 261–293. *In:* S. D. Gerking (ed.). The biological basis for freshwater fish production. Blackwell Scientific Publications, Oxford.

Bailey, W. M., and R. L. Boyd. 1971. A preliminary report on spawning of grass carp (*Ctenopharyngodon idella*) in Arkansas. Proceedings of the Southeastern Association of Game and Fish Commissioners 24(1970):560–569.

Balarin, J. D. 1979. Tilapia: A Guide to their biology and culture in Africa. University of Stirling, Stirling, Scotland.

Bardach, J. E. 1978. The growing science of aquaculture, 424–446. *In:* S. D. Gerking (ed.). Ecology of freshwater fish production. Blackwell Scientific Publications, London.

Bardach, J. E., J. H. Ryther, and W. O. McLarney. 1972. Aquaculture: The farming and husbandry of freshwater and marine organisms. Wiley-Interscience Publication, New York.

Bayne, D. R., D. Dunseth, and C. G. Ramirios. 1976. Supplemental feeds containing coffee pulp for rearing *Tilapia* in Central America. Aquaculture 7:133–146.

Behrends, L. L. 1977. Effects of three tilapias (*Sarotherodon*) and silver carp (*Hypophthalmichthys molitrix*) on phytoplankton communities and water quality in ponds with channel catfish (*Ictalurus punctatus*). M.S. thesis, Auburn University, Auburn, Alabama. 43 pp.

Behrends, L. L., J. J. Maddox, R. S. Pile, and J. C. Roetheli. 1979. Comparison of two methods of using liquid swine manure as an organic fertilizer in the production of filter-feeding fish. Aquaculture 19:35–39.

Bowen, J. T. 1970. A history of fish culture as related to the development of fishery programs, 71–93. *In:* N. G. Benson (ed.). A century of fisheries in North America. American Fisheries Society Special Publication 7, Washington, D.C.

Bryan, F. L. 1974. Diseases transmitted by foods contaminated by wastewater, 16–45. *In:* Wastewater use in the production of food and fiber—Proceedings. EPA Technical Series 660/2-74-041, Washington, D.C.

Buck, D. H., R. J. Baur, and C. R. Rose. 1978. Polyculture of Chinese carps in ponds with swine wastes, 144–155. *In:* R. O. Smitherman, W. L. Shelton, and J. H. Grover (eds.). Culture of exotic fishes symposium proceedings. American Fisheries Society, Fish Culture Section, Auburn, Alabama.

Buck, D. H., R. J. Baur, and C. R. Rose. 1979. Experiments in recycling swine manure in fish ponds, 489–492. *In:* T.V.R. Pillay and W. A. Dill (eds.). Advances in aquaculture. Fishing News Books, Surrey, England.

Buntz, J., and C. S. Manooch III. 1968. *Tilapia aurea* (Steindachner), a rapidly spreading exotic in South central Florida. Proceedings of the Southeastern Association of Game and Fish Commissioners 22(1968):495–501.

Burns, R. P., and R. R. Stickney. 1980. Growth of *Tilapia aurea* in ponds receiving poultry wastes. Aquaculture 20:117–121.

Calhoun, W. E. 1981. Progeny sex ratios from mass spawnings of sex-reversed brood stock of *Tilapia nilotica*. M.S. thesis, Auburn University, Auburn, Alabama. 44 pp.

Carlson, D. M., and P. S. Bonislawsky. 1981. The paddlefish (*Polyodon spathula*) fisheries of the midwestern United States. Fisheries 6(2):17–27.

Chadhuri, H., R. D. Chakrabarty, P. R. Sen, N.G.S. Rao, and S. Jena. 1975. A new high in fish production in India with record yields by composite fish culture in freshwater ponds. Aquaculture 6:343–355.

Chen, F. Y. 1969. Preliminary studies on the sex-determining mechanism of *Tilapia mossambica* Peters and *T. hornorum* Trewavas. Verhandlungen der Internationalen Vereinigung für Limnologie 17:719–724.

Chen, T. P., and Y. Li. 1980. Integrated agriculture-aquaculture studies in Taiwan, 239–241. *In:* R.S.V. Pullin and Z. H. Shehadeh (eds.). Integrated agriculture-aquaculture farming systems. ICLARM Conference Proceedings 4, Manila, Philippines.

Chervinski, J., and M. Lahav. 1976. The effect of exposure to low temperature on fingerlings of local tilapia (*Tilapia aurea*) and imported tilapia (*Tilapia vulcani*) and (*Tilapia nilotica*) in Israel. Bamidgeh 28:25–29.

Chervinski, J., and R. R. Stickney. 1981. Overwintering facilities for tilapia in Texas. Progressive Fish-Culturist 43:20–21.

Chimits, P. 1955. Tilapia and its culture, a preliminary bibliography. FAO Fishery Bulletin 8:1–33.

Chimits, P. 1957. The tilapias and their culture, a second review and bibliography. FAO Fishery Bulletin 10:1–24.

Clemens, H. P., and K. E. Sneed. 1957. The spawning behavior of the channel catfish, *Ictalurus punctatus*. U.S. Department of the Interior, Fish and Wildlife Service Special Scientific Report, Fisheries 219.

Coche, A. G. 1982. Cage culture of tilapia, 205–246. *In:* R.S.V. Pullin and R. H. Lowe-McConnell (eds.). The biology and culture of tilapias. ICLARM, Conference Proceedings 7, Manila, Philippines.

Coleman, M. S., J. P. Henderson, H. G. Chichester, and R. L. Carpenter. 1974. Aquaculture to achieve effluent standards, 199–214. *In:* Wastewater use in the production of food and fiber – Proceedings. EPA Technical Series 660/2-74-041, Washington, D.C.

Collis, W. J., and R. O. Smitherman. 1978. Production of tilapia hybrids with cattle manure or a commercial diet, 43–54. *In:* R. O. Smitherman, W. L. Shelton, and J. H. Grover (eds.). Culture of exotic fishes symposium proceedings, American Fisheries Society, Fish Culture Section, Auburn, Alabama.

Colt, J., S. Mitchell, G. Tchobanoglous, and A. Knight. 1979. The use and potential of aquatic species for waste water treatment. Appendix B. The environmental requirements of fish. California State Waste Resources Control Board Publication 65, Sacramento, California.

Conner, J. V., R. P. Gallagher, and M. F. Chatry. 1980. Larval evidence for natural reproduction of the grass carp (*Ctenopharyngodon idella* [Val.]) in the lower Mississippi River, 1–19. *In:* L. A. Fuiman (ed.). Proceedings of the Fourth Annual Larval Fish Conference, 27–28 February 1980, Oxford, Mississippi. FWS/OBS-80/43, 19 pp.

Courtenay, W. R., Jr., and D. A. Hensley. 1980. Special problems associated with monitoring exotic species, 281–307. *In:* C. H. Hocutt and J. R. Stauffer, Jr. (eds.). Biological monitoring of fish. Lexington Books, Lexington, Massachusetts.

Courtenay, W. R., Jr., and C. R. Robins. 1975. Exotic organisms: An unsolved complex problem. BioScience 25:306–313.

Crawford, K. W., D. R. Dunseth, C. R. Engle, M. L. Hopkins, E. W. McCoy, and R. O. Smitherman. 1978. Marketing tilapia and Chinese carps, 240–257. *In:* R. O.

Smitherman, W. L. Shelton, and J. H. Grover (eds.). Culture of exotic fishes symposium proceedings. American Fisheries Society, Fish Culture Section, Auburn, Alabama.

Crawford, K. W., and E. W. McCoy. 1977. Budgeting for selected aquacultural enterprises. Auburn University, Agricultural Experiment Station Bulletin 495, Auburn, Alabama.

Cremer, M. C., and R. O. Smitherman. 1980. Food habits and growth of silver and bighead carp in cages and ponds. Aquaculture 20:57–64.

Cruz, E. M., and Z. H. Shehadeh. 1980. Preliminary results of integrated pig-fish and duck-fish production tests, 225–238. *In:* R.S.V. Pullin and Z. H. Shehadeh (eds.). Integrated agriculture-aquaculture farming systems. ICLARM Conference Proceedings 4, Manila, Philippines.

Dabrowski, K. 1977. Protein requirements of grass carp fry (*Ctenopharyngodon idella* Val.). Aquaculture 12:63–73.

Dadzie, S. 1975. A preliminary report on the use of methallibure in tilapia culture. African Journal of Hydrobiology and Fisheries 4 (special issue) 3:127–140.

Dah-Shu, L. 1980. The method of cultivation of grass carp, silver carp and big-head carp. Translated by A. Jokela, Joint Subcommittee on Aquaculture, Translation 63, National Marine Fisheries Service.

Davis, A. T., and R. R. Stickney. 1978. Growth responses of *Tilapia aurea* to dietary protein quality and quantity. Transactions of the American Fisheries Society 107:479–483.

Delmendo, M. N. 1980. A review of integrated livestock-fowl-fish farming systems. 59–71. *In:* R.S.V. Pullin and Z. H. Shehadeh (eds.). Integrated agriculture-aquaculture farming systems. ICLARM Conference Proceedings 4, Manila, Philippines.

Denzer, H. W. 1968. Studies on the physiology of young tilapia, 357–366. *In:* T.V.R. Pillay (ed.). Proceedings of the world symposium on warm-water pond fish culture. FAO Fisheries Reports 4(44), Rome.

DeSilva, S. S., and D.E.M. Weerakoon. 1981. Growth, food intake and evacuation rates of grass carp, *Ctenopharyngodon idella,* fry. Aquaculture 25:67–76.

Donaldson, E. M., U.H.M. Fagerlund, D. A. Higgs, and J. R. McBride. 1979. Hormonal enhancement of growth, 455–597. *In:* W. S. Hoar, D. J. Randall, and J. R. Brett (eds.). Fish physiology. Vol. 8. Academic Press, New York.

Dunseth, D. R. 1977. Polyculture of channel catfish *Ictalurus punctatus,* silver carp *Hypophthalmichthys molitrix,* and three all-male tilapias *Sarotherodon* spp. Ph.D. dissertation, Auburn University, Auburn, Alabama. 62 pp.

Dunseth, D. R., and D. R. Bayne. 1978. Recruitment control and production of *Tilapia aurea* (Steindachner) with the predator, *Cichlasoma managuense* (Günther). Aquaculture 14:383–390.

Dunseth, D. R., and R. O. Smitherman. 1977. Pond culture of catfish, tilapia, and silver carp. Highlights of Agricultural Research, Auburn University, Auburn, Alabama, 24(3):4.

Fagerlund, U.H.M., and H. M. Dye. 1979. Depletion of radioactivity from yearling coho salmon (*Oncorhynchus kisutch*) after extended ingestion of anabolically effective doses of 17a-methyltestosterone-1,2-^{3}H. Aquaculture 18:303–315.

Fielding, J. R. 1968. New systems and new fishes for culture in the United States, 143–161. *In:* T.V.R. Pillay (ed.). Proceedings of the world symposium on warm-water pond fish culture. FAO Fisheries Reports 5(44), Rome.

Fischer, Z. 1972a. The elements of energy balance in grass carp (*Ctenopharyngodon idella* Val.). Part 2: Fish fed with animal food. Polskie Archiwum Hydrobiologii 19:65–82.

Fischer, Z. 1972b. The elements of energy balance in grass carp (*Ctenopharyngodon idella* Val.). Part 3: Assimilability of proteins, carbohydrates, and lipids by fish fed with plant and animal food. Polskie Archiwum Hydrobiologii 19:83–95.

Fischer, Z., and V. P. Lyakhnovich. 1973. Biology and bioenergetics of grass carp (*Ctenopharyngodon idella* Val.). Polskie Archiwum Hydrobiologii 20:521–557.

Fisheries Research Institute of Shanghai. 1978. Artificial propagation of grass carp, silver carp, bighead in China. Nursing fry culture of adult fish and technique about prevention and treatment of fish's diseases. Shanghai, China.

Fortes, R. D. 1979. Evaluation of tenpounder and tarpon as predators on java tilapia. Ph.D. dissertation, Auburn University, Auburn, Alabama. 185 pp.

Freeze, M., and S. Henderson. 1981a. The aquaculture industry in Arkansas in 1979. Proceedings of the Southeastern Association of Fish and Wildlife Agencies 34(1980):115–126.

Freeze, M., and S. Henderson. 1981b. Economic evaluation of the wild commercial fishery in Arkansas – 1 July 1978 to 30 June 1979. Proceedings of the Southeastern Association of Fish and Wildlife Agencies 34(1980):372–378.

Germany, R. D., and R. L. Noble. 1978. Population dynamics of blue tilapia in Trinidad Lake, Texas. Proceedings of the Southeastern Association of Fish and Wildlife Agencies 31(1977):412–417.

Glude, J. B. (ed.). 1977. NOAA aquaculture plan. National Oceanic and Atmospheric Administration, Washington, D.C.

Godfriaux, B. L., A. F. Eble, A. Farman Farmalan, C. R. Guerra, and C. A. Stephens (eds.). 1979. Power plant waste heat utilization in aquaculture. Alanheld, Osmun and Co., Montclair, New Jersey.

Gold, J. R. 1979. Cytogenetics, 353–405. *In:* W. S. Hoar, D. J. Randall, and J. R. Brett (eds.). Fish physiology, 8. Academic Press, New York.

Guerrero, R. D. 1982. Control of tilapia reproduction, 308–316. *In:* R.S.V. Pullin, and R. H. Lowe-McConnell (eds.). The biology and culture of tilapias. ICLARM, Conference Proceedings 7, Manila, Philippines.

Guillory, V., and R. D. Gasaway. 1978. Zoogeography of the grass carp in the United States. Transactions of the American Fisheries Society 107:105–112.

Halevy, A. 1979. Observations on polyculture of fish under standard farm pond conditions at the Fish and Aquaculture Research Station, Dor, during the years 1972–1977. Bamidgeh 31:96–104.

Hammerman, I. S., and R. R. Avtalion. 1979. Sex determination in *Sarotherodon (Tilapia)*. Part 2: The sex ratio as a tool for determination of genotype – a model of autosomal and gonosomal influence. Theoretical and Applied Genetics 55:177–187.

Henderson, S. 1978. An evaluation of the filter feeding fishes, silver and bighead carp, for water quality improvement, 121–143. *In:* R. O. Smitherman, W. L. Shelton, and J. H. Grover (eds.). Culture of exotic fishes symposium proceedings. American Fisheries Society, Fish Culture Section, Auburn, Alabama.

Henderson, S. 1980. Production potential of catfish grow-out ponds supplementally stocked with silver and bighead carp. Proceedings of the Southeastern Association of Fish and Wildlife Agencies 33(1979):584–590.

Henderson-Arzapalo, A., and R. R. Stickney. 1981. Effects of stocking density on two tilapia species raised in an intensive culture system. Proceedings of the Southeastern Association of Fish and Wildlife Agencies 34(1980): 379–387.

Hendricks, M. K., and R. L. Noble. 1980. Feeding interaction of three planktivorous fishes in Trinidad Lake, Texas. Proceedings of the Southeastern Association of Fish and Wildlife Agencies 33(1979): 324–330.

Hepher, B. 1978. Ecological aspects of warm-water fishpond management, 447–468. *In:* S. D. Gerking (ed.). Ecology of freshwater fish production. Blackwell Scientific Publications, London.

Hepher, B., and Y. Pruginin. 1981. Commercial fish farming, with special reference to fish culture in Israel. Wiley-Interscience Publication, New York.

Hickling, C. F. 1966. On the feeding process in the white amur, *Ctenopharyngodon idella*. Journal of Zoology 148:408–419.

Hida, T. S., J. R. Harada, and J. E. King. 1962. Rearing tilapia for tuna bait. U.S. Fish and Wildlife Service, Fishery Bulletin 198, 62:1–20.

Hida, T. S., and D. A. Thomson. 1962. Introduction of the threadfin shad to Hawaii. Progressive Fish-Culturist 24:159–163.

Hopkins, K. D. 1979. Production of monosex tilapia fry by breeding sex-reserved fish. Ph.D. dissertation, Auburn University, Auburn, Alabama. 44 pp.

Hopkins, K. D., E. M. Cruz, M. L. Hopkins, and K. C. Chong. 1981. Optimum manure loading rates in tropical freshwater fish ponds receiving untreated piggery wastes. 15–29. *In:* ICLARM Technical Reports 2, Manila, Philippines.

Hora, S. L., and T.V.R. Pillay. 1962. Handbook on fish culture in the Indo-Pacific region. FAO Fisheries Biology, Technical Paper 14.

Hubbs, C. L. 1968. An opinion on the effects of cichlid releases in North America. Transactions of the American Fisheries Society 97:197–198.

Huet, M. 1972. Textbook of fish culture: Breeding and cultivation of fish. Translated by H. Kahn. Fishing News Books, Surrey, England.

Huisman, E. A. 1981. Integration of hatchery, cage, and pond culture of common carp (*Cyprinus carpio* L.) and grass carp (*Ctenopharyngodon idella* Val.) in the Netherlands, 266–273. *In:* L. J. Allen and E. C. Kinney (eds.). Proceedings of the bioengineering symposium for fish culture. American Fisheries Society, Fish Culture Section Publication 1, Bethesda, Maryland.

Huisman, E. A., and P. Valentijn. 1981. Conversion efficiencies in grass carp (*Ctenopharyngodon idella* Val.) using a feed for commercial production. Aquaculture 22:279–288.

Hulata, G., S. Rothbard, and G. Wohlfarth. 1981. Genetic approach to the production of all-male progeny of tilapias. European Mariculture Society, Special Publication 6:181–190.

Jalabert, B., J. Mareau, D. Planquette, and R. Billard. 1974. Déterminisme du sexe chez *Tilapia macrochir* et *Tilapia nilotica*: Action de la méthyltestostérone dans l'alimentation des alevins sur la différentiation sexuelle: Proportion du sexes dans la descedance des mâles "inversés." Annales de Biologie Animale, Biochimie, Biophysique 14(4-B):729–739.

Jeffrey, N. B. 1970. Spawning of grass carp (*Ctenopharyngodon idella*). FAO Fish Culture Bulletin 2:3.

Jensen, G. L. 1979. Administration of methyltestosterone to direct the gonadal sex differentiation of female grass carp. Ph.D. dissertation, Auburn University, Auburn, Alabama. 202 pp.

Jensen, G. L., and W. L. Shelton. 1979. Effects of estrogens on *Tilapia aurea:* Implications for production of monosex genetic male tilapia. Aquaculture 16:233–242.

Jensen, G. L., W. L. Shelton, and L. O. Wilken. 1978. Use of methyltestosterone silastic implants to control sex in grass carp, 200–219. *In:* R. O. Smitherman, W. L. Shelton, and J. H. Grover (eds.). Culture of exotic fishes symposium proceedings. American Fisheries Society, Fish Culture Section, Auburn, Alabama.

Jhingran, V. G., and B. K. Sharma. 1980. Integrated livestock-fish farming in India, 135–142. *In:* R.S.V. Pullin and Z. H. Shehadeh (eds.). Integrated agriculture-aquaculture farming systems. ICLARM Conference Proceedings 4, Manila, Philippines.

Johnstone, R., T. H. Simpson, A. F. Youngson, and C. Whitehead. 1979. Sex reversal in salmonid culture, 2. The progeny of sex-reversed rainbow trout. Aquaculture 18:13–19.

Kelly, H. D. 1957. Preliminary studies on *Tilapia mossambica* Peters relative to experimental pond culture. Proceedings of the Southeastern Association of Fish and Game Commissioners 10(1956):139–149.

Kilgen, R. H. 1974. Mixed culture of catfish with blacktail redhorse sucker. Journal of the Alabama Academy of Science 45:139–143.

Kilgen, R. H., and R. O. Smitherman. 1971. Food habits of the white amur stocked in ponds alone and in combination with other species. Progressive Fish-Culturist 33:123–127.

Kirk, R. G. 1972. A review of recent development in *Tilapia* culture, with special reference to fish farming in heated effluents of power stations. Aquaculture 1:45–66.

Kohler, C., and F. A. Pagan-Font. 1978. Evaluation of rum distillation wastes, pharmaceutical wastes and chicken feed for rearing *Tilapia aurea* in Puerto Rico. Aquaculture 14:339–347.

Konradt, A. G. 1968. Methods of breeding the grass carp, *Ctenopharyngodon idella* (Val.) and silver carp, *Hypophthalmichthys molitrix,* 195–204. *In:* T.V.R. Pillay (ed.). Proceedings of the world symposium on warm-water pond fish culture. FAO Fisheries Reports 4(44).

Lauenstein, P. C. 1978. Intensive culture of tilapia with geothermally heated water, 82–85. *In:* R. O. Smitherman, W. L. Shelton, and J. H. Grover (eds.). Culture of exotic fishes symposium proceedings. American Fisheries Society, Fish Culture Section, Auburn, Alabama.

Lee, D. S., C. R. Gilbert, C. H. Hocutt, R. E. Jenkins, D. E. McAllister, and J. R. Stauffer, Jr. (eds.). 1980. Atlas of North American freshwater fishes. North Carolina Biological Survey 1980-12, North Carolina Museum of Natural History, Raleigh, North Carolina.

Lee, J. C. 1979. Reproduction and hybridization of three cichlid fish, *Tilapia aurea* (Steindachner), *Tilapia hornorum* (Trewavas), and *Tilapia nilotica* (Linnaeus) in aquaria and in plastic pools. Ph.D. dissertation, Auburn University, Auburn, Alabama. 94 pp.

Leventer, H. 1981. Biological control of reservoirs by fish. Bamidgeh 33:3–23.

Lin, S. Y. 1965. Induced spawning of Chinese carps by pituitary injection in Taiwan (a survey of technique and application). Chinese-American Joint Commission on Rural Reconstruction, Fisheries Series 5. Taipei, Taiwan.

Lin, S. Y. 1974. The dialectics of a proposal on biological control of eutrophication in sewage lagoons, 417–434. *In:* Wastewater use in the production of food and fiber — Proceedings. EPA Technical Series 660/2-74-041, Washington, D.C.

Lovell, R. T. 1979. Flavor problems in fish culture, 186–190. *In:* T.V.R. Pillay and W. A. Dill (eds.). Advances in aquaculture. Fishing News Books, Surrey, England.

Lovell, R. T., R. O. Smitherman, and E. W. Shell. 1978. Progress and prospects of fish farming, 261–291. *In:* A. M. Alschul and H. L. Wilcke (eds.). New protein foods. Academic Press, New York.

Lovshin, L. L. 1982. Tilapia hybridization, 279–308. *In:* R.S.V. Pullin and R. H. Lowe-McConnell (eds.). The biology and culture of tilapias. ICLARM, Conference Proceedings 7. Manila, Philippines.

McBay, L. G. 1962. The biology of *Tilapia nilotica* Linnaeus. Proceedings of the Southeastern Association of Game and Fish Commissioners 15(1961):208–218.

McCoy, E. W., and M. L. Hopkins. 1977. Establishing a market for an exotic fish species. Highlights of Agricultural Research, Auburn University, Auburn, Alabama 24(1):14.

Makeyeva, A. P., and B. V. Verigin. 1971. Use of pituitary injection in the propagation of silver carp and grass carp. Journal of Ichthyology 11:174–185.

Marian, T., and Z. Krasznai. 1978. Kariological investigations on *Ctenopharyngodon idella* and *Hypophthalmichthys nobilis* and their cross-breeding. Aquacultura Hungarica (Szarvas) 1:44–50.

Mensinger, G., and B. E. Brown. 1969. Live weight-dressed weight relationship for commercial fishes from four Oklahoma reservoirs. Proceedings of the Southeastern Association of Game and Fish Commissioners 22(1968):465–470.

Mgbenka, B. O. 1983. Intensive feeding of grass carp, *Ctenopharyngodon idella* Val., in earthen ponds. Ph.D. dissertation, Auburn University, Auburn, Alabama. 81 pp.

Moav, R. 1979. Genetic improvement in aquaculture industry, 610–622. *In:* T.V.R. Pillay and W. A. Dill (eds.). Advances in aquaculture. Fishing News Books, Surrey, England.

Moav, R., G. Wohlfarth, G. L. Schroeder, G. Hulata, and H. Barash. 1977. Intensive polyculture of fish in freshwater ponds. 1. Substitution of expensive feeds by liquid cow manure. Aquaculture 10:25–43.

Moriarty, C. M., and D.J.W. Moriarty. 1973. Quantitative estimation of the daily ingestion of phytoplankton by *Tilapia nilotica* and *Haplochromis nigrapinnis* in Lake George, Uganda. Journal of Zoology, London 1971:15–23.

Moyle, P. B. 1976. Inland fishes of California. University of California Press. Berkeley, California. 405 pp.

Murty, D. S., R. K. Dey, and P.V.G.K. Reddy. 1978. Experiments on rearing exotic carp fingerlings in composite fish culture in India. Aquaculture 13:331–337.

Nagy, A., M. Bercsenyi, and V. Csanyi. 1981. Sex reversal in carp (*Cyprinus carpio*) by oral administration of methyltestosterone. Canadian Journal of Fisheries and Aquatic Sciences 38:725–728.

Nagy, A., K. Rajki, L. Horvath, and V. Csanyi. 1978. Investigation on carp, *Cyprinus carpio* L. gynogenesis. Journal of Fish Biology 13:215–224.

Nair, R. R. 1968. A preliminary bibliography of the grass carp. FAO, Fisheries Circular 302.

National Academy of Sciences. 1978. Aquaculture in the United States—constraints and opportunities. Committee on Aquaculture, National Research Council, National Academy of Sciences, Washington, D.C.

Nerrie, B. L. 1979. Production of male *Tilapia nilotica* using pelleted chicken manure. M.S. thesis, Auburn University, Auburn, Alabama. 54 pp.

Newton, S. H. 1980. Catfish farming with Chinese carps. Arkansas Farm Research Journal 24:8.

Newton, S. H., J. C. Dean, and A. J. Handcock. 1978. Low intensity polyculture with Chinese carps, 137–143. *In:* R. O. Smitherman, W. L. Shelton, and J. H. Grover (eds.). Culture of exotic fishes symposium proceedings. American Fisheries Society, Fish Culture Section, Auburn, Alabama.

Noble, R. L., R. D. Germany, and C. R. Hall. 1975. Interactions of the blue tilapia and largemouth bass in a power plant cooling lake. Proceedings of the Southeastern Association of Game and Fish Commissioners 29(1975):247–251.

Omarov, M. O. 1970. The daily food consumption of silver carp (*Hypophthalmichthys molitrix* Val.). Journal of Ichthyology 10:425–426.

Opuszyński, K. 1972. Use of phytophagous fish to control aquatic plants. Aquaculture 1:61–74.

Opuszyński, K. 1981. Comparison of the usefulness of the silver carp and the bighead carp as additional fish in carp ponds. Aquaculture 25:223–233.

Osborn, M. F. 1977. Intensive culture of channel catfish with aeration and the use of silver carp as supplemental species. M.S. thesis, Auburn University, Auburn, Alabama. 63 pp.

Pagan-Font, F. A. 1975. Cage culture as a mechanical method for controlling reproduction of *Tilapia aurea* (Steindachner). Aquaculture 6:243–247.

Panov, D. A., Y. I. Sorokin, and L. G. Molenkova. 1969. Experimental study of the feeding of young silver carp (*Hypophthalmichthys molitrix*). Journal of Ichthyology 9:101–111.

Pearl River Fisheries Research Institute (PRFRI). 1980. Pond fish culture in China. China National Bureau of Aquatic Products. Guangzhou, China.

Perry, W. G., Jr., and J. W. Avault, Jr. 1976. Polyculture studies with channel catfish and buffalo. Proceedings of the Southeastern Association of Fish and Game Commissioners 29(1975):91–98.

Pillay, T.V.R. (ed.). 1972. Coastal aquaculture in the Indo-Pacific region. Fishing News Books, Surrey, England.

Pillay, T.V.R. 1979a. The state of aquaculture 1976, 1–10. *In:* T.V.R. Pillay and W. A. Dill (eds.). Advances in aquaculture. Fishing News Books, Surrey, England.

Pillay, T.V.R. 1979b. Aquaculture development in China. FAO Aquaculture Development and Coordination Programme Report 79/10.

Pretto-Malca, R. 1976. Polyculture systems with channel catfish as the principal species. Ph.D. dissertation, Auburn University, Auburn, Alabama. 190 pp.

Pruginin, Y., S. Rothbard, G. Wohlfarth, A. Halevy, R. Moav, and G. Hulata. 1975. All-male broods of *Tilapia nilotica* x *T. aurea* hybrids. Aquaculture 6:11–21.

Pullin, R.S.V., and C. M. Kuo. 1981. Developments in the breeding of cultured fishes, 899–978. *In:* J. T. Manassah and E. J. Briskey (eds.). Advances in food-producing systems for arid and semiarid lands. Part B. Academic Press, New York.

Pullin, R.S.V., and R. H. Lowe-McConnell (eds.). 1982. The biology and culture of tilapias. ICLARM, Conference Proceedings 7, Manila, Philippines.

Pullin, R.S.V., and Z. H. Shehadeh (eds.). 1980. Integrated agriculture-aquaculture farming systems. ICLARM Conference Proceedings 4, Manila, Philippines.

Rakocy, J. E., and R. Allison. 1981. Evaluation of a closed recirculating system for the culture of tilapia and aquatic macrophytes, 296–307. *In:* L. J. Allen and E. C. Kinney (eds.). Proceedings of the bio-engineering symposium for fish culture. American Fisheries Society, Fish Culture Section Publication 1, Bethesda, Maryland.

Ray, L. E. 1978. Production of tilapia in catfish raceways using geothermal waters, 86–89. *In:* R. O. Smitherman, W. L. Shelton, and J. H. Grover (eds.). Culture of exotic fishes symposium proceedings. American Fisheries Society, Fish Culture Section, Auburn, Alabama.

Redner, B. D., and R. R. Stickney. 1979. Acclimation to ammonia by *Tilapia aurea.* Transactions of the American Fisheries Society 108:383–388.

Refstie, T. 1981. Tetraploid rainbow trout produced by cytochalasin B. Aquaculture 25:51–58.

Refstie, T., V. Vassvik, and G. Gjedrem. 1977. Induction of polyploidy in salmonids by cytochalasin B. Aquaculture 10:65–74.

Regier, H. A. 1968. The potential misuse of exotic fish as introductions, 92–111. *In:* A symposium on introductions of exotic species. Ontario Ministry of Natural Resources, Research Report 82, Ottawa, Canada.

Robins, C. R., R. M. Bailey, C. E. Bond, J. R. Brooker, E. A. Lachner, R. N. Lea, and W. B. Scott. 1980. A list of common and scientific names of fishes from the United States and Canada, 4th ed. American Fisheries Society Special Publication 12. Bethesda, Maryland. 174 pp.

Romaire, R. P., C. E. Boyd, and W. J. Collis. 1978. Predicting nighttime dissolved oxygen decline in ponds used for tilapia culture. Transactions of the American Fisheries Society 107:804–808.

Rosen, R. A., and D. C. Hales. 1981. Feeding of paddlefish, *Polyodon spathula.* Copeia 1981:441–455.

Rosenthal, H. 1978. Bibliography on transplantation of aquatic organisms and its consequences on aquaculture and ecosystems. European Mariculture Society Special Publication 3. Bredene, Belgium.

Rothbard, S. 1981. Induced reproduction in cultivated cyprinids—the common carp and the group of Chinese carps: I. The technique of induction, spawning and hatching. Bamidgeh 33:103–121.

St. Amant, J. A. 1966. Progress report of the culture of *Tilapia mossambica* (Peters) hybrids in southern California. California Inland Fisheries Administration Report 66-9. Sacramento, California.

Sarig, S., and Y. Arieli. 1980. Growth capacity of tilapia in intensive culture. Bamidgeh 32:57–65.

Schoenen, P. 1982. A bibliography of important tilapias (Pisces: Cichlidae) for aquaculture. ICLARM, Bibliographies 3, Manila, Philippines.

Schroeder, G. L. 1980. Fish farming in manure-loaded ponds, 73–86. *In:* R.S.V. Pullin and Z. H. Shehadeh (eds.). Integrated agriculture-aquaculture farming systems. ICLARM Conference Proceedings 4, Manila, Philippines.

Schroeder, G., and B. Hepher. 1979. Use of agricultural and urban wastes in fish culture, 487–488. *In:* T.V.R. Pillay and W. M. Dill (eds.). Advances in aquaculture. Fishing News Books, Surrey, England.

Shelton, W. L., S. P. Boney, and E. M. Rosenblatt. 1982. Monosexing grass carp by sex reversal and breeding, 184–194. *In:* T. O. Robson (ed.). Proceedings and international symposium on herbivorous fish. European Weed Research Society, Wageningen, Netherlands.

Shelton, W. L., K. D. Hopkins, and G. L. Jensen. 1978. Use of hormones to produce monosex tilapia for aquaculture, 10–33. *In:* R. O. Smitherman, W. L. Shelton, and J. H. Grover (eds.). Culture of exotic fishes symposium proceedings. American Fisheries Society, Fish Culture Section, Auburn, Alabama.

Shelton, W. L., and G. L. Jensen. 1979. Production of reproductively limited grass carp for biological control of aquatic weeds. Office of Water Research Technology, U.S. Department of Interior, Water Resources Research Institute Bulletin 39.

Shelton, W. L., D. Rodriguez-Guerrero, and J. Lopez-Macias. 1981b. Factors affecting androgen sex reversal of *Tilapia aurea.* Aquaculture 25:59–65.

Shelton, W. L., R. O. Smitherman, and G. L. Jensen. 1981a. Density related growth of grass carp, *Ctenopharyngodon idella* (Val.), in managed small impoundments in Alabama. Journal of Fish Biology 18:45–51.

Shireman, J. V. 1974. Preliminary experiment on striped mullet culture in freshwater catfish ponds. Progressive Fish-Culturist 36:113.

Shireman, J. V. 1979. Proceedings of the grass carp conference. Aquatic Weeds Research Center, University of Florida, Gainesville, Florida.

Shireman, J. V., D. E. Colle, and R. W. Rottmann. 1978a. Growth of grass carp fed natural and prepared diets under intensive culture. Journal of Fish Biology 12:457–463.

Shireman, J. V., D. E. Colle, and R. W. Rottmann. 1978b. Size limits to predation on grass carp by largemouth bass. Transactions of the American Fisheries Society 107:213–215.

Sills, J. 1970. A review of herbivorous fish for weed control. Progressive Fish-Culturist 32:158–161.

Silvera, P.A.W. 1978. Factors affecting fry production in *Sarotherodon niloticus.* M.S. thesis, Auburn University, Auburn, Alabama. 39 pp.

Sinha, V. R., and M. V. Gupta. 1975. On the growth of grass carp, *Ctenopharyngodon idella* Val., in composite fish culture in Kalyani, West Bengal (India). Aquaculture 5:283–290.

Smith, P. L. 1971. Effects of *Tilapia aurea* (Steindachner), cage culture and aeration on channel catfish, *Ictalurus punctatus* (Rafinesque), production ponds. Ph.D. dissertation, Auburn University, Auburn, Alabama. 112 pp.

Smitherman, R. O., W. L. Shelton, and J. H. Grover (eds.). 1978. Culture of exotic fishes symposium proceedings. American Fisheries Society, Fish Culture Section, Auburn, Alabama.

Spataru, P., and M. Zorn. 1978. Food and feeding habits of *Tilapia aurea* (Steindachner) (Cichlidae) in Lake Kinneret (Israel). Aquaculture 13:67–79.

Stanley, J. G. 1976a. Reproduction of the grass carp (*Ctenopharyngodon idella*) outside its native range. Fisheries 1(3):7–10.

Stanley, J. G. 1976b. Production of hybrid, androgenetic, and gynogenetic grass carp and carp. Transactions of the American Fisheries Society 105:10–16.

Stanley, J. G. 1976c. Female homogamety in grass carp (*Ctenopharyngodon idella*) determined by gynogenesis. Journal of the Fisheries Research Board of Canada 33:1372–1374.

Stanley, J. G. 1979. Control of sex in fishes, with special reference to the grass carp, 201-242. *In:* J. V. Shireman (ed.). Proceedings of the grass carp conference. Aquatic Weeds Research Center, University of Florida, Gainesville, Florida.

Stanley, J. G., and W. M. Lewis (eds.). 1978. Special section: Grass carp in the United States. Transactions of the American Fisheries Society 107:104–124.

Stanley, J. G., and K. E. Sneed. 1973. Artificial gynogenesis and its application in genetics and selective breeding of fishes, 527–536. *In:* J. H. Blaxter (ed.). The early life history of fish. Springer-Verlag, New York.

Stanley, J. G., J. M. Martin, and J. B. Jones. 1975. Gynogenesis as a possible method for producing monosex grass carp (*Ctenopharyngodon idella*). Progressive Fish-Culturist 37:25–26.

Stanley, J. G., W. W. Miley II, and D. L. Sutton. 1978. Reproductive requirements and likelihood for naturalization of escaped grass carp in the United States. Transactions of the American Fisheries Society 107:119–128.

Stanley, J. G., and A. E. Thomas. 1978. Absence of sex reversal in unisex grass carp fed methyltestosterone, 194–199. *In:* R. O. Smitherman, W. L. Shelton, and J. H. Grover (eds.). Culture of exotic fishes symposium proceedings. American Fisheries Society, Fish Culture Section, Auburn, Alabama.

Stevenson, J. H. 1965. Observations on grass carp in Arkansas. Progressive Fish-Culturist 27:203–206.

Stickney, R. R. 1979. Principles of warmwater aquaculture. Wiley Interscience Publication, New York.

Stickney, R. R., J. H. Hesby, R. B. McGeachin, and W. A. Isbell. 1979. Growth of *Tilapia nilotica* in ponds with differing histories of organic fertilizer. Aquaculture 17:189–194.

Stickney, R. R., L. O. Rowland, and J. H. Hesby. 1978. Water quality – *Tilapia aurea* interaction in ponds receiving swine and poultry wastes. Proceedings of the World Mariculture Society 8(1977):55–71.

Stone, N. M. 1981. Growth of male and female *Tilapia nilotica* in ponds and cages. M.S. thesis, Auburn University, Auburn, Alabama. 19 pp.

Stott, B., and L. D. Orr. 1970. Estimating the amount of aquatic weed consumed by grass carp. Progressive Fish-Culturist 32:51–54.

Stroband, H.W.J. 1977. Growth and diet dependant [sic] structural adaptations of the digestive tract in juvenile grass carp (*Ctenopharyngodon idella* Val.). Journal of Fish Biology 11:167–174.

Sutton, D. L. 1977. Grass carp (*Ctenopharyngodon idella* Val.) in North America. Aquatic Botany 3:157–164.

Swingle, H. A. 1965. Growth rates of paddlefish receiving supplemental feeding in fertilized ponds. Progressive Fish-Culturist 27:220.

Swingle, H. S. 1957. Preliminary results on the commercial production of channel catfish in ponds. Proceedings of the Southeastern Association of Game and Fish Commissioners 10(1956):142–148.

Swingle, H. S. 1960. Comparative evaluation of two tilapias as pond fishes in Alabama. Transactions of the American Fisheries Society 89:142–148.

Swingle, H. S. 1968a. Biological means of increasing productivity in ponds, 243–247. *In:* T.V.R. Pillay (ed.). Proceedings of the world symposium on warm-water pond fish culture. FAO Fisheries Reports 4(44).

Swingle, H. S. 1968b. Estimation of standing crops and rates of feeding fish in ponds,

416–423. *In:* T.V.R. Pillay (ed.). Proceedings of the world symposium on warmwater pond fish culture. FAO Fisheries Reports 3(44).

Swingle, H. S. 1970. History of warmwater pond culture in the United States, 95–105. *In:* N. G. Benson (ed.). A century of fisheries in North America. American Fisheries Society Special Publication 7.

Sylvester, J. R. 1975. Biological considerations on the use of thermal effluents for finfish aquaculture. Aquaculture 6:1–10.

Tal, S., and I. Ziv. 1978. Culture of exotic fishes in Israel, 1–9. *In:* R. O. Smitherman, W. L. Shelton, and J. H. Grover (eds.). Culture of exotic fishes symposium proceedings. American Fisheries Society, Fish Culture Section, Auburn, Alabama.

Tamas, G., and L. Horvath. 1976. Growth of cyprinids under optimal zooplankton conditions. Bamidgeh 28:50–56.

Tang, Y. A. 1970. Evaluation of balance between fishes and available fish foods in multispecies fish culture ponds in Taiwan. Transactions of the American Fisheries Society 99:708–718.

Tapiador, D. D., H. F. Henderson, M. N. Delmendo, and H. Tsutsui. 1977. Freshwater fisheries and aquaculture in China. FAO Fisheries Technical Paper 168.

Theriot, R. F., and J. L. Decell. 1978. Large-scale operations management tests of the use of the white amur to control aquatic plants, 220–229. *In:* R. O. Smitherman, W. L. Shelton, and J. H. Grover (eds.). Culture of exotic fishes symposium proceedings. American Fisheries Society, Fish Culture Section, Auburn, Alabama.

Thys van den Audenaerde, D.F.E. 1968. An annotated bibliography of tilapia (Pisces, Cichlidae). Musée royale de l'Afrique centrale, Tervuren, Belgique, Documentation zoologique 14.

Trewavas, E. 1966. *Tilapia aurea* (Steindachner) and the status of *Tilapia nilotica exul, T. monodi,* and *T. lemassoni* (Pisces, Cichlidae). Israel Journal of Zoology 14(1965):258–276.

Trewavas, E. 1973. On the cichlid fish of the genus *Pelmatochromis* with proposal of a new genus for *P. congicus;* on the relationship between *Pelmatochromis* and *Tilapia* and the recognition of *Sarotherodon* as a distinct genus. Bulletin of the British Museum of Natural History (Zoology) 25:1–26.

Trewavas, E. 1982. Tilapias: Taxonomy and speciation, 3–13. *In:* R.S.V. Pullin and R. H. Lowe-McConnell (eds.). The biology and culture of tilapias, ICLARM Conference Proceedings 7, Manila, Philippines.

Tuten, J. S., and J. W. Avault, Jr. 1981. Growing red swamp crayfish (*Procambarus clarkii*) and several North American fish species together. Progressive Fish-Culturist 43:97–98.

Uchida, R. N., and J. E. King. 1962. Tank culture of tilapia. U.S. Fish and Wildlife Service, Fishery Bulletin 198, 62:21–51.

U.S. Department of Agriculture (USDA). 1981. Aquaculture: Outlook and situation. Economic and Statistic Service, AS-1, Washington, D.C.

Valenti, R. J. 1975. Induced polyploidy in *Tilapia aurea* (Steindachner) by means of temperature shock treatment. Journal of Fish Biology 7:519–528.

Webber, H. H., and P. F. Riordan. 1976. Criteria for candidate species for aquaculture. Aquaculture 7:107–123.

Welcomme, R. 1981. Register of international transfer of inland fish species. FAO Fisheries Technical Paper 213.

Wiliamovski, A. 1972. Structure of the gill apparatus and the suprabranchial organ of *Hypophthalmichthys molitrix* (silver carp). Bamidgeh 24:87–99.

Williamson, J., and R. O. Smitherman. 1976. Food habits of hybrid buffalofish, tilapia, Israeli carp and channel catfish in polyculture. Proceedings of the Southeastern Association of Game and Fish Commissioners 29(1975):86–91.

Winfree, R. A., and R. R. Stickney. 1981. Effects of dietary protein and energy on growth, feed conversion efficiency and body composition of *Tilapia aurea*. Journal of Nutrition 111:1001–1012.

Wohlfarth, G. W., and G. H. Hulata. 1981. Applied genetics of tilapias. ICLARM Studies and Reviews 6. Manila, Philippines.

Wohlfarth, G. W., and G. L. Schroeder. 1979. Use of manure in fish farming – a review. Agricultural Wastes 1:279–299.

Woodruff, V. C. 1978. Marketability of canned silver carp. M.S. thesis, Auburn University, Auburn, Alabama. 69 pp.

Woynarovich, E., and L. Horvath. 1980. The artificial propagation of warm-water finfishes – a manual for extension. FAO Fisheries Technical Paper 201.

Yamamoto, T. 1969. Sex differentiation, 117–175. *In:* W. S. Hoar and D. J. Randall (eds.). Fish Physiology, 3. Academic Press, New York.

Yashouv, A. 1968. Mixed fish culture – an ecological approach to increase pond productivity, 258–273. *In:* T.V.R. Pillay (ed.). Proceedings of the world symposium on warm-water pond fish culture. FAO Fisheries Reports 4(44).

Yashouv, A. 1971. Interaction between the common carp (*Cyprinus carpio*) and the silver carp (*Hypophthalmichthys molitrix*) in fish ponds. Bamidgeh 23:85–92.

Yashouv, A., E. Berner-Sansonov, and K. Reich. 1970. Forced spawning of silver carp. Bamidgeh 22:3–8.

Yashouv, A., and J. Chervinski. 1961. The food of *Tilapia nilotica* in ponds of the fish culture research station at Dor. Bamidgeh 13:33–39.

CHAPTER 14

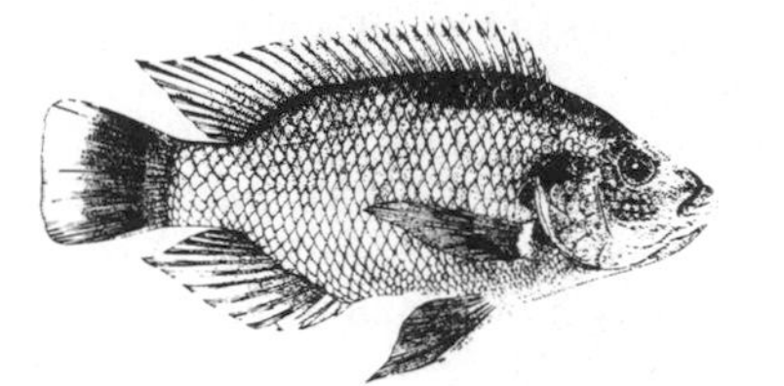

Control of Aquatic Weeds with Exotic Fishes

Jerome V. Shireman

Aquatic weeds in excessive amounts cause problems in waters that have many domestic, agricultural, industrial, and recreational uses. Formulation of aquatic plant management techniques is difficult because a number of user groups are involved, and each group has a different perception of the problem. In the past, most aquatic weed management strategies have involved chemical treatments, and the impacts of these treatments on the receiving environment were often unknown.

Because aquatic herbicides are expensive and might cause adverse effects, biological control organisms have been and are being investigated. The study of biological controls for aquatic plants received impetus from the discovery of the alligatorweed flea beetle (*Agasicles hygrophila*) during the 1960s for control of alligatorweed (*Alternanthera philoxeroides*). The use of fishes as biological controls gained popularity shortly after (Avault et al. 1968, Blackburn and Sutton 1971, Legner et al. 1975).

Aquatic Weed Problems

ALGAE

Algae are found in nearly all waters throughout the United States, and in certain water bodies reach nuisance proportions. In 1972, the United States Environmental Protection Agency initiated the National Eutrophication Survey to investigate nationwide the accelerated cultural eutrophication of lakes and streams. This survey investigated physical, chemical, and biological data on over 800 lakes and reservoirs throughout the United States. Research has shown a strong correlation between algal biomass (chlorophyll a) and total phosphorus concentrations (Edmondson 1971, Schindler 1975, Jones and Bachmann 1976, Canfield 1979). For this reason, there has been an effort on a nationwide scale to reduce phosphorus inputs into lakes and reservoirs.

NATIVE AQUATIC MACROPHYTES

Like algae, native aquatic macrophytes are found in nearly all inland waters and often reach nuisance proportions that require control.

To my knowledge, a survey has not been conducted to determine the extent of macrophyte infestations in the United States. Reports allude to problems caused by emergent, submersed, floating leafed and flowering plants; however, data are lacking pertaining to the factors that might limit the growth of these plants. Many of the submersed species would be amenable to biological controls.

NONNATIVE AQUATIC MACROPHYTES

Several nonnative species including water hyacinth (*Eichornia crassipes*), alligator weed (*Alternanthera philoxeroides*), Eurasian watermilfoil (*Myriophyllum spicatum*), and hydrilla (*Hydrilla verticillata*) are introduced plants that cause serious problems in the United States. Data are lacking on the factors that might limit their growth and distribution. Hydrilla is currently expanding its range (Haller 1979) and is difficult to control because it is a submersed plant and produces tubers that are unaffected by most control methods (Miller et al. 1976). This plant, however, is vulnerable to biological control by fishes.

Exotic Fishes for Weed Control

Numerous fishes are reported to consume aquatic vegetation (table 14–1). However, only a few of these fish have been investigated and show promise for weed control: the cichlids *Tilapia mossambica* and *Tilapia zilli* (Legner et al. 1975), *Tilapia melanoplura* (Hickling 1962, Avault et al. 1968) [*Tilapia rendalli,* see Thys van den Audenaerde 1968], *Tilapia nilotica* (Avault et al. 1968), the characoids *Metynnis roosevelti* and *Mylossoma argenteum* (Yeo 1967), and the cyprinids *Puntius gonionatus* (Hickling 1962), *Hypophthalmichthys molitrix* (Opuszyński 1979), and *Ctenopharyngodon idella* (Shireman and Smith 1981).

Many of these species may not be suitable for weed control because the individual has insufficient consumption (high stocking rates needed), they are prolific spawners (often cause overcrowding), or they are restricted to warm climates (must be overwintered in controlled environments).

Table 14–1. EXOTIC FISH SPECIES REPORTED TO CONSUME AQUATIC VEGETATION

Common name	Scientific name	Vegetation consumed[a]
Milkfish	*Chanos chanos*	Phytoplankton, epiphytes, and decayed macrophytes
Silver dollarfish	*Metynnis roosevelti*	Macrophytes
Silver dollarfish	*Mylossoma argenteum*	Macrophytes
Tambaqui	*Colossoma bidens*	Phytoplankton
Pirapitinga	*Mylossoma bidens*	Phytoplankton
Ayu	*Plecoglossus altivelis*	Algae
Common carp	*Cyprinus carpio*	Uproots vegetation
Crucian carp	*Carassius carassius*	Filamentous algae
Tawes	*Puntius javanicus*	Algae, grasses
Belinka	*P. belinka*	Epiphytes, soft macrophytes
Mata mera	*P. orphoides*	Phytoplankton, decayed macrophytes
Carnatic carp	*Barbus carnaticus*	Algae, macrophytes
Copper mahseer	*B. hexagonalis*	Macrophytes
Rohu	*Labeo rohita*	Macrophytes
Calbasu	*L. calbasu*	Epiphytes, macrophytes
Fingelipped carp	*L. fimbriatus*	Filamentous algae
Cauvary carp	*L. kontius*	Algae, pieces of macrophytes
Tambra	*Labeobarbus tambroides*	Phytoplankton, decayed plants
Mrigal	*Cirrhina mrigala*	Algae, pieces of macrophytes
Reba	*C. reba*	Algae, macrophyte parts
Catla	*Catla catla*	Algae, decayed macrophytes
Grass carp	*Ctenopharyngodon idella*	Macrophytes
Silver carp	*Hypophthalmichthys molitrix*	Algae
Caven	*Megalobrama bramula*	Macrophytes
Nilem	*Osteochilus hasselti*	Algae, soft macrophytes
Nagendram fish	*O. thomassi*	Filamentous algae
Sandkhol carp	*Thynnichthys sandkhol*	Phytoplankton
Gourami	*Osphronemus gouramy*	Plant leaves
Spotted gourami	*Trichogaster trichopterus*	Algae, macrophytes
Climbing perch	*Anabas testudineus*	Algae, soft macrophytes
Florida flagfish	*Jordanella floridae*	Filamentous algae
Mozambique tilapia	*Tilapia mossambica*	Macrophytes, filamentous algae
Redbreast tilapia	*T. rendalli*	Macrophytes
	T. melanopleura	Macrophytes
Nile tilapia	*T. nilotica*	Filamentous algae
Redbelly tilapia	*T. zilli*	Macrophytes
Blue tilapia	*T. aurea*	Phytoplankton, detritus
Pearlspot	*Etroplus suratensis*	Algae, macrophytes

[a] In some species, vegetation might be a small portion of the diet.

The use of phytophagous fishes for algae control has not been investigated to a great extent, but several species show promise as algae eaters. For example, *Tilapia aurea* (Leventer 1972), *Tilapia galilea,* and silver carp, *Hypophthalmichthys molitrix* (Leventer 1979) are used for algae control in Israeli reservoirs. Silver carp are used in polyculture situations to consume algae which result as a byproduct of other fish culture operations (Opuszyński 1979).

Tilapia spp. are also reported to control macrophytes. Legner et al. (1975) reported success in California irrigation canals with *T. zilli.* These fish suppressed growth of noxious aquatic weeds in California irrigation canals when stocked in high densities.

The grass carp, *Ctenopharyngodon idella,* has been successfully used for aquatic macrophyte control in a variety of situations and climates (Aliev 1963, Nikolsky and Aliev 1974, Bailey 1978, Shireman and Maceina 1981) and may provide the only practical control method for water bodies where herbicides cannot be used (Kobylinski et al. 1980).

Although the grass carp is a proven biological control, fish biologists and others have questioned its use because of possible adverse environmental impact. Reproduction has been documented in the Mississippi River (Conner et al. 1980), but the recruitment of these fish into larger-size classes has not been followed. In order to circumvent the escape of additional grass carp, a sterile hybrid (grass carp x bighead carp) has been touted as a viable replacement. This fish, however, is difficult to culture because of high mortality during the fry and larval stages and greater dependence upon zooplankton for a longer period (W. W. Miley II, personal communication). In Florida, grass carp hybrids, like grass carp, should not be restocked until they reach 300 mm total length to avoid predation (Shireman et al. 1978). The hybrid apparently does not consume vegetation in quantities as great as the grass carp, which necessitates higher stocking rates. Sutton (1980) reported that hybrids, fed upon vegetation in outdoor pools, first consumed vallisneria (*Vallisneria americana*) and in lesser amounts southern naiad (*Najas quadalupensis*) and chara (*Chara* spp.). He noted some feeding on Illinois pondweed (*Potamogeton illinoensis*), but hydrilla and Eurasian watermilfoil were not fed on by the hybrid carp. He also reported that growth of hybrid carp in pools and in an outdoor closed-circulation system was low.

Possible Environmental Impact

All plant control methods, if successful, will cause a change in the environment. The death and destruction of target plant species will eventually cause a change in animal populations by altering food availability, release of nutrients resulting in phytoplankton blooms, and water quality alteration including changes in the oxygen–carbon dioxide balance (Brooker and Edwards 1975). These changes are general and will be caused regardless of the control method. With biological control, however, these effects can be long lasting, especially if the fish reproduce in the area.

Tilapia zilli, when stocked in high numbers, is capable of removing vegetation. Legner et al. (1975) concluded that this species would control vegetation in irrigation canals in southern California without causing adverse impact to sport fishes. This view is apparently not universal as the California Fish and

Game Commission prohibits future introduction into central and northern California and is attempting to control the fish in the south (Pelzman 1973). There is considerable concern about the impact that this fish might have on other species because of its potential omnivorous feeding, competition for spawning sites, and destruction of habitat (Pelzman 1973). Noble et al. (1976) report that bass ceased to spawn in a Texas reservoir (Trinidad Lake) when blue tilapia numbers were high. In a pond experiment, Florida largemouth bass ceased to spawn successfully at a tilapia density of 2,240 kg/ha, but some spawning occurred at a tilapia density of 1,121 kg/ha. These data indicate that tilapia in high densities can have an impact on sport fish.

Grass carp impacts have been studied in ponds and natural lake systems by a number of researchers. Lembi et al. (1978) reported increased potassium levels in all her test ponds. Nitrogen and phosphorus levels were also higher, but not statistically different. Shireman et al. (1979) reported similar results for a 123 ha Florida lake stocked with grass carp. Potassium levels increased 300 to 400 percent when vegetation levels decreased. Michewicz et al. (1972) reported increases in water hardness and nitrogen but not phosphorus. Other authors (Terrell 1974, Forester and Lawrence 1978) did not find changes in water quality where grass carp were stocked. Increases in phytoplankton have been reported after grass carp stocking (Alikunhi and Sukumaran 1964, Nikolsky and Aliev 1974, Vinogradov and Zolotova 1974, Opuszyński 1979, Crisman and Kooijman 1980). Excessive phytoplankton blooms, however, might not occur if systems are not overstocked and some macrophytes are allowed to remain.

The effects of grass carp stocking on other animal populations is not well understood and the degree of change in these populations is usually related to stocking rate and amount of vegetation removed. Several studies have shown increases in invertebrate populations after grass carp removal (Aliev 1976, Lembi and Ritenour 1977, Kobylinski et al. 1980), ascribing increased nutrient input from decaying vegetation and utilization of these nutrients. At high stocking rates (45–69 kg/ha) vegetation refugia were eliminated causing reduction in invertebrates (Vinogradov and Zolotova 1974, Gasaway 1977). In other situations, zooplankton and benthic populations were unaffected (Rottmann 1976, Rottmann and Anderson 1976, Crisman and Kooijman 1980).

The impact of grass carp on other fish species occurs mostly through the elimination of vegetation. Russian studies have shown, for example, that in heavily stocked lakes pike (*Esox lucius*) and perch (*Lucioperca fluviatilis*) failed to spawn (Sutton et al. 1977). Vinogradov and Zolotova (1974) reported that grass carp introductions in Russia have adversely affected pike, perch, Crucian carp (*Carassius carassius*), and roach (*Rutilus rutilus*). Other studies have shown decreased bluegill populations and failure of bass to spawn (Forester 1975, Forester and Lawrence 1978). Several authors (Buck et al. 1975, Rottmann 1976, Rottmann and Anderson 1976) reported that the introduction of grass carp enhanced the production of cyprinids and centrarchids. Grass carp stockings in Arkansas lakes had little or no effect on shad (*Dorosoma* spp.), largemouth bass, or sunfish (*Lepomis* spp.) (Bailey 1978). Although the results from these studies seem conflicting, the impact is related to the dependence of the fish population on vegetation and the amount of vegetation control. The alternative is to ignore the vegetation and allow it to remain in the system. This alternative, however, is not a viable one where water is used for irrigation or transportation. In lakes managed for fish production, this might be a viable alternative; however, studies conducted in our laboratory indicate the degree of vegetation infestation (hydrilla) influences condition and subsequently growth of bluegill, redear sunfish (*Lepomis microlophus*), and largemouth bass (Colle and Shireman 1980). Condition values decreased as hydrilla increased in the water column. Harvestable bass, greater than 350 mm TL, exhibited low condition values when hydrilla coverage exceeded 30 percent, whereas condition factors of smaller bass were not adversely affected until coverage exceeded 50 percent. Sunfish maintained adequate condition until coverage exceeded 80 percent. From these data, we conclude that vegetation management plans for Florida fish management lakes should include vegetation removal.

Status of Weed Control and Possible Management Strategies with Exotic Fish

Of the fishes examined to date, the grass carp appears to be the best candidate for control of submersed plants. This fish is used by Arkansas and other states for this purpose in natural lakes and has been researched by a number of other states. Florida has conducted research on the grass carp and is currently investigating the use of the hybrid grass carp as a sterile replacement. To my knowledge, the hybrid is still in experimental stages and is not a proven biological control organism.

As stated previously, the grass carp does consume vegetation and if stocked in sufficient numbers can completely remove submersed plants from lake systems (Shireman and Maceina 1981). The degree of vegetation removal, therefore, is a function of the stocking rate. It does appear, however, that control is achieved in an all or none fashion. Florida studies indicate that if the fish does not remove vegetation completely in two to three years, control will not be achieved, especially in hydrilla-infested lakes. As an example, Lake Baldwin was stocked with fingerling grass carp in 1975. Before stocking, the lake was treated with herbicides, and vegetation control was evident for approximately eighteen months. In 1977, the lake was again hydrilla infested. A grass carp population estimate was conducted (Colle et al. 1978) and only 3 fish/ha remained. Larger grass carp, greater than 300 mm TL, were restocked and control was achieved (35 fish/ha hydrilla). This study indicates that if sufficient fish are present, all vegetation will be removed. Thus, as stated previously, adverse environmental impact might occur when all vegetation is removed. I must point out, however, that most Florida studies have not been carried to the extreme whereby the impact of complete vegetation removal could be studied.

In a recent paper (Shireman and Maceina 1981) we discussed three possible management strategies utilizing grass carp: (1) complete vegetation removal within two years with a heavy stocking rate, (2) winter stocking with fewer fish to maintain a lesser amount of vegetation in the system and adjust the grass carp population as needed, and (3) integrated control utilizing chemical treatments to obtain desired levels quickly and stock grass carp to maintain this level. Again, the grass carp population should be adjusted as needed. A word of caution is in order: it is much easier to stock additional grass carp than to remove unwanted fish from the system. We are currently studying the option of integrated control methods (Shireman and Haller 1980).

Verigin (1979) proposed the utilization of herbivorous fishes to reconstruct the ichthyofauna in water bodies undergoing anthropogenic eutrophication. In Russia, according to Verigin (1979), the native fish fauna consists primarily of relatively cold resistant, nonproductive fish species which are either zoophages or carnivores. In water bodies that are enriched by chemical and thermal effluents, Verigin (1979) proposed utilizing warmwater phytophagous fish (grass carp, silver carp, and bighead carp *Aristichthys nobilis*). He thinks that ichthyofauna reconstruction can increase not only economic productivity but also water quality.

Since 1968, Israeli scientists have utilized the concept of ichthyofauna reconstruction in reservoirs of the Israel National Water System (Leventer 1979). Silver carp were used to consume phytoplankton, bighead carp for zooplankton, grass carp for submerged plants, black carp (*Mylopharyngodon piceus*) for snails, common carp for filamentous algae, sea bass (*Dicentratchus punctatus*) for larval fishes, blue tilapia for bottom sediments, and mullet (*Mugil* spp.) for detritus. The results of this study were positive. *Tilapia aurea* and gray mullet decreased the organic sediment concentration, eliminating the bloom of the blue green alga *Oscillatoria chalybea;* grass carp eliminated vegetation; black carp and common carp reduced snail populations; sea bass were effective in controlling larval fishes; silver carp reduced phytoplankton by 25 percent and prevented blooms, and bighead carp were effective in keeping zooplankton numbers in check.

One of the major criticisms of utilizing biological control with fishes or any other aquatic plant treatment is the possible development of phytoplankton blooms. The Israeli study indicates that the intelligent utilization of a number of fish species for biological control can negate these problems. It might be possible to substitute endemic species for some of those used in Israel for total weed control (macrophytes and algae) in this country.

Conclusion

The grass carp is a proven biological control for submersed weeds. Other species have been tried, but show little potential because of temperature requirements, reproduction rate, low consumption rate, or adverse environmental impact.

The environmental impact of any exotic fish used for aquatic weed control should be determined before release. Although the grass carp has been researched for a number of years, biologists are still concerned about the establishment of wild populations that could reach numbers great enough to remove macrophytes from areas where they are desired. For this reason the hybrid grass carp is being researched as a sterile replacement for the grass carp. The hybrid does not consume vegetation as voraciously as the grass carp; therefore, its use might be restricted to areas where weeds are controlled prior to stocking.

Literature Cited

Aliev, D. S. 1963. Experience in the use of white amur in the struggle against the overgrowth of water bodies, 89–92. *In:* Problemy Rybokhozyaystvennogo Ispol' zovaniya Rastitel'noyadnykh Ryb v Vodoyemakh SSSR (Problems of the Fisheries Exploitation of Plant-Eating Fishes in the Water Bodies of the USSR). Ashkhabad, Akad, Nauk Turkmen, SSR.

Aliev, D. S. 1976. What's new in the use of the biological method for preventing the overgrowth and siltation of collecting and drainage network canals, 297–308. *In:* Gidrobiologiya Kanalov SSSR i Biologicheskiye Pomekhi v Tkh Ekspluatasiya (The Hydrobiology of Canals of the U.S.S.R. and Biological Intervention in Their Operation). Naukova Dumka.

Alikunhi, K. H., and K. K. Sukumaran. 1964. Preliminary observations in Chinese carps in India. Proceedings of the Indian Academy Sciences 60B(3):171–189.

Avault, J. W., Jr., R. O. Smitherman, and E. W. Shell. 1968. Evaluation of eight species of fish for aquatic weed control. *In:* T.V.R. Pillay (ed.). 1968. Proceedings of the world symposium of warm-water pond fish culture. Rome, 18-25 May 1966. Food and Agriculture Organization Fisheries Reports 5(4):109–122.

Bailey, W. M. 1978. A comparison of fish population before and after extensive grass carp stocking. Transactions of the American Fisheries Society 107(1):181–206.

Blackburn, R. D., and D. L. Sutton. 1971. Growth of the white amur (*Ctenopharyngodon idella* Val.) on selected species of aquatic plants, 87–93. Proceedings of the European Weeds Research Council, 3d International Symposium on Aquatic Weeds.

Brooker, M. P., and R. W. Edwards. 1975. Aquatic herbicide and the control of water weeds: A review paper. Water Research 9:1–15.

Buck, D. H., R. J. Bauer, and C. R. Rose. 1978. Polyculture of Chinese carps in ponds with swine wastes, 144–155. *In:* R. O. Smitherman, W. L. Shelton, and J. H. Grover (eds.). 1978. Symposium on Culture of Exotic Fishes presented at Aquaculture/Atlanta/'78, Atlanta, Georgia, 4 January 1978.

Canfield, D. E., Jr. 1979. Prediction of total phosphorus concentrations and trophic states in natural and artificial lakes: The importance of phosphorus sedimentation. Ph.D. dissertation, Iowa State University. Ames, Iowa. 93 pp.

Colle, D. E., and J. V. Shireman. 1980. Coefficients of condition for largemouth bass, bluegill, and redear sunfish in hydrilla-infested lakes. Transactions of the American Fisheries Society 109:521–531.

Colle, D. E., J. V. Shireman, R. D. Gasaway, R. L. Stetler, and W. T. Haller. 1978. Utilization of selective removal of grass carp (*Ctenopharyngodon idella*) from an 80-hectare Florida lake to obtain a population estimate. Transactions of the American Fisheries Society 107(5):724–729.

Conner, J. V., R. P. Gallagher, and M. F. Chatry. 1980. Larval evidence for natural reproduction of the grass carp (*Ctenopharyngodon idella*) in the lower Mississippi River. Proceedings of the Fourth Annual Larval Fish Conference. 1980. U.S. Fish and Wildlife Service, Biological Services Program, National Power Plant Team, Ann Arbor, Michigan, FWS/OBS-80/43:1–19.

Crisman, T. L., and F. M. Kooijman. 1980. Large-scale operations management test using the white amur at Lake Conway, Florida. Benthos, 298–304. *In:* J. E. Decell (ed.). 1980. Proceedings, 14th Annual Meeting, Aquatic Plant Control Research Planning and Operations Review. Vicksburg, Mississippi. U.S. Army Engineering and Waterways Experiment Station.

Edmondson, W. T. 1971. Changes in Lake Washington following an increase in the nutrient income. International Vereinigung für Theoretische und Angewandte Limnologie Verhandlungen. 14:167–175.

Forester, T. S. 1975. Effects of white amur, *Ctenopharyngodon idella* (Valenciennes), and common carp, *Cyprinus carpio* Linnaeus, on populations of pond fishes. M.S. thesis, Auburn University, Auburn, Alabama. 49 pp.

Forester, T. S., and J. M. Lawrence. 1978. Effects of grass carp and carp on populations of bluegill and largemouth bass in ponds. Transactions of the American Fisheries Society 107(1):172–175.

Gasaway, R. D. 1977. The effects of grass carp on community structure in four Florida lakes. *In:* The grass carp: A special research report to the governor and the cabinet. Florida Game and Fresh Water Fish Commission, Tallahassee, Florida.

Haller, W. T. 1979. Aquatic weeds in Florida, 1–14. *In:* J. V. Shireman (ed.). Proceedings of the Grass Carp Conference. Gainesville, Florida, Aquatic Weeds Research Center, University of Florida, Institute of Food and Agricultural Sciences.

Hickling, C. F. 1962. Fish Culture. Faber and Faber, London. 296 pp.

Jones, J. R., and R. W. Bachmann. 1976. Prediction of phosphorus and chlorophyll levels in lakes. Journal of the Water Pollution Control Federation 48:2176–2182.

Kobylinski, G. J., W. W. Miley II, J. M. Van Dyke, and A. J. Leslie. 1980. The effects of grass carp (*Ctenopharyngodon idella* Val.) on vegetation, water quality, zooplankton, and macroinvertebrates of Deer Point Lake, Bay County, Florida. Department of Natural Resources, Tallahassee, Florida. 114 pp.

Legner, E. F., W. J. Hauser, T. W. Fisher, and R. A. Medved. 1975. Biological aquatic weed control by fish in the lower Sonoran Desert of California. California Agriculture News 29(11):8–10.

Lembi, C. A., and B. G. Ritenour. 1977. The white amur as a biological control for aquatic weeds in Indiana. Purdue University Water Resources Research Center, Technical Report 95. 95 pp.

Lembi, C. A., B. G. Ritenour, E. M. Iverson, and E. C. Forss. 1978. The effects of vegetation removal by grass carp on water chemistry and phytoplankton in Indiana ponds. Transactions of the American Fisheries Society 107(1):161–171.

Leventer, H. 1972. Eutrophication control of Tsalmon Reservoir by the cichlid fish *Tilapia aurea.* Proceedings of the 6th International Water Pollution Research. Pergamon Press, New York. A/8/15:1-9.

Leventer, H. 1979. Biological control of reservoirs by fish. Report of the Mekoroth Water Company, Jordan District, Central Laboratory of Water Quality, Nazareth Elit, Israel. 71 pp.

Michewicz, J. E., D. L. Sutton, and R. D. Blackburn. 1972. Water quality of small enclosures stocked with white amur. Hyacinth Control Journal 10:22–25.

Miller, J. L., L. A. Garrard, and W. T. Haller. 1976. Some characteristics of hydrilla tubers in Lake Oklawaha. Hyacinth Control Journal 14:26–29.

Nikolsky, G. V., and D. D. Aliev. 1974. Role of Far Eastern herbivorous fish in ecosystems of natural water bodies used for acclimatization. Voprosy Ikhtiologii 14(6):974–979. (translation from Russian in Journal of Ichthyology 14(6):842–847.)

Noble, R. L., R. D. Germany, and C. R. Hall. 1975. Interactions of blue tilapia and largemouth bass in a power plant cooling reservoir. Proceedings of the Annual Conference of the Southeastern Association of Game and Fish Commissioners 29:247–251.

Opuszyński, K. 1979. Weed control and fish production, 103–138. *In:* J. V. Shireman (ed.). 1979. Proceedings of the Grass Carp Conference. Gainesville, Florida, Aquatic Weeds Research Center, University of Florida, Institute of Food and Agricultural Sciences.

Pelzman, R. J. 1973. A review of the life history of *Tilapia zillii* with a reassessment of its desirability in California. California Department of Fish and Game, Administrative Report no. 74-1.

Rottmann, R. W. 1976. Limnological and ecological effects of grass carp in ponds. M.S. thesis, University of Missouri, Columbia, Missouri. 62 pp.

Rottmann, R. W., and R. O. Anderson. 1976. Limnological and ecological effects of grass carp in ponds. Proceedings of the Southeastern Association of Game and Fish Commissioners 30:24–39.

Schlindler, D. W. 1975. Whole-lake eutrophication experiments with phosphorus, nitrogen and carbon. International Vereinigung für Theoretische und Angewandte Limnologie Verhandlungen 19:3221–3231.

Shireman, J. V., D. E. Colle, and R. G. Martin. 1979. Ecological study of Lake Wales, Florida, after introduction of grass carp (*Ctenopharyngodon idella*), 49–90. *In:* J. V. Shireman (ed.). Proceedings of the Grass Carp Conference. Gainesville, Florida, Aquatic Weeds Research Center, University of Florida, Institute of Food and Agricultural Sciences.

Shireman, J. V., D. E. Colle, and R. W. Rottmann. 1978. Size limits to predation on grass carp by largemouth bass. Transactions of the American Fisheries Society 107(1):213–215.

Shireman, J. V., and W. H. Haller. 1980. The ecological impact of integrated chemical and biological aquatic weed control. Annual Report to the Environmental Protection Agency, Gulf Breeze, Florida.

Shireman, J. V., and M. J. Maceina. 1981. The utilization of grass carp (*Ctenopharyngodon idella*) for hydrilla control in Lake Baldwin. Journal of Fish Biology 19:629–636.

Shireman, J. V., and C. R. Smith. 1981. Biological synopsis of grass carp (*Ctenopharyngodon idella*). Final Report, Contract no. 14-16-0009-78-912, U.S. Fish and Wildlife Service, National Fishery Research Laboratory, Gainesville, Florida. 226 pp.

Sutton, D. L. 1980. Aquatic plant references and feed conversion efficiencies of the hybrid grass carp (*Hypothalmichthys nobilis* Rich. x *Ctenopharyngodon idella* Val.). Annual Report to Florida Department of Natural Resources, Tallahassee, Florida. 23 pp.

Sutton, D. L., W. W. Miley II, and J. G. Stanley. 1977. On sight inspection of the grass carp in the U.S.S.R. and other European countries. Report to Florida Department of Natural Resources, Tallahassee, Florida. 48 pp.

Terrell, T. T. 1974. Effects of white amur (*Ctenopharyngodon idella* Val.) on plankton populations and eutrophication rate of Georgia ponds. Georgia Cooperative Fishery Unit, School of Forest Resources, University of Georgia, Athens, Georgia. 13 pp.

Verigin, B. V. 1979. The role of herbivorous fishes at reconstruction of ichthyofauna under the conditions of anthropogenic evolution of water bodies, 139–146. *In:* J. V. Shireman (ed.). 1979. Proceedings of the Grass Carp Conference. Gainesville, Florida. Aquatic Weeds Research Center, University of Florida, Institute of Food and Agricultural Sciences.

Vinogradov, V. K., and Z. K. Zolotova. 1974. The effect of the grass carp on the ecosystems of waters. Gidrobiologicheskiy zhurnal 10(2):90–98 (translated from Russian in Hydrobiological Journal 10(2):72–78).

Yeo, R. R. 1967. Silver dollar fish for biological control of submersed aquatic weeds. Weeds 15:27–31.

CHAPTER 15

Exotic Fishes and Sport Fishing

G. C. Radonski, N. S. Prosser, R. G. Martin, and R. H. Stroud

The introduction of nonindigenous fishes has been an important management tool for fisheries workers since the earliest days of recreational fishery management. Nonnative fishes introduced into other waters have been divided into two groups depending upon geographic origin, that is, exotic and transplants. At an invitational conference jointly sponsored by the American Fisheries Society (AFS) and the American Society of Ichthyologists and Herpetologists (ASIH), 18–19 February 1969, in Washington, D.C., an exotic was defined as a species "introduced from a foreign country." The attendees further accepted that species moved by man between watersheds within the country of origin would be designated as "transplants" (Stroud 1969a).

Any discussion of sport fish management in the United States would, of necessity, soon focus upon the introduction and management of nonindigenous fish species. If one applies the accepted (AFS/ASIH) definitions, transplants have been vastly more widely used for recreational fisheries purposes than have exotics. We need only mention rainbow trout east of the Pacific drainage, striped bass in inland lakes, and Pacific salmon in the Great Lakes to appreciate the overwhelming benefits that transplanted fishes have had on recreational fishing opportunities.

About 1,200 to 1,500 tropical aquarium species have been, and continue to be, imported from Africa, Ceylon, China, Japan, Malaysia, and Central and South America. A detailed nationwide survey of exotic fishes in U.S. waters determined that some 84 exotic species have been collected from the open waters of the nation and 39 have established breeding populations

(Anonymous 1979). Only a few temperate-zone species of exotic fishes have been brought into the United States.

This, of course, does not imply that the impacts of exotics on U.S. waters and recreational fisheries have been slight. Fisheries scientists have learned from long and sometimes bitter experience that introduction of exotic organisms within a given aquatic ecosystem cannot be accomplished without repercussion. The nature of the community response may vary from insignificant to highly beneficial or highly destructive, depending on the characteristics of the exotic species and the biological and physicochemical parameters of target environments. In relation to recreational fisheries, exotics of two basic types have been introduced: species directly targeted as sport fish and those species that have, through alteration of physical, chemical, and/or biological parameters, indirectly affected recreational fishing.

Historical Review

A brief review of the scope of exotic fish introductions of major impact on sport fisheries may be helpful for perspective.

Before 1875

Before organization of a federal fisheries agency and various state fisheries (and/or wildlife) agencies, only a few exotic species were imported. These include the common carp (*Cyprinus carpio*) and the goldfish (*Carassius auratus*).

1877–1900

During the first few years of its operation, the U.S. Fish Commission encouraged and assisted the states in organizing state fisheries agencies. Around 1875, attention was given to the possibilities of foreign fish introductions. During the following twenty-five years, several species of fish were imported from the temperate waters of Europe, including both the common carp and the brown trout (*Salmo trutta*).

1900–1950

During the first half of the twentieth century, the disastrous consequences of the introduction of common carp generated widespread fear of exotic fishes. Importation of temperate-zone fishes virtually ceased.

1950–present

Following World War II, fisheries biologists and some sportsmen again began to consider the importation of foreign fishes that might prove useful in some way. Purposeful introduction of exotic species has been made to satisfy a variety of well-predicted (for the most part) objectives. These objectives include the desire to provide greater diversity of sport and/or food fishes, to achieve aquatic weed control, and to promote and provide more effective utilization of new ecological niches created by man's growing propensity for habitat alteration (e.g., reservoir construction, thermal discharge, accelerated eutrophication, etc.).

In 1951, the first of several species of tilapia (*Tilapia* spp.) was imported, with others following quickly. In 1963, the initial importation was made by the U.S. Fish and Wildlife Service of the grass carp (*Ctenopharyngodon idella*) from Asia. The importation of silver carp (*Hypophthalmichthys molitrix*), bighead carp (*Aristichthys nobilis*), and mud carp (*Cirrhina molitorella*) quickly followed.

Several additional species were imported for experimental pond and lake use. These included two species of peacock fishes (*Cichla ocellaris, C. temensis,* perhaps others) from South America, Ohrid trout (*Salmo letnica*) from Yugoslavia, Amur pike (*Esox reicherti*) from Asia, and Nile perch (*Lates nilotica*) from Africa.

Introductions of Major Exotics

Common carp and brown trout, both imported from the temperate climate of Europe, represent the two most successful introductions of exotic fish species affecting recreational fisheries interests.

Carp were apparently first brought to the United States from France in 1831 and 1832 by a private citizen. Some of those fish were released into the Hudson River near Newburgh, New York (Bowen 1970). Forty years later, more carp were reportedly brought from Germany to California. In any event, 345 carp were received from Germany by the U.S. Fish Commission in mid-1877. They were placed in ponds at Baltimore, Maryland, and later transferred to ponds in Washington, D.C. During the fall of 1879, carp fingerlings were distributed to thirty-eight states and territories by the U.S. Bureau of Fisheries. By 1897, with widespread public disenchantment, carp culture was discontinued at national fish hatcheries.

Despite their popularity with discriminating European anglers, carp are considered a pest by many anglers and recreational fisheries workers in the United States. Millions of dollars have been spent, particularly in the upper Mississippi River drainage states, on carp control programs. Contrary to conventional wisdom that carp possess little potential as a sport species, regular and intensive angling effort is exerted for carp in parts of the country.

Overall, however, introduction of the common carp to U.S. waters would have to be considered a monumental mistake and one with which we must learn to live. As necessity is the mother of invention, so carp may find growing acceptance by the average angler as recreational fishing opportunities supported by the more "traditional" sport fish species become inadequate to meet angling demands.

Brown trout was the other clearly successful introduction of an exotic species. It generally is considered beneficial, although often competitive with native trout populations. This environmentally more exacting, temperate-zone species is a highly prized game fish, and its commercial sale is prohibited in all states where it occurs. The first brown trout eggs were received from Germany in 1883 and reared in New York Fish Commission hatcheries. Additional eggs of brown trout (Loch Leven strain) were received the following year from Scotland. By the dawn of the twentieth century, the industry of American fish culturists had assured their introduction, nationwide, into virtually all suitable trout waters. In the process, through nondiscriminating propagation activities, the identity of various European strains and/or subspecies became hopelessly lost.

Self-sustaining populations of brown trout in many U.S. waters support significant recreational fisheries. Additional recreational fishing is provided by fingerling and/or put-and-take brown trout stocking programs in appropriate waters.

Several additional exotics of value or concern to recreational fisheries interests have been introduced into U.S. waters. Goldfish were stocked in the Hudson River in 1843 by a private citizen of Newburgh, New York. They are now commonly cultured for ornamental use and as bait minnows for farm pond use in some states. They are found in natural waters throughout most of the United States, occasionally in sufficient number to contribute as an important prey-base for predatory gamefish.

Four introduced European cyprinids continue to survive over extremely limited range: the tench (*Tinca tinca*) and the ide (*Leuciscus idus*), both found in Connecticut; and the bitterling (*Rhodeus sericeus*) and rudd (*Scardinius erythrophthalmus*) that survive in the state of New York (Lachner et al. 1970).

Although various *Tilapia* spp. were brought into the United States as aquarium fishes, it was not until 1954 that experiments in using members of this genus as southern pondfish were begun at Auburn University, Alabama (Swingle 1960). The species studied at Auburn included the Mozambique tilapia (*Tilapia mossambica*), the Nile tilapia (*T. nilotica*), and Malacca hybrid tilapia (*T. hornorum* x *T. mossambica*).

In the early 1960s, the Alabama Division of Game and Fish experimented with *T. mossambica* and *T. nilotica,* received from Auburn. Tilapia were introduced into selected public fishing lakes, both to provide forage for warmwater predator fish and to serve as a new hook-and-line sport fish. The results held some promise, but due to winterkill problems and findings that the species used did not provide a significant sport fishery, the program was discontinued after 1981. A new study was initiated in 1981 in Alabama.

Tilapia mossambica, and to a lesser extent *T. zilli,* have become established in some southwestern states. Important hook-and-line sport fisheries are supported in the Gila River and surrounding irrigation systems in Arizona, and in the Salton Sea in California. To date, no adverse impacts of these introductions on resident sport fish communities have been documented in these waters. In central and southern Florida, tilapia have overwintered out-of-doors, and several species including blackchin tilapia (*T. melanotheron*), blue tilapia (*T. aurea*), and spotted tilapia (*T. mariae*) are established in Florida waters and are currently expanding their respective ranges. Although limited sport netting occurs, *Tilapia aurea* is the most widespread exotic in Florida, and populations of this species have expanded in certain Florida and Texas waters to the point of adversely influencing reproduction of desirable centrarchid game species.

Grass carp are now widely distributed in the United States, and it is believed that natural reproduction has occurred in the Mississippi River drainage. The response of various fisheries scientists and managers to the introduction of the grass carp has run the gamut from enthusiastic endorsement to utter rejection. This species, probably better than any other exotic on the current scene, epitomizes both the potential "opportunities" and "problems" posed by exotic introductions. On the one hand, the grass carp promises to offer a substitute for costly mechanical and chemical control of millions of acres of weed-infested waters (a potential boon to recreational fisheries). On the other hand this potentially prolific species raises the specter of an uncontrollable population increase at the expense of more desirable indigenous sport fish species and waterfowl. The jury is still out regarding the eventual disposition of the case for or against the introduction of the grass carp.

Several other exotic fishes such as the silver carp, bighead carp, and mud carp have been imported to the North American continent in recent years. These species are deemed, by some, to possess weed control potential when stocked as a complement to, or as hybridized crosses with, the grass carp.

Five exotic species, in addition to the brown trout, have been identified as possessing direct potential as game fish and have been introduced into selected waters of the United States. Although eggs of the Danube salmon (*Hucho hucho*), obtained by a New York fish culturist in 1864, were hatched successfully, the liberated fry failed to survive (Baird 1874).

Ohrid trout eggs were secured from Yugoslavia in an international fish-exchange program carried on by the U.S. Fish and Wildlife Service in 1964–65. Resulting fingerlings were reared in federal fish hatcheries and then planted in several small lakes in northern Minnesota in the fall of 1965. One thousand three-year-old Ohrid trout, averaging 25 cm in length, were planted in four other small Minnesota lakes in 1968. The remainder of the Ohrid trout hatchery stock was then released in 1969 as four-year-olds into a 574 ha Big Trout Lake in northern Minnesota. It was hoped that they might be able to establish themselves through natural reproduction. Survival was poor to negligible, growth was slow, ultimate size was small, and eating qualities were poor. In short, the recreational potential of the species proved marginal at best (Stroud 1969b).

Amur pike were first imported into the United States from the Soviet Union in 1965 by the U.S. Fish and Wildlife Service. A second shipment of about 660,000 Amur pike eggs was received in 1969. In each instance, the U.S. Fish and Wildlife Service gave the Amur pike eggs to the Pennsylvania Fish Commission (Stroud 1969c). The commission released several thousand resulting Amur pike fingerlings into Glendale Reservoir, in southwestern Pennsylvania, which flows into the Conemaugh River, a badly polluted tributary to the Allegheny River (Abele 1975).

Recent experimental introductions of a species of peacock fish (*Cichla temensis*) from Colombia and Brazil and the Nile perch (*Lates nilotica*) from Africa are under investigation in Texas. The latter species attains a much larger ultimate size than the tucanare (*C. ocellaris*) that was released as a game fish by the Florida Game and Fresh Water Fish Commission in 1964 in Dade County, Florida. All of the tucanare released in Florida were apparently killed during the cold winter of 1964–65 (Courtenay and Robins 1973). The Nile perch is being evaluated for possible use as a sport-predator fish in some of the forty-one existing and nine planned reservoirs in Texas subject to warming by thermal effluents from steam-electric plants. Studies over several years, both in Africa and Texas, into the life history of the Nile perch and other African fishes have been conducted to determine possible adverse impacts of their introduction on native sport fishes such as largemouth bass (*Micropterus salmoides*), crappie (*Pomoxis* spp.), and catfish (*Ictalurus* spp.). Because the Nile perch is believed unable to survive at water temperatures below 16 C, a final control is presumably built into the study in that the fish (assuming failure to acclimatize) will be confined to power-plant cooling lakes. Water temperatures in the lakes during the winter can be lowered by the temporary halting of power production and killing the Nile perch if the species proves to be harmful in some way (Hubbs 1973).

In November 1974, a Texas Parks and Wildlife Department (TPWD) fisheries biologist brought back 300 one-month-old peacock fish (*Cichla temensis*) from Brazil for evaluation as a sport-predator fish for thermal discharge reservoirs. Several hundred peacock fish eggs were also collected directly from wild fish in the Vaupés River (a tributary to the Amazon), Colombia, and brought to Texas in mid-December 1974 (Anonymous 1975). After several years of study and culture in a Texas hatchery, in 1980 the TPWD stocked 4,110 fingerling South American peacock fish into Coleto Creek Reservoir near Victoria, Texas. Also stocked were 50 one-year-old surplus brood fish. Coleto Creek Reservoir has not yet been opened for fishing. Previous stockings in Lake Bastrop failed because of winter power plant shutdowns that allowed the water temperature to drop below 17 C which is minimal for peacock fish.

Conclusion

Other species will undoubtedly be imported by various state, federal, and private agencies in the months and years immediately ahead. The principal concern is that any and all such activities be undertaken with great care and regard for possible harmful side-effects as well as possible benefits. Introducing a fish species to a body of water outside of its historic range should carry the same responsibility, whether the species is designated exotic or transplant. The present body of legal responsibilities centers on the AFS/ASIH definition of exotic.

The introduction of nonnative fishes, whether exotic or transplant, is an important sport fishery management tool. The responsibility for such introduction must not be taken lightly. The criterion that the fish is "valuable" is not sufficient reason to introduce an exotic species. For example, the European carp, when introduced into American waters, was considered valuable. The use of exotics in sport fishery management must be predicated on the ability to predict reasonably the impact of the introduced exotic species on existing biota and their environs.

Certainly, there is no need for haste in making introductions of exotic species. On the contrary, there is every reason to carry out exhaustive, long-term evaluations of the proposed introductions. The fact must be kept uppermost in mind that, once accomplished, the new introduction is virtually impossible to eradicate and, thereby, an unanticipated ecological disaster is virtually impossible to rectify.

In recognition of the absolute necessity for control of exotic fishes imported as sport fish or for any other purpose, the Board of Directors of the Sport Fishing Institute (SFI) adopted the following resolution at their 1973 Annual Meeting:

> RESOLVED, that the Directors of the Sport Fishing Institute, assembled in regular Annual Session at Montreal, on May 15, 1973, do herewith urge the U.S. Department of the Interior to prohibit the importation into the United States, except for well-controlled scientific study purposes, of all exotic fishes other than those that can be proven to lack harmful ecological effects upon the natural aquatic environments of the United States and the native fauna and flora found therein.

Tighter regulations governing importation of exotic fishes into the United States were proposed by the U.S. Fish and Wildlife Service as a direct consequence. Proposed regulations were first published in December 1973. A tremendous volume of related comment was subsequently received from interested and affected parties by the Department of the Interior.

The 1973 SFI board resolution was followed with another resolution on exotic fish at its 1976 Annual Meeting:

RESOLVED, that the Board of Directors of the Sport Fishing Institute, assembled in regular Annual Meeting this 11th day of May, 1976, at Kansas City, Missouri, herewith respectfully request the Secretary of the Interior to develop an appropriate protocol to govern procedures for evaluating probable effects on native fauna and habitats of importation from foreign countries and introduction into America of exotic fish species, and to examine the language of the Black Bass Act (16 U.S.C. 851-856) for its evident inadequacies with respect to needed authority for desirable control over interstate shipment of exotic fish species, and to propose suitable strengthening amendments to the Black Bass Act that would permit appropriate control over the interstate shipment of exotic fish species.

In large measure the preceding SFI resolution resulted in the establishment of the National Fishery Research (Exotic Fish) Laboratory in Gainesville, Florida. Recently (1982) the Black Bass Act received the critical review by the U.S. Congress called for in the resolution.

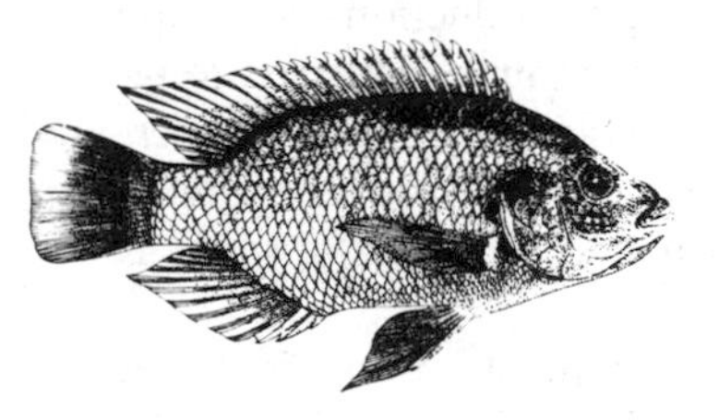

Literature Cited

Abele, R. 1975. Verbal communication of March 19, 1975. Director, Pennsylvania Fish Commission, Harrisburg.

Anonymous. 1975. Second batch of peacock bass in Texas. Texas Parks and Wildlife News (1/8/75). Texas Parks and Wildlife Department, Austin.

Anonymous. 1979. Fisheries and wildlife research 1979. Fish and Wildlife Service, U.S. Department of the Interior Annual Report.

Baird, S. F. 1874. Report of the Commissioner of Fish and Fisheries for 1872 and 1873. U.S. Department of Commerce, Washington, D.C.

Bowen, J. T. 1970. A history of fish culture as related to the development of fishery programs 71–94. *In:* A century of fisheries in North America. Special Publication 74, American Fisheries Society, Washington, D.C.

Courtenay, W. R., Jr., and C. R. Robins. 1973. Exotic aquatic organisms in Florida with emphasis on fisheries: A review and recommendations. Transactions of the American Fisheries Society 102(1):1–12.

Hubbs, C. 1973. Letter to Dr. Herbert R. Axelrod. (June 1). Department of Zoology, University of Texas, Austin.

Lachner, E. A., C. R. Robins, and W. R. Courtenay, Jr. 1970. Exotic fishes and other aquatic organisms introduced into North America. Smithsonian Contribution to Zoology 59:1–29.

Stroud, R. H. 1969a. Conference on exotic fishes and related problems. SFI Bulletin no. 203:1–4 (April). Sport Fishing Institute, Washington, D.C.

Stroud, R. H. 1969b. Ohrid trout tests, SFI Bulletin no. 210:7–8 (November-December). Sport Fishing Institute, Washington, D.C.

Stroud, R. H. 1969c. Exotic Amur pike. SFI Bulletin no. 205:7 (June). Sport Fishing Institute, Washington, D.C.

Swingle, H. S. 1960. Comparative evaluation of two tilapias as pond fishes in Alabama. Transactions of the American Fisheries Society 89(2):142–148.

CHAPTER 16

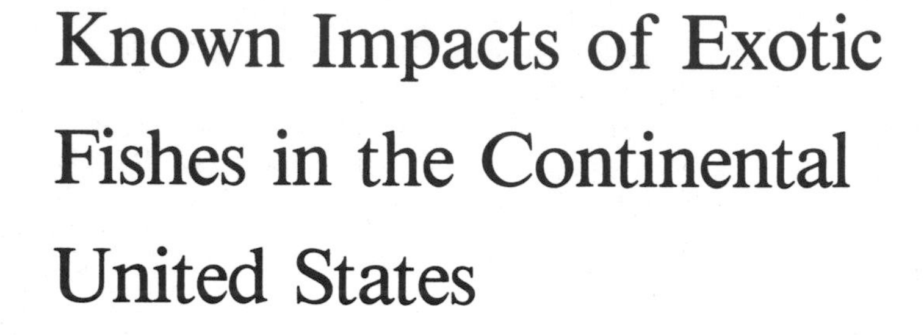

Known Impacts of Exotic Fishes in the Continental United States

Jeffrey N. Taylor,
Walter R. Courtenay, Jr., and
James A. McCann

Forty species of exotic fishes are presently known to have established one or more breeding populations in the fresh and coastal waters of the continental United States. Three species of temperate origins—brown trout (*Salmo trutta*), goldfish (*Carassius auratus*), and common carp (*Cyprinus carpio*)—are widely distributed, although in many states brown trout populations are maintained almost entirely by annual stocking. The other 37 species are predominantly representatives of families from tropical or semitropical regions (e.g., Cichlidae, Clariidae, Loricariidae, Poeciliidae) whose distributions in the United States are more restricted. Largely because of their limited tolerances for low temperatures, these species occur most frequently in states with mild climates, such as Florida (with 20 established species), California (17 species), Nevada (11 species), Texas (8 species), and Arizona (7 or more species) as reported by Courtenay et al. (chapter 4).

Since the establishment of the goldfish in the late 1600s (DeKay 1842), the rate at which exotic fishes have been introduced into U.S. waters has increased alarmingly (Courtenay and Hensley 1980); that knowledge of the distribution of existing populations is far from complete and that releases of exotic forms continue to occur are clearly indicated by recent discoveries of established populations of spotted tilapia (*Tilapia mariae*) and redbelly tilapia (*T. zilli*) in Nevada, Midas cichlid (*Cichlasoma citrinellum*) and redstriped eartheater (*Geophagus surinamensis*) in south Florida, and presently unidentified *Tilapia* species in California and Arizona.

Successful establishment of exotic fishes can result from either intentional or accidental releases (Courtenay and Robins 1973, 1975; Courtenay et al. 1974). Intentional introductions have been made for essentially four reasons: (1) to add new food, game, and forage species to native fish assemblages; (2) to attempt biological control of unwanted (often exotic) pest species; (3) for ornamental or other aesthetic purposes; and (4) as a means of releasing unwanted pets. Most accidental releases can be attributed to escapes of exotic fishes from private ponds, stock tanks, and culture facilities, either during periods of flooding or by way of improperly screened drainage pipes. Regardless of the means by which introductions are effected, successful establishment of an exotic species must necessarily precipitate changes in the physical and biological characteristics of the aquatic ecosystem receiving the introduction (Elton 1958, Laycock 1966, Magnuson 1976, Martin 1976, Courtenay and Hensley 1980, Li and Moyle 1981, Simberloff 1981). However, the exact nature of such changes is largely unpredictable (DeVos et al. 1956, Laycock 1966, Regier 1968, Magnuson 1976, Courtenay 1979, Shafland 1979, Philippart and Ruwet 1982).

Ecology of Exotic Introductions

Effects that exotic fish species can have in a novel aquatic environment depend on (a) the physiological, behavioral, and ecological potentialities of the exotic and (b) the physical and biological properties of the ecosystem receiving the introduction. Successful alien colonists often have physiological tolerances, feeding habits, or reproductive capabilities that are different from those of native forms (table 16–1) and that preadapt them for establishment in novel environments; such preadaptations may contribute to replacements of native species (resulting from differences in population growth rates), irrespective of the nature and severity of interactions that may also occur between an exotic and sympatric natives. Indeed, among intentionally introduced species, those that exhibit such attributes (e.g., the common carp and the tilapias) have routinely been selected for stocking purposes. The importance of such properties is well illustrated by the example of the brown trout which differs from native trouts in several important characteristics: it is more tolerant of high stream temperatures (Mather 1889, Cobb 1933, Needham 1938, Fenderson 1954, Dymond 1955, Sigler and Miller 1963, Harlan and Speaker 1969, Vincent and Miller 1969, Brynildson et al. 1973, Gard and Flittner 1974); feeds actively throughout the year (Millard and MacCrimmon 1972, Brynildson et al. 1973, Alexander 1979); experiences lower

Table 16–1. ATTRIBUTES OF EXOTIC SPECIES THAT PREADAPT THEM FOR SUCCESSFUL COLONIZATION AND POPULATION GROWTH IN NOVEL ENVIRONMENTS

I. Broad physiological tolerances for
   A. temperature extremes
   B. low oxygen levels
   C. fluctuations in salinity
   D. turbidity and pollution
   E. drought

II. Feeding habits and diet
   A. diet composition
   B. feeding schedules
   C. vulnerability to predation

III. Reproductive behavior
   A. rapid growth and maturation
   B. extended or continuous breeding
   C. multiple clutches
   D. advanced parental care

winter mortality (Smith and Smith 1945, Lemmien et al. 1957), and has a greater resistance to siltation, pollution, and other forms of stream modification (Smedley 1938, LaRivers 1962, Harlan and Speaker 1969, Marshall and MacCrimmon 1970, Nyman 1970, Brynildson et al. 1973, Moyle 1976a). Furthermore, survival of the brown trout is enhanced by its lower vulnerability to angling pressures than that of native trouts (Cobb 1933; Hubbs and Eschmeyer 1938; Needham 1938; Schuck 1942; Shetter and Hazzard 1942; Smith and Smith 1945; Wales 1946; Shetter 1947, 1948, 1968; Thorpe et al. 1947; Neave 1949; Cooper 1952; 1953a and b, 1959; Fenderson 1954; Hale 1954; Vestal 1954; Dymond 1955; Wales and German 1956; Lemmien et al. 1957; Miller 1957a and b; McCraig et al. 1960; La Rivers 1962; Pister 1962; Nelson 1965; Shetter and Alexander 1965; Staley 1966; Jenkins 1969; Vincent and Miller 1969; Marshall and MacCrimmon 1970; Nyman 1970; Millard and MacCrimmon 1972; Brynildson et al. 1973; Moyle 1976a; Allan and Roden 1978).

With regard to reproductive traits, many introduced cichlids exhibit short generation times, multiple clutches, and extended breeding seasons in the United States (table 16–1). For example, spotted tilapia and black acara (*Cichlasoma bimaculatum*) have been observed to breed during every month of the year in south Florida canals (Hogg 1976, personal observation [JNT]), whereas native species—geologically recent colonists of peninsular Florida from more temperate regions—mature more slowly and have strictly delimited breeding seasons. Guarding of free-swimming young, as seen in cichlids and the walking catfish (*Clarias batrachus*), may also enhance survivorship over that of natives with less advanced (or no) parental care. Our observations in Florida suggest that spotted tilapia, black acaras, and firemouth cichlids (*Cichlasoma meeki*) are all less prone to nest desertion and subsequent losses to egg predators than are native sunfishes of the genus *Lepomis*. Such anecdotes indicate that effects of differential reproductive capabilities may play a key role in accounting for patterns of replacement of native fishes by exotics in the United States.

The "structure" of a community receiving an introduction is important in determining whether an exotic becomes established. Transfer of an exotic species typically involves removal of individuals from conditions under which the species evolved and adapted, thereby releasing them, to some unpredictable extent, from the effects of factors that determine the species' density and distribution within its native range (e.g., climatic regime, predators, competitive interactions, diseases and parasites). Release from these constraints is often compounded by the absence of comparable limitations in the community into which an exotic is introduced; an alien species may then express broader flexibility in characteristics that may enhance its chances for survival in a new environment (i.e., habitat utilization, feeding behavior, diet, reproductive schedules, or growth potential). Probabilities of successful colonization are relatively high both in simple communities with low diversities of native species (Minckley and Deacon 1968, Shafland 1979) and in those subject to physical or biological disturbances (Elton 1958, Sigler 1958, Courtenay and Robins 1973, Moyle and Nichols 1974, Moyle 1976a and b, Parsons 1982).

Given the characteristics of an alien species and of the ecosystem into which it is released, the introduction of an exotic may result in one of several outcomes. Survival requires that initial pulse size be adequate to ensure location and procurement of suitable mates; if this prerequisite is not met, failure to establish is inevitable. Even if enough individuals are introduced to provide for continued recruitment, extinction may still occur if the introduced form is incapable of adapting to ecological conditions in its new environment; indeed, attempted introductions of fishes – and other animal groups – have failed far more frequently than succeeded (Larkin 1979, Simberloff 1981). If extinction is avoided, a small, highly localized deme may develop, held in check at low densities by one or more limiting factors (perhaps, but not necessarily, similar to those that delimited the species' distribution within its native range). Alternatively, if such factors are not operative, local increases in density occur as reproduction proceeds. As population size increases, density-dependent regulatory factors may begin to manifest themselves, providing an upper limit to population growth. However, in addition to density increases, dispersal and range expansion should occur, unless confinement within a closed system precludes this possibility. As establishment, population growth, and range expansion take place, the nature, frequency, and complexity of interactions possible with different components of the native community multiply rapidly. Clearly, the effects that an exotic species can or will have in a new environment depend directly on opportunities for interaction with native species, which in turn are a function of the relative densities of the species concerned. In this regard, extremely high densities, as characterized by population "explosions" (Elton 1958), would be expected to result in the most far-reaching effects on native communities.

The point of the preceding discussion is simple. The ecological principles that govern the outcomes of exotic introductions (whether accidental or intended) are identical to those that operate in natural colonization events or instances of range expansions involving native forms. In the same vein, "impact studies" are basically ecological analyses of the nature and results of interactions between introduced exotics and native species. As used here, the *impact* of an exotic fish species on a freshwater community is defined as any effect attributable to that exotic that causes – directly or indirectly – changes in the density, distribution, growth characteristics, condition, or behavior of one or more native populations within that community. In the remainder of the chapter, we attempt to review and evaluate the nature, degree, and frequency of impacts that exotic fishes are believed to have had on freshwater fish assemblages in U.S. waters.

Before we proceed, the definition proposed above requires further comment. First, the definition is independent of human judgments as to the "value" of an introduction; impacts are nothing more than *changes* in native populations brought about by the activities of an exotic species. Benefits of introductions, in human terms, are discussed at length in accompanying chapters on aquaculture, vegetation control, and sport fishing. Second, a demonstration of impact, as employed here, implies verification of a causal relationship between changes in a native population and the presence of an exotic. Rigorous documentation of a cause-effect relation requires an experimental design in which appropriate controls and replications are used (see DeBach 1974 for a discussion of experimental methods for evaluating biological control of exotic insect pests). Although the strongest possible evidence for impact is derived from experimental manipulations, the utility of such an approach has been compromised by a number of practical considerations. For example, the development and application of protocols and methodologies for evaluating proposed introductions before the release of the animals into open waters have only too recently become a concern in the United States (Courtenay and Robins 1973, Li and Moyle 1981; see also chapter 18). Also, the nature of interactions between exotics and natives may vary along a continuum from simple to complex, and impacts on natives may be direct or secondary. Although some effects (e.g., predation) are readily documented, more subtle changes, such as shifts in activity schedules or decreased growth rates, may be easily overlooked or difficult to measure. Moreover, indirect effects on natives may occur secondarily as results of changes in habitat structure precipitated by activities of introduced species – an occurrence that further complicates efforts to establish causality. Finally, conditions under which experiments are conducted may be oversimplified and even artificial, compared with conditions in nature, and the extent

Table 16–2. CLASSIFICATION OF ECOLOGICAL EFFECTS OF INTRODUCED, EXOTIC FISHES ON NATIVE AQUATIC COMMUNITIES

I. Habitat alterations
  A. Removal of vegetation
    1. by consumption
    2. by uprooting
    3. by increasing turbidity
  B. Degradation of water quality
    1. by siltation
    2. by substrate erosion
    3. by eutrophication

II. Introduction of parasites and diseases

III. Trophic alterations
  A. Forage supplementation
  B. Competition for food
  C. Predation

IV. Hybridization

V. Spatial alterations
  A. Aggressive effects
  B. Overcrowding

to which experimental results can be applied to actual communities may be open to question. To counteract this possibility, efforts to reproduce conditions in nature should routinely be incorporated in the design of impact-oriented experiments. Ideally, manipulations should be carried out in natural settings; if this is not practical, large outdoor ponds are the best alternative.

In the absence of experimental analyses, other approaches have been applied to studies of impacts of exotic fish introductions. Comparative methods, involving simultaneous observations of differences in densities or behaviors of natives in the presence of an exotic and nearby areas where the exotic does not yet occur, can provide evidence for impact to the degree that alternative causal agents can be eliminated as explanations for observed patterns of differences. Evidence of impact derived from comparative studies is strengthened whenever a relation between a native and the presence or absence of an exotic is documented repeatedly in independent circumstances, or when changes in the exotic's population structure (whether induced or fortuitous) are accompanied by coordinate fluctuations in a native population (e.g., see below the account of interactions between blue tilapia, *Tilapia aurea,* and largemouth bass, *Micropterus salmoides,* in Trinidad Lake Reservoir, Texas, under section on Overcrowding). Comparative approaches usually are not capable of revealing *mechanisms* underlying patterns of differences; causality may be obscured by multifaceted interactions in which a number of levels or effects (see table 16–2) operate concurrently. In such cases, appropriate experimentation is necessary to verify causal interrelations.

Less convincing as a form of evidence for impact—because of failure to establish causality—is the occurrence of a correlation between the introduction of an exotic and a subsequent change in a native population. Unfortunately, as we illustrate later, much of the proposed documentation of impact put forward to date is of exactly this nature. Such a situation stems in large part from the ad hoc nature of many nonexperimental impact studies. Initiation of efforts to document impact at a site *after* an exotic has already been released and become established not only lacks requisite controls, but is certain to suffer from the criticism that the operation of alternative causal agents—other than the presumed influence of the exotic—has not been taken into consideration. On a positive note, establishment of a correlation between some characteristic of a native population and the presence of an exotic, though insufficient as a demonstration of impact, can be useful as a means of identifying productive avenues of experimentation aimed at verifying causality.

Ecological Classification of Impacts of Exotic Fishes

Moyle (1976b) grouped impacts of introduced exotic fishes into two classes: ecosystem alteration and reduction or elimination of native fish populations. Impacts of the first type included habitat modifications precipitated by activities of introduced forms that in turn led to effects on native populations; those of the second class (e.g., competition, predation, and hybridization) affected native populations directly. In the context of terminology employed here, Moyle's ecosystem or habitat alterations are viewed as "effects" of exotic fishes that may cause, but do not themselves constitute, "impacts" on native fishes. Other classes of possible effects are outlined in table 16–2; each category (except diseases and parasites, treated in chapters 11 and 12) is briefly discussed here.

HABITAT ALTERATIONS

Modifications of habitat structure by exotic fishes relate primarily to removal of aquatic vegetation and degradation of water quality. Vegetation levels can be measurably altered in three ways: consumption of plants by herbivorous species; uprooting of macrophytes as a result of digging activities associated with foraging or nesting; and roiling, organic enrichment, and other activities that increase turbidity and thereby reduce light penetration and photosynthesis. Alterations in levels of aquatic plant populations by the actions of exotic fishes may have significant consequences for native fishes, as well as for other components of aquatic communities. Plants provide oxygen (reviewed by Hubbs and Eschmeyer 1938), substrate and nutrients for an invertebrate community that in turn serves as important prey resources for fishes, egg-deposition sites for certain species, protective refugia for early life-history stages of many indigenous species, and foraging habitat for waterfowl and aquatic mammals. Barnett and Schneider (1974) found that, during the growing season, dense submerged aquatic plant communities in several central Florida lakes supported higher numbers of individuals and total biomass per unit area – particularly of forage species and juvenile game fishes – than did most other aquatic habitats; as vegetation died in winter, large numbers of these fishes became available to native predators. Savino and Stein (1982) provided experimental evidence that vegetation effectively reduces predation by largemouth bass on juvenile bluegills (*Lepomis macrochirus*). The limnological roles of macrophytes in primary production and food webs in aquatic ecosystems were reviewed by Boyd (1971). Further information on the biological significance of aquatic plants in biotic relations was summarized by Wilson (1939), Hotchkiss (1941), Sculthorpe (1967), Provine (1975), Ware et al. (1975), Gasaway and Drda (1977), Nall and Schardt (1978), and Shireman and Maceina (1981). Although excessive development of macrophytes can have detrimental effects on aquatic ecosystems (Hubbs and Eschmeyer 1938, Hogan 1946, Vinogradov and Zolotova 1974), examples of extreme overabundance seem to be most frequent in man-made bodies of water into which (ironically) exotic plant species have been introduced, particularly in southern states. Even in such instances, the importance of aquatic vegetation is reflected by recent recognition that management efforts should be focused on control rather than on elimination of plant life (Stott and Orr 1970, Sneed 1971, Pardue 1973, Terrell and Terrell 1975, Addor and Theriot 1977; see also chapter 14).

Reductions in macrophytes (whether through activities of exotic fishes or by other means) can precipitate increases in turbidity through wave-mediated erosion and continual mixing of silt previously stabilized by rooted aquatics. Deterioration in water quality, particularly a reduction in clarity, can also result directly from the actions of introduced fishes. Roiling of shallow littoral zones commonly accompanies foraging behavior of such bottom-feeding species as the common carp. Nesting and spawning activities, especially in species that form dense aggregations during reproductive periods, can also lead to at least temporary declines in clarity. Where extremely high densities of exotic fishes occur, accumulation and subsequent decay of organic wastes may stimulate phytoplankton blooms that in turn may cause imbalances in mineral and nutrient cycling within the ecosystem. Hestand and Carter (1978) proposed that increased phytoplankton production resulting from recycling of nutrients released through rapid destruction of macrophytes might be more damaging to water quality than the presence of dense stands of weeds.

Effects of increased turbidity on native fishes could be manifested in various ways. Wallen (1951) and Cordone and Kelley (1961) provided extensive literature reviews of the detrimental effects of turbidity on fishes and other organisms, citing examples of disruptions in reproductive activities (e.g., desertion of nests, elimination of spawning, egg mortality) and physiological stresses involving interference with normal respiratory and secretory functions. Schneberger and Jewell (1928) observed a negative correlation between turbidity and fish production in a series of twenty farm ponds. By exposing fishes to differing levels of turbidity in experimental ponds, Buck (1956) demonstrated lower production, slower growth, and less recruitment in centrarchids in ponds with high turbidities; similar trends were noted in comparisons of populations from turbid and clear lakes. Saunders and Smith (1965) found that low-standing crops of brook trout (*Salvelinus fontinalis*), associated with silting in a stream in Prince Edward Island, recovered markedly after the stream was scoured to reduce siltation. In laboratory experiments, Grandall and Swenson (1982) recorded differential behavioral responses to silt levels in brook trout and creek chubs (*Semotilus atromaculatus*) which could result in spatial segregation between the two species in natural habitats; similarly, Heimstra et al. (1969) observed differential effects of silt turbidity on behavioral activities in juvenile largemouth bass and green sunfish (*Lepomis cyanellus*) in aquaria. Concern over deleterious effects that introductions of exotic fishes might have on water quality is further warranted by historical records that repeatedly document a correlation between habitat degradation (largely attributable to increases in siltation and turbidity) and declines in population densities and species diversity in native freshwater fish faunas (see, for example, Trautman 1981).

TROPHIC ALTERATIONS

Introductions of exotic fishes may alter trophic relationships within aquatic communities in at least three ways, all of which can effect changes in native fish populations. First, successful establishments of exotics may result in quantitatively significant increases in forage available to native predator species; however, even if an overall increase in forage biomass can be demonstrated, a decline in prey quality could conceivably occur if replacement of native forage species accompanies establishment of an exotic (see discussion by Shafland and Pestrak 1984). Second, feeding activities of exotics can alter prey resources available to native species. Many introduced species, particularly of the family Cichlidae, express broad flexibility and opportunism in their feeding habits in newly occupied environments, leading to considerable dietary overlap with native fishes. If food resources are in short supply, such overlap provides a basis for competitive interactions that may lead to changes in diet, growth, condition, or fecundity in the less adept competitor. More subtle trophic effects may occur as a result of the removal of vegetation through activities of exotic species. Weed beds often harbor diverse communities of organisms that are commonly used as food by many native fishes; consumption or uprooting of vegetation by exotics effectively removes these communities, forcing natives to locate and exploit alternate sources of food. And third, population structures of native species can be altered by foraging activities of introduced predators, though the precise outcomes of such interactions may vary, depending on such factors as the effectiveness of the piscivore, relative densities of prey and predator (plus other factors that regulate their population levels), and availability of alternate prey resources and refugia.

HYBRIDIZATION

Successful establishment of an exotic species can lead to interactions during reproductive activities that result in impacts on native fish populations. Hybridization with subsequent deterioration of native genetic stocks is the most clear-cut example of such occurrences; however, probabilities of hybrid-cross formation depend largely on close phylogenetic affinities, an exigency not often met, given the largely neotropical origins of most species introduced into temperate North America. Hybridization among introduced exotics (e.g., tilapias), however, remains a problem to those of us trying to identify introduced fishes.

SPATIAL ALTERATIONS

Given sufficiently high densities, interactions between exotic and native populations may become frequent enough to affect directly the distribution, density, and survival of native fishes. Interspecific competition mediated by aggressive dominance or territoriality may lead to displacement from preferred microhabitats (e.g., feeding or resting stations, spawning sites, and refugia) or activity periods. Also, more subtle (though equally consequential) effects on natives may occur in response to overcrowding by exotics. Cessation of reproduction in the presence of high densities of sunfishes (genus *Lepomis*) has been repeatedly demonstrated in populations of largemouth bass and other native species (Swingle 1957c, Chew 1972, Barwick and Holcomb 1976, Smith 1976, Dean and Bailey 1977). Diminished growth and reproduction in native populations resulting from overcrowding by exotics capable of building high population densities (e.g., common carp and tilapias) are distinct possibilities deserving considerably greater attention in future research than they have received in the past.

Examples of Known Impacts of Exotic Fishes on Native Aquatic Communities

In reviewing evidence now available for the occurrence of impacts of exotic fishes on aquatic communities in the United States, we again follow the classification outlined in table 16–2. For each of the categories listed, examples of impact studies are presented on a species-by-species basis. We attempt to distinguish among documented impacts, anecdotes, and speculations; comments on methodologies used (e.g., experimental, comparative, or correlational) are appended whenever pertinent. Although no exhaustive review has been attempted, we have included a number of references to research on exotic fishes done outside the United States. Since individual examples are rarely discussed in detail, the reader is encouraged to consult original literature for more comprehensive accounts of specific studies.

HABITAT ALTERATIONS

Removal of vegetation

Several species of herbivorous fishes have been intentionally released into U.S. waters for the purpose of controlling nuisance growths of (often exotic) aquatic vegetation (Courtenay and Robins 1973). Foremost among these is the grass carp, *Ctenopharyngodon idella,** which is now established and reproducing in the lower Mississippi River (Conner et al. 1980; see chapter 4), although high population densities have not yet been reported. Though both observational and experimental studies have been conducted under highly variable conditions, they substantiate that grass carp when stocked at appropriate densities are effective in controlling aquatic macrophytes (Hickling 1961, 1965; Stroganov 1963; Avault 1965a and b; Pentelow and Stott 1965; Stevenson 1965; Avault et al. 1968; Nair 1968; Cross 1969; Crowder and Snow 1969; Prowse 1969, 1971; Sills 1970; Stott and Orr 1970; Stott and Robson 1970; Cagni et al. 1971; Kilgen and Smitherman 1971; Sneed 1971; Bailey 1972, 1978; Michewicz et al. 1972a; Opuszyński 1972; Greenfield 1973; Johnson and Laurence 1973; Sutton and Blackburn 1973; Terrell and Fox 1974; Vinogradov and Zolotova 1974; Buck et al. 1975; Terrell and Terrell 1975; Ware et al. 1975; Beach et al. 1976; Drda 1976; Rottmann and Anderson 1976; Gasaway and Drda 1977, 1978; Lembi and Ritenour 1977; Fowler and Robson 1978; Hestand and Carter 1978; Kilgen 1978; Lembi et al. 1978; Lewis 1978; Mitzner 1978; Osborne 1978; Miley et al. 1979; Kobylinski et al. 1980; Schardt and Nall 1981; Shireman and Maceina 1981; see also chapter 14); however, less accord has been reached with regard to the impacts that weed removal by grass carp has had on native aquatic communities. Given that the extent of impact should be related to the degree of vegetation control desired, complete eradication of macrophytes could have serious ramifications, resulting in loss of food sources, shelter, and spawning substrates for certain native fishes and other aquatic organisms (Hubbs and Eschmeyer 1938, Provine 1975, Gasaway and Drda 1977). However, in recent weed removal programs, management of vegetation at particular levels, rather than eradication, has been the primary objective of biological control efforts

* For a comprehensive bibliography covering all aspects of research on the grass carp in the United States and abroad, see Smith and Shireman (1981).

involving grass carp (Stott and Orr 1970, Sneed 1971, Pardue 1973, Terrell and Terrell 1975, Addor and Theriot 1977, Miley et al. 1979, Shireman and Maccina 1981). Toward this end, selection of initial stocking densities and properly sized individuals is geared to desired levels of control. Such adjustments are most readily managed in monocultures, occurrences that typically characterize weed infestations (often by exotic plants) in man-made waterways. Weed problems in natural ecosystems, however, usually involve a heterogeneous community of aquatic macrophytes, though one or a few species may predominate. In these circumstances, feeding selectivity in grass carp becomes a vital consideration in stocking to control the density of vegetation. When exposed to a range of plant species simultaneously, grass carp commonly express definite dietary choices or preferences (Stroganov 1961, 1963; Avault et al. 1968; Cross 1969; Prowse 1969; Stott and Robson 1970; Cagni et al. 1971; Michewicz et al. 1972b; Opuszyński 1972; Sutton 1972; Sutton and Blackburn 1973; Provine 1975; Lembi and Ritenour 1977; Colle et al. 1978; Fowler and Robson 1978; Hestand and Carter 1978; Lembi et al. 1978; Mitzner 1978; Miley et al. 1979; Kobylinski et al. 1980; Schardt and Nall 1981; Smith and Shireman 1981). Consequently, given appropriate stocking densities, the fish consumes weed species in a particular sequence, and that sequence may be independent of the relative densities of the plants available for consumption. Therefore, management problems arise if the plant species targeted for control are not preferred by grass carp. At low stocking densities, subsistence on preferred (but nontarget) species may actually allow target plant species to increase in biomass; conversely, stocking densities required for control of problem species may result in elimination of a number of nontarget species before foraging on targeted forms begins (Ware et al. 1975, Fowler and Robson 1978, Kobylinski et al. 1980, Smith and Shireman 1981).

In addition to decimation of vegetation, effects of grass carp herbivory may include alterations in water quality, which may in turn lead to impacts on native fishes. Hestand and Carter (1978:44) pointed out that

> since aquatic macrophytes act as a reservoir for inorganic nutrients, their rapid destruction results in the recycling of nutrients into solution making them available for increased phytoplankton production . . . which could be more damaging to water quality than dense stands of macrophytes. . . . Stimulation of the phytoplankton community generally results in an increase in the number of individuals, changes in phyla, and the numbers of phyla present. One symptom of eutrophication is the shifting of the phytoplankton association towards one which contains large numbers of a few undesirable species such as blue-green algae. Thus it would appear that the introduction of grass carp into an aquatic ecosystem could cause major changes in that system.

Increased primary productivity and associated effects on the structure of phytoplankton communities would be expected to stimulate concomitant changes in populations of zooplankton, macroinvertebrates, fishes, and other vertebrate populations as a result of alterations in water quality characteristics, food webs, and energy-flow relations.

As witnessed by references cited previously, habitat alteration through grass carp herbivory has been amply demonstrated. To what extent have such alterations led to changes in native fish populations?

As herbivores, adult grass carp display diet specializations not shared by native North American fishes; thus, impacts on natives should, to a large degree, be indirect, resulting from removal of submerged macrophytes and concurrent changes in water quality (Bailey 1978). When effects on fishes have been secondarily induced by actions of grass carp, establishing causality has been difficult; in fact, to date, evidence of effects of vegetation removal by grass carp on native fish communities is largely correlational. For example, Bailey (1978), who summarized the effects of stocking of grass carp in thirty-one Arkansas lakes, reported changes in total standing crop of fishes; in biomass of shad (*Dorosoma*); in numbers of catchable largemouth bass, sunfishes, and crappies (*Pomoxis*); and in numbers of young-of-the-year largemouth bass and sunfishes. Observed effects varied unpredictably from lake to lake, with no overall trends. This observation is not surprising, given extensive differences in stocking levels and initial vegetational coverages, in addition to potential contributions of unexplored factors that are capable of influencing fish population structures (e.g., weather, water level fluctuations, fertilization, and fishing pressure). That such descriptive approaches are inconclusive as evidence for impact is further illustrated by conflicting interpretations of data gathered after the release of grass carp in four Florida lakes. Ware and Gasaway (1976) concluded that "biologically significant deleterious effects were determined" for native fish populations, involving reductions in growth and population levels of game species, evidence of over-crowding, and elimination of several forage species — all (presumably) attributable to the effects of vegetation removal by grass carp. However, Beach et al. (1976) argued that these same changes were artifacts of prestocking population estimation techniques and could not be "blamed" on activities of grass carp. These and similarly designed examples of studies of effects of grass carp on native fishes (e.g., Crowder and Snow 1969, Sutton 1972, Montegut et al. 1976, Mitzner 1978, Land 1981, Hardin et al. 1982) all share shortcomings of descriptive, nonexperimental approaches to demonstrating impacts. Without controls and replicates, conclusions drawn are always subject to alternative, untested explanations.

Results of experimental studies of the effects of grass carp herbivory on native fishes have also varied. Stanley (1972) found that yields of native fishes were as much as ten times higher in ponds stocked with grass carp fingerlings (22–56 per acre) than in control ponds without grass carp. Growths of *Pithophora, Najas,* and duckweed were also completely controlled at these densities. Buck et al. (1975) reported nonsignificant increases in production of bluegills and golden shiners (*Notemigonus crysoleucas*) in the presence of grass carp in small pools, and Kilgen (1978) observed that growth of channel catfish (*Ictalurus punctatus*) and striped bass (*Morone saxatilis*) in 0.04-hectare ponds was not affected by grass carp herbivory. Similarly, Lembi and Ritenour (1977) reported that production of adult sport fishes (largemouth bass, bluegills, and redear sunfish [*Lepomis microlophus*]) in 0.2 to 0.3-hectare ponds with and without grass carp did not differ significantly; however, in ponds where vegetation had been eliminated by grass carp, recruitment of young-of-the-year fishes was significantly reduced, presumably because the reduction in cover increased predation by larger natives. Similarly, Baur et al. (1979) concluded that severe reduction of vegetative cover by grass carp on 0.4-hectare ponds increased the vulnerability of juvenile fishes to predation by largemouth bass. Experimental evidence for decreased production of bluegills and largemouth bass in the presence of grass carp was also presented by Smitherman (in Greenfield 1973:50) and Forester and Lawrence (1978). In the absence of predators, Rottmann and Anderson (1976) found significantly higher densities of fathead minnows (*Pimephales promelas*) or young-of-the-year bluegills in (0.03 to 0.08-hectare) ponds from which vegetation had been cleared by grass carp than in weed-infested control ponds; however, final biomass of young bluegills was not significantly different between treatments because growth rates in the carp-free control ponds were the faster.

Evidence for impact of grass carp herbivory on non-fishes is also equivocal. Changes in zooplankton and macroinvertebrate faunas, correlated with vegetation removal by grass carp in Florida and Indiana lakes, were discussed by Beach et al. (1976), Rottmann and Anderson (1976), Shireman (1976), Lembi and Ritenour (1977), Gasaway and Drda (1977), Osborne (1978), Fry and Osborne (1980), Kobylinski et al. (1980), Crisman and Kooijman (1981), and others (see Smith and Shireman 1981). Gasaway and Drda (1977), Land (1981), and Hardin et al. (1982) further speculated that consumption of waterfowl food plants by grass carp and concurrent reductions in associated invertebrate communities contributed to observed declines in migratory waterfowl in Florida. Finally, Forester and Avault (1978) demonstrated statistically higher yields—in terms of both numbers and weight—of harvestable-sized crayfish (*Procambarus clarki*) in control ponds (0.01 hectare) than in those with grass carp, whereas Rottmann and Anderson (1976) found no differences in densities of glass shrimp (*Palaemonetes kadiakensis*) between ponds with and without grass carp.

Unlike the grass carp, the common carp was not released into United States waters for aquatic weed control. Nevertheless, the common carp's foraging behavior results in vegetation removal both by direct consumption and by uprooting associated with its habit of digging through the substrate in search of food (Cole 1905, 1906; Smallwood and Struthers 1928; Cahn 1929; Struthers 1929, 1930, 1931, 1932; Hubbs and Eschmeyer 1938; Frey and Vike 1941; Ricker and Gottschalk 1941; Sharp 1942; Black 1946; Chamberlain

1948; Anderson 1950; Threinen 1952; Cahoon 1953; Lagler and Latta 1954; Threinen and Helm 1954; Tryon 1954; Sigler 1955, 1958; McCrimmon 1956, 1968; Swingle 1957a; Miller et al. 1959; Ensign 1960; Jessen and Kuehn 1960a and b; Grizzell and Neely 1962; Shell 1962; MacKay 1963; Moyle and Kuehn 1964; Walden 1964; Foye 1965; Mathis 1965; Burns 1966; Swee and McCrimmon 1966; King and Hunt 1967; Avault et al. 1968; Sills 1970; Scott and Crossman 1973; Moyle 1976a; Leventer 1981). Further deleterious effects on growth and survival of rooted aquatics stem from increased turbidities and associated declines in light penetration that accompany foraging activities. That carp are directly responsible for such effects on vegetation and water quality has been established experimentally in controlled tests in which exclosures were used, as well as in carp removal efforts (Ricker and Gottschalk 1941, Anderson 1950, Cahoon 1953, Rose and Moen 1953, Threinen and Helm 1954, Tryon 1954, Jessen and Kuehn 1960b, Robel 1961, Mathis 1965, King and Hunt 1967). Deterioration in native fish populations has often accompanied the spread and buildup of carp populations. Although vegetation removal attributable to carp may have led to declines in native populations, we are aware of no experimental studies documenting a direct causal relation, in spite of valid bases for expecting dependencies on macrophytes for shelter, forage, and spawning sites. Nonetheless, unlike situations involving recently introduced species, considerable time and effort have been expended in investigating the effects of common carp on native fishes. Evidence for impact is strong in the sense that, in numerous independent studies, increases in carp populations and concomitant changes in habitat structure have been repeatedly associated with declines in or displacement of native assemblages (Cahn 1929; Ricker and Gottschalk 1941; Sharp 1942; Aldrich 1943; O'Donnell 1943; Holloway 1948; Threinen 1952; Cahoon 1953; Rose and Moen 1953; Sigler 1955, 1958; Buck 1956; Miller et al. 1959). However, the multiplicity of effects possible—given the complex manner in which carp interact with virtually every physical and biological component of an ecosystem—has made it difficult to pinpoint simple cause-effect relationships. Yet inability to establish causality hardly lessens concern for the repeated occurrence of detrimental effects on native fishes in areas where carp have been introduced, no matter how circumstantial the evidence for impact remains. It has become increasingly clear that the results of carp buildups in U.S. waters are predictably detrimental to native fishes, irrespective of whether underlying causal mechanisms have been specified for particular examples. Further support for the occurrence of impacts is provided by studies where the removal of common carp was followed by a recovery in native populations (e.g., Ricker and Gottschalk 1941, Cahoon 1953, Rose and Moen 1953, Miller et al. 1959), though, again, the exact nature of significant interactions has not been reported.

A comparable situation exists with regard to a proposed relation between effects of carp feeding behavior on aquatic vegetation and deterioration of waterfowl habitats. Destruction of aquatic vegetation through carp feeding activities has been interpreted as a contributing factor in declining waterfowl populations in many states by many authors (Cole 1905; Chambcrlain 1948; Holloway 1948; Moyle 1949; Cahoon 1953; McCrimmon 1956, 1968; Sigler 1958; Jessen and Kuehn 1960a and b; Robel 1961; Moyle and Kuehn 1964; Foye 1965; King and Hunt 1967); however no causality has been established.

Of the cichlid fishes established in United States waters, the redbelly tilapia is largely herbivorous, showing marked preferences for macrophytic vegetation whenever it is available (Ben-Tuvia 1959, Fryer 1961, Welcomme 1966, Avault et al. 1968, Phillippy 1969, Abdel-Malek 1972a and b, Alkholy and Abdel-Malek 1972, Pelzman 1973, Hauser et al. 1976, Moyle 1976a, Fitzpatrick et al. 1981, Leventer 1981). As a result, the species has been widely used in weed control programs (Lowe 1955; Swingle 1957a; Hickling 1961; Shell 1962; Avault 1965a; Pierce and Yawn 1965; Avault et al. 1968; Buntz and Manooch 1969; Phillippy 1969; Prowse 1969; Legner and Medved 1972, 1973a and b; Legner et al. 1973; Pelzman 1973; Legner et al. 1975; Hauser et al. 1976; Moyle 1976a; Legner and Pelsue 1977; Harris 1978; Legner 1978, 1979; Black 1980; Fitzpatrick et al. 1981), though most such efforts have been confined to closed, man-made bodies of water. No effects on native communities have yet been attributed to vegetation removal by redbelly tilapia, though concern has been voiced over the potential destruction of habitats critical to the survival of endangered faunal elements of aquatic communities in the southwestern United States, should redbelly tilapia or other cichlids be released into open waters (Deacon 1979; Courtenay and Deacon 1982, 1983).

Three other tilapia species that have been investigated for use in weed and algal control programs include the blue tilapia (often referred to as the Nile perch, *T. nilotica,* in early reports; McBay 1961, Shell 1962, Avault 1965a, Pierce and Yawn 1965, Avault et al. 1968, Habel 1975, Germany 1977, Harris 1978); the Mozambique tilapia (*T. mossambica;* Swingle 1957a and b; Shell 1962; Avault 1965a; Childers and Bennett 1967; Lahser 1967; Avault et al. 1968; Legner and Medved 1972, 1973a and b; Legner et al. 1973; Legner et al. 1975; Hauser et al. 1976); and the Wami tilapia (*T.hornorum;* Legner and Medved 1972, Legner et al. 1973, Legner et al. 1975). Effects on native fishes of vegetation removal by any of these species have not been examined.

Degradation of water quality

Ideally, impact studies concerned with how exotic fishes affect native populations by way of water quality changes should satisfy two requirements: (1) the presence of the exotic must first be shown to be responsible for observed alterations in water quality; and (2) these alterations, in turn, must cause changes in the structure or dynamics of at least one native population. As in vegetation removal studies, the first requirement has often been demonstrated; satisfying the second requirement has proved difficult (when attempted), and often effects on natives are inferred, if not asserted. An exception is Buck's (1956) studies of the effects of carp-induced turbidity on production of several species of native fishes in Oklahoma. Stocking of six small hatchery ponds with common carp produced higher turbidities than were recorded in two carp-free control ponds. Each pond was then stocked with 125 young-of-the-year largemouth bass, 100 yearling bluegills, and 50 yearling channel catfish. Draining of the ponds after one growing season (summer 1955) yielded the following results: (1) clear (control) ponds produced higher growth rates in all three native species and greater total weights of bass and bluegills than did turbid ponds with carp; and (2) a higher total weight of channel catfish was produced in turbid ponds than in controls, in spite of a slower growth rate, because of greatly increased survival. Similar trends in production of native fishes were observed in ponds in which turbid conditions were artificially established and maintained through application of clay sodium silicate treatments—adding credence to the role assigned to turbidity in interpreting results of the carp experiments. Further evidence of the effects of turbidity on native fishes was provided by concurrent experimental studies in farm ponds and comparative observations in reservoirs, yielding conclusions similar to those obtained from the hatchery ponds. Though high turbidities were not attributable to carp in the farm pond or reservoir studies, reports that link foraging and spawning activities in the common carp to "roiling" of aquatic habitats and increased turbidities are abundant (Cole 1905, 1906; Smallwood and Struthers 1928; Cahn 1929; Struthers 1930, 1931, 1932; Hubbs and Eschmeyer 1938; Ricker and Gottschalk 1941; Sharp 1942; Chamberlain 1948; Anderson 1950; Threinen 1952; Cahoon 1953; Moen 1953b; Lagler and Latta 1954; Threinen and Helm 1954; Sigler 1955, 1958; McCrimmon 1956, 1968; Mraz and Cooper 1957; Miller et al. 1959; Ensign 1960; Jessen and Kuehn 1960a and b; Grizzell and Neely 1962; Shell 1962; MacKay 1963; Moyle and Kuehn 1964; Walden 1964; Foye 1965; Mathis 1965; Burns 1966; Swee and McCrimmon 1966; King and Hunt 1967; Avault et al. 1968; Kimsey and Fisk 1969; Sills 1970; Scott and Crossman 1973; Moyle 1976a; Forester and Lawrence 1978; Leventer 1981). These reports, plus others that illustrate detrimental effects of increasing turbidities on fish populations (e.g., Schneberger and Jewell 1928, Wallen 1951), provide circumstantial evidence for the contention that changes in water quality caused by foraging activities in common carp can have harmful effects on native aquatic communities. However, as in vegetation removal studies discussed earlier, causality is difficult to verify in natural settings; therefore, proposed cause-effect relations between carp-induced turbidities and such occurrences as displacements of native species, disruption or cessation of reproductive activities, deterioration of spawning areas, and damage to shallow-water communities (examples of which are presented in later sections) must be viewed as conjectural until more rigorous documentation is obtained.

Two principal types of studies have addressed the effects of grass carp introductions on water quality characteristics (e.g., dissolved oxygen; pH;

alkalinity; turbidity; primary productivity; tannin-lignin content; nitrogen content; and levels of phosphorus, potassium, calcium, and other minerals): (1) controlled, experimental manipulations in ponds, pools, or aquaria, and (2) observations of changes in natural aquatic systems after they are stocked with grass carp. Those of the second category (e.g., Crowder and Snow 1969, Courtenay and Robins 1975, Beach et al. 1976, Gasaway and Drda 1978, Mitzner 1978, Miley et al. 1979, Shireman et al. 1979, Kobylinski et al. 1980, Kaleel 1981) often lack baseline data collected before carp introduction, and thus can provide no more than correlational evidence for the occurrence of grass carp influences on variables monitored. Differing stocking rates, enclosure sizes, monitoring schedules, and other starting conditions have all contributed unexplained variability in results obtained in the numerous experimental studies that have been done (e.g., Avault et al. 1968, Michewicz et al. 1972a, Stanley 1974, Buck et al. 1975, Terrell 1975, Rottmann and Anderson 1976, Lembi and Ritenour 1977, Fowler and Robson 1978, Hestand and Carter 1978, Lembi et al. 1978, Osborne 1978). For example, responses of phytoplankton communities and levels of organic nitrogen differed significantly among four small lakes in Florida after grass carp were stocked (Beach et al. 1976); analogously, Lembi and Ritenour (1977) showed that responses in a number of variables fluctuated widely and could even be reversed from one year to the next in individual ponds in Indiana that contained grass carp. Overall, few repeated patterns of response in particular variables emerge when available evidence is viewed en masse, and most authors reported little or no change in at least some water quality characteristics examined after grass carp introductions (see review in Smith and Shireman 1981). Exceptions include tendencies toward decreases in pH (Michewicz et al. 1972a, Buck et al. 1975, Beach et al. 1976, Rottmann and Anderson 1976, Fowler and Robson 1978, Hestand and Carter 1978, Kobylinski et al. 1980); decreases in dissolved oxygen (Michewicz et al. 1972a, Fowler and Robson 1978, Mitzner 1978); increases in potassium (Avault et al. 1968, Lembi and Ritenour 1977, Shireman et al. 1979); tannin-staining of water (Avault et al. 1968, Michewicz et al. 1972a, Beach et al. 1976, Hestand and Carter 1978); increases in levels of nitrogenous compounds (Michewicz et al. 1972a, Stanley 1974, Beach et al. 1976, Miley et al. 1979, Shireman et al. 1979, Kobylinski et al. 1980; see Mitzner 1978 to the contrary); increases in orthophosphates (Stanley 1974, Terrell 1975, Shireman et al. 1979); and increased organic enrichment or plankton blooms (Crowder and Snow 1969, Buck et al. 1975, Courtenay and Robins 1975, Gasaway and Drda 1978, Miley et al. 1979; see Mitzner 1978 to the contrary). To date, modifications in water quality resulting from grass carp introductions have not been shown to cause changes in native fish populations.

Increases in turbidity in ponds have been attributed to digging activities in the blue tilapia (Noble et al. 1975) and to organic enrichment through fecal decomposition by redbelly tilapia (Hickling 1961, Phillippy 1969). Though blue tilapia have been observed to thrive in eutrophic lakes in Florida (Buntz and Manooch 1968, Ware 1973, Shafland 1979) and reservoirs in Texas (Gleastine 1974), no direct evidence exists that their presence has caused increases in turbidity or algal blooms in these habitats. Alterations in water quality by tilapia, like those by grass carp, have not yet been linked to changes in native populations.

TROPHIC ALTERATIONS

Forage supplementation

Only one fish, the wakasagi (*Hypomesus nipponensis*), in California (Wales 1962, Moyle 1976a) has been released in U.S. waters solely as a forage supplement; however, the bairdiella (*Bairdiella icistia*) has provided forage for the orangemouth corvina (*Cynoscion xanthulus*) in addition to being a popular angling species, since its introduction into the Salton Sea (Walker et al. 1961, Moyle 1976a, Courtenay et al., chapter 4). Theoretically, the introduction of any exotic species into a novel environment should alter the forage base; however, the nature and extent of changes that result may be complex and unpredictable, depending on such factors as (1) native predators' abilities to locate, capture, and ingest the exotic; (2) the size, density, and prey-quality of the exotic relative to that of other forage species present; and (3) effects of interactions (competitive or otherwise) between the exotic and native forage species. Few field evaluations of the actual use of introduced fishes (versus native species) as forage have yet been conducted in natural systems. Gasaway (1977) documented predation by largemouth bass on grass carp during the week after stocking in Lake Baldwin, Florida, and numerous authors have observed that stocking of fingerling grass carp often results in heavy losses to native predators (Montegut et al. 1976, Shireman 1976, Lembi and Ritenour 1977, Lembi et al. 1978, Kobylinski et al. 1980, Shireman and Maceina 1981). As a result, Shireman et al. (1978) have recommended stocking grass carp larger than 450 mm to eliminate predation on fish released for aquatic weed control. In pond studies in Illinois, Childers and Bennett (1967) found that population levels of Mozambique tilapia, at the end of a growing season, were inversely related to initial densities of largemouth bass. Shafland and Pestrak (1984) showed that largemouth bass in experimental ponds in Florida strongly preferred smaller-sized blue tilapia when presented with fish of a broad range of lengths; they suggested that the forage value of tilapia, evaluated as a function of general body shape, length-depth relations, and presence of spiny-rayed fins, may lie between those of gizzard shad (*Dorosoma cepedianum*) and bluegills. No rigorous efforts have been made to quantify effects on predator-prey relations in native communities, even though occurrences of predation by native piscivores have been observed for a number of exotic species in both natural habitats and artificial systems. Examples are the common carp (Cole 1905, 1906; Shebley 1917; Struthers

1931; Hubbs and Eschmeyer 1938; Frey and Vike 1941; Sharp 1942; Wales 1942; Aldrich 1943; Jonez and Sumner 1954; Lagler and Latta 1954; Sigler 1958; Grizzell and Neely 1962; Shell 1962; Lewis and Helms 1964; Avault et al. 1968; McCrimmon 1968; Sills 1970), blue tilapia (Swingle 1960, Crittenden 1962, Lewis and Helms 1964, Pierce and Yawn 1965, Germany 1977), Mozambique tilapia (Swingle 1960, Legner and Medved 1973a, Legner et al. 1975, Lobel 1980), and redbelly tilapia (Avault et al. 1968, Legner and Medved 1973a, Legner et al. 1975, Legner 1979, Fitzpatrick et al. 1981). The lack of quantification of these effects requires immediate rectificiation. Over the past two decades, introduced tilapias have repeatedly demonstrated capacities for rapid range expansion and explosive population increases in several states: Arizona (Minckley 1973, personal communication; personal observation [WRC]), California (Knaggs 1977; G. F. Black, personal communication; personal observation [WRC]), Florida (Ware 1973, Ware et al. 1975, Courtenay and Hensley 1979, Shafland 1979; personal observation [JNT]); and Texas (Gleastine 1974, Germany 1977; C. Hubbs, personal communication). Such occurrences must be accompanied by substantial changes in the composition—in terms of both diversity and biomass—of native communities; these changes, in turn, may have significant consequences for food-web structure and stability of sport fish populations. Indeed, in evaluating potential effects of the blue tilapia in Florida, Shafland and Pestrak (1984) speculated that its major impact on native fishes may well be exerted through alterations in the forage base available to the largemouth bass, a fish that supports an annual multimillion dollar industry and tourist trade in Florida.

Competition for food

In addition to reducing prey availability through defoliational effects, feeding activities of introduced species may directly diminish food resources required by native fishes. The degree of overlap in prey utilization, and hence the potential for trophic competition, should depend on dietary habits, preferences, and potentialities of the introduced species in relation to those of native species present, and on the variety of prey available. Efforts to predict the outcomes of trophic interactions between introduced and native fishes are complicated by generalist feeding habits and trophic opportunism (i.e., abilities—even in so-called stenophagous species—to shift or expand diet composition in new environments) typically exhibited by successful introductions. Although frequently implicated as an explanation for modifications in native aquatic communities, competition—for food as well as other resources—between introduced and native fishes has rarely been conclusively demonstrated as a cause of observed changes—as pointed out by several investigators (McCrimmon 1968, Moyle 1976b, Schoenherr 1981, Simberloff 1981). Perhaps the best example as regards trophic resources relates to interactions between blue tilapia and shad in Texas and Florida, though the evidence is hardly unequivocal. Replacement of native shads (*Dorosoma* spp.) by blue tilapia in eutrophic lakes and reservoirs has been documented in both Florida (Horel 1969; Babcock and Chapman 1973; Ware 1973; Shafland et al. 1980; Wattendorf 1981, 1982) and Texas (Gleastine 1974, Germany 1977, Hubbs et al. 1978, Hendricks and Noble 1979). Hendricks and Noble (1979) presented evidence—based on stomach content analyses of blue tilapia, gizzard shad, and threadfin shad (*D. petenense*) taken from Trinidad Lake, Henderson County, Texas in 1975—for extensive dietary overlap among these three planktivores and suggested that decreasing shad populations in the lake may have resulted from interspecific competition for food with blue tilapia. However, small samples (15 fish or less) were used in the stomach analyses and no evidence was presented to indicate that food (i.e., plankton) was in short supply; further, sampling techniques (gill-netting and seining) utilized to track changes in abundance of shad over time may not have provided reliable estimates of population densities. More accurate censusing data (based on samples collected with rotenone) are available to evidence replacement of shad by blue tilapia in several Florida lakes (Horel 1969; Babcock and Chapman 1973; Ware 1973; Shafland et al. 1980; Wattendorf 1981, 1982). In addition, stomach content analyses of large samples of blue tilapia and shad from Lake Lena, Polk County, revealed substantial amounts of detritus in the diets of blue tilapia and gizzard shad, whereas threadfin shad depended largely on phytoplankton (Wattendorf et al. 1980). As the blue tilapia population increased after its appearance in Lake Lena in 1970 (Buntz and Chapman 1971), the population of gizzard shad declined through 1979 and that of threadfin shad increased (Shafland et al. 1980). Unfortunately, in subsequent years gizzard and threadfin shads were not sorted to species in the field. Although the combined density of the two species declined further in 1980 (Wattendorf 1981), the density estimated in 1981 (Wattendorf 1982) evidenced an overall increase in shad. Whether these estimates accurately reflect changes in population densities of each individual species of shad is not known.

Food studies that document diet overlap between introduced and native freshwater fishes have been carried out for many species: brown trout (Clemens 1928; Sibley 1928; Metzelaar 1929; Rimsky-Korsakoff 1930; Pate 1933; Needham 1938; Raney and Lachner 1942; Idyll 1942; Webster and Little 1942, 1944; Wales 1946; Maciolek and Needham 1951; Evans 1952; Reimers et al. 1955; Tebo and Hassler 1963; Nyman 1970; Wiley and Varley 1978; Hiscox 1979); common carp (Sibley 1928, 1929; Smallwood and Struthers 1928; Struthers 1929, 1930, 1931; Harrison 1950; Moen 1953a and b; Sigler 1958; Rehder 1959; McCrimmon 1968; Whitaker 1974); pike killifish, *Belonesox belizanus* (Belshe 1961; Miley 1978; Anderson 1981, 1982); shortfin molly, *Poecilia mexicana* (Hardy 1981); several introduced cichlids in Florida (Finucane and Rinckey 1964; Hogg 1976; Anderson 1981, 1982); convict cichlid, *Cichlasoma nigrofasciatum* (Hardy 1981); blue tilapia (McBay 1961); Mozambique tilapia (Kelly 1956, Moyle 1976a); redbelly tilapia (Cox 1972, Black 1980), and two species of Asiatic gobies, *Acanthogobius flavimanus* and *Tridentiger trigonocephalus* (Haacker 1979). None of these studies, however, has attempted to link dietary overlap to competitive effects on native fishes.

Predation

Instances of predation provide perhaps the best examples of impacts (as defined earlier) of exotic species on native fishes, in that consumption of individuals of a native population by an introduced predator immediately alters that population's composition, though long-term effects on prey populations may vary appreciably from one community to another. Although viewed in most circles as a beneficial introduction in the United States, the brown trout, particularly at larger sizes, is well known as a piscivore, feeding on a wide variety of native species (Bean 1906; Metzelaar 1929; Sibley and Rimsky-Korsakoff 1931; Hubbs and Eschmeyer 1938; Needham 1938; Smedley 1938; Idyll 1942; Wales 1946; Evans 1952; Fenderson 1954; Reimers et al. 1955; Wales and German 1956; Sharpe 1957, 1960, 1962; McCraig et al. 1960; O'Donnall and Churchill 1960; Sigler and Miller 1963; Walden 1964; Staley 1966; Hannuksela 1969; Harlan and Speaker 1969; Nyman 1970; Shetter and Alexander 1970; Jones 1972; Brynildson et al. 1973; Moyle and Marciochi 1975; Moyle 1976a and b; Alexander 1977, 1979; Allan and Roden 1978; Wiley 1978; Wiley and Varley 1978; Hiscox 1979; Garman and Nielsen 1982). Furthermore, piscivorous habits of brown trout have been implicated as a major factor in reductions in a number of native populations, particularly other salmonids. Specific examples of declines in populations of native trouts attributable to brown trout predation – acting either alone or in combination with competition, habitat alteration, or greater vulnerability to angling pressures in native trouts – have been documented in California, Colorado, and Michigan. Brown trout predation in the upper Kern River basin currently poses a threat to the survival of the native golden trout (*Salmo aguabonita*), the official state fish of California. In fact, the California Department of Fish and Game is eradicating brown trout populations in an effort to preserve the golden trout (E. P. Pister, personal communication). Predation by brown trout has probably also contributed to declines in populations of cutthroat trout (*Salmo clarki*), and McCloud River Dolly Varden (*Salvelinus malma*) in California (Moyle 1976a and b) and Gila trout (*Salmo gilae*) in New Mexico (J. N. Rinne, personal communication). Predator pressures exerted by lake trout (*Salvelinus namaycush*) and introduced brown trout are believed to have prevented stocked brook trout from establishing a self-sustaining population in Castle Lake, California. Chemical removal of the lake's heterogeneous trout community, followed by stocking with brook trout, led to successful establishment of a reproducing population (Wales 1946, Wales and German 1956, Staley 1966). This example is of further interest relative to results of efforts to introduce North American trouts into Europe. Attempts to establish brook trout and rainbow trout (*Salmo gairdneri*) in France often failed in the presence of brown trout, but have been successful in waters lacking this European native (Vivier 1955). In Shadow Mountain Reservoir, Colorado, two-thirds of the fish eaten by brown trout were found to be young rainbow trout, as judged by stomach content analyses from samples taken during the summers of 1957–59 (Sharpe 1957, 1960, 1962). In Michigan, removal of about 60 percent of the brown trout longer than 300 mm from a portion of the North Branch of the Au Sable River resulted in an increase in numbers of large brook trout, but no significant change in the numbers of smaller trout (Shetter and Alexander 1970), even though brown trout were later identified as the most significant natural predator in the Au Sable in terms of total trout consumed (Alexander 1977, 1979). In discussing the role of the brown trout in the Au Sable, Alexander (1977:1) concluded:

> Brown trout predation is probably detrimental to brook trout populations where brook trout fisheries are the primary management objective. However, brown trout predation is probably beneficial in controlling slow growth in high population densities of small trout. Further, if older brown trout lacked smaller trout as prey they would grow slower and fewer large "trophy" size fish would be produced. Also, the predation of the brown trout on coarse fish populations is beneficial to all trout.

Finally, Moyle and Marciochi (1975) noted that the negative correlation demonstrated between the abundance of the rare Modoc sucker (*Catostomus microps*) and the abundance of large brown trout in the upper Pit River drainage of California is most likely attributable to brown trout predation on the sucker; and Garman and Nielsen (1982) demonstrated that the abundance of torrent suckers (*Moxostoma rhothoecum*) decreased in a section of a Virginia stream stocked with large brown trout, but remained constant in a control section without trout.

In Florida, field observations, experimental manipulations, and stomach content analyses have verified that predation by pike killifish is a major cause of mortality in native forage populations (Belshe 1961; Miley 1978; Turner 1981; Anderson 1981, 1982; personal observation [JNT]). In experimental feeding studies involving pike killifish and four native cyprinodontoids, Miley (1978) showed that the mosquitofish (*Gambusia affinis*) and sailfin molly (*Poecilia latipinna*) were preyed upon much more heavily than the least killifish (*Heterandria formosa*) and the bluefin killifish (*Lucania goodei*) because of the surface-dwelling habits of the first two species. However, predation on all species was significantly reduced in the presence of aquatic vegetation (*Hydrilla verticillata*) as a refuge. In a heavily vegetated Dade County canal, native poeciliids and cyprinodontids constituted 98 percent of the number and nearly 60 percent of the biomass of all native fishes. In a collection made five months after vegetation was removed from the same site, these figures dropped to 15 percent and 23 percent, respectively, the most notable reduction occurring in the numbers of mosquitofish: 210 individuals, composing 2.4 percent of the native fish biomass, were taken in the first collection, but none were collected after removal of *Hydrilla*. Further, in a series of collections from Dade County, native poeciliids and cyprinodontids made up 54 percent of the number and 34 percent of the biomass of samples taken at sites where pike killifish occurred, compared with 97 percent of the number and 58 percent of the biomass at nearby sites not occupied by the predator (Miley 1978). Recent rotenone samples taken from Dade County ditches inhabited by pike killifish (and few other piscivores) have occasionally failed to yield a single individual of any native poeciliid or cyprinodontid species (personal observation [JNT]), further attesting to the voracious, efficient predatory habits of the pike killifish.

Additional occurrences of predation on native fishes by introduced species have been noted for the oscar (*Astronotus ocellatus*) (Hogg 1976), in Florida, and the redbelly tilapia (Schoenherr 1979, 1981; Black 1980) and the bairdiella (Quast 1961) on the desert pupfish, *Cyprinodon macularius,* in the Salton Sea, California; however, the effects of predation on the population structures of native species have not been investigated.

Though frequently cited as a potential threat of considerable consequence, predation on eggs or young by exotic fishes has not been demonstrated to be a common occurrence in freshwater communities in the United States. Greeley (1932) reported the presence of rainbow trout eggs in the stomachs of brown trout, but doubted that egg predation was a significant source of mortality. Though much maligned as an egg predator even as early as the turn of the century (see Cole 1905), the common carp has not been shown in any rigorous fashion to feed—either commonly or systematically—on eggs or spawn of native fishes. Indeed, evidence to the contrary comes from results of stomach content analyses that reveal the presence of only small numbers of eggs, if any, in the diets of carp from a number of different populations (Cole 1905, 1906; Struthers 1929, 1930, 1931; Harrison 1950; Moen 1953a and b; Lagler and Latta 1954; Ensign 1960). Jonez and Sumner (1954) reported one of the few apparent exceptions to this generalization; they observed large groups of carp foraging in spawning areas of the razorback sucker (*Xyrauchen texanus*) in Lakes Mead and Mojave; subsequent verification (through stomach content analyses) that egg predation was occurring (W. L. Minckley, personal communication) suggests that egg predation by carp may have been a key factor in the recent decline of the razorback sucker in the Colorado River basin.

The possibility has been proposed that egg and larval mortality in native fishes can result as consequences of (1) roiling or vegetation removal associated with carp feeding activities or (2) nest desertion in response to disturbances caused by feeding or spawning carp. The same possibility exists as regards the effects of grass carp herbivory on egg and larval mortality, but has not been investigated in the United States. Among introduced cichlids, schools of juvenile black acara have been observed to rob eggs from nests left unattended by male bluegills in Florida (Hogg 1976). Schoenherr (1979) stated that substrate-spawning redbelly tilapia "probably also eat fish eggs"; similar behavior has been occasionally observed in juvenile spotted tilapia in temporarily vacated sunfish nests in Florida (personal observation [JNT]). Haacker (1979) reported the presence of 478 fish eggs in the gut of a chameleon goby in California, though eggs were not identified to species. Finally, though not known as egg predators, young pike killifish (14.4–17.7 mm in standard length at birth) are able to eat newborn mosquitofish (and presumably fry of other species) soon after birth (Belshe 1961, Turner 1981).

HYBRIDIZATION

Though clearly a threat to the genetic integrities of native gene pools, hybridization events involving crosses between exotic and native fishes in open waters of the United States are apparently rare. The only introduced species known to hybridize with native forms is the brown trout (see reviews in Schwartz 1972, 1981; Dangel et al. 1973; Chevassus 1979), and reports of natural crosses in wild populations are uncommon (Brown 1966, Nyman 1970, Allan 1977).

SPATIAL ALTERATIONS

Spatial alterations refer to density-related effects on native fishes that result directly from behavioral encounters between individual members of exotic and native populations. Interactions may occur in association with any activity involving the use of space. Immediate effects of these interactions are usually expressed as spatial displacements from occupied microhabitats or temporal shifts in activity schedules. Although such effects themselves constitute impacts, they may, in addition, precipitate changes in native population structures through alterations in patterns of growth, survivorship, and reproduction. Spatial alterations, as behaviorally mediated phenomena, are expected to be density related (i.e., the severity of an effect should be a function of the frequency of behavioral contacts that occur, which in turn depends on the relative densities of the interacting populations). Effects may be grouped into two general classes on the basis of the nature of the behavioral interaction underlying an alteration: (1) aggressive effects, including competitively based interactions often mediated by systems of aggressive dominance or territoriality; and (2) overcrowding, density-related phenomena in which behaviors (especially reproductive activities) in affected populations are turned off or altered in response to the mere presence of the population responsible for the effect. Unlike changes brought about by direct, observable aggressive encounters, effects attributable to overcrowding appear to be triggered solely by sensory stimuli, probably visual or olfactory in nature, received by members of the affected population. Causal mechanisms involved in prospective examples of effects resulting from overcrowding are obscure, and their investigation deserves immediate, behaviorally based experimentation. To date, most studies describing changes attributed to overcrowding have been carried out in decidedly nonbehavioral contexts; not surprisingly, verification of behavioral interactions as causal agents underlying impacts will eventually require studies involving observations of behavior!

Evidence for impacts of exotics through spatial alterations is reviewed below; as in previous classes of effects, demonstrating changes in native populations in the presence of an exotic is much simpler and more commonplace than identifying the exotic as the cause of observed changes. Attempts to attribute the change to the exotic are complicated by behavioral interactions that result when an exotic becomes established in a novel environment. Effects on natives may be caused by any combination of changes in habitat structure, resource availability, susceptibility to predation, or microhabitat utilization, all of which may result (simultaneously) from activities of the exotic; partitioning out variance in native populations attributable to each such exotic-induced change, though admittedly a difficult task, should be the ultimate objective in the design of any impact study.

Aggressive effects

Temporal overlap in use of space can provide a context for competition between exotic and native species, if one assumes that the spatial "commodity" in question (e.g., feeding stations, spawning sites or territories, resting stations, and refugia) is in short supply or of variable quality. Salmonid fishes are well known for their sedentary habits (Shetter 1937, 1968; Schuck 1945; Allen 1951; Miller 1957a; Kalleberg 1958; Chapman and Bjornn 1969; Jenkins 1969; Lewis 1969; Mense 1975; Moyle 1976a) and their propensities to form territorial spacing systems in association with feeding, reproduction, and other activities (Lindroth 1955, Newman 1956, Kalleberg 1958, Chapman 1966, Chapman and Bjornn 1969, Jenkins 1969, Fausch and White 1981). In addition, similarities in agonistic displays and the occurrence of interspecific defense of territories have been repeatedly noted among salmonids (Lindroth 1955, Newman 1956, Kalleberg 1958, Jenkins 1969, Fausch and White 1981); benefits of social status realized by territorial individuals have been postulated by Miller (1958), Chapman (1966), Egglishaw (1967), and Fausch and White (1981). Studies of the life history and ecology of brown trout, based on introduced populations in the United States, have provided a number of examples of spatial displacements of native salmonids that have been linked to aggressive interactions with brown trout; unfortunately, as reviewed below, the occurrence of competition as the driving force behind observed displacements is more often assumed than rigorously validated.

Similarities in habitat use between brown trout and native species have been cited as a basis for competitive interactions leading to patterns of spatial segregation observed in natural communities. For example, Nyman (1970: 343) stated that "the ecological demands of brook trout and brown trout are almost identical," and Scott and Crossman (1973:199) wrote that the brown trout's requirements "are essentially the same as for the native brook trout, with which it is usually in competition."

Though descriptive characterizations of microhabitat utilization among sympatric salmonid populations reveal distributional differences relative to such variables as current, depth, temperature, and cover (Chapman and Bjornn 1969, Lewis 1969, Vincent and Miller 1969, Brynildson et al. 1973, Gard and Flittner 1974, Moyle 1976a), such studies are incapable of distinguishing between segregational effects because of (1) species-specific preferences for differing environmental conditions and (2) displacements mediated by aggressive behavioral interactions. The same is true, in general, for studies of trophic relations between brown trout and native species. Stream-dwelling salmonids are frequently characterized as trophic opportunists—feeding largely on drift—and availability is the most important determinant of diet composition (Needham 1929, Tebo and Hassler 1963, Egglishaw 1967, Jenkins 1969, Nyman 1970). Defense of stable feeding territories or stations from which excursions are made to sample food items in the drift is also a widespread habit within the group (Newman 1956, Kalleberg 1958, Jenkins 1969, Fausch and White 1981).

The co-occurrence of trophic opportunism and defense of feeding stations has too often been viewed as a priori evidence for the existence of competition among salmonids; indeed, an extensive rationale has been developed to explain how competition for space (e.g., feeding stations) has been substituted, evolutionarily, for direct competition for trophic resources (see Chapman 1966). Unfortunately, in spite of the intuitive appeal (and perhaps the factual validity) of such a hypothesis, competition for feeding sites among salmonids has not been conclusively demonstrated in any given case as a cause of displacement of native forms by brown trout. In fact, in the only experimental study in which such a demonstration was attempted, Fausch and White (1981) found that removal of brown trout from an enclosed length of stream failed to result in changes in feeding-site occupation in resident brook trout; in this stream, fairly uniform distributions of water velocities and invertebrate drift were postulated as providing an excess of equally desirable feeding stations, and little competition occurred. However, the same authors were able to document a spatial shift in brook trout to daytime *resting* positions with more favorable water velocity characteristics and a greater degree of cover after elimination of comparably sized brown trout, indicating that brown trout had been actively excluding brook trout "from preferred resting positions, a critical and scarce resource." Though earlier authors (e.g., Newman 1956, Kalleberg 1958, Jenkins 1969) discussed the occurrence and potential significance of resting stations and nocturnal refugia in stream-dwelling salmonids, no earlier attempts had been made to relate competition for such resources to displacements of native species. Finally, as regards reproductive activities, spawning requirements (i.e., shallow, gravelly headwaters) are basically the same for the brook trout and brown trout, and both species breed from late autumn to early winter (Greeley 1932, Hubbs and Eschmeyer 1938, Nyman 1970, Scott and Crossman 1973, Moyle 1976a); whether overlap in breeding activities is detrimental to the reproductive success of either species has not been investigated.

Evidence for competitively induced spatial displacements of native fishes by exotic species other than the brown trout is largely inferential. With regard to reproductive interactions, roiling resulting from spawning activities of the common carp has been generally cited as detrimental to shallow water areas utilized for spawning by native species (Cole 1905, 1906; Shebley 1917; Struthers 1931, 1932; Sharp 1942; McCrimmon 1968; Deacon et al. 1972); however, attempts to demonstrate direct behavioral interference in spawning activities of natives have proved unsuccessful in a number of instances (Cole 1905, 1906; Struthers 1929; Mraz and Cooper 1957; Sigler 1958; Boyer and Vogele 1971). Nevertheless, effects on natives – particularly centrarchids – may be expressed indirectly through nest desertion and brood mortality precipitated by disturbances and increased siltation resulting from carp spawning activities (Struthers 1929, 1932; Mraz and Cooper 1957; Sigler 1958; Foye 1965). In experimental studies done in 0.1-hectare ponds, Forester and Lawrence (1978) documented interference in bluegill recruitment by both carp and grass carp, relative to levels of production in control ponds lacking either cyprinid species; in addition, largemouth bass failed to reproduce in four of six ponds stocked with either grass carp or common carp, though young were produced in all of three control ponds. Unfortunately the behavioral mechanisms underlying observed disruptions in breeding habits were not investigated.

The wide range in spawning-site preferences exhibited by introduced cichlid fishes provides a basis for extensive overlap – and potential interference – with breeding habits of native species. Representatives of the family established in the United States include hole-nesting species (e.g., the Midas cichlid; R. S. Anderson, personal communication), solitary substrate-spawners with preferences for solid, often vertical, objects as egg-deposition sites (the black acara and spotted tilapia, Hogg 1976); and both substrate-spawners (the firemouth cichlid, Hogg 1976; personal observation [JNT]; and the redbelly tilapia, Daget 1952, Loiselle 1977) and mouthbrooders (the blue tilapia, McBay 1961, Ware 1973, Gleastine 1974, Noble et al. 1975; and the Mozambique tilapia, Lobel 1980) that nest colonially in shallow habitats over mud or sand substrates. As such, overlap in spawning-substrate use with a wide variety of native nest-building taxa, particularly centrarchids, is a distinct possibility. In spite of the potential for interference and displacement of native fishes by introduced cichlids during reproductive periods, studies of overlap in breeding requirements, spawning interactions, and their impacts on natives are uncommon. Indeed, though parallels in breeding habits and competition for spawning sites between native centrarchids and mouthbrooding tilapias in Florida have been repeatedly cited (Buntz and Manooch 1968,

Babcock and Chapman 1973, Ware 1973, Courtenay et al. 1974, Harris 1978), in the only experimental study done to examine such possibilities, Shafland and Pestrak (unpublished data) demonstrated clear segregation in nest placement between largemouth bass and blue tilapia stocked together in 0.01-hectare ponds; bass nested predominantly in shallower areas over gravel, whereas tilapia pits were concentrated in deeper water over sand. Limited data suggest that greater variability in nest placement was expressed in each species when stocked singly, but small sample sizes precluded statistical testing for differences between "sympatric" and "allopatric" treatments. Given the recent growth and expansion of tilapia populations observed in areas of the United States with mild climates, coupled with declines in recruitment exhibited by associated native populations, intensive studies of spawning interactions between natives and introduced tilapias appear to be a prerequisite to understanding mechanisms underlying impacts on native populations.

Overcrowding

Evidence that overcrowding by exotic fishes can lead to impacts on native species is sparse and inconclusive; however, studies of the negative effects of overcrowding by native forage species on reproduction of such important game species as largemouth bass indicate that the possibility that comparable interactions can result where populations of exotics are large deserves closer scrutiny. Swingle (1957c), who investigated the effects of density on reproduction in fishes stocked in experimental ponds, found that at high densities little or no recruitment occurred. Results from experiments with a number of cyprinids, including common carp and goldfish, revealed that the production of a "repressive factor" excreted by the fish inhibited spawning, even though mature gametes were developed. When a similar relation between reproduction and density was observed in centrarchids, the repressive-factor interpretation was interpolated to this group, though experimental verification was not provided. Later, Chew (1972) observed that retention (and eventual resorption) of ripe ova in female largemouth bass was the basis for nonrecruitment in eutrophic lakes overcrowded with bluegills and other forage species in Florida; similar results were obtained in hatchery ponds in which hypereutrophic, overcrowded conditions were simulated. Like Swingle (1957c), Chew (1972) attributed inhibition of spawning in largemouth bass to the accumulation of a repressive factor excreted by the dense forage populations, though no efforts were made to evaluate or verify such an interpretation. Evidence against the operation of a repressive factor, at least of an olfactory nature, in centrarchids was provided by studies in experimental ponds in Florida in which both normal and anosmic bass in Florida were used (Barwick and Holcomb 1976); further, in the absence of egg predation by sunfishes, these authors concluded (p. 246) that "lack of reproduction . . . was apparently associated with a physical factor that in some way interrupted normal bass breeding activity," and, though behavioral observations were not attempted, that harassment of spawning bass by the sunfishes could be the basis for nonrecruitment. Smith (1976) attempted to pinpoint this relation; he concluded (p. 684) that, based on results of introducing high densities of sunfishes at different stages of bass reproductive activities, "reduced or complete inhibition of spawning of largemouth bass is linked to aggressive interaction with other species, primarily affects the male bass, and is effective in suppressing the spawning behavior sequence even before nest building." Unfortunately, behavioral data to substantiate the occurrence of such interactions were not collected.

Recent studies in the United States have suggested that high population densities in introduced exotic fishes may have debilitating effects on reproductive activities in native species through overcrowding. In Trinidad Lake, Texas, largemouth bass populations declined with no evidence of recruitment as densities of blue tilapia rose to approximately 2,240 kg/ha during the period 1972–75; bass also failed to reproduce in a 0.2-hectare earthen pond stocked with tilapia at 2,240 kg/ha (Noble et al. 1975). In the pond, male bass did not build nests and females eventually resorbed their eggs. Spawning of largemouth bass and other species resumed after a complete winterkill of blue tilapia in Lake Trinidad in 1975–76 (Noble 1977). As in the centrarchid studies outlined above, negative effects on reproduction are clearly density related; however, underlying mechanisms have yet to be elucidated. In an effort to determine whether competition for nest sites caused suppression of spawning, Shafland and Pestrak (unpublished data) observed breeding behavior in largemouth bass stocked in 0.01-hectare ponds with blue tilapia at similar densities but variable sex ratios. Production of young-of-the-year bass in ponds with tilapia was only 10-24 percent of that in control ponds without tilapia; however, since differences in bass production among all-male, all-female, and mixed male-female tilapia treatments were not statistically significant, the authors concluded that competition for nest sites between male bass and male tilapia was not a sufficient explanation for the extent and level of reproductive suppression observed. They proposed, as an alternative, that inhibition of spawning resulted from behavioral interactions associated with the frequency of incidental bass-tilapia encounters. Further testing of this alternative hypothesis is required to establish whether overcrowding is involved or the observed interference resulted from aggressive interactions.

With regard to exotics other than blue tilapia, displacements of native fishes by common carp can sometimes be linked to suppression of spawning (Cahn 1929, Cahoon 1953, Forester and Lawrence 1978); in addition, removal of carp has led to reversals of suppression and recovery of native populations in a number of instances (Ricker and Gottschalk 1941, Cahoon 1953, Rose and Moen 1953). Nevertheless, the role of overcrowding per se has not been investigated in relation to impacts of carp populations, and observed effects on natives are equally likely to result from other causes (e.g., vegetation removal, increased turbidity, aggressive interference during nesting periods). Indeed, the "reality" of overcrowding as a discrete mechanism capable of causing changes in native populations is yet to be established. With reference to the bass-tilapia example discussed above, controlled experiments designed to distinguish between density effects — as a function of, for example, frequency of visual contacts — and other behavioral interactions are required in order to determine whether overcrowding itself can alter reproductive activities in bass. Clearly, such experiments must either incorporate or eliminate the possibilities of effects resulting from aggressive interference. Although the present status of the phenomenon remains obscure, possible effects of overcrowding on native fishes deserve further investigation. In recent years, explosive growth of introduced tilapia populations has been repeatedly documented in California, Arizona, Texas, and Florida; the possibility that high densities, in and of themselves, can exert negative effects on native game fishes, such as largemouth bass, represents a sufficiently grave threat to demand rigorous evaluation.

Conclusion

Defining "impact" as any effect on a native population attributable to the presence of an exotic species deems its occurrence a foregone conclusion; that no effects should result from such perturbations strains one's confidence in ecological principles. Though effects may not be in doubt, at least on theoretical grounds, rigorous documentation of specific examples of impacts is largely lacking because investigators have failed to accommodate the consequences of multiple causality in the design of impact studies. The need for controlled, experimental manipulations aimed at testing particular hypotheses about cause-effect relationships involving interactions between exotic and native species is paramount. Further, wherever behavioral interactions are implicated as prospective causal mechanisms, behaviorally oriented experiments will be required to investigate such hypotheses.

By and large, though the long-term effects of specific introductions remain unpredictable, successful establishments of exotic fishes appear most likely in (1) aquatic habitats with mild year-round temperatures; (2) disturbed or altered habitats, or (3) communities of low diversity (i.e., "unsaturated"; see Diamond and Veitch 1981). The last of these three categories includes both natural communities, such as those that characterize aquatic systems of the arid Southwest, and artificial (i.e., man-made) systems, such as heated reservoirs and canals. Indeed, introduced fishes from neotropical regions may well be better suited to such habitats than naturalized colonists from contiguous areas with more temperate climates. Perhaps the major exception to these predicted patterns relates to prospective introductions of exotic species, such as certain Old World cyprinids, from temperate regions where climates and aquatic habitats closely match those that occur across much of the United States. The successful reproduction of the grass carp in the Mississippi River (Conner et al. 1980) provides an important lesson as to the ability of such species to become established in what both fisheries biologists and ecologists would agree is a mature, diverse community.

Finally, the applicability of the same ecological principles that structure dispersal and colonization in natural communities has been stressed in evaluating the impacts of exotic introductions. Though a full discussion would be beyond the scope of the present chapter, it should be emphasized that the same parallels hold with regard to artificial transplantation of native species by man within the United States; indeed, given that communities receiving transplants are likely to include closely related species, the prospects for interaction and interference would seem to be increased, implying that consequences of transplantation may be even more serious than those of exotic introductions.

Acknowledgments

The cooperation of the interlibrary loan service staffed by G. L. Parsons, D. E. Moore, and S. L. Pompey, in S. E. Wimberly Library, Florida Atlantic University, was invaluable in conducting the literature search required to complete the manuscript. We also thank B. J. Rice for typing the literature cited section.

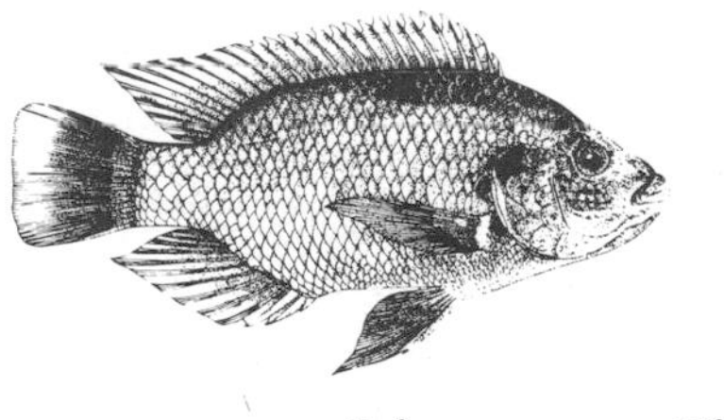

Literature Cited

Abdel-Malek, S. A. 1972a. Food and feeding habits of some Egyptian fishes in Lake Quarun. Part I. *Tilapia zillii* (Gerv.). B. According to different length groups. Bulletin of the Institute of Oceanography and Fisheries, Alexandria, Egypt 2:203–213.

Abdel-Malek, S. A. 1972b. Food and feeding habits of some Egyptian fishes in Lake Quarun. Part I. *Tilapia zillii* (Gerv.). C. According to different sexes. Bulletin of the Institute of Oceanography and Fisheries, Alexandria, Egypt 2:239–259.

Addor, E. E., and R. F. Theriot. 1977. Test plan for the large-scale operations management test of the use of the white amur to control aquatic plants. Instruction Report A-77-1, U.S. Army Engineer Waterways Experiment Station, CE, Vicksburg, Mississippi.

Aldrich, A. D. 1943. Natural production of pond fish in waters near Tulsa, Oklahoma. Transactions of the North American Wildlife Conference 8:165–166.

Alexander, G. R. 1977. Consumption of small trout by large predatory brown trout in the North Branch of the Au Sable River, Michigan. Michigan Department of Natural Resources, Fisheries Research Report no. 1855:1–26.

Alexander, G. R. 1979. Predators of fish in coldwater streams, 153–170. *In:* H. Clepper (ed.). Predator-prey systems in fisheries management. Sport Fishing Institute, Washington, D.C.

Alkholy, A. A., and S. A. Abdel-Malek. 1972. Food and feeding habits of some Egyptian fishes in Lake Quarun. Part I. *Tilapia zillii* (Gerv.). A. According to different localities. Bulletin of the Institute of Oceanography and Fisheries, Alexandria, Egypt 2:185–201.

Allan, J. H. 1977. First report of the tiger trout hybrid, *Salmo trutta* Linneaus x *Salvelinus fontinalis* (Mitchill) in Alberta. Canadian Field-Naturalist 91(1):85–86.

Allan, R. C., and D. L. Roden. 1978. Fish of Lake Mead and Lake Mojave. Biology Bulletin No. 7. Nevada Department of Wildlife, Reno, Nevada. 105 pp.

Allen, K. R. 1951. The Horokiwi stream. A study of a trout population. Fisheries Bulletin, Wellington, New Zealand 10:1–231.

Anderson, J. H. 1950. Some aquatic vegetation changes following fish removal. Journal of Wildlife Management 14:206–209.

Anderson, R. S. 1981. Food habits of selected non-native fishes: Stomach contents. First annual performance report, Non-Native Fish Research Laboratory, Boca Raton, Florida Game and Fresh Water Fish Commission. 16 pp.

Anderson, R. S. 1982. Food habits of selected non-native fishes: Stomach contents. Second annual performance report, Non-Native Fish Research Laboratory, Boca Raton, Florida Game and Fresh Water Fish Commission. 22 pp.

Avault, J. W., Jr. 1965a. Biological weed control with herbivorous fish. Proceedings of the Annual Meeting of the Southern Weed Conference 18:590–591.

Avault, J. W., Jr. 1965b. Preliminary studies with grass carp for aquatic weed control. Progressive Fish-Culturist 27:207–209.

Avault, J. W., Jr., R. O. Smitherman, and E. W. Shell. 1968. Evaluation of eight species of fish for aquatic weed control. Proceedings of the world symposium on warm-water fish culture, Rome, Italy. Food and Agriculture Organization of the United Nations, Fisheries Report no. 44, 5 (VII/E-3):109–122.

Babcock, S., and P. Chapman. 1973. Description and status of the commercial fishery for blue tilapia in Polk County with recommendations to the Commission on methods of obtaining additional revenue from this fishery. Mimeographed report, Florida Game and Fresh Water Fish Commission. 35 pp.

Bailey, W. M. 1972. Arkansas' evaluation of the desirability of introducing the white amur (*Ctenopharyngodon idella* Val.) for control of aquatic weeds. Arkansas Fish and Game Commission mimeo. 59 pp.

Bailey, W. M. 1978. A comparison of fish populations before and after extensive grass carp stocking. Transactions of the American Fisheries Society 107(1):181–206.

Barnett, B. S., and R. W. Schneider. 1974. Fish populations in dense submerged plant communities. Hyacinth Control Journal 12:12–14.

Barwick, D. H., and D. E. Holcomb. 1976. Relation of largemouth bass reproduction to crowded sunfish populations in Florida ponds. Transactions of the American Fisheries Society 105(2):244–246.

Baur, R. J., D. H. Buck, and C. R. Rose. 1979. Production of age-0 largemouth bass, smallmouth bass, and bluegills in ponds stocked with grass carp. Transactions of the American Fisheries Society 108(5):496–498.

Beach, M. L., W. W. Miley II, J. M. van Dyke, and D. M. Riley. 1976. The effects of the Chinese grass carp (*Ctenopharyngodon idella* [Val.]) on the ecology of four Florida lakes and its use for aquatic weed control. Mimeographed final report, Florida Department of Natural Resources, Tallahassee, Florida. 246 pp.

Bean, T. H. 1906. Discussion period of the 35th annual meeting of the American Fisheries Society. Transactions of the American Fisheries Society 35:137.

Belshe, J. F. 1961. Observations of an introduced tropical fish (*Belonesox belizanus*) in southern Florida. M.S. thesis, University of Miami, Coral Gables, Florida. 71 pp.

Ben-Tuvia, A. 1959. The biology of the cichlid fishes of Lakes Tiberias and Huleh. Bulletin of the Research Council of Israel 8B:153–188.

Black, G. F. 1980. Status of the desert pupfish, *Cyprinodon macularius* (Baird and Girard), in California. Inland Fisheries Endangered Species Program, special publication 80-1, California Department of Fish and Game. 42 pp.

Black, J. D. 1946. Nature's own weed killer, the German carp. Wisconsin Conservation Bulletin 11, no. 4.

Boyd, C. E. 1971. The limnological role of aquatic macrophytes and their relationship to reservoir management, 153–166. *In:* G. E. Hall (ed.). Reservoir fisheries and limnology. Special publication no. 8, American Fisheries Society.

Boyer, R. L., and L. E. Vogele. 1971. Longear sunfish behavior in two Ozark reservoirs, 13–25. *In:* G. E. Hall (ed.). Reservoir fisheries and limnology. Special publication no. 8, American Fisheries Society.

Brown, C.J.D. 1966. Natural hybrids of *Salmo trutta* and *Salvelinus fontinalis*. Copeia 1966(3):600–601.

Brynildson, O. M., V. A. Hacker, and T. A. Klick. 1973. Brown trout: Life history, ecology and management. Publication no. 234, Wisconsin Department of Natural Resources. 16 pp.

Buck, D. H. 1956. Effects of turbidity on fish and fishing. Transactions of the North American Wildlife Conference 21:249–261.

Buck, D. H., R. J. Baur, and C. R. Rose. 1975. Comparison of the effects of grass carp and the herbicide Diuron in densely vegetated pools containing golden shiners and bluegills. Progressive Fish-Culturist 37:185–190.

Buntz, J., and P. Chapman. 1971. Fish population survey and comparison of Lakes Lena and Hancock. Mimeographed report, Florida Game and Fresh Water Fish Commission, Tallahassee. 22 pp.

Buntz, J., and C. S. Manooch III. 1968. *Tilapia aurea* (Steindachner), a rapidly spreading exotic in south central Florida. Proceedings of the Annual Conference of the Southeastern Association of Game and Fish Commissioners 22:495–501.

Buntz, J., and C. S. Manooch III. 1969. A brief summary of the cichlids in the south Florida region. Mimeographed report, Florida Game and Fresh Water Fish Commission. 3 pp.

Burns, J. W. 1966. Carp, 510–515. *In:* A. Calhoun (ed.). Inland fisheries management. California Fish and Game Department, Sacramento, California.

Cagni, J. E., D. L. Sutton, and R. D. Blackburn. 1971. A review of the amur (*Ctenopharyngodon idella*) as a biological control agent for aquatic weeds. Appendix C (15 pp.). Aquatic Plant Control Research Planning Conferences. Interagency Research Advisory Committee, Aquatic Plant Control Program. Office of the Chief of Engineers, Department of the Army.

Cahn, A. R. 1929. The effect of carp on a small lake: The carp as a dominant. Ecology 10(3):271–274.

Cahoon, W. G. 1953. Commercial carp removal at Lake Mattamuskeet, North Carolina. Journal of Wildlife Management 17(3):212–217.

Chamberlain, E. B., Jr. 1948. Ecological factors influencing the growth and management of certain waterfowl food plants on Back Bay National Wildlife Refuge. Transactions of the North American Wildlife Conference 13:347–356.

Chapman, D. W. 1966. Food and space as regulators of salmonid populations in streams. American Naturalist 100:345–357.

Chapman, D. W., and T. O. Bjornn. 1969. Distribution of salmonids in streams, with special reference to food and feeding, 153–176. *In:* T. G. Northcote (ed.). Symposium on salmon and trout in streams. H. R. McMillan Lectures in Fisheries, University of British Columbia, Vancouver, Canada.

Chevassus, B. 1979. Hybridization in salmonids: Results and perspectives. Aquaculture 17:113–128.

Chew, R. L. 1972. The failure of largemouth bass, *Micropterus salmoides floridanus* (LeSueur), to spawn in eutrophic, over-crowded environments. Proceedings of the Annual Conference of the Southeastern Association of Game and Fish Commissioners 26:306–319.

Childers, W. F., and G. W. Bennett. 1967. Experimental vegetation control by largemouth bass–tilapia combinations. Journal of Wildlife Management 31:401–407.

Clemens, W. A. 1928. The food of trout from the streams of Oneida County, New York State. Transactions of the American Fisheries Society 58:183–197.

Cobb, E. W. 1933. Results of trout tagging to determine migrations and results from plants made. Transactions of the American Fisheries Society 63:308–318.

Cole, J. L. 1905. The German carp in the United States. Report of the U.S. Bureau of Fisheries (1904):525–641.

Cole, J. L. 1906. The status of the carp in America. Transactions of the American Fisheries Society 34:201–207.

Colle, D. E., J. V. Shireman, and R. W. Rottmann. 1978. Food selection by grass carp fingerlings in a vegetated pond. Transactions of the American Fisheries Society 107(1):149–152.

Conner, J. V., R. P. Gallagher, and M. F. Chatry. 1980. Larval evidence for natural reproduction of the grass carp (*Ctenopharyngodon idella*) in the lower Mississippi River. Proceedings of the Fourth Annual Larval Fish Conference, Biological Services Program, National Power Plant Team, Ann Arbor, Michigan, FWS/OBS-80/43:1–19.

Cooper, E. L. 1952. Rate of exploitation of wild eastern brook trout and brown trout populations in the Pigeon River, Otsego County, Michigan. Transactions of the American Fisheries Society 81:224–234.

Cooper, E. L. 1953a. Returns from plantings of legal-sized brook, brown and rainbow trout in the Pigeon River, Otsego County, Michigan. Transactions of the American Fisheries Society 82:265–280.

Cooper, E. L. 1953b. Growth of brook trout (*Salvelinus fontinalis*) and brown trout (*Salmo trutta*) in the Pigeon River, Otsego County, Michigan. Papers of the Michigan Academy of Scicncc, Arts and Letters 38:151–162.

Cooper, E. L. 1959. Trout stocking as an aid to fish management. Pennsylvania State University Agricultural Experimental Station Bulletin 663.

Cordone, A. J., and D. W. Kelley. 1961. The influences of inorganic sediment on the aquatic life of streams. California Fish and Game 47(2):189–228.

Courtenay, W. R., Jr. 1979. The introduction of exotic organisms, 237–252. *In:* H. P. Brokaw (ed.). Wildlife and America. U.S. Government Printing Office, Washington, D.C.

Courtenay, W. R., Jr., and J. E. Deacon. 1982. The status of introduced fishes in certain spring systems in southern Nevada. Great Basin Naturalist 42(3):361–366.

Courtenay, W. R., Jr., and J. E. Deacon. 1983. Fish introductions in the American Southwest: A case history of Rogers Spring, Nevada. Southwestern Naturalist 28(2):221–224.

Courtenay, W. R., Jr., and D. A. Hensley. 1979. Range expansion in southern Florida of the introduced spotted tilapia, with comments on its environmental impress. Environmental Conservation 6(2):149–151.

Courtenay, W. R., Jr., and D. A. Hensley. 1980. Special problems associated with monitoring exotic species, 281–307. *In:* C. H. Hocutt and J. R. Stauffer, Jr. (eds.). Biological monitoring of fish. Lexington Books, Lexington, Massachusetts.

Courtenay, W. R., Jr., and C. R. Robins. 1973. Exotic aquatic organisms in Florida with emphasis on fishes: A review and recommendations. Transactions of the American Fisheries Society 102(1):1–12.

Courtenay, W. R., Jr., and C. R. Robins. 1975. Exotic organisms: An unsolved, complex problem. BioScience 25(5):306–313.

Courtenay, W. R., Jr., H. F. Sahlman, W. W. Miley II, and D. J. Herrema. 1974. Exotic fishes in fresh and brackish waters of Florida. Biological Conservation 6(4):292–302.

Cox, T. J. 1972. The food habits of the desert pupfish (*Cyprinodon macularius*) in Quitobaquito Springs, Organ Pipe National Monument, Arizona. Journal of the Arizona Academy of Sciences 7(1):25–27.

Crisman, T. L., and F. M. Kooijman. 1981. Large-scale operations management test. The plankton and benthic invertebrate response. Miscellaneous Paper A81-3, 361–368. U.S. Army Engineer Waterways Experiment Station, CE, Vicksburg, Mississippi.

Crittenden, E. 1962. Status of *Tilapia nilotica* Linnaeus in Florida. Proceedings of the Annual Conference of the Southeastern Association of Game and Fish Commissioners 16:257–262.

Cross, D. G. 1969. Aquatic weed control using grass carp. Journal of Fish Biology 1:27–30.

Crowder, J. P., and J. R. Snow. 1969. Use of grass carp for weed control in ponds. Food and Agriculture Organization of the United Nations, Fish Culture Bulletin 2(1):6.

Daget, J. 1952. Observations sur la ponte de *Tilapia zillii* (Gervais), poisson de la famille des Cichlidae. Annals and Magazine of Natural History, London 12:309–310.

Dangel, J. R., P. T. Macy, and F. C. Withler. 1973. Annotated bibliography of interspecific hybridization of fishes of the subfamily Salmoninae. U.S. Department of Commerce, National Oceanic and Atmospheric Administration (NOAA) Technical Memorandum NMFS NWFC-1. 48 pp.

Deacon, J. E. 1979. Endangered and threatened species of the West. *In:* The endangered species: A symposium. Great Basin Naturalist Memoirs 3:41–64.

Deacon, J. E., L. J. Paulson, and C. O. Minckley. 1972. Effects of Las Vegas wash effluents upon bass and other game fish reproduction and success, and quantitative and qualitative analysis of stomach samples of major game and forage fishes of Lake Mead. Mimeographed final report, Nevada Department of Fish and Game, Las Vegas, Nevada. 74 pp.

Dean, W. J., and W. H. Bailey. 1977. Reproductive repression of largemouth bass in a heated reservoir. Proceedings of the Annual Conference of the Southeastern Association of Fish and Wildlife Agencies 31:463–470.

DeBach, P. 1974. Biological control by natural enemies. Cambridge University Press, London, England. 323 pp.

DeKay, J. E. 1842. Zoology of New York – IV: Fishes. W. and A. White and J. Visscher, Albany, New York.

DeVos, A., R. H. Manville, and R. G. Van Gelder. 1956. Introduced mammals and their influence on native biota. Zoologica 41:163–194.

Diamond, J. M., and C. R. Veitch. 1981. Extinctions and introductions of the New Zealand avifauna: Cause and effect? Science 211:499–501.

Drda, T. F. 1976. Grass carp project vegetation study. Mimeographed completion report, Florida Game and Fresh Water Fish Commission. 33 pp.

Dymond, J. R. 1955. The introduction of foreign fishes in Canada. Proceedings of the International Association of Theoretical and Applied Limnology 12:543–553.

Egglishaw, H. J. 1967. The food, growth and population structure of salmon and trout in two streams in the Scottish Highlands. Freshwater and Salmon Fisheries Research (Department of Agriculture and Fisheries, Edinburgh, Scotland) 38:1–32.

Elton, C. S. 1958. The ecology of invasion by plants and animals. Chapman and Hall, London, England. 181 pp.

Ensign, A. R. 1960. Not what – how! Wisconsin Conservation Bulletin 25(4):1–3.

Evans, H. E. 1952. The food of a population of brown trout, *Salmo trutta* Linn., from central New York. American Midland Naturalist 47(2):413–420.

Fausch, K. D., and R. J. White. 1981. Competition between brook trout (*Salvelinus fontinalis*) and brown trout (*Salmo trutta*) for positions in a Michigan stream. Canadian Journal of Fisheries and Aquatic Sciences 38:1220–1227.

Fenderson, C. N. 1954. The brown trout in Maine. Maine Department of Inland Fish and Game, Bulletin no. 2:1–16.

Finucane, J. H., and G. R. Rinckey. 1964. A study of the African cichlid, *Tilapia heudeloti* Dumeril, in Tampa Bay, Florida. Proceedings of the Annual Conference of the Southeastern Association of Game and Fish Commissioners 18:259–269.

Fitzpatrick, L. A., B. W. Rickel, M. O. Saeed, and C. O. Ziebell. 1981. Factors influencing the effectiveness of *Tilapia zillii* in controlling aquatic weeds. Arizona Cooperative Fisheries Research Unit, Research Report 81–1:1–19.

Forester, J. S., and J. W. Avault, Jr. 1978. Effects of grass carp on freshwater red swamp crawfish in ponds. Transactions of the American Fisheries Society 107(1): 156–160.

Forester, T. S., and J. M. Lawrence. 1978. Effects of grass carp and carp on populations of bluegill and largemouth bass in ponds. Transactions of the American Fisheries Society 107(1):172–175.

Fowler, M. C., and T. O. Robson. 1978. The effects of the food preferences and stocking rates of grass carp (*Ctenopharyngodon idella* Val.) on mixed plant communities. Aquatic Botany 5(3):261–276.

Foye, R. E. 1965. The carp menace in Merrymeeting. Maine Fish and Game (spring): 26–27.

Frey, D. G., and L. Vike. 1941. A creel census of Lakes Waubesa and Kegonsa, Wisconsin, in 1939. Transactions of the Wisconsin Academy of Sciences, Arts and Letters 33:339–362.

Fry, D. L., and J. A. Osborne. 1980. Zooplankton abundance and diversity in central Florida grass carp ponds. Hydrobiologia 68(2):145–155.

Fryer, G. 1961. Observations of the biology of the cichlid fish *Tilapia variabilis* Boulenger in the northern waters of Lake Victoria (East Africa). Revue de Zoologie et de Botanique africaine 64:1–33.

Gard, R., and G. A. Flittner. 1974. Distribution and abundance of fishes in Sagehen Creek, California. Journal of Wildlife Management 38(2):347–358.

Garman, G. C., and L. A. Nielsen. 1982. Piscivority by stocked brown trout (*Salmo trutta*) and its impact on the nongame fish community of Bottom Creek, Virginia. Canadian Journal of Fisheries and Aquatic Sciences 39(6):862–869.

Gasaway, R. D. 1977. Predation on introduced grass carp (*Ctenopharyngodon idella*) in a Florida lake. Florida Scientist 40(2):167–173.

Gasaway, R. D., and T. F. Drda. 1977. Effects of grass carp introduction on waterfowl habitat. Transactions of the North American Wildlife and Natural Resources Conference 42:73–85.

Gasaway, R. D., and T. F. Drda. 1978. Effects of grass carp introduction on macrophytic vegetation and chlorophyll content of phytoplankton in four Florida lakes. Florida Scientist 41(2):101–109.

Germany, R. D. 1977. Population dynamics of the blue tilapia and its effects on the fish populations of Trinidad Lake, Texas. Ph.D. dissertation, Texas A & M University, College Station, Texas. 85 pp.

Gleastine, B. W. 1974. A study of the cichlid *Tilapia aurea* (Steindachner) in a thermally modified reservoir. M.S. thesis, Texas A & M University, College Station, Texas. 258 pp.

Grandall, K. S., and W. A. Swenson. 1982. Responses of brook trout and creek chubs to turbidity. Transactions of the American Fisheries Society 111(3):392–395.

Greeley, J. R. 1932. The spawning habits of brook, brown and rainbow trout, and the problem of egg predators. Transactions of the American Fisheries Society 62:239–248.

Greenfield, D. W. 1973. An evaluation of the advisability of the release of the grass carp, *Ctenopharyngodon idella,* into the natural waters of the United States. Transactions of the Illinois Academy of Sciences 66(1/2):47–53.

Grizzell, R. A., Jr., and W. W. Neely. 1962. Biological controls for water-weeds. Transactions of the North American Wildlife Conference 27:107–113.

Haacker, P. L. 1979. Two Asiatic gobiid fishes, *Tridentiger trigonocephalus* and *Acanthogobius flavimanus,* in southern California. Bulletin of the Southern California Academy of Sciences 78(1):56–61.

Habel, M. 1975. Overwintering of the cichlid, *Tilapia aurea,* produces fourteen tons of harvestable sized fish in a south Alabama bass-bluegill public fishing lake. Progressive Fish-Culturist 37:31–32.

Hale, J. G. 1954. Investigations on trout catch and populations in the West Branch of the Split Rock River, Lake County, Minnesota, 1951–53. Minnesota Fish and Game Investigations, Fisheries Series 148. 14 pp.

Hannuksela, P. R. 1969. Food habits of brown trout in the Anna River, Michigan. Michigan Department of Natural Resources Report 1759.

Hardin, S., R. Land, G. Morse, and M. Spelman. 1982. Large-scale operations management test of use of the white amur for control of problem aquatic plants. Report 2. First year poststocking results. Vol. 2. The fish, mammals, and waterfowl of Lake Conway, Florida. Technical Report A-78-2, prepared by the Florida Game and Fresh Water Fish Commission, Orlando, Florida, for the U.S. Army Engineer Waterways Experiment Station, CE, Vicksburg, Mississippi.

Hardy, T. 1981. Ecological status of the aquatic community within the outflow of Ash Springs, Nye County, Nevada, with special reference to the Pahranegut roundtail chub (*Gila robusta jordani*). Final report, contract #14-16-001-6319 FS, amendment 6, U.S. Fish and Wildlife Service. 72 pp.

Harlan, J. R., and E. B. Speaker. 1969. Iowa fish and fishing. Iowa State Conservation Commission, Des Moines, Iowa. 365 pp.

Harris, C. 1978. Tilapia: Florida's alarming alien menace. Florida Sportsman 9(11): 12, 15, 17–19.

Harrison, H. M. 1950. The foods used by some common fish of the Des Moines River drainage. Proceedings of the Iowa Conservation Commission Biological Seminar, July 1950:31–44 (mimeo).

Hauser, W. J., E. F. Legner, R. A. Medved, and S. Platt. 1976. *Tilapia*—a management tool for biological control of aquatic weeds and insects. Fisheries (Bethesda) 1(2):15–16.

Heimstra, N. W., D. K. Damkot, and N. G. Benson. 1969. Some effects of silt turbidity on behavior of juvenile largemouth bass and green sunfish. Technical Paper no. 20, Bureau of Sport Fisheries and Wildlife, Fish and Wildlife Service, U.S. Department of the Interior, Washington, D.C. 9 pp.

Hendricks, M. K., and R. L. Noble. 1979. Feeding interactions of three planktivorous fishes in Trinidad Lake, Texas. Proceedings of the Annual Conference of the Southeastern Association of Fish and Wildlife Agencies 33:324–330.

Hestand, R. S., and C. C. Carter. 1978. Comparative effects of grass carp and selected herbicides on macrophyte and phytoplankton communities. Journal of Aquatic Plant Management 16:43–50.

Hickling, C. F. 1961. Tropical inland fisheries. John Wiley and Sons, New York. 287 pp.

Hickling, C. F. 1965. Biological control of aquatic vegetation. Pest Articles and News Summaries 11(3):237–244.

Hiscox, J. I. 1979. Feeding habits and growth of stocked salmonids in a California reservoir. California Department of Fish and Game Report 79-3.

Hogan, J. 1946. The control of aquatic plants with fertilizers in rearing ponds at the Lonoke hatchery, Arkansas. Transactions of the American Fisheries Society 76:183–189.

Hogg, R. G. 1976. Ecology of fishes of the family Cichlidae introduced into the fresh waters of Dade County, Florida. Ph.D. dissertation, University of Miami, Coral Gables, Florida. 142 pp.

Holloway, A. D. 1948. Twelve years of fishing records from Lake Mattamuskeet. Transactions of the North American Wildlife Conference 13:474–480.

Horel, G. 1969. Federal Aid Project F-12-10. Lake management, research and development. Study 3. Fish population studies. Mimeographed report, Florida Game and Fresh Water Fish Commission. 70 pp.

Hotchkiss, N. 1941. The limnological role of higher plants, 152–162. *In:* A symposium on hydrobiology. University of Wisconsin Press, Madison, Wisconsin.

Hubbs, C., T. Lucier, G. P. Garrett, R. J. Edwards, S. M. Dean, E. Marsh, and D. S. Belk. 1978. Survival and abundance of introduced fishes near San Antonio, Texas. Texas Journal of Science 30(4):369–376.

Hubbs, C. L., and R. W. Eschmeyer. 1938. The improvement of lakes for fishing. Michigan Department of Conservation, Institute of Fisheries Research, Bulletin 2:1–233.

Idyll, C. 1942. Food of rainbow, cutthroat and brown trout in the Cowichan River system, B.C. Journal of the Fisheries Research Board of Canada 5(5):448–458.

Jenkins, T. M. 1969. Social structure, position choice and microdistribution of two trout species (*Salmo trutta* and *Salmo gairdneri*) resident in mountain streams. Animal Behavior Monographs 2(2):57–123.

Jessen, R. L., and J. H. Kuehn. 1960a. When the carp are eliminated. Minnesota Department of Conservation, Official Bulletin 23(134):46–50.

Jessen, R. L., and J. H. Kuehn. 1960b. A preliminary report on the effects of the elimination of carp on submerged vegetation. Minnesota Fish and Game Investigations, Department of Conservation 2:1–12.

Johnson, M., and J. M. Laurence. 1973. Biological weed control with the white amur. Appendix E, 12 pp. *In:* Herbivorous fish for aquatic plant control. Mimeographed technical report 4, Aquatic Plant Control Program. U.S. Army Corps of Engineers, Vicksburg, Mississippi.

Jones, A. C. 1972. Contributions to the life history of the Piute sculpin in Sagehen Creek, California. California Fish and Game 58(4):285–290.

Jonez, A., and R. C. Sumner. 1954. Lakes Mead and Mojave investigations. Mimeographed final report, Nevada Fish and Game Commission (D-J Project F-1-R), Reno, Nevada. 186 pp.

Kaleel, R. T. 1981. Large-scale operations management test. Water quality analysis of Lake Conway. Miscellaneous paper A-81-3, 340–360. U.S. Army Engineer Waterways Experiment Station, CE, Vicksburg, Mississippi.

Kalleberg, H. 1958. Observations in a stream tank of territoriality and competition in juvenile salmon and trout. Institute of Freshwater Research, Drottningholm, Report no. 39:55–98.

Kelly, H. D. 1956. Preliminary studies on *Tilapia mossambica* Peters relative to experimental pond culture. Proceedings of the Annual Conference of the Southeastern Association of Game and Fish Commissioners 10:139–149.

Kilgen, R. H. 1978. Growth of channel catfish and striped bass in small ponds stocked with grass carp and water hyacinths. Transactions of the American Fisheries Society 107(1):176–180.

Kilgen, R. H., and R. O. Smitherman. 1971. Food habits of the white amur (*Cteno-*

pharyngodon idella) stocked in ponds alone and in conjunction with other species. Progressive Fish-Culturist 33(3):123–127.

Kimsey, J. B., and L. O. Fisk. 1969. Freshwater nongame fishes of California. California Department of Fish and Game. 54 pp.

King, D. R., and G. S. Hunt. 1967. Effect of carp on vegetation in a Lake Erie marsh. Journal of Wildlife Management 31(1):181–188.

Knaggs, E. H. 1977. Status of the genus *Tilapia* in California's estuarine and marine waters. California-Nevada Wildlife Transactions 1977:60–67.

Kobylinski, G. J., W. W. Miley II, J. M. van Dyke, and A. J. Leslie, Jr. 1980. The effects of grass carp (*Ctenopharyngodon idella* Val.) on vegetation, water quality, zooplankton, and macroinvertebrates of Deer Point Lake, Bay County, Florida. Mimeographed final report, Bureau of Aquatic Plant Research and Control, Florida Department of Natural Resources, Tallahassee, Florida. 114 pp.

Lagler, K. F., and W. C. Latta. 1954. Michigan fish predators. Michigan Department of Conservation, pamphlet no. 12, 17 pp.

Lahser, C. W., Jr. 1967. *Tilapia mossambica* as a fish for aquatic weed control. Progressive Fish-Culturist 29(1):48–50.

Land, R. 1981. Large-scale operations management test. Fish, mammals, and waterfowl. Miscellaneous Paper A-81-3, 325–339. U.S. Army Engineer Waterways Experiment Station, CE, Vicksburg, Mississippi.

LaRivers, I. 1962. Fishes and fisheries of Nevada. Nevada State Fish and Game Commission, Nevada. 782 pp.

Larkin, P. A. 1979. Predator-prey relations in fishes: An overview of the theory, 13–22. *In:* H. Clepper (ed.). Predator-prey systems in fisheries management. Sport Fishing Institute, Washington, D.C.

Laycock, G. 1966. The alien animals. Natural History Press, Garden City, New York. 240 pp.

Legner, E. F. 1978. Efforts to control *Hydrilla verticillata* Royle with herbivorous *Tilapia zillii* (Gervais) in Imperial County irrigation canals. Proceedings of the California Mosquito Control Association 46:103–104.

Legner, E. F. 1979. Considerations in the management of *Tilapia* for biological aquatic weed control. Proceedings of the California Mosquito Control Association 47:44–45.

Legner, E. F., T. W. Fisher, and R. A. Medved. 1973. Biological control of aquatic weeds in the lower Colorado River basin. Proceedings of the California Mosquito Control Association 41:115–117.

Legner, E. F., W. J. Hauser, T. W. Fisher, and R. A. Medved. 1975. Biological aquatic weed control by fish in the lower Sonoran Desert of California. California Agriculture News 29(11):8–10.

Legner, E. F., and R. A. Medved. 1972. Predators investigated for the biological control of mosquitoes and midges at the University of California, Riverside. Proceedings of the California Mosquito Control Association 40:109–111.

Legner, E. F., and R. A. Medved. 1973a. Predation of mosquitoes and chironomid midges in ponds by *Tilapia zillii* (Gervais) and *T. mossambica* (Peters) (Teleosteii: Cichlidae). Proceedings of the California Mosquito Control Association 41:119–121.

Legner, E. F., and R. A. Medved. 1973b. Influence of *Tilapia mossambica* (Peters), *T.*

zillii (Gervais) (Cichlidae) and *Mollienisia latipinna* LeSueur (Poeciliidae) on pond populations of *Culex* mosquitoes and chironomid midges. Mosquito News 33(3): 354–364.

Legner, E. F., and F. W. Pelsue. 1977. Adaptations of *Tilapia* to *Culex* and chironomid midge ecosystems in south California. Proceedings of the California Mosquito Control Association 45:95–97.

Lembi, C. A., and B. G. Ritenour. 1977. The white amur as a biological control for aquatic weeds in Indiana. Mimeographed technical report no. 95, Purdue University Water Resources Research Center, West Lafayette, Indiana. 95 pp.

Lembi, C. A., B. G. Ritenour, E. M. Iverson, and E. C. Forss. 1978. The effects of vegetation removal by grass carp on water chemistry and phytoplankton in Indiana ponds. Transactions of the American Fisheries Society 107:161–171.

Lemmien, W. A., P. I. Tack, and W. F. Morofsky. 1957. Results from planting brown trout and rainbow trout in Augusta Creek, Kalamazoo County, Michigan. Michigan Agricultural Experimental Station Quarterly Bulletin 40:242–249.

Leventer, H. 1981. Biological control of reservoirs by fish. Bamidgeh 33(1):3–23.

Lewis, S. L. 1969. Physical factors influencing fish populations in pools of a trout stream. Transactions of the American Fisheries Society 98(1):14–19.

Lewis, W. M. 1978. Observations on the grass carp in ponds containing fingerling channel catfish and hybrid sunfish. Transactions of the American Fisheries Society 107(1):153–155.

Lewis, W. M., and D. R. Helms. 1964. Vulnerability of forage organisms to largemouth bass. Transactions of the American Fisheries Society 93(3):315–318.

Li, H. W., and P. B. Moyle. 1981. Ecological analysis of species introductions into aquatic systems. Transactions of the American Fisheries Society 110(6):772–782.

Lindroth, A. 1955. Distribution, territorial behaviour and movements of sea trout fry in the river Indalsälven. Report of the Institute of Freshwater Research, Drottningholm 36:104–119.

Lobel, P. S. 1980. Invasion by the Mozambique tilapia (*Sarotherodon mossambicus;* Pisces; Cichlidae) of a Pacific atoll marine ecosystem. Micronesica 16(2):349–355.

Loiselle, P. V. 1977. Colonial breeding by an African substratum-spawning cichlid fish, *Tilapia zillii* (Gervais). Biology of Behaviour 2:129–142.

Lowe, R. H. 1955. Species of *Tilapia* in East African dams, with a key for their identification. East African Agriculture Journal 20(4):256–262.

McBay, L. G. 1962. The biology of *Tilapia nilotica* Linnaeus. Proceedings of the Annual Conference of the Southeastern Association of Game and Fish Commissioners 15:208–218.

McCraig, R. S., J. W. Mullan, and C. O. Dodge. 1960. Five-year report on the development of the fishery of a 25,000 acre domestic water supply reservoir in Massachusetts. Progressive Fish-Culturist 22:15–23.

McCrimmon, H. R. 1956. Fishing in Lake Simcoe. Publication of the Ontario Department of Lands and Forests. 137 pp.

McCrimmon, H. R. 1968. Carp in Canada. Fisheries Research Board of Canada, Bulletin no. 165, 93 pp.

Maciolek, J. A., and P. R. Needham. 1951. Ecological effects of winter conditions on trout and trout foods in Convict Creek, California. Transactions of the American Fisheries Society 81:202–217.

MacKay, H. H. 1963. Fishes of Ontario. Publication of the Ontario Department of Lands and Forests, Ottawa, Canada, 166–175.

Magnuson, J. J. 1976. Managing with exotics—a game of chance. Transactions of the American Fisheries Society 105(1):1–9.

Marshall, T. L., and H. R. MacCrimmon. 1970. Exploitation of self-sustaining Ontario stream populations of brown trout (*Salmo trutta*) and brook trout (*Salvelinus fontinalis*). Journal of the Fisheries Research Board of Canada 27:1087–1102.

Martin, R. 1976. Exotic fish problems and opportunities in the Southeast. Proceedings of the Southeastern Association of Game and Fish Commissioners 30:1–6.

Mather, F. 1889. Brown trout in America. Bulletin of the U.S. Fish Commission 7(1887):21–22.

Mathis, W. P. 1965. Observations of control of vegetation in Lake Catherine using Israeli carp and a fall and winter drawdown. Southwestern Weed Science Society Proceedings 18:197–205.

Mense, J. B. 1975. Relation of density to brown trout movement in a Michigan stream. Transactions of the American Fisheries Society 104(4):688–694.

Metzelaar, J. 1929. The food of the trout in Michigan. Transactions of the American Fisheries Society 59:146–152.

Michewicz, J. E., D. L. Sutton, and R. D. Blackburn. 1972a. Water quality of small enclosures stocked with white amur. Hyacinth Control Journal 10:22–25.

Michewicz, J. E., D. L. Sutton, and R. D. Blackburn. 1972b. The white amur for aquatic weed control. Weed Science 20(1):106–110.

Miley, W. W., II. 1978. Ecological impact on the pike killifish, *Belonesox belizanus* Kner (Poeciliidae), in southern Florida. M.S. thesis, Florida Atlantic University, Boca Raton, Florida. 55 pp.

Miley, W. W., II, A. J. Leslie, Jr., and J. M. van Dyke. 1979. The effects of grass carp (*Ctenopharyngodon idella* Val.) on vegetation and water quality in three central Florida lakes. Mimeographed final report, Bureau of Aquatic Plant Research and Control, Florida Department of Natural Resources, Tallahassee, Florida. 119 pp.

Millard, T. J., and H. R. MacCrimmon. 1972. Evaluation of the contribution of supplemental plantings of brown trout *Salmo trutta* (L.) to a self-sustaining fishery in the Sydeham River, Ontario, Canada. Journal of Fish Biology 4:369–384.

Miller, H. J., C. L. Brynildson, and C. W. Threinen. 1959. Rough fish control. Wisconsin Conservation Department Publication no. 229, 15 pp.

Miller, R. B. 1957a. Permanence and size of home territory in stream-dwelling cutthroat trout. Journal of the Fisheries Research Board of Canada 14:687–691.

Miller, R. B. 1957b. Have the genetic patterns of fishes been altered by introductions or by selective fishing? Journal of the Fisheries Research Board of Canada 14:797–806.

Miller, R. B. 1958. The role of competition in the mortality of hatchery trout. Journal of the Fisheries Research Board of Canada 15:27–45.

Minckley, W. L. 1973. The fishes of Arizona. Arizona Game and Fish Department, Phoenix, Arizona. 293 pp.

Minckley, W. L., and J. E. Deacon. 1968. Southwestern fishes and the enigma of "endangered species." Science 159(3822):1424–1432.

Mitzner, L. 1978. Evaluation of biological control of nuisance aquatic vegetation by grass carp. Transactions of the American Fisheries Society 107(1):135–145.

Moen, T. 1953a. Food habits of the carp in northwest Iowa lakes. Iowa Academy of Science 60:665–686.

Moen, T. 1953b. Carp are not vegetarians. Iowa Conservationist 12(5):129–134.

Montegut, R. S., R. D. Gasaway, D. F. DuRant, and J. L. Atterson. 1976. An ecological evaluation of the effects of grass carp (*Ctenopharyngodon idella*) introduction in Lake Wales, Florida. Mimeographed report, Florida Game and Fresh Water Fish Commission. 113 pp.

Moyle, J. B. 1949. Fish population concepts and management of Minnesota lakes for sport fishing. Transactions of the North American Wildlife Conference 14:283–294.

Moyle, J. B., and J. H. Kuehn. 1964. Carp, a sometimes villain. Reprinted from Waterfowl Tomorrow, U.S. Department of Interior. 8 pp.

Moyle, P. B. 1976a. Inland fishes of California. University of California Press, Berkeley, California. 405 pp.

Moyle, P. B. 1976b. Fish introductions in California: History and impact on native fishes. Biological Conservation 9:101–118.

Moyle, P. B., and A. Marciochi. 1975. Biology of the Modoc sucker, *Catostomus microps* (Pisces: Catostomidae) in northeastern California. Copeia 1975(3): 556–560.

Moyle, P. B., and R. Nichols. 1974. Decline of the native fish fauna of the Sierra-Nevada foothills, Central California. American Midland Naturalist 92:72–83.

Mraz, D., and E. L. Cooper. 1957. Reproduction of carp, largemouth bass, bluegills, and black crappies in small rearing ponds. Journal of Wildlife Management 21(2): 127–133.

Nair, K. K. 1968. A preliminary bibliography of the grass carp, *Ctenopharyngodon idella* Valenciennes. Food and Agriculture Organization of the United Nations, Fisheries Circular no. 302:1–15.

Nall, L. E., and J. D. Schardt. 1978. Large-scale operations management test of use of the white amur for control of problem aquatic plants. Report 1. Baseline studies. Vol. 1. The aquatic macrophytes of Lake Conway, Florida. Technical Report A-78-2, U.S. Army Engineer Waterways Experiment Station, CE, Vicksburg, Mississippi.

Neave, F. 1949. Game fish populations of the Cowichan River. Bulletin of the Fisheries Research Board of Canada, no. 84, 32 pp.

Needham, P. R. 1929. Quantitative studies of the fish food supply in selected areas. *In:* A biological survey of the Erie-Niagara system. Supplement to the 18th Annual Report of the New York Conservation Department 1928:220–232.

Needham, P. R. 1938. Trout streams. Comstock Publishing Co., Ithaca, New York. 233 pp.

Nelson, J. S. 1965. Effects of fish introductions and hydroelectric development on fishes in the Kananaskis River System, Alberta. Journal of the Fisheries Research Board of Canada 22(3):721–753.

Newman, M. A. 1956. Social behaviour and interspecific competition in two trout species. Physiological Zoology 29(1):64–81.

Noble, R. L. 1977. Response of reservoir fish populations to tilapia reduction. Mimeographed progress report to the Sport Fishery Research Foundation. 4 pp.

Noble, R. L., R. D. Germany, and C. R. Hall. 1975. Interactions of blue tilapia and largemouth bass in a power plant cooling reservoir. Proceedings of the Annual Conference of the Southeastern Association of Game and Fish Commissioners 29:247–251.

Nyman, O. L. 1970. Ecological interaction of brown trout, *Salmo trutta* L. and brook trout, *Salvelinus fontinalis* (Mitchill) in a stream. Canadian Field-Naturalist 84:343–350.

O'Donnall, J. D., and W. S. Churchill. 1960. Certain physical, chemical and biological aspects of the Brule River, Wisconsin. Transactions of the Wisconsin Academy of Sciences, Arts and Letters 43:172–189.

O'Donnell, D. J. 1943. The fish population in three small lakes in northern Wisconsin. Transactions of the American Fisheries Society 72:187–196.

Opuszyński, K. 1972. Use of phytophagous fish to control aquatic plants. Aquaculture I(1):61–74.

Osborne, J. A. 1978. Management of emergent and submergent vegetation in stormwater retention ponds using grass carp. Mimeographed final report, Florida Department of Natural Resources, Florida Technical University, Orlando, Florida.

Pardue, G. B. 1973. Production response of bluegill sunfish, *Lepomis macrochirus* Rafinesque, to added attachment surface for fish food organisms. Transactions of the American Fisheries Society 102(3):622–626.

Parsons, P. A. 1982. Adaptive strategies of colonizing animal species. Geological Review 57:117–148.

Pate, V.S.L. 1933. Studies on fish food in selected areas. *In:* A biological survey of the Upper Hudson watershed. Supplement to the 22d Annual Report of the New York Conservation Department 1932:130–156.

Pelzman, R. J. 1973. A review of the life history of *Tilapia zillii* with a reassessment of its desirability in California. California Department of Fish and Game, Inland Fisheries Administrative Report 74-1:1–9.

Pentelow, F.T.K., and B. Stott. 1965. Grass carp for weed control. Progressive Fish-Culturist 27:210.

Philippart, J-Cl., and J-Cl. Ruwet. 1982. Ecology and distribution of tilapias, 15–59. *In:* R.S.V. Pullin and R. H. Lowe-McConnell (eds.). The biology and culture of tilapias. ICLARM Conference Proceedings 7, 432 pp. International Center for Living Aquatic Resource Management, Manila, Philippines.

Phillippy, C. L. 1969. *Tilapia melanopleura* as a control for aquatic vegetation. Mimeographed report, Florida Game and Fresh Water Fish Commission. 13 pp.

Pierce, P. C., and H. M. Yawn. 1965. Six field tests using two species of tilapia for controlling aquatic vegetation. Proceedings of the Southern Weed Control Conference 18:582–583.

Pister, E. P. 1962. The brown trout. Outdoor California 23(6):7–10.

Provine, W. C. 1975. The grass carp. Mimeographed special report, Texas Parks and Wildlife Department, Inland Fisheries Research. 51 pp.

Prowse, G. A. 1969. The role of cultured pond fish in the control of eutrophication in lakes and dams. Verhandlungen der internationalen Vereinigung für theoretische und angewandte Limnologie 17:714–718.

Prowse, G. A. 1971. Experimental criteria for studying grass carp feeding in relation to weed control. Progressive Fish-Culturist 33:128–133.

Quast, J. C. 1961. The food of the bairdiella. *In:* B. W. Walker (ed.). The ecology of the Salton Sea, California, in relation to the sportfishery. California Department of Fish and Game, Fishery Bulletin 113:153–164.

Raney, E. C., and E. A. Lachner. 1942. Autumn food of recently planted young brown trout in small streams of central New York. Transactions of the American Fisheries Society 71:106–111.

Regier, H. A. 1968. The potential misuse of exotic fish as introductions, 92–111. *In:* K. H. Loftus (ed.). A symposium on introductions of exotic species. Research Report 82, Ontario Department of Lands and Forests, Ottawa, Canada.

Rehder, D. D. 1959. Some aspects of the life history of the carp, *Cyprinus carpio,* in the Des Moines River, Boone County, Iowa. Iowa State College Journal of Science 34:11–26.

Reimers, N., J. A. Maciolek, and E. P. Pister. 1955. Limnological study of the lakes in Convict Creek Basin, Mono County, California. U.S. Fish and Wildlife Service, Fisheries Bulletin 56:437–503.

Ricker, W. E., and J. Gottschalk. 1941. An experiment in removing coarse fish from a lake. Transactions of the American Fisheries Society 70:382–390.

Rimsky-Korsakoff, V. N. 1930. The food of certain fishes of the Lake Champlain watershed. *In:* A biological survey of the Champlain watershed. Supplement of the 19th Annual Report of the New York Conservation Department 1929:88–104.

Robel, R. J. 1961. The effects of carp populations on the production of waterfowl food plants on a western waterfowl marsh. Transactions of the North American Wildlife Conference 26:147–159.

Rose, E. T., and T. Moen. 1953. The increase in game fish populations in East Okoboji Lake, Iowa, following intensive removal of rough fish. Transactions of the American Fisheries Society 82:104–114.

Rottmann, R. W., and R. O. Anderson. 1976. Limnological and ecological effects of grass carp in ponds. Proceedings of the Annual Conference of the Southeastern Association of Fish and Wildlife Agencies 30:24–39.

Saunders, J. W., and M. W. Smith. 1965. Changes in stream populations of trout associated with increased silt. Journal of the Fisheries Research Board of Canada 22(2):395–404.

Savino, J. F., and R. A. Stein. 1982. Predator-prey interaction between largemouth bass and bluegills as influenced by simulated, submersed vegetation. Transactions of the American Fisheries Society 111(3):255–266.

Schardt, J. D., and L. E. Nall. 1981. Large-scale operations management test. Aquatic macrophytes. Miscellaneous Paper A-81-3, 302–324. U.S. Army Engineer Waterways Experiment Station, CE, Vicksburg, Mississippi.

Schneberger, E., and M. E. Jewell. 1928. Factors affecting pond fish production. Kansas Forestry, Fish and Game Commission, Bulletin no. 9:5–14.

Schoenherr, A. A. 1979. Niche separation within a population of freshwater fishes in an irrigation drain near the Salton Sea, California. Bulletin of the Southern California Academy of Sciences 78:46–55.

Schoenherr, A. A. 1981. The role of competition in the replacement of native fishes by introduced species, 173–203. *In:* R. J. Naiman and D. L. Soltz (eds.). Fishes in North American deserts. John Wiley and Sons, New York.

Schuck, H. A. 1942. The effect of population density of legal-sized trout upon the yield per standard fishing effort in a controlled section of stream. Transactions of the American Fisheries Society 71:236–248.

Schuck, H. A. 1945. Survival, population density, growth and movement of the wild brown trout in Crystal Creek. Transactions of the American Fisheries Society 73:209–230.

Schwartz, F. J. 1972. World literature to fish hybrids with an analysis by family, species, and hybrid. Publication no. 3, Gulf Coast Research Laboratory and Museum, Ocean Springs, Mississippi. 328 pp.

Schwartz, F. J. 1981. World literature to fish hybrids with an analysis by family, species, and hybrid. Supplement 1. National Oceanic and Atmospheric Administration Technical Report no. 750, National Marine Fisheries Service, Special Scientific Report – Fisheries. U.S. Dept. of Commerce, Washington, D.C. 507 pp.

Scott, W. B., and E. J. Crossman. 1973. Freshwater fishes of Canada. Fisheries Research Board of Canada. Bulletin 184. 966 pp.

Sculthorpe, C. C. 1967. The biology of aquatic vascular plants. Edward Arnold, London, England. 610 pp.

Shafland, P. L. 1979. Non-native fish introductions with special reference to Florida. Fisheries (Bethesda) 4(3):18–24.

Shafland, P. L., D. S. Levine, and R. J. Wattendorf. 1980. An evaluation of interspecific interactions between blue tilapia and certain native fishes in Florida: Fish population analysis. Second annual performance report, Non-Native Fish Research Laboratory, Boca Raton, Florida Game and Fresh Water Fish Commission. 31 pp.

Shafland, P. L., and J. M. Pestrak. 1984. Predation of blue tilapia by largemouth bass in experimental ponds. Proceedings of the 35th Annual Conference of the Southeastern Association of Fish and Wildlife Agencies, 18–21 October 1981, Tulsa, Oklahoma, 443–448.

Sharp, R. W. 1942. Some studies of the distribution and ecology of the German carp in Minnesota, with suggested control measures. Fisheries Research Investigational Report no. 45, Minnesota Department of Conservation. 25 pp.

Sharpe, F. P. 1957. Investigations of the feeding habits of the German brown trout, *Salmo trutta,* in Shadow Mountain Reservoir. Colorado Cooperative Fisheries Research Unit quarterly report 4(1):49–63.

Sharpe, F. P. 1960. Investigation of Shadow Mountain trout fishery: Distribution, migration and growth rates. Colorado Cooperative Fisheries Research Unit 6:36–50.

Sharpe, F. P. 1962. Some observations of the feeding habits of brown trout. Progressive Fish-Culturist 24(2):60–64.

Shebley, W. H. 1917. History of the introduction of food and game fishes into the waters of California. California Fish and Game 3(1):3–12.

Shell, E. W. 1962. Herbivorous fish to control *Pithophora* sp. and other aquatic weeds in ponds. Weeds 10:326–327.

Shetter, D. S. 1937. Migration, growth rate, and population density of brook trout in the North Branch of the Au Sable, Michigan. Transactions of the American Fisheries Society 66:203–210.

Shetter, D. S. 1947. Further results from spring and fall plantings of legal-sized hatchery-reared trout in streams and lakes in Michigan. Transactions of the American Fisheries Society 74:35–58.

Shetter, D. S. 1948. The relationship between the legal-sized trout population and the catch by anglers in portions of two Michigan trout streams. Papers of the Michigan Academy of Science, Arts and Letters 34:97–107.

Shetter, D. S. 1968. Observations on the movements of wild trout in two Michigan stream drainages. Transactions of the American Fisheries Society 97(4):472–480.

Shetter, D. S., and G. R. Alexander. 1965. Results of angling under special and normal trout fishing regulations in a Michigan trout stream. Transactions of the American Fisheries Society 94(3):219–226.

Shetter, D. S., and G. R. Alexander. 1970. Results of predator reduction on brook trout and brown trout in 4.2 miles (6.76 km) of the north branch of the Au Sable River. Transactions of the American Fisheries Society 99(2):312–319.

Shetter, D. S., and A. S. Hazzard. 1942. Planting "keeper" trout. Michigan Conservation 11(4):3–5.

Shireman, J. V. 1976. Ecological study of Lake Wales, Florida after introduction of grass carp. Mimeographed annual report, Florida Department of Natural Resources, Tallahassee, Florida. 63 pp.

Shireman, J. V., D. E. Colle, and R. G. Martin. 1979. Ecological study of Lake Wales, Florida after introduction of grass carp (*Ctenopharyngodon idella*), 49–90. *In:* J. V. Shireman (ed.). Proceedings of the Grass Carp Conference. University of Florida, Gainesville, Florida.

Shireman, J. V., D. E. Colle, and R. W. Rottmann. 1978. Size limits to predation on grass carp by largemouth bass. Transactions of the American Fisheries Society 107(1):213–215.

Shireman, J. V., and M. J. Maceina. 1981. The utilization of grass carp, *Ctenopharyngodon idella* Val., for hydrilla control in Lake Baldwin, Florida. Journal of Fish Biology 19:629–636.

Sibley, C. K. 1928. Food of some Oswego drainage fishes. *In:* A biological survey of the Oswego River. Supplement to the 17th Annual Report of the New York Conservation Department 1927(1928).

Sibley, C. K. 1929. The food of certain fishes of the Lake Erie drainage basin. *In:* A biological survey of the Erie-Niagara system. Supplement to the 18th Annual Report of the New York Conservation Department 1928:180–188.

Sibley, C. K., and V. N. Rimsky-Korsakoff. 1931. Food of certain fishes in the watershed. *In:* A biological survey of the St. Lawrence watershed. Supplement to the 20th Annual Report of the New York Conservation Department 1930:109–120.

Sigler, W. F. 1955. An ecological approach to understanding Utah's carp populations. Utah Academy Proceedings 32:95–104.

Sigler, W. F. 1958. The ecology and use of carp in Utah. Bulletin 405, Agricultural Experiment Station, Utah State University, Logan, Utah. 62 pp.

Sigler, W. F., and R. R. Miller. 1963. Fishes of Utah. Utah State Department of Fish and Game, Salt Lake City, Utah. 203 pp.

Sills, J. B. 1970. A review of herbivorous fish for weed control. Progressive Fish-Culturist 32(3):158–161.

Simberloff, D. 1981. Community effects of introduced species, 53–81. *In:* M. H. Nitecki (ed.). Biotic crises in ecological and evolutionary time. Academic Press, Inc., New York.

Smallwood, W. M., and P. H. Struthers. 1928. Carp control studies in Oneida Lake. *In:* A biological survey of the Oswego River system. Supplement to the 17th Annual Report of the New York Conservation Department 1927:67–83.

Smedley, H. H. 1938. Trout of Michigan. Privately published in Muskegon, Michigan. 49 pp.

Smith, C. R., and J. V. Shireman. 1981. Grass carp bibliography. Job completion report for the U.S. Corps of Engineers. A contribution of the Center for Aquatic Weeds, University of Florida, Gainesville, Florida.

Smith, L. L., Jr., and B. S. Smith. 1945. Survival of seven- to ten-inch planted trout in two Minnesota streams. Transactions of the American Fisheries Society 73:108–116.

Smith, S. L. 1976. Behavioral suppression of spawning in largemouth bass by interspecific competition for space within spawning areas. Transactions of the American Fisheries Society 105(6):682–685.

Sneed, K. 1971. The white amur: A controversial biological control. American Fish Farmer and World Aquaculture News 2(6):6–9.

Staley, J. 1966. Brown trout, 233–242. *In:* A. Calhoun (ed.). Inland fisheries management. California Department of Fish and Game.

Stanley, J. G. 1972. Unisex studies on the white amur, 89–95. Mimeographed annual report, Bureau of Sport Fisheries and Wildlife, Fish Farming Experimental Station, Stuttgart, Arkansas.

Stanley, J. G. 1974. Nitrogen and phosphorus balance of grass carp, *Ctenopharyngodon idella* fed elodea, *Egeria densa.* Transactions of the American Fisheries Society 103(3):587–592.

Stevenson, J. H. 1965. Observations on grass carp in Arkansas. Progressive Fish-Culturist 27:203–206.

Stott, B., and L. D. Orr. 1970. Estimating the amount of aquatic weed consumed by grass carp. Progressive Fish-Culturist 32:51–54.

Stott, B., and T. O. Robson. 1970. Efficiency of grass carp (*Ctenopharyngodon idella* Val.) in controlling submerged water weeds. Nature (London) 226(5248):870.

Stroganov, N. S. 1961. Food preferences of grass carp, 181–191. *In:* Data of the All-Union Conference on culture of the herbivores *Ctenopharyngodon idella* and *Hypophthalmichthys molitrix* in Russian waters. Ashkhabad Academy of Sciences, Turkmen, U.S.S.R.

Stroganov, N. S. 1963. The food selectivity of the Amur fishes, 1–12. *In:* Symposium on problems of fishery exploitation of plant eating fishes in water bodies of the U.S.S.R. Ashkhabad Academy of Sciences, Turkmen, U.S.S.R.

Struthers, P. H. 1929. Carp control studies in the Erie Canal. *In:* A biological survey of the Erie-Niagara system. Supplement to the 18th Annual Report of the New York Conservation Department 1928:208–219.

Struthers, P. H. 1930. Carp control studies in the Cayuga and Owasco Lake basin. *In:* A biological survey of the Champlain watershed. Supplement to the 19th Annual Report of the New York Conservation Department 1929:261–280.

Struthers, P. H. 1931. Carp control studies in the Seneca, Canandaigua and Keuka lake basins. *In:* A biological survey of the St. Lawrence watershed. Supplement to the 20th Annual Report of the New York Conservation Department 1930:217–229.

Struthers, P. H. 1932. A review of the carp control problems in New York waters. *In:* A biological survey of the Oswegatchie and Black River systems. Supplement to the 21st Annual Report of the New York Conservation Department 1931:272–289.

Sutton, D. L. 1972. Control of aquatic plant growth in earthen ponds by the white amur. Mimeographed annual report, Florida Department of Natural Resources, Fort Lauderdale, Florida. 77 pp.

Sutton, D. L., and R. D. Blackburn. 1973. Feasibility of the amur (*Ctenopharyngodon idella* Val.) as a biocontrol of aquatic weeds. Appendix D, 42 pp. *In:* Herbivorous fish for aquatic plant control. Mimeographed technical report 4, Aquatic Plant Control Program. U.S. Army Corps of Engineers, Vicksburg, Mississippi.

Swee, U. B., and H. R. McCrimmon. 1966. Reproductive biology of the carp, *Cyprinus carpio* L., in Lake St. Lawrence, Ontario. Transactions of the American Fisheries Society 95(4):372–380.

Swingle, H. S. 1957a. Control of pond weeds by the use of herbivorous fishes. Proceedings of the Southern Weed Control Conference 10:11–17.

Swingle, H. S. 1957b. Further experiments with *Tilapia mossambica* as a pondfish. Proceedings of the Annual Conference of the Southeastern Association of Game and Fish Commissioners 11:152–154.

Swingle, H. S. 1957c. A repressive factor controlling reproduction in fishes. Proceedings of the Pacific Science Congress 8(1953):865–871.

Swingle, H. S. 1960. Comparative evaluation of two tilapias as pondfishes in Alabama. Transactions of the American Fisheries Society 89(2):142–148.

Tebo, L. B., Jr., and W. W. Hassler. 1963. Food of brook, brown and rainbow trout from streams in western North Carolina. Journal of the Elisha Mitchell Scientific Society 79(1):447–453.

Terrell, J. W., and A. C. Fox. 1974. Food habits, growth and catchability of grass carp in the absence of aquatic vegetation. Proceedings of the Annual Conference of the Southeastern Association of Game and Fish Commissioners 28:251–258.

Terrell, J. W., and T. T. Terrell. 1975. Macrophyte control and food habits of the grass carp in Georgia ponds. Verhandlungen der internationalen Vereinigung für theoretische und angewandte Limnologie 19:2515–2520.

Terrell, T. T. 1975. The impact of macrophyte control by the white amur (*Ctenopharyngodon idella*). Verhandlungen der internationalen Vereinigung für theoretische und augewandte Limnologie 19:2510–2514.

Thorpe, L. M., H. J. Rayner, and D. A. Webster. 1947. Population depletion in brook, brown, and rainbow trout stocked in the Blackledge River, Connecticut, in 1942. Transactions of the American Fisheries Society 74:166–187.

Threinen, C. W. 1952. Carp in the ecology of southern Wisconsin lakes. Wisconsin Conservation Bulletin 17(7):14–15.

Threinen, C. W., and W. T. Helm. 1954. Experiments and observations designed to show carp destruction of aquatic vegetation. Journal of Wildlife Management 18(2):247–250.

Trautman, M. B. 1981. The fishes of Ohio. Ohio State University Press, Columbus, Ohio. 782 pp.

Tryon, C. A., Jr. 1954. The effects of carp exclosures on growth of submerged aquatic vegetation in Pymatuning Lake, Pennsylvania. Journal of Wildlife Management 18(2):251–254.

Turner, J. S. 1981. Population structure and reproduction in the introduced Florida population of the pike killifish, *Belonesox belizanus* (Pisces: Poeciliidae). M.S. thesis, University of Central Florida, Orlando, Florida. 56 pp.

Vestal, E. H. 1954. Creel returns from Rush Creek test stream, Mono County, California, 1947–1951. California Fish and Game 40(2):89–104.

Vincent, R. E., and W. H. Miller. 1969. Altitudinal distribution of brown trout and other fishes in a headwater tributary of the south Platte River, Colorado. Ecology 50(3):464–466.

Vinogradov, V. K., and Z. K. Zolotova. 1974. The influence of the grass carp on aquatic ecosystems. Hydrobiology Journal (U.S.S.R.) 10(2):72–78.

Vivier, P. 1955. Sur l'introduction des Salmonides exotiques en France. Verhandlungen der internationalen Vereinigung für theoretische und angewandte Limnologie 12:527–534.

Walden, H. T. 1964. Familiar freshwater fishes of America. Harper and Row, New York. 324 pp.

Wales, J. B. 1962. Introduction of pond smelt from Japan into California. California Fish and Game 48(2):141–142.

Wales, J. H. 1946. Castle Lake trout investigation. First phase: Interrelationships of four species. California Fish and Game 32(3):109–143.

Wales, J. H., and E. R. German. 1956. Castle Lake investigation. Second phase: Eastern brook trout. California Fish and Game 42(2):93–108.

Walker, B. W., R. R. Whitney, and G. W. Barlow. 1961. The fishes of the Salton Sea, 77–91. *In:* B. W. Walker (ed.). The ecology of the Salton Sea, California, in relation to the sportfishery. California Department of Fish and Game, Fishery Bulletin 113.

Wallen, I. E. 1951. The direct effect of turbidity on fishes. Bulletin of the Oklahoma Agricultural and Mechanical College, Biological Series no. 2, 48(2):1–27.

Ware, F. J. 1973. Status and impact of *Tilapia aurea* after 12 years in Florida. Mimeographed report, annual meeting of the American Fisheries Society, Disney World, Florida. 7 pp.

Ware, F. J., and R. D. Gasaway. 1976. Effects of grass carp on native fish populations in two Florida lakes. Proceedings of the Annual Conference of the Southeastern Association of Fish and Wildlife Agencies 30:324–355.

Ware, F. J., R. D. Gasaway, R. A. Martz, and T. F. Drda. 1975. Investigations of herbivorous fishes in Florida, 79–84. *In:* Proceedings of a Symposium on Water Quality Management through Biological Control. Report no. ENV-07-75-1, University of Florida, Gainesville, Florida.

Wattendorf, R. J. 1981. Lake Lena blue tilapia investigations: Fishery analyses. First annual performance report, Non-Native Fish Research Laboratory, Boca Raton, Florida Game and Fresh Water Fish Commission. 16 pp.

Wattendorf, R. J. 1982. Lake Lena blue tilapia investigations: Fishery analyses. Second annual performance report, Non-Native Fish Research Laboratory, Boca Raton, Florida Game and Fresh Water Fish Commission. 18 pp.

Wattendorf, R. J., D. S. Levine, and P. L. Shafland. 1980. An evaluation of interspecific interactions between blue tilapia and certain native fishes in Florida: Foods, feeding and selectivity analysis. Second annual performance report, Non-Native Fish Research Laboratory, Boca Raton, Florida Game and Fresh Water Fish Commission. 18 pp.

Webster, D. A., and G. S. Little. 1942. Food of newly planted adult brown trout (*Salmo trutta*) in Six Mile Creek, New York. Copeia 1942(3):192.

Webster, D. A., and G. S. Little. 1944. Further observations on the food of newly planted brown trout (*Salmo trutta*) in Six Mile Creek near Ithaca, New York. Copeia 1944(1):56–57.

Welcomme, R. L. 1966. Recent changes in the stocks of *Tilapia* in Lake Victoria. Nature (London) 212:52–54.

Whitaker, J. O., Jr. 1974. Foods of some fishes from the White River at Petersburg, Indiana. Proceedings of the Indiana Academy of Science 84:491–499.

Wiley, R. W. 1978. Trends in fish population, 1963 through 1976 – 14 years of gill-netting, Flaming Gorge Reservoir. Fishery Research Report no. 1, Flaming Gorge Reservoir Monograph Series, Wyoming Game and Fish Department, Cheyenne, Wyoming. 13 pp.

Wiley, R. W., and J. D. Varley. 1978. The diet of rainbow and brown trout from Flaming Gorge Reservoir, 1964 through 1969. Fishery Research Report no. 1, Flaming Gorge Reservoir Monograph Series, Wyoming Game and Fish Department, Cheyenne, Wyoming. 16 pp.

Wilson, L. R. 1939. Rooted aquatic plants and their relation to the limnology of fresh-water lakes, 107–122. *In:* Problems of lake biology. Publication no. 10, American Association for the Advancement of Science.

CHAPTER 17

Toward the Development of an Environmental Ethic for Exotic Fishes

Charles H. Hocutt

During the 1970s a reevaluation of man's role in the environment and the ecological impact of his activities was accompanied in the United States by a political and social awareness (e.g., the National Environmental Policy Act [NEPA] and Earth Day). Terms like "environment," "ecology," "conservation," and "pollution" became common words in the nation's vocabulary; today they are supplemented with "PCB," "Three Mile Island," "OPEC," and "snail darter," to name a few. During this renaissance of environmental concern it was reemphasized that environmental degradation occurred not only through point and non-point sources of physicochemical alterations in water quality, but also through biological contaminants.

The Invitational Conference on Exotic Fishes and Related Problems held on 18–19 February 1969, in Washington, D.C., cosponsored by the American Fisheries Society and the American Society of Ichthyologists and Herpetologists (Stroud 1969), formally addressed species involved, country of origin, mode of transportation, where introduced, and the motive of such introductions. This information was subsequently summarized by Lachner et al. (1970).

The environmental movement and the concern over exotic species have both evolved over the past decade, and are intimately related in many aspects.

Definition of Exotic

A foremost difficulty in approaching the subject of exotic fishes is defining the term *exotic.* The term has been used inconsistently throughout the literature (e.g., Lachner et al. 1970, Courtenay and Robins 1973, Courtenay 1979). In the context of the continental United States, the organizers of this symposium have defined exotic fish as those forms not indigenous to North America. How does one define North America? North America is geographically and politically recognized as encompassing Mexico; however, such a definition for ichthyofaunal purposes is self-defeating in that Mexico is a major transition area. Drs. Robert R. Miller and Michael L. Smith are preparing a chapter entitled "Zoogeography of Nearctic Mexican Fishes" that will appear in a comprehensive work on zoogeography of North American freshwater fishes (Hocutt and Wiley, in press). It defines North American waters as those south into Mexico to "20° N latitude on the Atlantic slope [where there is a natural volcanic barrier] and to somewhat below 17° N latitude on the Pacific versant. The southern boundary in Mexico is the southern limit of the family Cyprinidae." This is a sound zoogeographic definition of North America.

CONTEMPORARY APPLICATION

The stated intention of our national water policy as premised in the Federal Water Pollution Control Act of 1972 (PL92–500) was to "restore and maintain the chemical, physical, and biological integrity of the nation's waters." A specific objective of PL92–500 was to set effluent limitations on industrial point source discharges, with an ultimate goal of achieving zero-discharge by 1985. To achieve this goal, PL92–500 specified that the discharge of any pollutant would be prohibited unless authorized by a National Pollutant Discharge Elimination System (NPDES) permit, as administered under Section 402 of the act; the U.S. Environmental Protection Agency (USEPA) was created under the act to monitor man-related alterations on water quality and to regulate the NPDES permit system. However, subsection 316(a) of the Federal Water Pollution Control Act allows for a less stringent effluent requirement for one category of point source discharges, that being thermal pollution. Under Section 316(a) of PL92–500 a less stringent effluent requirement (i.e., variance) for a thermal effluent can be obtained if it can be demonstrated with "representative important species" (RIS) that the regulations are more stringent than necessary to assure the protection of the aquatic ecosystem receiving the discharge. It is conceivable to employ RIS to receive variances for other effluent classes as well.

The USEPA (1977) designates RIS by reviewing the existing literature for the respective site and applying a sixfold definition to potential candidates. Generally, qualification specifically includes those species which are: (1) commercially or recreationally available (i.e., within the top ten species landed by dollar value); (2) threatened and endangered; (3) critical to the function of the ecological system; (4) potentially capable of becoming localized nuisance species; (5) necessary in the food chain for the well-being of the species determined in 1–4; or (6) representative of the thermal requirements of important species, but which themselves may not be important. RIS are those species which are "representative in the terms of their biological requirements of a balanced, indigenous community of shellfish, fish, and wildlife in the body of water into which the discharge is made" (USEPA 1977). However, nonnative species can be chosen as RIS (USEPA 1977) even though the term *indigenous* is used in the law.

Therefore, the USEPA or its designated enforcement organization in each state can select species as RIS that may be indigenous neither to the drainage nor conceivably to North America. The brown trout is a good example. This discussion brings to bear the dilemma of the development of an environmental ethic for exotic fishes. How are exotics to be considered in our society?

On the one hand, introduction of exotics has been opposed for a variety of ecological and public health considerations. For instance, exotics can facilitate dispersal of new parasites or diseases to indigenous fishes and other wildlife, occupy a different trophic level than anticipated, and so on. On the other hand, exotics are often purposefully introduced for sport or food, aquarium resources, or as agents of biological control. Another consideration is that exotics are often established accidentally or unintentionally by the public. Thus enforcement agencies can exert little control other than at centers of importation into the country. These thoughts have been ably discussed throughout this book and elsewhere in the literature; thus I do not intend to dwell upon the pros and cons of exotics.

I am concerned, however, with how established exotics are treated in view of the legal mandate of our nation's water quality laws. In an attempt to discern national views toward exotics, I addressed a 10-point questionnaire to the fifty state fish and wildlife agencies. Some 80 percent (40 of 50 states) responded and the data are presented in table 17–1.

Results of Questionnaire

1. Please provide the definition of exotic employed.

Some 65 percent of the states (26) responded to this question by defining "exotic" as not native to the state. Only five states (12.5 percent) defined "exotic" as not indigenous to North America. New Mexico and Oregon were the only states to declare an exotic fish as one outside its native range or basin, which is by far the purest definition.

2. Has your state adopted legislation specifically addressing exotic species, especially fish?

Thirty-one states (77.5 percent) responded that they had adopted formal legislation or informal departmental resolutions regarding the import of exotics. Where limited to a list of a few species, these species invariably included grass carp, walking catfish, and piranha.

3. Does your agency have an interdepartmental program aimed specifically toward exotic fish management and control?

Only eleven states (27.5 percent) responded affirmatively to this question, and one state abstained. It is expected, however, that many of the states responding "No" did not actively consider their own definition of "exotic," which would have included transplanted indigenous fauna of North America, especially salmonids, esocids, centrarchids, and large percids.

4. Has your state enacted legislation or is there an informal policy addressing restrictions on holding, culturing, and distribution of exotic fishes at commercial fish farms?

Thirty-one states (77.5 percent) responded "Yes" to this question. Of these, four state policies addressed particular species: Alaska, brook trout; Arkansas, silver and bighead carp; Illinois, walking catfish and grass carp; and Mississippi, piranha, walking catfish, and grass carp.

5. Has your state enacted legislation or is there an informal policy addressing import, culture, and release of aquarium fishes?

Some twenty-seven states (67.5 percent) responded "Yes" to this question, but 37 percent of this total (10 states) had a restricted policy limited almost exclusively to prohibiting release, which of course cannot be enforced. Potentially, tropical fishes pose a tremendous threat to certain southern states from California to Florida. The latest figures I could obtain indicate that aquarium fishes account for 50.1 percent of all live animal sales in the United States, with some 250 million specimens distributed among 10 million households.

6. Is an eradication program in force or planned within your state for established exotic fishes or those recently introduced?

Only eleven states (27.5 percent) indicated that any steps were being taken to eradicate undesirable exotics. Of this total three states (27.3 percent) had species-specific programs: Illinois and New York, grass carp; New Hampshire, goldfish. Six of eleven states (54.5 percent) had limited programs, which were being conducted only under specific conditions and when feasible.

7. Are established exotic fishes (e.g., brown trout, carp, tilapia) regarded as "native indigenous fauna" with regard to preparation of environmental impact assessment?

Table 17–1. PERSPECTIVE OF STATE FISH AND WILDLIFE AGENCIES ON EXOTIC FISHES (NUMBERS CORRESPOND TO QUESTIONS IN TEXT)

State	Question									
	1	2	3	4	5	6	7	8	9	10
AL	Not native to state	Yes, particularly walking catfish and piranha	No	No	No	No	Yes	No	Yes	AFS (1975)
AK	Not native to state	Yes	No	Yes, for brook trout	Yes	No	Yes, brook trout	No	No	Never applied
AZ	–	–	–	–	–	–	–	–	–	–
AR	Not native to state	Yes, for 4 species	No	Yes, for silver and bighead carp	No	No	Yes	Yes, striped bass	Yes	AFS (1975)
CA	Not native to state	Yes	No	Yes	Yes	Yes, where appropriate	Policy inconsistent	Yes, striped bass	Yes	AFS (1980)
CO	Not native to state	Yes	Yes	Yes	No	No	Yes	Yes, game fish	Yes	$10/fish
CT	Specific list of North American and "exotic" fishes	Yes	No	Yes	Yes, prohibits release	Yes, where feasible	Yes	No	No	No replacement costs
DE	Walking catfish and grass carp	Yes	No	No	Only for walking catfish	No	Yes	Yes, brown trout	Yes	AFS (1975)
FL	Not native to state	Yes	Yes	Yes	Yes	Yes	Policy inconsistent	No	Yes	State price list
GA	Not native to North America	Yes	No	Yes	Yes	No	Yes	–	Yes	AFS (1975)
HI	–	–	–	–	–	–	–	–	–	–
ID	Not native to state	Yes	Yes	Yes	Yes	Where practical	Yes	Yes	Responded no, but probably yes	AFS (1978)

IL	Not native to state	Yes	Yes	Yes, for walking catfish and grass carp	Yes	Yes, for grass carp	Yes	Yes	Yes	AFS (1975)
IN	Not native to state	Yes, for grass carp and walking catfish	No	Yes	No	No	No	—	No	—
IA	Not native to state	Yes	No	Yes	Yes, prohibits its release	No	Yes	Yes, brown trout	Yes	AFS (1978)
KS	Outside indigenous range, generally extracontinental	Yes, but limited	No	No	Yes (limited)	No	Yes	Under consideration	Yes	AFS (1978)
KY	Not native to state	Yes	Yes	Yes	Yes	No	Yes	Yes	Yes, excluding white amur	AFS (1975)
LA	Not native to state	Yes	No	Yes	Yes	Yes, where practical	No	Yes	Yes	AFS (1975)
ME	Not native to state	Yes	Yes	Yes	Yes, for walking catfish, piranha, grass carp	No	Yes	No	No	Based on hatchery costs
MD	—	—	—	—	—	—	—	—	—	—
MI	—	—	—	—	—	—	—	—	—	—
MN	Not native to state	Yes, informally	Yes	Yes	No	Yes, where feasible	Yes	Yes	Yes	AFS (1978)
MS	Not native to North America	Yes	No	Only for piranha, walking catfish, grass carp	—	No	Yes	No	No	AFS (1975)
MO	All introduced fishes	Yes	No	No	No	No	Yes	Yes	Yes	AFS (1978)
MT	Not native to state	No	No	Yes	Yes, prohibits its release	No	Inconsistent	Yes, brown trout	No	Fish not used to assess damage

Table 17-1. PERSPECTIVE OF STATE FISH AND WILDLIFE AGENCIES ON EXOTIC FISHES (NUMBERS CORRESPOND TO QUESTIONS IN TEXT) *(continued)*

State	Question									
	1	2	3	4	5	6	7	8	9	10
NE	–	–	–	–	–	–	–	–	–	–
NV	Not found in contiguous 48 states and Alaska, but some native North American forms included	Yes	Yes	Yes	Yes	–	Yes	–	–	–
NH	Not native to state	No	No	Yes	Yes, for walking catfish, grass carp, caribe, piranha	Yes, for goldfish	Yes	No	Yes	AFS (1975) guide, plus inflation
NJ	Not native to state	No	No	No	No	No, generally	Yes	Yes	Yes	AFS (1978)
NM	Outside indigenous range, or basin	No	Manage fisheries of desirable species	Yes	For release only	No	Yes	Yes	Yes	$1/specimen
NY	Not native to state and undesirable	Yes	No	No	No	Yes, for grass carp	Yes	No	Responded no, but probably yes	Not available
NC	Not native to North America	Yes, for piranha, grass carp, walking catfish	No	No	No	No	Yes	No	Yes	AFS (1975)
ND	–	–	–	–	–	–	–	–	–	–
OH	Not native to state	Yes	No	Yes	Yes	No	Yes	No	Yes	State price list

OK	Not native to North America	Yes	No	Yes	Yes	No	Yes	No	Yes	AFS (1975)
OR	Not indigenous to water body	Yes, for walking catfish	Only for some species	Yes	Yes	No	No	No	Inconsistent	AFS (1975)
PA	Not native to state	Yes	No	Yes	Yes	No	Yes	No	No	State price list
RI	Not native to state	No	No	Not applicable	Yes, prohibits its release	No	Yes	Yes	No	Not assessed monetary damage for any fish kill
SC	Not native to state	Yes	No	Yes	Yes	No	Inconsistent	Probably (see 9)	Yes, for carp and brown trout	AFS (1975)
SD	Not native to North America	No	No	Yes	No	Yes on individual waters, not statewide	Regarded as native for examples given	Yes, brown trout	Inconsistent; carp have commercial value; trout, sport fish value	AFS (1978) plus commercial prices
TN	Not native to state	Yes	No	No	No	No	Yes	No	Yes	AFS (1975)
TX	Not native to United States	Yes	No	Yes	Yes	No	–	No	No	State price list
UT	Not native to United States	No	No	Yes	No	No	Yes	Yes, game species	No	AFS (1978) guide, and hatchery costs
VT	–	–	–	–	–	–	–	–	–	–
VA	–	–	–	–	–	–	–	–	–	–
WA	–	–	–	–	–	–	–	–	–	–
WV	Not native to state	No	No	Yes	Yes	No	Yes	Yes	Yes	AFS (1975)
WI	Not native to state	Yes, indirectly	Yes	Yes	Yes, prohibits its release	Yes	Yes	No	No	State price list
WY	Not native to state	Yes	–	Yes	Yes	—	–	–	–	

Thirty-two states replied "Yes" to this question, and four other states had inconsistent policies. Thus, some 90 percent of responding states considered established exotic fishes as indigenous fauna, at least in part. This is not surprising since, as mentioned earlier, the majority of states (65 percent) defined "exotic" as not native to the state. Many states such as Rhode Island and New Mexico have nearly 80 percent of their fauna as transplants.

8. The Federal Water Pollution Control Act of 1972 suggests using "Representative Important Species" to assess structural and functional integrity of aquatic systems receiving point source discharges. Does your state employ exotic species as "Representative Important Species"?

Only eighteen states (45 percent) employ exotics as RIS. Of this total, a surprisingly large percent (44.4 percent) restrict their use of RIS to particular species: Arkansas, California, striped bass; Delaware, Iowa, Montana, South Dakota, brown trout; Colorado, Utah, game species. A restricted use to "game" species eliminates the possible inclusion of exotics representative of various trophic levels.

9. In the event a fish kill occurs, are all or some exotic species used to assess monetary damage to the ecosystem?

Twenty-four states (60 percent) applied a monetary value to exotic fishes in the event of a fish kill. Two states, Idaho and New York, responded "No," but I consider, from answers to other questions, that they misinterpreted the question. Kentucky responded "Yes," but excluded the grass carp. Two states, Oregon and South Dakota, were inconsistent in their policy.

10. How is replacement cost of exotic fish determined, and is the replacement cost of exotic fishes equivalent to that placed on native fauna?

The response to this question will certainly reinforce the value of the American Fisheries Society's committee on Monetary Values of Fish. Twenty-four states (60 percent) followed either the 1975 guide of the Southern Division (15 states), 1978 guide of the North Central Division (7 states), or the 1980 guide of the American Fisheries Society (2 states). Six states (15 percent) employed their own price list, while one state used hatchery costs to assess values of fish. Almost inconceivably, four states (Alaska, Connecticut, Montana, Rhode Island) either do not or have never had the opportunity to assess monetary damage to aquatic communities in the event of a fish kill.

Discussion

There is a clear legal mandate to maintain the biological integrity of this nation's waters. Exotic fishes pose a distinct potential source of disruption to these waters, but unfortunately there is a significant inconsistency in how states or scientists define exotic. Also, there is no consistent manner of treating exotics in response to the legal mandate. Therefore, the following recommendations are proposed.

Definition of exotic: The co-organizers of this symposium defined exotic fishes as those not indigenous to North America. This definition is fine for this symposium; however, the utility of the definition is not realistic except on a federal level. The argument put forth by Dr. Clark Hubbs certainly has much more merit on a regional basis, and has ecological merit as well. Dr. Hubbs (1977) argues that an exotic is "any species introduced to a location outside its natural geographic range." This concept coincides with the definition employed by the majority of states surveyed (i.e., not native to the state), and can be brought to public attention through fishing license sellers and distributors of tropical (aquarium) fishes and supplies. In my view, the key to minimizing potential introduction of exotics is through education, as well as regulation.

As mentioned, Dr. Hubbs's definition of exotic has ecological merit as well. Species closely related phylogenetically are (with exception) functionally more similar (Stauffer and Hocutt 1980), and will have similar ecological requirements and thresholds. Thus, species "A" transplanted into a drainage outside its native range, but where sibling species "B" occurs, may replace species "B" through competition for feeding, spawning, or nursery sites. Additionally, closely related species are more likely to hybridize, thus decreasing the genetic integrity of each.

Certainly, the potential effects of introduction of truly foreign species should not be minimized, but introductions on a regional basis are surely more common (e.g., minnow bucket transplants). Also, interdrainage transfers have far greater potential for succeeding in nature since they are already acclimated to regional climate patterns and general water quality conditions. Editors of the recently published *Atlas of North American Freshwater Fishes* (Lee et al. 1980) constantly had to contend with transplanted indigenous North American fauna in verifying distribution maps for some 775 species north of the Rio Grande.

Indigenous Fauna: Suffice it to say that innumerable exotics and transplants have been established in the past century in waters outside their native range and it has become commonplace to think of these species as native in their new environs. In almost all instances, eradication is impractical if not impossible, and often is totally undesired if the species in question provides recreational or economic benefits. It seems completely sensible to me to continue to think of these species as potential candidates for use as "representative important species" (USEPA 1977), that is, to evaluate the potential effects of point and non-point discharge sources on them, and to determine if a man-related alteration on a water body will cause consequential damage by making conditions more favorable for their propagation and/or enhancing their potential to become nuisance species. This approach does not mean that appropriate steps should not be taken to limit importation, culture, distribution, and elimination of fauna not indigenous to North America. It does mean that the legal mandate provides an excellent rationale for studying exotics in detail.

Monetary Value of Fish: In the event of a fish kill involving exotics, it seems reasonable to assign a dollar value to exotics or transplants defined as representative important species. Moreover, state or federal regulatory agencies have every right to assess a fee to a discharger for clean-up and disposition of fishes, even though the kill is largely composed of undesirable species. This cost is not intangible. Furthermore, an ecosystem largely composed of undesirable species is nonetheless functioning, and a disruption can cause irreparable damage to that system, including its biological assimilative capacity.

Conclusion

The invitation by Drs. Courtenay and Stauffer to discuss this subject has placed me in a "no win" position. As an ecologist and ichthyologist, I am opposed to wholesale introductions of exotics and transplants into waters outside their native ranges. However, as a fisheries biologist and a member of the sport fishing public, I can support this activity under certain conditions.

In closing, I emphasize that the view toward exotic and transplanted fishes is no different from the philosophy on which the environmental movement of this nation was founded. The confusion over conservation policy is largely a result of the intrusion of economics and politics into what is fundamentally a question of ethics. Our society has demanded in the past, is doing so today, and will continue to demand certain environmental tradeoffs within reasonable bounds.

Mankind is living in a relatively closed system which is evolving as we would have it. Durwood Allen (1962), among others, defined conservation simply as "wise use." If we consider the limited resources we have at hand, conservation sometimes includes changing the environment to reap benefits otherwise not available. State fish and wildlife agencies have responded by establishing fisheries from transplanted populations to provide the public with an aesthetically pleasing opportunity not otherwise available to them. I think that it is important for the participants of this symposium to remember that, while many of them represent a conservative element who resist change in the biological integrity of fish communities (in the sense of exotics or transplants), the society that sponsored this symposium has a membership that is predominately oriented toward management of resources for the benefit of the public. This benefit in many instances includes managing exotic or transplanted fishes. Admittedly, state and federal agencies have not always made the best choice in establishing nonnative species, but their concern has genuinely been to provide their constituents with the best possible natural resource.

I am extremely critical, however, of two specific areas at the state level. First, and a matter related to the role of the U.S. Environmental Protection Agency and federal policy, many state agencies lack persons either trained sufficiently to make decisions regarding import and management of exotics or capable of selecting "representative important species." Indeed, several states have agencies other than a "Fish and Wildlife Department" that are responsible for NPDES regulatory decisions and these agencies in many instances may lack a staff biologist. Second, I am constantly amazed at how state fish and wildlife departments ignore the advice of university persons. Every effort is made to separate "management" philosophy from "research," while in fact the two are inseparable.

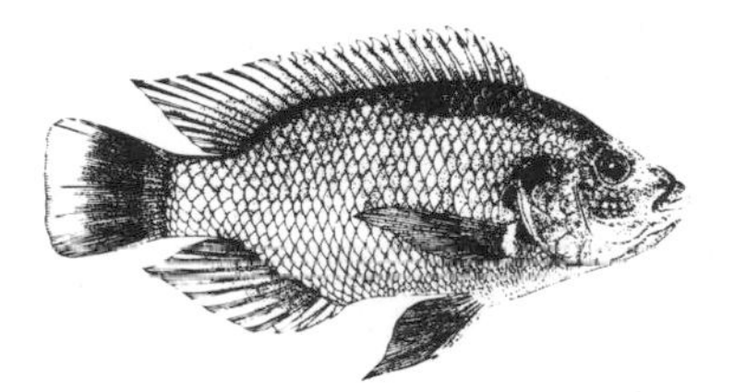

Literature Cited

Allen, D. L. 1962. Our wildlife legacy. Funk and Wagnalls, New York. 422 pp.

American Fisheries Society. 1975. Monetary values of fish. Special Publication of the Southern Division, American Fisheries Society.

American Fisheries Society. 1978. Monetary values of fish. Special Publication of the North Central Division, American Fisheries Society.

American Fisheries Society. 1980. Monetary values of fish. American Fisheries Society, Bethesda, Maryland.

Courtenay, W. R., Jr. 1979. The introduction of exotic organisms, 237–252. *In:* H. P. Brokaw (ed.). Wildlife and America. U.S. Government Printing Office, Washington, D.C.

Courtenay, W. R., Jr., and C. R. Robins. 1973. Exotic aquatic organisms in Florida with emphasis on fishes: A review and recommendations. Transactions of the American Fisheries Society 102(1):1–12.

Hocutt, C. H., and E. O. Wiley (eds.). In press. Zoogeography of North American freshwater fishes. John Wiley & Sons, New York.

Hubbs, C. 1977. Possible rationale and protocol for faunal supplementations. Fisheries 2(2):12–14.

Lachner, E. A., C. R. Robins, and W. R. Courtenay, Jr. 1970. Exotic fishes and other aquatic organisms introduced into North America. Smithsonian Contributions in Zoology 59:1–29.

Lee, D. S., C. R. Gilbert, C. H. Hocutt, R. E. Jenkins, D. E. McAllister, and J. R. Stauffer, Jr. 1980. Atlas of North American freshwater fishes. North Carolina State Museum of Natural History, Raleigh, North Carolina. 854 pp.

Stauffer, J. R., Jr., and C. H. Hocutt. 1980. Inertia and recovery: An approach to stream classification and stress evaluation. Water Resources Bulletin 16(1):72–78.

Stroud, R. H. 1969. Conference on exotic fishes and related problems. Sport Fishing Institute Bulletin 203:1–4.

U.S. Environmental Protection Agency. 1977. Interagency 316a Technical Guidance Manual and guide for thermal effects sections of nuclear facilities environmental impact statements. U.S. Environmental Protection Agency, Washington, D.C. 79 pp.

CHAPTER 18

A Suggested Protocol for Evaluating Proposed Exotic Fish Introductions in the United States

Christopher C. Kohler and Jon G. Stanley

Natural resources administrators are routinely called upon to make decisions regarding the well-being of ecosystems under their jurisdiction. Not the least of these are decisions concerning proposed introductions of exotic species. With respect to fishes, exotic species may be introduced for fisheries management or aquaculture, or they may be ornamental fishes intended for the home aquarist market.

The establishment of an exotic fish in an open waterway rarely occurs without consequences. Decisions regarding each exotic fish introduction should be based on information that elucidates its potential benefits and risks. Because of the current global movement of exotic fishes, it would not be practical to impose inflexible guidelines. Consequently, the protocol suggested here is a *tiered* approach. There are several levels of review, each requiring progressively more information.

A *protocol* may be defined as a formal procedure for systematically generating a data base; in this case, the introduction of an exotic fish, with the data base being necessary for rendering a regulatory decision. An *exotic fish* is any fish originating from a foreign territory; it is not synonymous with a nonnative or a transplanted fish. A *nonnative* fish is a fish that has been introduced from outside of its native range, regardless of geographic boundaries. On the other hand, *transplanted fishes* are those that have been transferred outside of their native range within the same country. Although the discussion that follows is directed at exotic fish introductions, it also applies to transplanted fishes.

Before the protocol is presented, the numerous factors that were incorporated in its synthesis are described. It was not our intention to render a review of the literature. Such a review would require a complete volume. Several authors (Andrews 1980, Axelrod 1973, Courtenay and Robins 1973, Mann 1979, Scanlon et al. 1978) have made recommendations relative to the introduction of exotic fishes. The protocol herein described represents the views of the authors, but not necessarily those of their employers or of the Exotic Fish Section of the American Fisheries Society.

Feasibility

Any authority considering introduction of an exotic species must evaluate many complex factors. The first step is to determine feasibility. Will the exotic fulfill some need at the receiving site that cannot be fulfilled by a native fish? Is it available from transplantation? Permits for transfer and safeguards against accompanying diseases and parasites may be needed. Will the exotic escape? We believe consideration of the factors below will serve as a basis for an approval, rejection, or deferral of a proposed introduction.

VALIDITY OF THE INTRODUCTION

First the authority approving an introduction should review the purported need for an exotic species. Native fishes may be available to satisfy requirements for commercial or recreational fisheries. Hence, we suggest a conservative approach toward introductions into natural waters. For the ornamental or hobby trade, however, the uniqueness of each species alone justifies need. We, therefore, recommend a less vigorous review of aquarium fishes. Fishes for aquaculture fall between the two extremes.

Second, it should be determined whether the proposed introduction adequately fulfills the need. Information on the species' characteristics should be reviewed to predict whether the objectives of the proposed introduction would be accomplished. In short, does the fish match the need?

STATUS IN THE NATIVE RANGE

Whether an exotic fish introduction is feasible will depend on stock availability. Information should be obtained on the native distribution and abundance. Who will collect the fish, how will they be collected, and does col-

Table 18–1. EXOTIC FISHES ON THE ENDANGERED SPECIES LIST

Common name	Scientific name	Historic range
Ala Balik (trout)	*Salmo platycephalus*	Turkey
Ayumodoki (loach)	*Hymenophysa (= Botia) curta*	Japan
Blindcat, Mexican	*Prietella phreatophila*	Mexico
Bonytongue, Asian	*Scleropages formosus*	Thailand, Indonesia, Malaysia
Catfish (no common name)	*Pangasius sanitwongsei*	Thailand
Catfish, giant	*Pangasianodon gigas*	Thailand
Cicek (minnow)	*Acanthorutilus handlirschi*	Turkey
Nekogigi	*Coreobagrus ichikawai*	Japan
Tango, Miyako (Tokyo bitterling)	*Tanakia tangago*	Japan
Temolek, Ikan (minnow)	*Probarbus jullieni*	Thailand, Cambodia, Vietnam, Malaysia, Laos
Topminnow, Gila	*Poeciliopsis occidentalis*	United States (AZ, NM), Mexico

Source: Department of the Interior 1980.

lection require approval of the country of origin? Is the species so rare that collecting will adversely affect native populations? Can collecting a restricted number of wild fish and culturing them to obtain commercial numbers suffice to provide stocks?

Only abundant, healthy fish stocks should be considered as sources for exotic introductions. Species that in their native range are endangered, threatened, or rare should not be introduced elsewhere, unless the aim of the introduction is to establish additional breeding populations in an effort to preserve the species. The Department of Interior's list of endangered species includes only 11 species that are not native to the United States (table 18–1). The list would be longer if additional information were available.

SITE OF PROPOSED INTRODUCTION

An exotic fish introduction might be proposed for either a closed or an open system. Closed systems include artificial ponds, raceways, aquaria, and other systems not connecting to open waters; however, it should be appreciated that accidental or intentional releases do occur. Open systems include those riverine and lacustrine systems, as well as oceans, where species emigration is not constrained. The intended destination of the exotic species should be decisive in determining whether introduction would be permitted with little review or pending an exhaustive evaluation. With the exception of some species, fishes destined for home aquaria, laboratory use, or other closed systems might be approved with little review. Conversely, the advisability of stocking exotics into open systems should be closely scrutinized, particularly if the organism would have access to a major drainage.

A wide range of situations exists between the extremes of closed and open systems: ponds fitted with screens, streams entering sea water, cages held in open water, and closed systems located along major waterways. These other circumstances may temper judgment concerning the security of the holding system. Each case should be decided on its own merits, but as a general guideline we recommend approval of introduction if there is reasonable assurance against escape and if relatively small numbers are involved. If massive numbers could escape should the system malfunction, we take a more conservative stand and recommend the proposed introduction be studied fully.

SAFEGUARDS AGAINST DISEASE INTRODUCTION

Diseases caused by bacteria or viruses (see chapter 11) or parasites (see chapter 12) may be conveyed along with the introduced exotic fish species. Transfer of diseased exotic fish was no doubt responsible for introduction of whirling disease into North America from Europe and infectious pancreatic necrosis into Europe from North America. Recent range extensions of parasites were shown to have accompanied introductions of fishes (Bauer and Hoffman 1976, Bohl 1979).

A protocol should contain provisions for taking reasonable precautions to avoid introduction of diseased fishes. Therapeutic treatment during shipment may lessen the chance of an accompanying disease. Treatment of fishes is an excellent way to eliminate certain parasites (Brown and Gratzek 1980). Antibiotics may or may not rid fishes of bacteria. Treatments are generally ineffective against viruses in fishes. Most fish diseases can be avoided by treatment of fertilized eggs; hence, egg shipment is the surest method of transfer of disease-free fishes.

The probability of disease transfer also would be lowered by a program for the certification of the fishes or hatchery as disease free. In Europe, twenty of twenty-two countries operate a certification program (Dill 1972). In the United States, imported salmonids must be certified free of whirling disease and viral hemorrhagic septicemia (Anonymous 1978). In Canada, the hatchery must be certified free of all major diseases for the preceding two years, based on semiannual inspection. Whichever type certification is used, the inspection should be by a fish pathologist if it is to be effective or legal. An international convention outlines precautions that will reasonably assure against importation of communicable diseases (FAO/OIE 1977).

Diseases may enter despite inspection. Several hundred grass carp (*Ctenopharyngodon idella*) were examined for *Bothriocephalus gowkongensis* (= *B. opsalichthydis*) and none was found (Bauer and Hoffman 1976). Nevertheless, the cestode must have been present in small numbers or at a sexually immature stage because later it was rampant in fish farms receiving these fish. Detection of a pathogen or parasite is enhanced if exotic fishes are held under quarantine. This allows the agent to go to maturity or multiply to detectable levels before the final decision is made to release the fish.

In this protocol we suggest that every practical measure be taken to prevent translocation of fish diseases. The exact methods will depend on the fish species, life stage, point of origin, and use of the fish at the receiving site.

LEGAL AND POLICY CLEARANCE

Regulations and legislation by state and federal governments restrict the importation of exotic species (see Bowden 1979, Brown 1979). We herein review some of the types of state regulations and cite federal law. It is not our intention to render a legal interpretation of laws on introduction of exotics.

In most states, a conservation department has constitutional or regulative authority over wildlife and fisheries, including the planting of fishes. Exotic species generally fall under this authority, and the states usually accomplish control by requiring permits for private persons or companies to possess exotic species. Stocking into the wild is not allowed. In most states the statutes are more specific. Exotic aquarium fish usually are not regulated closely.

Introductions of exotic fishes are regulated under three federal acts (Anonymous 1978). For example, the Lacey Act prohibits importation of any Clariidae. Also prohibited are viable eggs or fishes of the family Salmonidae unless certified to be free of whirling and Egtved disease (viral hemorrhagic septicemia). Import of all other species must be through designated ports. Release into the wild of any imported animal or its progeny is prohibited except by state wildlife conservation agencies or with their written permission. A 1981 amendment to the Lacey Act (Public Law 97-79) makes violation of any state law regarding possession or transportation of fish or wildlife a federal offense.

The Black Bass Act of 1935 prohibits importation of black bass (*Micropterus* spp.) or certain other fishes. The purpose of this legislation was to prevent commerce in sport fishes, and it has rarely been enforced in regard to importation of exotic fishes.

The Endangered Species Act of 1973 included sections that prohibited importation of exotics. The rationale for this act was to prevent depletion of endangered species in their native range (table 18-1). A permit may be obtained for scientific purposes or for propagation.

State and federal agencies and employees operate under various policies that prohibit introductions of exotics. Executive Order 11987 (of 1977) restricts the stocking of exotics by federal agencies or by those receiving federal funds. Introduction of an exotic requires the same Environmental Impact Statement (EIS) as any other federal project. Such an EIS was written for planting grass carp in the Panama Canal.

Acclimatization Potential

Acclimatization potential refers to the likelihood that a particular exotic species could form a self-sustaining population within the range of potential habitats. The probability of an exotic species becoming acclimatized is high if the receiving habitat resembles the native one. Welcomme (1981) documented over 100 cases of international transfers of fishes, and determined that approximately 30 of these became established in the wild. Of those established, 11 are considered as pests.

The aim of the protocol is to reduce the chances of making erroneous decisions. In this section we describe the kinds of information needed to decide whether acclimatization is possible. A thorough review would be in order whenever an exotic species has a high acclimatization potential and the proposed site for introduction is an open system.

NICHE REQUIREMENTS

Physicochemical parameters unsuitable for life or reproduction may preclude survival in the wild at the site of introduction. For example, if the local environmental temperatures fall outside the range of tolerance, then the fish could be cleared for introduction with minimal review. The lower lethal tolerance level is the parameter relevant for most introductions. The preponderance of proposed importations deals with tropical fishes that are unable to tolerate winter temperatures of the temperate zone. For 14 non-native fishes, the lower lethal temperature ranged from 5.0 to 12.9° C (Shafland and Pestrak, 1982), sufficient to cause mortality in temperate waters. Tropical fishes, however, could avoid mortality by congregating in industrial thermal effluents (Noble et al. 1975). Thus, the release site must be considered in relation to nearby artificial or natural thermal outfalls.

The opposite temperature consideration might arise in planting coldwater species in southern waters. Conceivably, exotic salmonids might be proposed for introduction into southern waters for a seasonal fishery. Salmonids will perish at temperatures above approximately 25 C (McCauley 1958, MacCrimmon and Campbell 1969). An interesting observation is that successful reproduction in rainbow trout (*Salmo gairdneri*) requires a minimum temperature of 13 C (MacCrimmon 1971). Probably the lack of sufficient seasonal temperature fluctuation limits the acclimatization potential of many cool-water fishes in the warmer latitudes.

Salinity tolerance is another example of an important dimension that may partially define the niche of an organism. Most fishes are capable of survival in either fresh water or the sea, but not both. Stringent evaluation is relevant only if the site of introduction involves water with salinities compatible with life of the exotic.

Euryhaline fishes may migrate via avenues not available to stenohaline fishes. Reproduction, however, is nearly always limited to one habitat. Generally, adults that migrate into fresh water to spawn are capable of spending their entire life cycle in fresh water; striped bass, *Morone saxatilis* (Bigelow and Schroeder 1953) and Pacific salmon, *Oncorhynchus* spp. (Stewart et al. 1981) are now acclimatized to freshwater habitats. In evaluating prospective introductions, we recommend assuming that adults will tolerate the salinity of the site if the eggs and larvae normally occur at such a salinity. Further, we recommend obtaining data on salinity tolerance of all proposed introductions and, if not available, assuming the species is euryhaline if any family members are euryhaline (e.g., see Robins et al. 1980).

Obviously, depending upon the proposed species, many other physicochemical parameters may need to be investigated.

Acclimatization potential depends on whether "unfilled" niches are present near the release site. An introduction is likely to be successful if an acceptable niche is available; otherwise the exotic must compete with established populations. Ironically, one of the chief rationales for introducing an exotic is to utilize more effectively some facet of the habitat. Fisheries productivity would be enhanced by an assemblage of species that fully exploited all food resources. Particular habitats might be depauperate because zoogeographical barriers prevented entry or because native species were reduced by man's activities. We recommend that exotic species not be introduced if a suitable native species is available for transplantation or if a fish that occupies the same or similar niche already exists in the waters.

Unfortunately, we are not optimistic about rigorous application of niche theory to an exotic species protocol. Detailed knowledge of the habitat and niche exists for only a few North American species (Werner and Hall 1977, Keast 1978). For most exotics the niche and habitat requirements are not known sufficiently to predict exactly how the exotic will fit into fish communities in receiving waters. The available information should be evaluated in terms of the perceived niche vacancy and the characteristics of the exotic.

REPRODUCTIVE POTENTIAL

Crucial to any evaluation of acclimatization potential is assessment of reproduction. We recommend the approach taken by Stanley et al. (1978) in which the spawning of grass carp was predicted for the Mississippi River. The requirements for maturation, spawning, and egg maturation were compared to the conditions found in the United States. Although spawning of grass carp occurred in the Mississippi as predicted (Conner et al. 1980), large populations have not been found. As Stanley et al. (1978) predicted, the habitat requirement for the larvae apparently proved to be limiting; only a few second-generation grass carp have been captured to date.

We believe that accurate predictions of reproductive potential can be made if adequate information is available about spawning and nursery habitat. Because life histories of exotic species are rarely published in American literature (e.g. Islam and Talbot 1968), the needed information may be available only through compendia (Breder and Rosen 1966, Donaldson 1977).

The conservative approach to introductions is to assume that reproduction will occur unless otherwise indicated by studies of factors that restrict reproduction. Absence of vegetation for phytophilous species, absence of suitable river current for fish with semibuoyant eggs, and absence of an estuary for species requiring brackish nurseries are examples of factors that might restrict reproduction. Many fishes, however, are adaptable and may reproduce in nontraditional habitats; e.g., the alewife (*Alosa pseudoharengus*) and smelt (*Osmerus mordax*) spawn on the shore bottom of the Great Lakes instead of in streams.

We believe that fecundity is not an overriding issue in evaluating reproductive potential of exotic fish. High fecundity usually is accompanied by high mortality and low fecundity with parental care or protection. However, predation may be less intense in the nonnative habitat giving a highly fecund exotic an advantage over a species with lower fecundity. Any factor that reduces average survival relative to that in the native range may result in failure of the planted fish in becoming established (see chapter 2).

The objective of evaluating reproduction in the protocol is to attempt to identify exotic species that might develop pest populations. For the exotic to become abundant, reproductive conditions would have to be met with sufficient spawning and nursery habitat available to sustain high population levels. The protocol requires focusing attention on critical spawning requirements so that exotics are not placed in waters where they may become pests.

MIGRATION AND MOVEMENTS

Movement away from the stocking site enhances the probability of locating additional suitable habitat. Dispersal, however, might reduce the chances of members of the opposite sex finding each other. Thus, migratory behavior may either favor or discourage acclimatization. We believe that, if large numbers of exotics are stocked, then migration would enhance reproductive potential because optimum spawning grounds would more likely be found. If few exotics are released, or escape, then migration is likely to decrease encounters between pairs. Thus, for species that migrate it is more difficult to predict acclimatization.

Migratory fishes may be less desirable because they may not stay in the location stocked. Further, eradication of escaped exotics would be more difficult because they might disperse.

We recommend that highly migratory species be evaluated more thoroughly before widespread or massive stocking than nonmigratory species. Although alewife have long been known to be migratory, they were recently transplanted to several mainstream reservoirs in the southeastern United States, giving them access to several major drainages, potentially even to the Mississippi River (Kohler and Ney 1982).

Control Potentials

Several control techniques are available that could alter the ability of an exotic species to establish populations. Techniques fall into two categories: the use of toxicants that kill escaped fish before they have a chance to multiply and spread, and sex control of stocked individuals so that reproductive potential is reduced. The availability of control techniques might lead to approval for a species that was otherwise likely to be disapproved.

TOXICANTS

Some forty chemical agents have been used to kill fishes (Lennon et al. 1970). Several of these toxicants are selective for fishes, leaving other biota unharmed. Among fish species, there is a wide range of sensitivities to toxicants so that potentially an unwanted species might be excised. In one case, grass carp were largely eliminated by 0.1 mg/liter $^{-1}$ rotenone, whereas most sport fishes survived (Colle et al. 1978). Another toxicant, antimycin, selectively kills scaled fish in channel catfish (*Ictalurus punctatus*) ponds (Burress and Luhning 1969). However, in practice, a toxicant should be applied at a concentration to assure complete kill of all the target fish throughout the water body, thus killing desirable fishes as well. The desirable fishes can often be replaced.

For most exotic fishes, specific toxicants are not available and general fish poisons must be used. Nevertheless, toxicants have the potential to eliminate an exotic from a local water in which it was experimentally stocked, or to which it escaped. The ability to kill experimental populations later would allow more realistic testing of environmental impacts or viability because the exotic could be stocked in natural closed systems.

If toxicants are to be employed in a protocol for a particular introduction, the applicant must have the financial, legal, and technical capability to apply the agent. Specifically, the toxicant must be covered in the budget, permits must be obtained for the water in question, and the trained personnel and application equipment must be available.

Toxicants may also serve as a backup to a protocol. If the protocol failed to predict an adverse impact and an exotic fish became locally established, the introduction might be terminated by poisoning.

Finally, toxicants are useful in enforcement. Many states have regulations controlling introductions that, if not followed, will result in action in which the private or public waters are poisoned, and the responsible individual or company is billed for the cost.

SEX CONTROL

Fish populations may be rendered incapable of reproduction by stocking exclusively one sex, by artificial gynogenesis to produce monosexes, or by stocking polyploids (Stanley 1981). Sex control techniques would be most applicable to test introductions of exotic fishes. Monosex grass carp were stocked in Lake Conway, Florida, to assess efficacy (Nall and Schardt 1980) and ecological effects (Land 1980, Kaleel 1980). The chief advantage of using nonreproducing populations is that the introduction of the exotic is reversible. If the exotic proved to be less useful than predicted or caused a severe impact, no further stockings would be made, and the population might simply be allowed to die.

Sex control offers a practical benefit for commercial producers of exotic fishes. Unusual species or special varieties of aquarium fishes or food fishes can be rendered incapable of reproduction, assuring that competitors would be unable to reproduce the fish. There would also be more incentive to develop special varieties if reproduction could be controlled.

Sex control does not offer an absolute guarantee against reproduction. Monosex fish may encounter the opposite sex that escaped from other sources or were stocked by other parties. Every individual may not be affected by the sex control technique; e.g., broods of "all-male" tilapia hybrids contain a few females (Mires 1977). Some triploid animals are capable of limited reproduction (Stanley 1981). For many exotics, reduced fecundity or restricted sexual capability could lower reproductive potential below that needed to sustain a wild population. Sunfish hybrids are fertile, but the reproductive potential is below that needed to support population growth in the face of predation (Ellison and Heidinger 1978). Only for the potentially most damaging exotic would absolute sterility be required in field tests.

Sex control has significance to protocols because the establishment of exotics can be potentially controlled and accurate information on benefits and risks can be obtained. If the proposed introduction involves only sex-controlled fish, then the decision to approve introduction could be made early in the review process.

Potential Impact

BENEFITS

Several benefits may accrue from the introduction of an exotic species. For example, it may serve as a biological control of such nuisances as aquatic weeds or mosquitoes. When compared to the potential environmental hazards of chemical controls, the risks of a biological agent may well be acceptable. Another benefit of an exotic species could be its functioning as a forage base for commercially or recreationally valuable native species. Moreover, the exotic species may, itself, develop a new commercial or recreational fishery. The potential benefits of exotic fishes are real. Pessimism is as counterproductive as unfounded optimism.

Exotic species are used in experimental and commercial aquaculture (see chapter 13). For example, various tilapias and carps have been introduced on a global scale for aquaculture, particularly in the developing countries. Once the procedures for culturing a species have been developed, its introduction outside its native range is often rationalized, both for commercial and humanitarian reasons.

Several species of exotic fishes have been used in other countries to control nuisance aquatic vegetation (see chapter 14). The species most often considered for this purpose is the grass carp. In addition to being an effective control agent for submerged weeds, the flesh of the grass carp is considered delectable and has commercial value (National Academy of Sciences (1976). Much interest has recently been generated for introducing grass carp x bighead carp (*Hypophthalmichthys* [*Aristichthys*] *nobilis*) F_1 hybrids for aquatic weed control because they are sterile; an AFS symposium on this hybrid was held concurrent with the present Exotic Fish Symposium. Other herbivorous fishes used for aquatic weed control include various tilapias, silver dollar fish (*Metynnis roosevelti*), and silver carp (*Hypophthalmichthys molitrix*).

Exotic fishes are used in public health programs. Mosquitofish (*Gambusia affinis*), native to the southern United States, has been widely introduced as a mosquito control measure but with limited success (Rosenthal 1980) and not without creating problems (Myers 1965). Redear sunfish (*Lepomis microlophus*) were introduced to Puerto Rico as one of many population control measures for the snail vector of schistosomiasis (Francisco A. Pagan-Font, personal communication).

Exotic species are used in fisheries management to establish or expand commercial and/or recreational fisheries. The brown trout (*Salmo trutta*) is exotic to North America but has long been established and is esteemed as a trophy. Conversely, North American salmonids have been introduced to several other continents, as have centrarchids. For expanding forage bases, transplantation of native species is the prevailing practice.

The use of exotic fishes for aquaria far exceeds all other uses. Simply stated, the ornamental fish trade is big business. Conroy (1975) estimates that in the United States the extent of the trade in ornamental fishes and aquarium accessories is broadly comparable to that involving cats and dogs combined. The trade involves both salt- and freshwater fishes; many of the freshwater species are cultured. Florida is the world's largest breeding center with over 150 tropical fish farms (Brown and Gratzek 1980).

RISKS

Unfortunately, most intentional stockings of exotic fishes have been undertaken on a trial-and-error basis, and many of these introductions have been expensive failures (Courtenay and Robins 1973). As Magnuson (1976) has aptly stated, managing with exotics can be a "game of chance."

A major concern of exotic species introductions is the inadvertent transfer of diseases and parasites. The ramifications of such an occurrence are obvious and were noted in a previous section.

An exotic species may displace one or more native species (see Courtenay 1979 and Rosenthal 1980 for reviews on this subject) through interspecific interactions; displacement may occur because the exotic is competing for resources such as food and breeding sites, exerting severe predation on native species, or has altered the habitat. The common carp (*Cyprinus carpio*), with its proclivity for uprooting vegetation and muddying waters, is the most often cited nuisance exotic fish species in North America. Millions of dollars have been expended on control and eradication programs, but with little success (Laycock 1966, Courtenay and Robins 1973).

An exotic may adversely affect nonaquatic wildlife or agricultural interests. Such has been the fear with regard to potential grass carp impacts on waterfowl habitat and rice fields (Courtenay and Robins 1973, Roberts et al. 1973). There is a good probability that the exotic might be ecologically incompatible with an endangered, threatened, or rare native species. Another complication is the possibility that, after introduction, an exotic may hybridize with closely related species. The consequences of an introduction cannot be appraised without life history data for the exotic species and the indigenous populations.

The potential for unintentional introduction of waterborne human diseases is a critical threat which must also be considered. A disease such as schistosomiasis, caused by a blood fluke, could concurrently be introduced if its intermediate snail host was accidentally transported with the exotic fish. Precautionary procedures to avoid this and other such transfers are obviously in order.

The Protocol

In previous sections we have described factors to be considered with regard to exotic fish introductions. These factors are incorporated into a protocol that we recommend be used for systematically evaluating proposed exotic fish introductions. The protocol employs a "Review and Decision Model" (figure 18–1), and is a decision tree in which a hierarchy of factors is considered in successive levels of review. The protocol requires establishment of an Exotic Fish Protocol Committee to evaluate proposals for each exotic fish introduc-

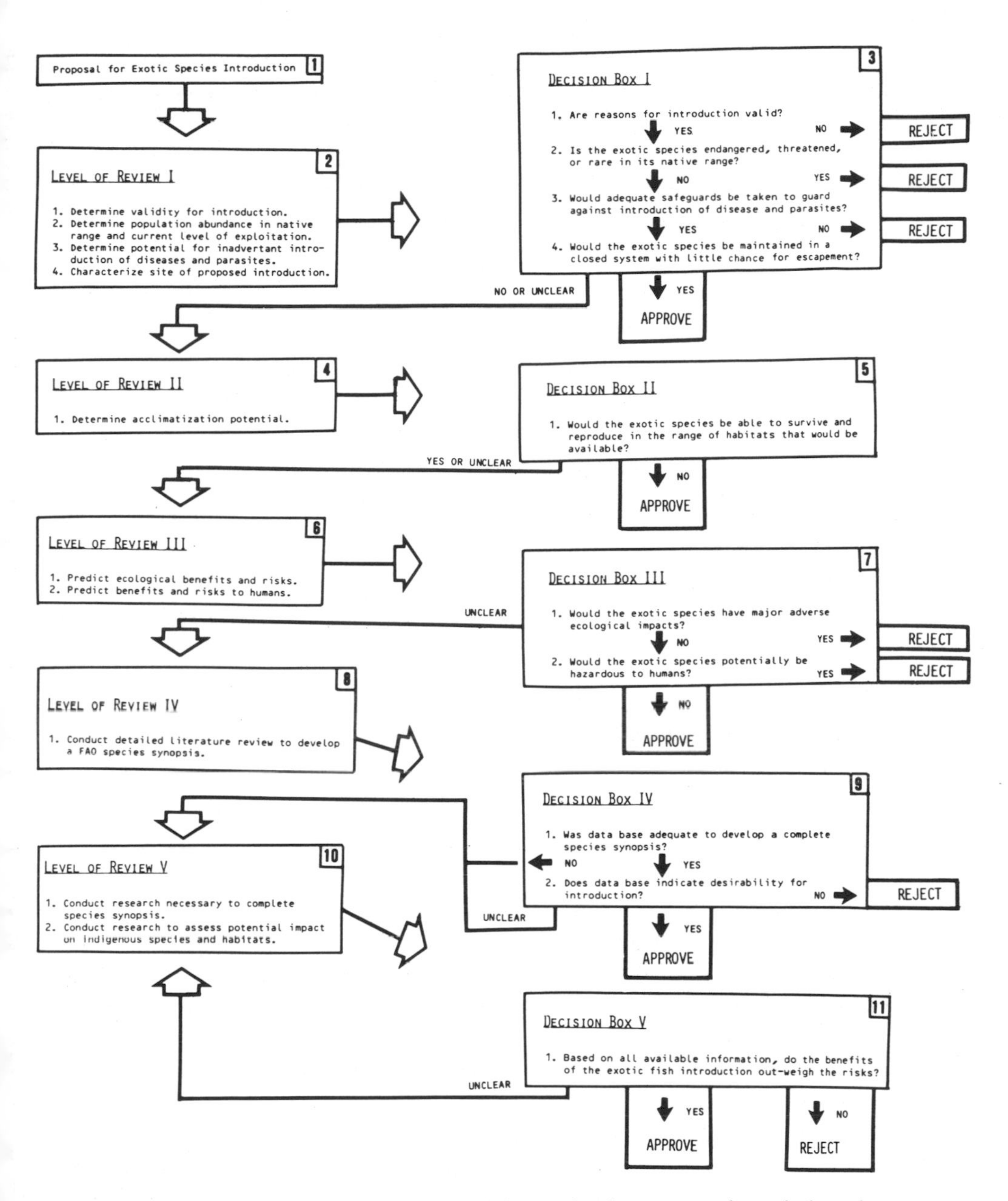

Figure 18–1. Review and decision model for evaluating proposed exotic introductions.

tion (excluding those exotics already widely established and most ornamental fishes). We do not suggest how, or by whom, this committee would be composed.

PROPOSAL FOR EXOTIC FISH INTRODUCTION

An entity desiring to introduce an exotic fish would prepare a proposal that responds to the following:

1. What exotic species do you propose to introduce (common and scientific name)?

2. Where is its native range? What is the present range?

3. What is the purpose of the introduction?

4. Where, how many, and into what type of system would this exotic species be introduced?

5. What precautions have or will be taken to ensure that the exotic species is not harboring diseases and parasites?

6. If the exotic species is to be maintained in a closed system, what measures would be taken to guard against accidental escape to open waters?

7. What is your knowledge of the acclimatization potential?
a. Habitat/niche requirements
b. Reproduction

A bibliography of pertinent literature should be appended to the proposal.

EVALUATION PROCESS

The evaluation process is based on a "Review and Decision Model" (figure 18–1) comprised of five levels of review and five corresponding "Decision Boxes." Each level of review mandates progressively greater scrutiny of the proposed exotic fish introduction. However, not all proposed exotic fish introductions would be subjected to the entire review process; a verdict could often be rendered during the initial stage of review. The evaluation process is as follows:

1. *Level of Review I*
a. Purpose of introduction. Does the proposing entity have valid reasons for introducing an exotic species, and could native species serve the same function?

b. Abundance in native range. Knowledge of the population abundance of the exotic species in its native range is an important aspect of the evaluation. Is the exotic species endangered, threatened, or rare? Is it exploited from the wild or under culture? Will importation preserve the species?

c. Diseases and parasites. The evaluation would include assessing the safeguards for avoiding transmission of diseases and parasites to the proposed receiving system(s).

d. Site of introduction. It is important to discern from the outset whether the exotic species would be stocked in an open or a closed system. Would the exotic species be stocked in or have potential access to a major drainage? If the exotic species is to be maintained in a closed system, the proposing entity must identify steps it would take to guard against accidental escape.

2. *Decision Box I*

A proposal for an exotic fish introduction would be rejected if (1) reasons for introduction were not deemed valid; (2) the species is endangered, threatened, or rare in its native range (except for species survival); or (3) the proposing entity has not established that adequate safeguards would be taken to avoid introduction of diseases and parasites. The proposal would be approved at this stage if the above criteria are properly met, and provided that the introduction would be limited to a closed system. If this last condition is not fully met, the evaluation process would proceed to the next level of review.

3. *Level of Review II*

This level of review is directed at determination of the exotic species' acclimatization potential. Should pertinent information be unavailable, the Protocol Committee might grant the proposing entity permission to conduct research with a limited number of specimens under confined conditions for the purpose of obtaining required data. If the proposing entity is not qualified to conduct the research, then it would be its responsibility to contract the research to a qualified laboratory. In some cases, the Protocol Committee may require that all research be conducted within the exotic species' native range.

4. *Decision Box II*

The proposal for the exotic fish introduction would be approved if the probability for establishment of a self-sustaining population were negligible. Alternatively, further evaluation would be mandated for those species that might produce self-sustaining populations or if evidence is insufficient for making a reasonable prediction.

5. *Level of Review III*

This level of review is based on predicting the potential impact of the exotic species on the ecological integrity of the system(s) where it is proposed for introduction. In addition, the benefit/risk analysis would include assessing the array of potential impacts on man. Review at this level requires detailed knowledge on the ecological role of the exotic species in its native range, as well as considerable information on the community structure of the proposed receiving system(s).

6. *Decision Box III*

The introduction would be approved if the available information strongly suggests that the exotic species would not exert a major adverse impact on the receiving system(s) or to man.

7. *Level of Review IV*

Level of Review IV requires evaluation of a literature review based on the format for a FAO Species Synopsis. However, additional sections concerning impacts of transplantation (documented or potential) would also be required.

8. *Decision Box IV*

The Protocol Committee would either approve or reject the proposed exotic fish introduction, based on analysis of a complete species synopsis. Additional review (Level V) would be necessary if the current data base is not sufficient for formulation of a complete species synopsis or if it is still unclear whether the introduction is desirable.

9. *Level of Review V*

This level of review requires further research to complete the species synopsis and/or specific research to assess the potential impact of the introduction to the indigenous species and habitats. Research might be conducted under controlled conditions in a location near to where the introduction is contemplated, or the Protocol Committee may require that all studies be carried out within the species' native range. In either case, the qualifications of the staff and research facilities would be evaluated by the Protocol Committee before the conduct of studies. Topics would be investigated as specified by the Protocol Committee, and would be based on research guidelines that are currently being prepared by a committee of the Exotic Fish Section of the American Fisheries Society.

10. *Decision Box V*

The Protocol Committee may find it necessary to specify additional research if important questions remain to be resolved. In such a case, the fifth and final evaluation stage would become a loop of the "Review and Decision" modes until a ruling could be made.

Conclusion

We have described factors to be considered in evaluating proposed exotic fish introductions. The protocol we suggest requires consideration of these factors in a logical sequence so that many decisions can be made early in the review process. All proposed introductions need not be subjected to the entire evaluation. The protocol is simply the vehicle by which the evaluation would proceed; the committee employing the protocol would have to interpret the generated data base and render an informed decision.

Many will likely consider this protocol too stringent and just as many will feel the opposite. Such conflicts are unavoidable, regardless of our concerted attempt to take a "middle-of-the-road" stance. Following this or a similar protocol should ensure that exotic species are used in a wise manner.

Literature Cited

Andrews, J. D. 1980. A review of introductions of exotic oysters and biological planning for new importations. Marine Fisheries Review 42:1–11.

Anonymous. 1978. Code of federal regulations. Wildlife and fisheries. Special edition of the Federal Register. Title 50. U.S. Government Printing Office, Washington.

Axelrod, H. R. 1973. Report of Exotic Fishes Committee. Transactions of American Fisheries Society 102:239–248.

Bauer, O. N., and G. L. Hoffman. 1976. Helminth range extension by translocation of fish, 163–172. *In:* L. A. Page (ed.). Wildlife diseases. Plenum Publishing Corporation, New York, New York.

Bigelow, H. B., and W. C. Schroeder. 1953. Fishes of the Gulf of Maine. U.S. Fish and Wildlife Service Fishery Bulletin 53:1–577.

Bohl, M. 1979. Disease control of reproduction of grass carp in Germany, 243–251. *In:* J. V. Shireman (ed.). Proceedings of the grass carp conference. University of Florida Institute of Food and Agricultural Sciences, Gainesville, Florida.

Bowden, G. 1979. Law, politics and biology: Aquaculture in the primordial ooze, 306–330. *In:* R. Mann (ed.). Exotic species in mariculture. MIT Press, Cambridge, Massachusetts.

Breder, C. M., Jr., and D. E. Rosen. 1966. Modes of reproduction in fishes. Natural History Press, Garden City, New York.

Brown, W. Y. 1979. The federal role in regulating species introductions into the United States, 258–264. *In:* R. Mann (ed.). Exotic species in mariculture. MIT Press, Cambridge, Massachusetts.

Brown, E. E., and J. B. Gratzek. 1980. Fish farming handbook, food, bait, tropicals and goldfish. AVI Publishing Company, Westport, Connecticut.

Burress, R. M., and C. W. Luhning. 1969. Field trials of antimycin as a selective toxicant in channel catfish ponds. Investigations in Fish Control 25:1–12.

Colle, D. E., J. V. Shireman, R. D. Gasaway, R. L. Stetler, and W. T. Haller. 1978. Utilization of selective removal of grass carp (*Ctenopharyngodon idella*) from an 80-hectare Florida lake to obtain a population estimate. Transactions of the American Fisheries Society 107:724–729.

Conner, J. V., R. P. Gallagher, and M. F. Chatry. 1980. Larval evidence for natural reproduction of the grass carp (*Ctenopharyngodon idella*) in the lower Mississippi River. Proceedings of the Fourth Annual Larval Fish Conference. Biological Services Program, National Power Plant Team, Ann Arbor, Michigan, FWS/OBS-80/43:1–19.

Conroy, D. A. 1975. An evaluation of the present state of world trade in ornamental fish. Food and Agriculture Organization Fisheries Circular 335, 120 pp.

Courtenay, W. R., Jr. 1979. Biological impacts of introduced species and management policy in Florida, 237–257. *In:* R. Mann (ed.). Exotic species in mariculture. MIT Press, Cambridge, Massachusetts.

Courtenay, W. R., Jr., and C. R. Robins. 1973. Exotic aquatic organisms in Florida with emphasis on fishes: A review and recommendations. Transactions of the American Fisheries Society 102:1–12.

Department of the Interior. 1980. Republication of lists of endangered and threatened species and corrections of technical errors in final rules. Federal Register 45:33768–33781.

Dill, W. A. (ed.). 1972. Report of the symposium on the major communicable diseases in Europe and their control. European Inland Fisheries Advisory Commission Technical Paper 71. Food and Agriculture Organization of the United Nations. Rome, Italy.

Donaldson, E. M. 1977. Bibliography of fish reproduction 1963–1974. Fisheries and Environment Canada, Fisheries and Marine Service Technical Report 732. West Vancouver, British Columbia.

Ellison, D. G., and R. C. Heidinger. 1978. Dynamics of hybrid sunfish in southern Illinois farm ponds. Proceedings of the Annual Conference of the Southeastern Association of Fish and Wildlife Agencies 30:82–87.

FAO/OIE (Food and Agriculture Organization of the United Nations/International Office of Epizootics). 1977. Control of the spread of major communicable fish diseases. FAO Fisheries Reports 192. Rome, Italy.

Islam, B. N., and G. B. Talbot. 1968. Fluvial migration, spawning, and fecundity of Indus River hilsa, *Hilsa ilisha*. Transactions of the American Fisheries Society 97:350–355.

Kaleel, R. T. 1980. Large-scale operations management test using the white amur at Lake Conway, Florida; water quality. Proceedings of the 14th Annual Meeting, Aquatic Plant Control Research Planning and Operations Review, Miscellaneous Paper A-80-3: 285–297.

Keast, A. 1978. Trophic and spatial interrelationships in the fish species of an Ontario temperate lake. Environmental Biology of Fishes 3:7–31.

Kohler, C. C., and J. J. Ney. 1982. Suitability of alewife as a pelagic forage fish for southeastern reservoirs. Proceedings 34th Annual Conference Southeastern Association of Fish and Wildlife Agencies 34:171–286.

Land, R. 1980. Large-scale operations management test using the white amur at Lake

Conway, Florida; fish, mammals, and waterfowl. Proceedings of the 14th Annual Meeting, Aquatic Plant Control Research Planning and Operations Review, Miscellaneous Paper A-80-3:273–284.

Laycock, G. 1966. The alien animals. Natural History Press. Garden City, New York. 240 pp.

Lennon, R. E., J. B. Hunn, R. A. Schnick, and R. M. Burress. 1970. Reclamation of ponds, lakes, and streams with fish toxicants: A review. Food and Agriculture Organization Technical Paper 100.

McCauley, R. M. 1958. Thermal relations of geographic races of *Salvelinus*. Canadian Journal of Zoology 36:655–662.

MacCrimmon, H. R. 1971. World distribution of rainbow trout (*Salmo gairdneri*). Journal of the Fisheries Research Board of Canada 28:663–704.

MacCrimmon, H. R., and J. S. Campbell. 1969. World distribution of brook trout, *Salvelinus fontinalis*. Journal of the Fisheries Research Board of Canada 26:1699–1725.

Magnuson, J. J. 1976. Managing with exotics – a game of chance. Transactions of the American Fisheries Society 105:1–9.

Mann, R. 1979. Exotic species in aquaculture: An overview of when, why, and how, 331–357. *In:* R. Mann (ed.). Exotic species in mariculture. MIT Press, Cambridge, Massachusetts.

Mires, D. 1977. Theoretical and practical aspects of the production of all male tilapia hybrids. Bamidgeh 29:94–101.

Myers, G. S. 1965. *Gambusia,* the fish destroyer. Tropical Fish Hobbyist 13:31–32, 53–54.

Nall, L. E., and J. D. Schardt. 1980. Large-scale operations management test using the white amur at Lake Conway, Florida; aquatic macrophytes. Proceedings of the 14th Annual Meeting, Aquatic Plant Control Research Planning and Operations Review, Miscellaneous Paper A-80-3:249–272.

National Academy of Sciences. 1976. Making aquatic weeds useful: Some perspectives for developing countries. National Academy of Sciences. Washington, D.C.

Noble, R. L., R. D. Germany, and C. R. Hall. 1975. Interactions of blue tilapia and largemouth bass in a power plant cooling reservoir. Proceedings of the Annual Conference of the Southeastern Association of Gamc and Fish Commissioners 29:247–251.

Roberts, T. R., C. H. Park, and R. Straus. 1973. Book review of Chinese Freshwater Fish Culture. Transactions of the American Fisheries Society 102:668–669.

Robins, C. R., R. M. Bailey, C. E. Bond, J. R. Brooker, E. A. Lachner, R. N. Lea, and W. B. Scott. 1980. A list of common and scientific names of fishes from the United States and Canada. Special Publication no. 12, American Fisheries Society, Bethesda, Maryland. 174 pp.

Rosenthal, H. 1980. Implications of transplantations to aquaculture and ecosystems. Marine Fisheries Review. May(1980):1–14.

Scanlon, P. F., T. R. Teitt, and G. H. Cross. 1978. An overview of problems of introduced species and approaches by states to controlling exotics. Proceedings of the Annual Conference of the Southeastern Association of Game and Fish Commissioners 30:674–679.

Shafland, P. L., and J. M. Pestrak. 1982. Lower lethal temperatures for fourteen non-native fishes in Florida. Environmental Biology of Fishes 7:149–156.

Stanley, J. G. 1981. Manipulation of developmental events to produce monosex and sterile fish. Rapports et Procès-verbaux des Réunions Conseil International pour l'Exploration de la Mer 178:485–491.

Stanley, J. G., W. W. Miley II, and D. L. Sutton. 1978. Reproductive requirements and likelihood for naturalization of escaped grass carp in the United States. Transactions of the American Fisheries Society 107:119–128.

Stewart, D. J., J. F. Kitchell, and L. B. Crowder. 1981. Forage fishes and their salmonid predators in Lake Michigan. Transactions of the American Fisheries Society 110:751–763.

Welcomme, R. L. 1981. Register of international transfers of inland fish species. Food and Agriculture Organization Fisheries Technical Paper no. 213. 120 pp.

Werner, E. G., and D. J. Hall. 1977. Competition and habitat shift in two sunfishes (Centrarchidae). Ecology 58:869–876.

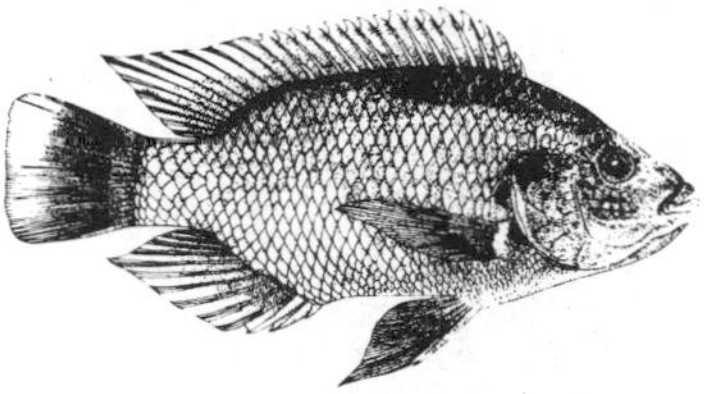

Summary

The successful colonization of introduced fishes has proceeded since World War II on a global scale. It was the intent of this work to document both the benefits and consequences that have resulted from the numerous transfers of fishes, and through examples, biological data, and ecological rationale, to urge further caution with candidates for future introductions.

From a review of the chapters included herein, it is clear that fish introductions are far more complex than simply releasing something, and should be so viewed. Impacts should not be judged solely on what the exotic can do *for* man but also on what the exotic can do *to* water quality, native biota, and habitat. Introductions of biota pose a perpetual threat to the integrity of our ecosystems. Once these introductions become established, it is not possible to "turn off the tap." Even the most persistent chemical pollutants do not reproduce, grow, and disperse.

Benefits can be derived from the introduction of exotic fishes. Haste and perceived objectives, however, are no substitute for caution, proper thought, and careful research. Within the past few decades, people of industrialized nations have become concerned with the conservation and preservation of endangered and threatened gene pools, water quality, and such concepts as biological integrity. Many developing nations, however, are more concerned with providing food for their human population than in conservation and preservation of irreplaceable resources; while such concern is laudable and understandable, we would urge those nations to examine more thoroughly the long-term consequences to native resources.

Finally, we hope this effort will awaken those who have (or should have) ultimate responsibility in the wise use of biotic resources – namely, biologists and resource managers, worldwide – to the need for establishing biologic as well as economic cost/benefit ratios before and not after introductions. Otherwise, our mistakes could prove far more costly in biologic and economic terms than any benefits perceived or derived.

Contributors

Salvador Contreras-B., Escuela de Graduados en Ciencias Biológicas, Universidad Autónoma de Nuevo León, Apartado Postal 732, San Nicolás de los Garza, Nuevo León, Mexico

Walter R. Courtenay, Jr., Department of Biological Sciences, Florida Atlantic University, Boca Raton, Florida 33431–0991

E. J. Crossman, Department of Ichthyology and Herpetology, Royal Ontario Museum, Toronto, Ontario M5S 2C6 Canada

Donald S. Erdman, CODREMAR, P.O. Box 2639, San Juan, Puerto Rico 00903

Marco A. Escalante-C., Escuela Ciencias del Mar, Universidad Autónoma de Sinaloa, Apartado Postal 714, Mazatlán, Sinaloa, Mexico

John B. Gratzek, Department of Medical Microbiology, College of Veterinary Medicine, University of Georgia, Athens, Georgia 30602

Dannie A. Hensley, Department of Marine Sciences, University of Puerto Rico, Mayaguez, Puerto Rico 00708

Charles H. Hocutt, Horn Point Environmental Laboratory, University of Maryland, Box 775, Cambridge, Maryland 21613–0775

Glenn L. Hoffman, U.S. Fish and Wildlife Service, Fish Farming Experimental Station, P.O. Box 860, Stuttgart, Arkansas 72160–0860

Christopher C. Kohler, Fisheries Research Laboratory, Southern Illinois University, Carbondale, Illinois 62901

James A. McCann, National Fisheries Center, U.S. Fish and Wildlife Service, P.O. Box 700, Kearneysville, West Virginia 25430–0700

R. M. McDowall, Fisheries Research Division, Ministry of Agriculture and Fisheries, Private Bag, Christchurch, New Zealand

J. A. Maciolek, National Fishery Research Center, U.S. Fish and Wildlife Service, Building 204, Naval Support Activity, Seattle, Washington 98115

Roland J. McKay, Queensland Museum, Gregory Terrace, Fortitude Valley, Queensland, Australia 4006

R. G. MARTIN, Sport Fishing Institute, 608 Thirteenth Street, N.W., Washington, D.C. 20005

N. S. PROSSER, Sport Fishing Institute, 608 Thirteenth Street, N.W., Washington, D.C. 20005

G. C. RADONSKI, Sport Fishing Institute, 608 Thirteenth Street, N.W., Washington, D.C. 20005

GOTTFRIED SCHUBERT, Arbeitsgruppe Biologie der Fische, Universität Hohenheim, Stuttgart, West Germany

WILLIAM L. SHELTON, Zoology Department, University of Oklahoma, 730 Van Vleet Oval, Norman, Oklahoma 73019

JEROME V. SHIREMAN, Center for Aquatic Weeds, University of Florida, Gainesville, Florida 32601

EMMETT B. SHOTTS, JR., Department of Medical Microbiology, College of Veterinary Medicine, University of Georgia, Athens, Georgia 30602

R. ONEAL SMITHERMAN, Department of Fisheries and Allied Aquacultures, Auburn University, Alabama 36849

JON G. STANLEY, Maine Cooperative Fishery Research Unit, University of Maine, Orono, Maine 04469.

JAY R. STAUFFER, JR., School of Forestry, The Pennsylvania State University, University Park, Pennsylvania 16802

R. H. STROUD, Sport Fishing Institute, 608 Thirteenth Street, N.W., Washington, D.C. 20005

JEFFREY N. TAYLOR, Department of Biological Sciences, Florida Atlantic University, Boca Raton, Florida 33431-0991

R. L. WELCOMME, Food and Agriculture Organization, United Nations, Via delle Terme di Caracalla, 00100 Rome, Italy

Index of Names

Abdel-Malek, S. A., 337
Abele, R., 318
Addor, E. E., 329, 333
A. E. Woods State Fish Hatchery, 65
Ahmed, A.T.A., 235
Ahuja, S. K., 185
Alabama Division of Game and Fish, 316
Al-Daham, N. K., 283
Aldrich, A. D., 336, 341
Alexander, G. R., 323, 324, 343, 344
Aliev, D. S., 305, 306
Alikunhi, K. H., 306
Alkholy, A. A., 337
Allan, J. H., 346
Allan, R. C., 324, 343
Allanson, B. R., 275
Allen, A. W., 49, 108
Allen, D. L., 384
Allen, G. H., 277
Allen, K. R., 347
Allendorf, F. W., 14
Allison, R., 276, 282
Alvarez, J., 106, 107, 108, 113
American Fisheries Society (AFS), 1-7, 313
American Society of Ichthyologists and Herpetologists, 1, 313
Anacker, R. L., 220
Anderson, J. H., 336, 338
Anderson, R. O., 306, 332, 335, 339
Anderson, R. S., 342, 344, 349
Andrews, C. W., 80, 83
Andrews, J. D., 388
Andriasheva, M. A., 273
Appelbaum, S., 268
Arieli, Y., 275
Arizona Department of Game and Fish, 62, 65
Arkansas Game and Fish Division, 49, 267, 269
Aro, K. V., 80
Arredondo, L., 102, 106
Arthington, A. H., 186
Atton, F. M., 89
Auburn University, 49, 62, 266, 269, 272, 273, 280, 284, 316
Australian Federation of Aquarium Fish Importers and Traders, 189
Australian Fisheries Council, 195
Australian National Parks and Wildlife Service, 182
Avault, J. W., Jr., 263, 274, 275, 285, 286, 302, 303, 332, 333, 335, 336, 337, 338, 339, 341
Avtalion, R. R., 283
Axelrod, H. R., 3, 4, 388

Babaev, B., 241
Babcock, S., 342, 350
Bachmann, R. W., 303
Backiel, T., 269
Bade, E., 50
Bailey, R. M., 3, 54, 55, 150
Bailey, W. H., 331
Bailey, W. M., 267, 305, 307, 332, 334
Baird, S. F., 49, 50, 52, 317
Balarin, J. D., 264, 274, 275
Balon, E. K., 22, 49
Banarescu, P., 108, 109
Barajas, M. L., 110, 118
Barbour, C. D., 115, 116
Bardach, J. E., 262, 263, 264, 267, 274, 275, 282
Barnett, B. S., 329
Barron, J. C., 53

Bartlett, K. H., 215, 228
Barwick, D. H., 331, 350
Bauer, B. H., 119, 390
Bauer, O. N., 233, 237, 238, 240, 247
Baughman, J. L., 52
Baur, R. J., 335
Baxter, G. R., 55
Bayne, D. R., 276, 277, 282
Beach, M. L., 332, 334, 335, 339
Beam, M., 205
Bean, T. H., 50, 343
Becker, C. D., 243
Becker, G. C., 51
Beckman, W. C., 51
Behnke, R. J., 62, 65, 106
Behrends, L. L., 272, 277
Bell, G. R., 97
Belshe, J. F., 54, 342, 344, 346
Bennett, G. W., 65, 337, 340
Ben-Tuvia, A., 337
Berdegue, A. J., 57, 58
Berg, L. S., 48, 49, 50, 51, 52, 67
Berra, T. M., 187
Best, E., 211
Bigelow, H. B., 393
Bird, E. A., 170
Bjornn, T. O., 347, 348
Black, G. F., 64, 65, 337, 341, 343, 345
Black, J. D., 335
Blackburn, R. D., 302, 332, 333
Bohl, M., 390
Bond, C. E., 52
Bonislawsky, P. S., 285
Bonnet, L. C., 165
Booke, H. E., 51
Bounds, R. L., 61
Bowden, G., 391
Bowen, J. T., 264
Bowler, J. M., 179
Boyd, C. E., 329
Boyd, R. L., 267
Boyer, R. L., 349
Branson, B. A., 108, 111
Breder, C. M., Jr., 394
Brittan, M. R., 53, 67
Brock, V. E., 131, 133, 138, 139, 143, 146, 150, 151
Brooker, M. P., 305
Brown, B. E., 285
Brown, C.J.D., 3, 54, 56, 57, 65, 346
Brown, E. E., 221, 390, 397
Brown, W. H., 64
Brown, W. Y., 391
Brunson, W. D., 243
Bryan, F. L., 277
Bryan, W. A., 161
Brynildson, O. M., 323, 324, 343, 348
Buchanan, R. E., 220
Buck, D. H., 272, 273, 277, 279, 306, 330, 332, 334, 336, 338, 339
Buntz, J., 62, 63, 275, 337, 340, 342, 349
Bureau of Fisheries, U.S., 163, 167, 169, 170, 315
Burgess, G. H., 57, 106, 113, 117
Burgess, J. E., 53
Burke, C. N., 221
Burns, J. W., 336, 338
Burns, R. P., 227
Burress, R. M., 395
Bussing, W. A., 53
Bykhovskaya-Pavlovskaya, I. E., 235

Cadwallader, P. L., 183, 184, 185, 206
Cagni, J. E., 332, 333
Cahn, A. R., 51, 335, 336, 338, 351
Cahoon, W. G., 336, 337, 338, 351
Caldwell, D. K., 54, 185
Calhoun, W. E., 276, 283
California Department of Fish and Game, 48, 57, 58, 343
California Fish and Game Commission, 305–6
Campbell, C. J., 3
Campbell, H. J., 32
Campbell, J. S., 392
Canadian Department of Fisheries and Oceans, 97
Canaris, A. G., 245
Canella, I. R., 234
Canella, M. F., 234
Canfield, D. E., Jr., 303
Carl, G. C., 86, 87, 89, 90, 92
Carlander, K. D., 61
Carlson, D. M., 285
Carter, C. C., 329, 332, 333, 339
Carter, J., 3
Cashner, R. C., 117
Catt, J., 83, 84
Cavuilati, S., 147
Chadhuri, H., 264
Chamberlain, E. B., Jr., 335, 337, 338
Chapman, D. W., 347, 348
Chapman, P., 342, 350
Cházari, E., 102

Chen, F. Y., 283
Chen, T. P., 278
Chernoff, B., 108, 112, 113, 117
Cherry, D. S., 185
Chervinski, J., 275
Chevassus, B., 346
Chew, R. L., 331, 350
Childers, W. F., 65, 337, 340
Chimits, P., 274
Christie, W. J., 79, 81
Churchill, W. S., 343
Clarke, T. A., 161
Clemens, H. P., 264
Clemens, W. A., 342
Coad, B. W., 89
Cobb, E. W., 323, 324
Coche, A. G., 276
Cole, J. L., 335, 337, 338, 340, 345, 349
Coleman, M. S., 276
Colle, D. E., 307, 333, 395
Collins, V. G., 228
Collis, W. J., 277, 279
Colt, J., 275, 285
Colwell, R. K., 15
Connell, D. W., 186
Conner, J. V., 48, 266, 282, 305, 332, 353, 393
Conroy, D. A., 28, 229, 230, 234, 397
Contreras, A. J., 118, 120
Contreras, S., 102, 107, 108, 109, 110, 112, 113, 114, 115, 116, 118, 119, 120, 123, 124, 408
Conway, G., 19
Cooper, E. L., 3, 50, 324, 338, 349
Cordone, A. J., 329
Cornell, J. H., 62, 64, 65
Cortés, M. T., 108, 120
Corujo-Flores, I. N., 163
Courtenay, W. R., Jr., 2, 3, 4, 9, 17, 18, 19, 29, 33, 45, 48, 53, 54, 55, 56, 57, 58, 59, 60, 62, 63, 65, 66, 97, 108, 111, 113, 114, 115, 122, 180, 187, 263, 264, 274, 318, 322, 323, 325, 326, 332, 337, 339, 340, 341, 350, 375, 388, 398, 408
Cox, T. J., 343
Craig, E. G., 82, 92
Crawford, K. W., 281
Cremer, M. C., 272, 273
Cressey, R. F., 246
Crisman, T. L., 306, 335
Crittenden, E., 62, 341
Crombie, A. C., 11
Cross, D. G., 332, 333
Crossman, E. J., 49, 78, 79, 80, 82, 83, 86, 87, 92, 94, 95, 336, 338, 347, 349, 408
Crowder, J. P., 332, 334, 339
Cruz, E. M., 278
Culley, D. D., 186

Dabrowski, K., 268
Dadzie, S., 283
Daget, J., 60, 349
Dahl, A. H., 133
Dah-Shu, L., 267
Dangel, J. R., 346
Darlington, P. J., 163
Dashu, -N., 234
Dashwood, J., 148
Davis, A. T., 276
Deacon, J. E., 2, 3, 32, 45, 53, 55, 56, 57, 59, 63, 66, 186, 325, 337, 349
Dean, W. J., 331
DeBach, P., 15, 16, 17, 326
De Buen, F., 102, 120
Decell, J. L., 283
Deevey, E. S., 11, 12
DeKay, J. E., 41, 48, 49, 322
Delgadillo, T., 103, 108, 121, 122
Delmendo, M. N., 277
Dence, W. A., 50
Denzer, H. W., 275
Department of Interior, U.S., 389
DeSilva, S. S., 268
Devambez, L. C., 131
DeVos, A., 323
Dial, R. S., 56, 60, 63, 64
Diamond, J. M., 352
Dill, W. A., 390
Dobzhansky, T., 10
Donaldson, E. M., 282, 394
Drda, T. F., 329, 332, 335, 339
Dubinina, M. N., 241
Dunham, A. E., 16
Dunseth, E. R., 271, 272, 275, 276, 280, 281, 282
Dye, H. M., 282
Dykova, E., 236
Dymond, J. R., 80, 82, 84, 87, 92, 94, 323, 324

Edmondson, W. T., 302
Edwards, D. J., 207, 242
Edwards, R. W., 305
Egglishaw, H. J., 347, 348

Egusa, S., 238
Ellison, D. G., 396
Elton, C. S., 323, 325, 326
Emery, A. R., 91
Emlen, S. T., 18
Endler, J. A., 55
Ensign, A. R., 336, 338, 345
Environmental Protection Agency, U.S., 302, 375, 376, 385
Erdman, D. S., 162, 163, 167, 168, 408
Escalante, M. A., 102, 109, 408
Eschmeyer, R. W., 324, 328, 329, 332, 335, 338, 341, 343, 349
Essbach, A. R., 61, 62
Evans, H. E., 342, 343
Evelyn, T.P.T., 248
Evermann, B. W., 163
Exotic Fish Section, AFS, 2-3, 4, 5, 79, 81

Fagerlund, U.H.M., 282
Fago, D., 51
Farrin, A. E., 240
Fausch, K. D., 347, 348
Fedoruk, A. N., 87
Fenderson, C. N., 323, 324, 343
Ferguson, D. E., 186
Fernando, C. H., 27, 242
Fielding, J. R., 284
Fijan, N., 242
Finucane, J. H., 63, 342
Fischer, Z., 268
Fish, G. R., 212
Fish and Wildlife Service (FWS), U.S., 3, 65, 166, 167, 169, 317, 318, 319
Fish Commission, U.S., 47, 49, 50, 52, 88, 314, 315
Fish Culture Section, AFS, 5, 265
Fish Culture Station, U.S. FWS, 65, 267
Fisk, L. O., 338
Fitzpatrick, L. A., 337, 341
Fleming, C. A., 200
Flittner, G. A., 323, 348
Florida Game and Fresh Water Fish Commission, 54, 58, 60, 62, 66-67, 318
Foissner, W., 234, 236
Follett, W. I., 108, 109, 113, 118, 120, 121
Food and Agriculture Organization (FAO), 23
Foote, K. J., 44, 61
Ford, D. M., 217
Forester, J. S., 335
Forester, T. S., 306, 335, 338, 349, 351
Fortes, R. D., 276
Fouquet, D., 234
Fowler, H. W., 50, 53, 54, 58, 67, 146
Fowler, M. C., 332, 333, 339
Fox, A. C., 65, 332
Foye, R. E., 336, 337, 338, 349
Freeze, M., 285
Frey, D. G., 335, 341
Frost, N., 83
Fry, D. L., 14, 15, 335
Fryer, G., 246, 248, 337
Fuentes, E. R., 15
Fujita, T., 235, 245
Furtado, J. Y., 242

Gard, R., 323, 348
Garman, G. C., 343, 344
Gasaway, R. D., 49, 267, 281, 306, 329, 332, 334, 335, 339, 340
German, E. R., 324, 343
Germany, R. D., 62, 275, 337, 341, 342
Ghittino, P., 240
Gibbons, N. E., 220
Gilbert, C. R., 116
Gilhen, J., 83
Gill, D. E., 10, 12, 16
Gilmore, R. G., 60
Gleastine, B. W., 340, 341, 342, 349
Glodek, G. S., 111, 112
Glucksman, J., 180, 187
Glude, J. B., 264
Godfriaux, B. L., 277
Gold, J. R., 283
González de la Rosa, C., 113
Goode, G. B., 47
Goodman, D., 9, 10, 11
Goodyear, C. P., 187
Gosline, W. A., 53
Gosse, J. P., 60
Gots, B. L., 79
Gottschalk, J., 335, 336, 338, 351
Grandall, K. S., 330
Gratzek, J. B., 221, 233, 390, 397, 408
Greeley, J. R., 50, 51, 345, 349
Greene, C. W., 51
Greenfield, D. W., 49, 332, 335
Greenwood, P. H., 137
Grizzell, R. A., Jr., 336, 338, 341
Guerrero, R. D., 276, 282
Guillory, V., 49, 108, 267, 281
Gupta, M. V., 264, 272

Haacker, P. L., 67, 343, 346
Habel, M., 337

Hadley, W. F., 93
Hahn, D. E., 55
Hairston, N. G., 11
Haldar, D. P., 237
Hale, J. G., 324
Hales, D. C., 285
Halevy, A., 264, 270
Hall, D. J., 393
Haller, W. T., 303
Halliwell, D. B., 53, 58
Hambrick, P. S., 15
Hamlyn-Harris, R., 182
Hammerman, I. S., 283
Hannuksela, P. R., 343
Hardin, S., 334, 335
Hardy, T., 342
Harlan, J. R., 323, 324, 343
Harris, C., 62, 337, 350
Harrison, E. J., 93
Harrison, H. M., 342, 345
Hassler, W. W., 342, 348
Hastings, R. W., 51
Hauser, W. J., 62, 337
Havelka, J., 33
Hawaii Cooperative Fishery Research Unit, 143
Hawaii Department of Health, 142
Hazzard, A. S., 324
Heidinger, R. C., 396
Heimstra, N. W., 330
Heinze, K., 245
Helm, W. T., 336, 338
Helms, D. R., 341
Henderson, S., 272, 273, 277, 279, 285
Henderson-Arzapalo, A., 276
Hendricks, M. K., 275
Hendricks, M. L., 107, 342
Hendrickson, D. A., 106, 107, 108, 111, 113, 118, 119, 120
Hensley, D. A., 17, 18, 29, 45, 48, 63, 108, 111, 113, 114, 122, 263, 274, 322, 323, 341, 408
Hepher, B., 264, 269, 273, 274, 276, 277
Herre, W.C.T., 133
Herwig, N., 246
Hestand, R. S., 329, 332, 333, 339
Hickling, C. F., 268, 303, 332, 337, 339
Hida, T. S., 148, 150, 265, 275
Hildebrand, S. F., 162, 163, 164, 165, 166, 167, 168, 185
Hills, E. S., 179
Hine, P. M., 207, 242, 246
Hirschmann, H., 236
Hiscox, J. I., 342, 343
Hochachka, P. W., 13
Hocutt, C. H., 15, 375, 383, 408
Hoffman, G. L., 33, 233, 235, 236, 237, 238, 239, 240, 241, 242, 243, 246, 247, 390, 408
Hogan, J., 329
Hogg, R. G., 58, 59, 63, 66, 324, 342, 345, 349
Holčík, J., 50, 51
Holcomb, D. E., 331, 350
Holloway, A. D., 336, 337
Hoover, F. G., 52, 56, 57, 64, 65
Hopkins, K. D., 277, 283
Hopkins, M. L., 281
Hora, S. L., 267, 268
Horel, G., 342
Horn, H. S., 17
Horn, M. H., 52, 56
Horvath, L., 267, 268
Hoshina, T., 234
Hosmer, A., 147
Hotchkiss, N., 329
Hubbell, T. R., 154
Hubbs, C. L., 3, 53, 54, 55, 56, 57, 59, 61, 62, 65, 66, 67, 103, 108, 109, 110, 112, 115, 117, 119, 265, 318, 324, 328, 329, 332, 335, 338, 341, 342, 343, 349, 383
Huet, M., 262, 263, 267, 274
Huisman, E. A., 268
Hulata, G. J., 32, 35, 282
Hunt, G. S., 336, 337, 338
Hunter, J. S., 202
Hurlbert, S. H., 186
Hutchinson, G. E., 15
Hutton, R. F., 3

Idyll, C., 342, 343
Iltis, A., 60
Iñigo, F., 165, 170
Isbrücker, I.J.H., 54
Islam, B. N., 394
Ivanova, N. S., 234

Jackson, P.B.N., 31
Jackson, P. D., 184
Jahn, T. L., 234
Jalabert, B., 283
Jeffrey, N. B., 266
Jeney, G., 236
Jenkins, R. M., 3
Jenkins, T. M., 324, 347, 348
Jensen, G. L., 283, 284

Jessen, R. L., 336, 337, 338
Jewell, M. E., 330, 338
Jhingran, V. G., 36, 278
Johnson, M., 332
Johnson, R. E., 3, 52
Johnson, R. P., 80, 85, 87, 94
Johnson, R. S., 163
Johnstone, R., 284
Jones, A. C., 343
Jones, J. R., 303
Jones, R. A., 50
Jonez, A., 341, 345
Jubb, R. A., 63

Kaleel, R. T., 339, 396
Kalleberg, H., 347, 348
Kallman, K. D., 56
Kanayama, R. K., 138, 142, 149, 150
Karlin, S., 11
Keast, A., 393
Kelly, H. D., 275, 342
Kilgen, R. H., 268, 284, 332, 335
Kimsey, J. B., 338
King, D. R., 336, 337, 338
King, F. W., 55
King, J. E., 275
King, W., 1, 3, 4
Kiribati Department of Natural Resources, 148
Kirk, R. G., 277
Kirkegard, D., 234
Kirtisinghe, P., 246
Knaggs, E. H., 62, 64, 65, 66, 341
Kobylinski, G. J., 305, 332, 333, 335, 339, 340
Koch, L., 48
Kohler, C., 277, 395, 408
Koneman, E. W., 220
Konradt, A. G., 267
Kooijman, F. M., 306, 335
Körting, W., 242
Koslov, A. P., 13
Koster, W. J., 51
Krasznai, Z., 273, 282, 283
Kritsky, D. C., 239
Krotas, B. A., 233, 242
Krueger, W. H., 58
Krumholz, L. A., 146
Kucas, S. T., 110
Kudo, R. R., 235, 236
Kuehn, J. H., 336, 337, 338
Kukowski, G. E., 67
Kulakovskaya, O. P., 233, 242
Kuo, C. M., 267
Kushlan, J. A., 58

LaBounty, J., 121, 124
Lachner, E. A., 1, 28, 54, 58, 342, 374, 375
Lagler, K. F., 336, 338, 341, 345
Lahav, M., 275
Lahser, C. W., Jr., 337
Laird, M., 161
Lake, J. S., 180, 182, 184
Land, R., 334, 335, 396
LaRivers, I., 56, 324
Larkin, P. A., 325
Latta, W. C., 336, 338, 341, 345
Lauenstein, P. C., 276, 281
Laurence, J. M., 332
Lavier, G., 236
Lawrence, J. M., 306, 335, 338, 349, 351
Laycock, G., 47, 49, 201, 323, 398
Lea, R. N., 58, 67
LeCren, E. D., 269
Lee, D. S., 109, 110, 111, 113, 116, 117, 118, 119, 120, 121, 265, 274, 287, 383
Lee, J. C., 275
Legendre, V., 81
Legner, E. F., 62, 64, 65, 66, 302, 303, 305, 337, 341
Lembi, C. A., 306, 332, 333, 335, 339, 340
Lemmien, W. A., 324
Lenette, E. H., 220
Lennon, R. E., 395
Leventer, H., 273, 304, 308, 336, 337, 338
Levine, D. S., 60
Levins, R., 11, 16
Lewis, S. L., 347, 348
Lewis, W. M., 267, 332, 341
Lewis, W. M., Jr., 185
Li, H. W., 323, 326
Li, Y., 278
Lichatowich, T., 147
Lien-Siang, L., 234
Lin, S. Y., 267, 272
Lindroth, A., 347
Little, G. S., 342
Lobel, P. S., 146, 349
Loiselle, P. V., 65, 349
Lom, J., 236, 237
Lovell, R. T., 272, 281
Lovshin, L. L., 275, 276, 282
Lowe, R. H., 337
Lowe-McConnell, R. H., 274

Lucký, Z., 235, 238, 247
Luhning, C. W., 395
Lyakhnovich, V. P., 268

McAllister, D. E., 47, 90
MacArthur, R. H., 10, 11, 16, 17
McBay, L. G., 274, 275, 337, 342, 349
McCann, J. A., 3, 57, 408
McCauley, R. M., 392
McCoy, E. W., 281
McCraig, R. S., 324, 343
MacCrimmon, H. R., 23, 47, 78, 79, 82, 85, 87, 88, 89, 163, 165, 323, 324, 336, 337, 338, 341, 342, 349, 392
McDaniel, D., 247
McDowall, R. M., 178, 180, 182, 184, 200, 201, 205, 206, 209, 211, 408
Maceina, M. J., 305, 307, 308, 329, 332, 333, 340
McGrenna, M., 246
Maciolek, J. A., 133, 149, 150, 151, 153, 342, 408
MacKay, H. H., 84, 336, 338
MacKay, R. J., 180, 182, 183, 186, 188, 408
McLaren, P., 33
McNelly, J. L., 51
Maglio, V. J., 186, 187
Magnuson, J. J., 323, 398
Mainland, G. B., 139–40
Maitland, J., 83
Maitland, P. A., 240
Maitland, P. S., 29
Makeyeva, A. P., 272
Malevitskaya, M. A., 233, 241
Malone, J. M., 266
Mann, R., 97, 388
Manooch, C. S., III, 62, 63, 275, 337, 340, 349
Manter, H. W., 240
Marciochi, A., 344
Marian, T., 273, 282, 283
Marsh, M. C., 163
Marshall, T. L., 47, 80, 82, 85, 87, 94, 165, 324
Martin, R., 323, 409
Mather, F., 47, 324
Mathis, W. P., 336, 338
Matthews, W. J., 110
Mazumbar, H. K., 52
Mearns, A. J., 56
Medved, R. A., 337, 341
Meehean, O. L., 170
Meek, S. E., 102, 108, 115
Meier-Brook, C., 240
Mense, J. B., 347
Mensinger, G., 285
Metzelaar, J., 342, 343
Meyer, F. P., 248
Mgbenka, B. O., 268
Michewicz, J. E., 306, 332, 333, 339
Migaki, G., 234
Mihálik, J., 50, 51
Miley, W. W., II, 53, 54, 55, 305, 332, 333, 339, 342, 344, 345
Millard, T. J., 323, 324
Miller, D. L., 58, 67
Miller, H. J., 336, 338
Miller, J. L., 303
Miller, R. B., 323, 324, 347
Miller, R. R., 54, 56, 58, 59, 102, 106, 108, 109, 112, 113, 117, 133, 145, 162, 184, 343, 375
Miller, W. H., 323, 324, 348
Minckley, W. L., 52, 53, 55, 57, 59, 61, 63, 64, 65, 66, 116, 122, 186, 325, 341
Mires, D., 396
Mirzoeva, L. M., 238
Mišik, V., 49
Missouri Department of Conservation, 285
Mongeau, J. -R., 81, 87
Mitchell, A. J., 248
Mitchell, C. P., 207
Mitzner, L., 332, 333, 334, 339
Mizelle, J. D., 239, 240
Moav, R., 270, 277, 283
Moen, T., 336, 338, 342, 345, 351
Molnar, K., 234, 235, 236, 237, 238, 239, 240, 242
Montegut, R. S., 334, 340
Mookerjee, H. K., 52
Moore, E., 207
Moravec, F., 246
Moriarty, C. M., 275
Moriarty, D.J.W., 275
Morita, C. M., 163
Morris, R. A., 148
Morrissy, N., 184
Moulton, J. C., 50
Moyle, P. B., 44, 47, 49, 51, 55, 56, 57, 66, 67, 109, 184, 287, 323, 324, 325, 326, 336, 337, 338, 340, 341, 342, 343, 344, 347, 348, 349
Moznov, A., 233

Mraz, D., 338, 349, 396
Mulley, J. C., 180
Murphy, G. I., 148
Murty, D. S., 268
Musick, J. A., 50
Musselius, V. A., 238, 242
Myers, G. S., 50, 51, 133, 150, 162, 185, 397

Nagy, A., 283, 284
Nair, K. K., 332
Nair, R. R., 267
Nakagawa, P., 142, 150
Nall, L. E., 329, 332, 333
National Fishery Research Laboratory, U.S. FWS, 3, 4, 5, 54, 320
Navarro, L., 106, 107, 108, 118, 120
Neave, F., 86, 324
Needham, P. R., 141, 323, 324, 342, 343, 348
Neely, W. W., 336, 338, 341
Nelson, J. S., 82, 87, 90, 91, 94, 324
Nerrie, B. L., 277, 279
Newman, M. A., 347, 348
Newton, S. H., 271
New Zealand Freshwater Fisheries Advisory Council, 208
Ney, J. J., 395
Neyman, J., 16
Nichols, J. T., 163, 166, 325
Nielsen, L. A., 343, 344
Nikolsky, G. V., 305, 306
Nikulina, V. N., 243
Noble, R. G., 275
Noble, R. L., 62, 275, 306, 339, 342, 349, 351, 392
North Carolina Wildlife Resources Commission, 65
Norton, S. E., 152
Nunogawa, J. H., 150
Nusbaum, K. E., 229
Nyman, O. L., 324, 342, 343, 346, 347, 348, 349

Obregón, F., 109
Oda, D. K., 149
O'Donnall, J. D., 343
O'Donnell, D. J., 336
Odum, E. P., 9
Odum, H. T., 185
Ogawa, K., 238
Okada, Y., 48, 67
Oliver-Gonzalez, J., 168
O'Malley, H., 163, 169
Omarov, M. O., 272
Ontario Department of Fish and Game, 78
Opuszyński, K., 268, 273, 303, 304, 306, 332, 333
Ordal, E. J., 220
Oring, L. W., 18
Orr, L. D., 268, 329, 332, 333
Ortiz-Carrasquillo, W., 167, 168, 169
Osborn, M. F., 271, 272
Osborne, J. A., 332, 335, 339
Ossiander, F. J., 247
Otto, R. G., 185

Paetz, M. J., 82, 87, 90, 94
Pagan-Font, F. A., 171, 277, 282, 397
Paine, R. T., 30, 184
Pandian, T. J., 187
Panov, D. A., 272
Paperna, I., 238, 246
Pardue, G. B., 329, 333
Park, T., 16
Parrish, J. D., 138, 149
Parsons, P. A., 325
Partsch, K., 236
Pate, V.S.L., 342
Patterson, C., 178
Pelren, D. W., 61
Pelsue, F. W., 62, 65, 66, 337
Pelzman, R. J., 66, 306, 337
Pentelow, F.T.K., 332
Perrone, M., Jr., 18
Perry, W. G., Jr., 286
Pestrak, J. M., 55, 330, 340, 341, 350, 351, 392
Pet Industry Joint Advisory Council, 215, 216
Phelps, S. R., 14
Philipartt, J-Cl., 323
Philipp, A., 147
Phillipps, W. J., 211
Phillipy, C. L., 67, 337, 339
Pianka, E., 10, 11, 16
Pierce, P. C., 337, 341
Pigg, J., 61, 62
Pillay, T.V.R., 262, 264, 267, 268
Pister, E. P., 324, 343
Platania, S. P., 111
Poss, S., 112
Preciado, A. S., 30
Pretto-Malca, R., 271, 273, 280, 285

Price, C. E., 239, 240
Prosser, N. S., 409
Prost, M., 238
Provine, W. C., 49, 329, 332, 333
Prowse, G. A., 332, 333, 337
Pruginin, Y., 264, 269, 273, 274, 276, 283
Puerto Rico Department of Agriculture, 164, 165
Puerto Rico Department of Natural Resources, 165
Pullin, R.S.V., 267, 274, 277

Quast, J. C., 345
Queensland Fisheries Service, 183, 195

Radforth, I., 87
Radonski, G. C., 409
Raj, H. D., 220
Rakocy, J. E., 276
Randall, J. E., 138, 139, 149
Raney, E. C., 342
Ray, L. E., 276, 281
Reddy, S. R., 187
Redner, B. D., 275
Refstie, T., 284
Regan, C. T., 58
Regier, H. A., 96, 184, 263, 323
Rehder, D. D., 342
Reichenbach-Klinke, H.-H., 221, 233, 240, 247
Reimers, N., 342, 343
Reynolds, W. D., 91
Ricker, W. E., 335, 336, 338, 351
Rimler, R. M., 220
Rimsky-Korsakoff, V. N., 342, 343
Rinckey, G. R., 63, 342
Ringuelet, R. A., 58
Rinne, J. N., 61, 343
Riordan, P. F., 268
Ritenour, B. G., 306, 332, 333, 335, 339, 340
Rivas, L. R., 54, 58, 59, 60
Rivera-González, J. E., 165, 170
Robel, R. J., 336, 337
Roberts, R. J., 247
Roberts, T. R., 178, 398
Robins, C. R., xiv, 33, 43, 50, 51, 54, 55, 57, 59, 60, 62, 97, 106, 133, 137, 146, 163, 264, 274, 318, 323, 325, 326, 332, 339, 375, 388, 398
Robson, T. O., 332, 333, 339
Roden, D. L., 324, 343
Rogers, W. A., 239, 242
Romaire, R. P., 277
Roman, E., 239, 240
Roman-Chiriac, E., 240
Romero, H., 108
Rosas-M., M., 103, 106, 107, 108, 115, 120, 124
Rose, E. T., 336, 351
Rosen, D. E., 44, 54, 55, 56, 57, 163, 178, 394
Rosen, D. W., 186, 187
Rosen, R. A., 285
Rosenthal, H., 32, 262, 397, 398
Rothbard, S., 267
Rottmann, A. W., 332, 335, 339
Rottmann, R. W., 306
Roughley, T. C., 180, 182
Rougier, E., 146, 147, 150
Rovozzo, G. C., 221
Rowell, T. W., 97
Rubec, P. J., 89
Rudomentova, N. K., 246
Ruwet, J-Cl., 323
Ryder, R. A., 88

Sahara, Y., 234
St. Amant, J. A., 44, 52, 54, 55, 56, 57, 64, 65, 275
Sarig, S., 275
Saunders, J. W., 330
Savino, J. F., 329
Scanlon, P. F., 388
Schäperclaus, W., 236, 247
Schardt, J. D., 329, 332, 333, 396
Schindler, D. W., 302
Schmidt, G. D., 245
Schmidt, R. E., 50
Schneberger, E., 330, 338
Schneider, R. W., 329
Schoenen, P., 274
Schoener, T. W., 15, 16
Schoenherr, A. A., 341, 345
Schroeder, G. L., 277
Schroeder, W. C., 393
Schubert, G., 229, 234, 236, 237, 242, 409
Schuck, H. A., 324, 347
Schulte, T. S., 165, 170
Schultz, E. E., 52
Schultz, R. J., 54, 133
Schwartz, F. J., 50, 51, 346
Scott, D., 201

Scott, W. B., 49, 79, 80, 83, 86, 87, 92, 94, 95, 336, 338, 347, 349
Sculthorpe, C. C., 329
Seaman, E. A., 3
Sehgal, K. L., 36
Semmens, K., 286
Shafland, P. L., 44, 45, 60, 66, 323, 325, 330, 340, 341, 342, 350, 351, 392
Shapovalov, L., 44, 45, 57
Sharma, B. K., 278
Sharp, I., 54, 55, 57
Sharp, R. W., 335, 336, 338, 341, 349
Sharpe, F. P., 343, 344
Shearer, K. D., 180
Shebley, W. H., 340, 349
Shehadeh, Z. H., 277, 278
Shell, E. W., 274, 336, 337, 338, 341
Shelton, W. L., 66, 268, 269, 276, 282, 283, 284, 409
Shetter, D. S., 324, 343, 344, 347
Shiffer, C. N., 50, 58
Shima, S., 150
Shireman, J. V., 267, 268, 284, 303, 305, 306, 307, 308, 329, 332, 333, 335, 339, 340, 409
Shomura, R. S., 148, 150
Shotts, E. B., Jr., 220, 221, 228, 229, 409
Shute, J. R., 110
Sibley, C. K., 342, 343
Sigler, W. F., 323, 325, 336, 337, 338, 341, 342, 343, 349
Sills, J., 266, 332, 338, 341
Silvera, P.A.W., 275
Simberloff, D., 16, 323, 325, 341
Simon, J. R., 55
Simpson, J. C., 51, 55
Sindermann, C. J., 240
Sinha, V. R., 264, 272
Smallwood, W. M., 335, 338, 342
Smedley, H. H., 324, 343
Smith, B. S., 324
Smith, C. L., 51
Smith, C. R., 303, 332, 333, 335, 339
Smith, H. M., 50, 67
Smith, L. L., Jr., 324
Smith, M. L., 52, 102, 375
Smith, M. W., 330, 351
Smith, P. L., 280
Smith, S. L., 331
Smitherman, R. O., 2, 5, 267, 268, 272, 273, 274, 277, 279, 280, 281, 285, 332, 409
Smith-Vaniz, W. F., 61, 62, 64, 66
Sneed, K., 49, 264, 283, 329, 332, 333
Snow, J. R., 332, 334, 339
Socolof, R. B., 3, 44, 45
Somero, G. N., 13
Sopuk, R. D., 80
Spataru, P., 275
Speaker, E. B., 323, 324, 343
Spotte, S., 247
Springer, V. G., 63
Ŝrámek-Huŝek, R., 237
Staley, J., 324, 343
Stanley, J. G., 267, 282, 283, 284, 334, 339, 393, 396, 409
Starnes, L. B., 50
Stauffer, J. R., Jr., 2, 13, 383, 409
Stein, R. A., 329
Stephanides, T., 186
Sterba, G., 52
Stevenson, J. C., 3
Stevenson, J. H., 266, 332
Stewart, D. J., 393
Stickney, R. R., 262, 263, 275, 276, 277, 279
Stokell, G., 201, 202
Stone, N. M., 286
Stott, B., 268, 329, 332, 333
Strelkov, Y. A., 233
Stroband, H.W.J., 268
Stroganov, N. S., 332, 333
Stroud, R. H., 1, 2, 3, 313, 318, 374, 409
Struthers, P. H., 335, 338, 340, 342, 345, 349
Stumpp, M., 245
Sukumaran, K. K., 306
Sumner, R. C., 341, 345
Sutton, D. L., 283, 302, 305, 306, 332, 333, 334
Swain, D. P., 89
Swee, U. B., 336, 338
Swenson, W. A., 330
Swift, C. C., 109
Swingle, H. A., 285, 286
Swingle, H. S., 263, 264, 269, 274, 275, 276, 282, 331, 336, 337, 341, 350
Sylvester, J. R., 277

Tabata, R., 148, 149
Tal, S., 269, 272, 273
Talbot, G. B., 394
Tamas, G., 268
Tang, Y. A., 270

Tapiador, D. D., 267, 273
Tarplee, W. H., Jr., 61, 64, 66
Taylor, J. N., 4, 324, 341, 344, 345, 346, 349, 409
Taylor, L., 148
Tebo, L. B., Jr., 342, 348
Teleki, G., 91
Terrell, J. W., 329, 332, 333
Terrell, T. T., 306, 329, 332, 333, 339
Texas Parks and Wildlife Department, 318
Theinemann, A., 22
Theriot, R. F., 283, 329, 333
Thomas, A. E., 283, 284
Thomerson, J. E., 3
Thompson, S., 234
Thomson, D. A., 150, 200, 201, 202, 265
Thorpe, L. M., 324
Threinen, C. W., 336, 338
Thys van den Audenaerde, D.F.E., 62, 63, 65, 122, 274, 303
Tikhomirova, V. A., 246
Tilzey, R.D.J., 184
Timbol, A. S., 149, 151, 153
Tomiyama, I., 67
Trautman, M. B., 330
Treviño-Robinson, D., 108
Trewavas, E., 61, 62, 274
Trivers, R., 18
Trust, T. J., 215, 228
Tryon, C. A., Jr., 336
Turner, H. J., 97
Turner, J. S., 344, 346
Tuten, J. S., 284, 286

Uchida, R. N., 275
Udvardy, M.D.F., 14
Uland, B., 268
University of California, Riverside, 62
University of Georgia, 215, 216, 217, 218
U.S. Army, 146
Usui, C. A., 67
Uyeno, T., 106

Valenti, R. J., 283
Valentijn, P., 268
Van Dine, D. L., 150
Van Leeuwen, B. H., 36
Van Pel, H., 147
Varley, J. D., 342, 343
Veitch, C. R., 352
Vergara, M., 112
Verigin, B. V., 272, 308
Vestal, E. H., 324
Vike, L., 335, 341
Vincent, R. E., 323, 324, 348
Vinogradov, V. K., 306, 329, 332
Vismania, K. O., 243
Vivier, P., 344
Vogele, L. E., 349
Vojtek, J., 239, 240
Volovik, S. V., 233

Wainright, S. C., 56, 60, 63, 64
Walden, H. T., 336, 338, 343
Wales, J. B., 48, 340, 341
Wales, J. H., 324, 342, 343
Walker, B. W., 57, 58, 340
Walker, P. G., 50, 51
Wallace, R. L., 51, 55
Wallach, J. D., 233
Wallen, I. E., 329, 338
Ware, F. J., 329, 332, 333, 334, 340, 341, 342, 349, 350
Wattendorf, R. J., 60, 342
Waugh, G. D., 203, 207
Weatherley, A. H., 180, 182, 184
Webber, H. H., 268
Webster, D. A., 342
Wedemeyer, G., 247
Weerakoon, D.E.M., 268
Welch, J. P., 141
Welcomme, R. L., 23, 29, 32, 37, 243, 262, 337, 392, 409
Wellborn, T. L., Jr., 237, 248
Wells, J. G., 215
Werner, E. G., 393
West, G. J., 180, 187
Wharton, J. C. F., 180, 233, 246
Wheeler, A., 29
Whitaker, A. H., 200
Whitaker, J. O., Jr., 342
White, R. J., 347, 348
Whitney, R. R., 51, 92
Whittaker, R. H., 9, 10, 11
Whitworth, W. R., 50
Wilbert, N., 236
Wilde, C. W., 51
Wiley, E. O., 375
Wiley, R. W., 342, 343
Williamovski, A., 272
Williams, J., 55, 64, 66, 285
Williams, V. R., 161

Wilmott, S., 78
Wilson, E. O., 11, 16
Wilson, L. R., 329
Winfree, R. A., 276
Wisconsin Conservation Department, 51
Wodzicki, K. A., 201
Wohlfarth, G. W., 32, 35, 277, 282
Woodruff, V. C., 281
Woynarovich, 267
Wright, B. H., 80
Wydoski, R. S., 51, 92

Yamada, R., 133
Yamaguti, S., 246
Yamamoto, T., 283
Yashouv, A., 263, 268, 272, 275
Yawn, H. M., 337, 341
Yukimenko, S. S., 241

Zaret, T. M., 18, 30, 184
Ziv, I., 269, 272, 273
Zolotova, Z. K., 306, 329, 332
Zorn, M., 275

Index of Fishes

Acanthogobius flavimanus (yellowfin goby), 18, 42, 67
Alburnus alborella (bleak), 30
Alburnus alburnus (bleak), 29, 33, 38, 202
Algansea lacustris (acumara), 104, 107
Alosa sapidissima (American shad), 92
Ambloplites rupestris (rock bass), 105, 117
Amphilius platychir (mountain catfish), 30
Anchoa compressa (deep body anchovy), 134, 138
Aplocheilus lineatus (striped panchax), 135, 142
Aristichthys nobilis (bighead carp)
  aquaculture of, 266, 270, 271, 273-74, 280
  as beneficial, 38
  hybridization of, 273-74
  marketability of, 281
  and parasite transfer, 235, 237, 238, 239
  and reproduction control, 282, 284
Astatoreochromis alluadi, 28
Astronotus ocellatus (oscar)
  colonization by, 18
  in continental United States, 42, 58
  in Hawaii, 137, 142, 152
  native range of, 58
  in North Pacific, 144, 147
  in Puerto Rico, 170

Bairdiella icistia (bairdiella), 42, 57-58
Barbus barbus (barbel), 201
Barbus conchonius (rosy barb), 104, 107, 165-66, 181, 182-83
Barbus semifasciolatus (half-banded barb), 134, 141, 152
Barbus titteya (cherry barb), 104, 107
Basilichthys bonariensis (Argentine pejerry), 31, 39
Belonesox belizanus (pike killifish), 54, 344-45
Betta brederi (Java betta), 144, 147
Betta splendens (Siamese fightingfish), 91
Blicca bjoerkna (silver bream), 29, 33, 38
Brycon melanopterus, 239

Carassius auratus (goldfish)
  and aquarium industry, 28-29
  in Australia, 181, 182, 183
  in Canada, 79, 80, 86-87
  colonization by, 17, 18
  in continental United States, 42, 48, 314, 316
  in Hawaii, 134, 139, 152
  in Mexico, 104, 108
  mixed view of, 39
  native range of, 48, 108
  in New Zealand, 201, 204, 209
  and parasite transfer, 235, 236-37, 239, 243
  in Puerto Rico, 166
  in South Pacific, 145, 147
  and sport fishing, 314, 316
Carassius carassius (Crucian carp), 181, 182, 246, 306
Carpiodes carpio (river carpsucker), 104, 111
Cephalopholis urodelus (roi), 136, 138, 149
Chanos chanos (milkfish), 4
Chelon engeli, 137, 138, 149
Chirostoma spp., 102
Chirostoma estor (blanco de Pátzcuaro), 30, 105, 115, 116
Chirostoma grandocule (charal blanco), 105, 115
Chirostoma jordani (charal comun), 105, 115, 116

Chirostoma labarcea (charal de Chapala), 105, 116
Chirostoma sphyraena (blanco de Chapala), 105, 116
Chromileptes altivelis (polka-dot grouper), 138
Cichla ocellaris (tucunare or peacock bass), 170–71, 172, 318
  in Hawaii, 137, 142, 152
  impact on native fish, 30
  in North Pacific, 144, 147
  and parasite transfer, 239
Cichlasoma spp., 124–25, 137, 142, 152
Cichlasoma bimaculatum (black acara), 18, 42, 58, 324
Cichlasoma citrinellum (Midas cichlid), 42, 58, 323
Cichlasoma cyanoguttatum (Rio Grande cichlid), 105, 121, 124
Cichlasoma managuense (jaguar guapote), 38
Cichlasoma meeki (firemouth cichlid)
  colonization by, 18
  in continental United States, 42, 59
  in Hawaii, 137, 141, 152
  native range of, 59
  preadaptations of, 325
Cichlasoma nigrofasciatum (convict cichlid)
  in Australia, 181, 183
  in Canada, 79, 90
  colonization by, 18
  in continental United States, 42, 59
  native range of, 59
Cichlasoma octofasciatum (Jack Dempsey), 18, 42, 59–60, 181, 183
Cichlasoma salvini (yellowbelly cichlid), 59
Cichlasoma urophthalmos (mojarra criolla), 105, 121
Cichla timensis (striped tucanare), 318
Clarias batrachus (walking catfish)
  colonization by, 17, 18
  in continental United States, 42, 52–53
  dispersal of, 14
  native range of, 52
  in North Pacific, 144, 147
  as pest, 38
  preadaptations of, 324–25
Clarius fuscus (Chinese catfish), 135, 139, 152
Colisa labiosa (thicklip gourami), 239
Coregonus albus (= *C. clupeaformis,* lake whitefish), 202
Coregonus laeveraetus (powan), 38
Coregonus peled (northern whitefish), 38
Crenichthys baileyi (White River springfish), 32
Ctenobrycon spilurus (silver tetra), 239
Ctenopharyngodon idella (grass carp), 4, 25
  and accidental introductions, 29
  aquaculture of, 266–69, 270, 271, 283–84
  breeding habits of, 28
  colonization by, 17, 18
  in continental United States, 42, 48–49, 317
  in Hawaii, 135, 142
  hybridization of, 305, 307
  impact of, 38, 306–7, 332–35, 339–40
  marketability of, 281
  in Mexico, 104, 108
  native range of, 48, 108
  in New Zealand, 207, 208–9, 210
  and parasite transfer, 33, 234, 236, 238, 239, 240, 241–42
  in Puerto Rico, 166
  and reproduction control, 281–84
  in South Pacific, 144, 147
  and sport fishing, 317
  weed control by, 28, 207, 208, 303, 304, 305, 306–9
Cynoscion xanthulus (orangemouth corvina), 42, 58
Cyprinodon alvarezi, 124
Cyprinus carpio (common carp), 25
  and accidental introductions, 29
  aquaculture of, 285
  in Australia, 180–81, 184–85
  in Canada, 78, 79, 80, 87–89, 94
  colonization by, 17
  in continental United States, 43, 49, 314, 315, 335–37, 338, 345, 349
  early transfer of, 22, 263
  egg predation by, 345
  in Hawaii, 134, 139, 152
  impacts of, 33–36, 88, 94, 109, 184–85, 335–37, 338, 345, 349
  in Mexico, 104, 108–9
  mixed view of, 39
  native range of, 108
  in New Zealand, 201, 207, 210
  and overcrowding, 351–52
  and parasite transfer, 32–33, 234, 236, 238, 241–42, 243, 246
  in Puerto Rico, 166
  in South Pacific, 144, 145, 147

and sport fishing, 314, 315
stunting of, 33
survivorship curve of, 12

Dallia pectoralis (Alaska blackfish), 79, 86
Dorosoma spp. (shads), 334, 341–42
Dorosoma petenense (threadfin shad)
as beneficial, 38
competition with *Tilapia aurea*, 341–42
in Hawaii, 135, 142, 152
impact of, 106
in Mexico, 104, 106
native range of, 106
in Puerto Rico, 165

Esox lucius (northern pike), 30, 39, 85, 306
Esox masquinongy (muskellunge), 93
Esox niger (chain pickerel), 92
Esox reicherti (Amur pike), 4, 318
Etheostoma spectabile (orangethroat darter), 14
Etroplus suratensis (green chromide), 38

Fundulus diaphanus (banded killifish), 167, 172
Fundulus fonticola, 162
Fundulus grandis (gulf killifish), 135, 140
Fundulus zebrinus (plains killifish), 104, 112

Galaxias spp., impact of exotic fish on, 30, 178, 179, 211–12
Galaxias nigrostriatus, 185
Galaxias olidus (mountain galaxia), 184
Gambusia affinis (mosquitofish), 25
in Australia, 181, 182, 185–87
in Canada, 79, 89–90
hardiness of, 185–87
in Hawaii, 135, 140, 143, 150, 152
impact of, 113, 185–87
in Mexico, 104, 113
mixed view of, 39
and mosquito control, 28, 185
native range of, 113
in New Zealand, 205, 206, 210
in Oceania, 144, 145, 146
and parasite transfer, 241, 242
in Puerto Rico, 168
Gambusia dominicensis (Dominican mosquitofish), 181, 182, 186
Gambusia panuco (Panuco gambusia), 105, 113
Geophagus surinamensis (redstriped eartheater), 43, 60, 323
Gila bicolor mohavensis (Mohave tui chub), 104, 109
Gila orcutti (arroyo chub), 104, 109
Gnathonemus petersi (elephant-nose mormyrid), 239
Gobio gobio (gudgeon), 201
Goodeidae (goodeids), 30
Gymnocorymbus ternetzi (black tetra), 46, 239

Helostoma rudolfi (= *H. temmincki;* see below), 239
Helostoma temmincki (kissing gourami), 46, 147
Hemibarbus maculatus (chi ha yu), 38
Hemichromis bimaculatus (jewelfish), 18, 42, 60, 91
Hemiculter eigenmanni, 38
Hemiculter leucisculus, 38
Herklotsichthys quadrimaculatus (goldspot herring), 161
Hesperoleucas navarroensis, 239
Heterotis niloticus (African bonytongue), 38
Hucho hucho (huchen or Danube salmon), 79, 81, 318
Hypomesus nipponensis (wakasagi), 43, 47–48
Hypophthalmichthys molitrix (silver carp)
aquaculture of, 266, 269–73, 280
as beneficial, 38
hybridization of, 273–74
marketability of, 281
in Mexico, 104, 109
native range of, 109
in New Zealand, 207, 210
and parasite transfer, 234, 235, 237, 238
in Puerto Rico, 166
and reproduction control, 282, 284
for weed control, 207, 210, 303, 304
Hypostomus spp. (suckermouth catfishes), 42, 53
Hypseleotris swinhoris, 38

Ictalurus spp. (bullhead catfishes), 29, 33
Ictalurus catus (white catfish), 167
Ictalurus furcatus (blue catfish), 104, 111
Ictalurus melas (black bullhead), 38, 92, 104, 111–12, 167
Ictalurus nebulosus (brown bullhead)
in Canada, 92
in Hawaii, 135, 139, 161
in New Zealand, 202, 204, 206, 209

Ictalurus nebulosus (continued)
and parasite transfer, 238
as pest, 38
in Puerto Rico, 163–64, 167
Ictalurus punctatus (channel catfish)
aquaculture of, 269
in Hawaii, 135, 142, 152
impact of, 112
in Mexico, 104, 112
native range of, 112
and parasite transfer, 234, 238, 242
in Puerto Rico, 167
in South Pacific, 145, 147
Ictiobus spp. (buffalofish), 264, 285

Kuhlia rupestris (nato), 136, 138

Labeo rohita (rohu), 4, 38
Lates nilotica (Nile perch), 4, 39, 195–96, 318
Lepidogalaxias salamandroides (Australian mudminnow), 178
Lepomis spp. (sunfishes), 30, 334
Lepomis auritus (redbreast sunfish)
in Mexico, 105, 118
native range of, 117
as pest, 38
in Puerto Rico, 168–69, 172
stunting of, 33
Lepomis cyanellus (green sunfish), 118
Lepomis gibbosus (pumpkinseed), 33, 38, 92, 239–40
Lepomis gulosus (warmouth), 105, 118, 169, 172
Lepomis macrochirus (bluegill)
in Hawaii, 136, 141, 152
impact of *Ctenopharyngodon idella* on, 334–35
in Mexico, 105, 118–19
native range of, 118
in Puerto Rico, 164, 169, 172
Lepomis macrochirus x *Micropterus salmoides,* 150
Lepomis megalotis (longear sunfish), 105, 119
Lepomis microlophus (redear sunfish), 105, 119, 169, 172
Lepomis punctatus (spotted sunfish), 105, 119
Leuciscus idus (ide), 43, 49–50, 316
Leuciscus leuciscus (Eurasian dace), 38, 201
Limnothrissa miodon (roughtongue kapenta), 24, 27, 38
Lucioperca fluviatilis (= *Perca flavescens;* European perch), 306
Lutjanus fulvus (toau), 136, 138, 149
Lutjanus kasmira (taape), 136, 138, 149

Macquaria colonorum, 144, 147
Macquaria novemaculeata (Australian bass), 201
Madigania unicolor (= *Liopotheropon unicolor;* spangled perch), 144, 147
Megupsilon aporus (El Potosí pupfish), 124
Melanotaenia australis (crimsonspotted rainbowfish), 242
Melaniris chagrensis, 30
Menidia beryllina (tidewater silverside), 105, 116
Mesopristes argenteus (silver grunter), 144, 147
Metynnis roosevelti, 303, 304
Micropterus coosae (redeye bass), 169, 172
Micropterus dolomieui (smallmouth bass)
in Hawaii, 136, 142, 151, 152
in Mexico, 105, 120
native range of, 119
in South Pacific, 144, 147
Micropterus salmoides (largemouth bass), 23, 329
in Canada, 92
and forage fish changes, 340–41
in Hawaii, 136, 140, 152
impact of *Ctenopharyngodon idella* on, 334, 335
impact of *Tilapia zilli* on, 306
impacts of, 30, 102, 120, 124
in Mexico, 102, 105, 120, 124
mixed view of, 39
native range of, 120
in New Zealand, 205
and overcrowding, 350–51
and parasite transfer, 243, 245, 246
in Puerto Rico, 164, 170, 172
in South Pacific, 145, 147
Misgurnus anguillicaudatus (oriental weatherfish)
as beneficial, 38
and commercial fishery, 28
in continental United States, 42, 52
in Hawaii, 134, 139, 152
in Mexico, 104, 111
native range of, 52, 111

Monopterus albus (ricefield eel), 136, 139, 152
Morone chrysops (white bass), 105, 117, 168
Morone saxitalis (striped bass), 92, 105, 117, 136, 138
Mylossoma argenteum, 303, 304

Neoceratodus forsteri (Australian lungfish), 178
Nocturus insignis (margined madtom), 79, 89
Notemigonus crysoleucas (golden shiner), 104, 109–10, 242
Nothobranchius guentheri (nothobranchus), 135, 142
Notropis lutrensis (red shiner), 104, 110

Oncorhynchus gorbuschka (pink salmon), 39
Oncorhynchus kisutch (Coho salmon), 92, 205
Oncorhynchus masou (cherry salmon), 79, 81
Oncorhynchus nerka (sockeye salmon), 202, 204, 209, 210
Onchorynchus tshawytscha (Chinook salmon)
in Australia, 181, 182
as beneficial, 38
in Hawaii, 134, 138
in New Zealand, 201, 202, 204, 207, 209
Ophicephalus striatus (snakehead), 135, 139, 144, 147, 152
Opsarichthys uncirostris, 38
Oreodaemion quathlambae (maluti minnow), 30
Orestias spp., 31
Orthodon microlepidotus (Sacramento blackfish), 285–87
Oryzias latipes (medaka), 136, 141
Osphronemus gouramy (gourami), 28, 39, 137, 142, 145, 147

Perca flavescens (yellow perch), 92
Perca fluviatilis (European perch), 30, 39
in Australia, 180, 181, 185–86
in New Zealand, 201, 210
Percottus glehni, 38
Percina oxyrhyncha (sharpnose darter), 14–15
Petenia splendida (tenhuaycaa), 105, 121–22
Phoxinus phoxinus (European minnow), 201
Phractocephalus hemibiopterus (redtail catfish), 239
Pimephales promelas (fathead minnow), 104, 110, 166, 242
Pimephales vigilax (bullhead minnow), 104, 110
Platichthys flesus (European flounder), 79, 91
Plecoglossus altivelis (ayu), 134, 138
Poecilia spp. (mollies), 30
Poecilia latipinna (sailfin molly)
in Australia, 181, 182, 183
in Canada, 79, 90
hardiness of, 185–86
in Hawaii, 135, 140, 150, 152
in Mexico, 105, 113
native range of, 105
in New Zealand, 206, 210
in Oceania, 144, 146
as pest, 38
Poecilia mexicana (shortfin molly)
colonization by, 18
in continental United States, 42, 54–55
in Hawaii, 133, 135, 142, 143, 152
native range of, 54
in New Zealand, 206, 210
in Oceania, 144, 145, 146
and parasite transfer, 246
Poecilia reticulata (guppy)
in Australia, 181, 182–83
in Canada, 79, 90
in continental United States, 43, 55
in England, 29
in Hawaii, 135, 141, 143, 152, 161
impact of, 32
in Mexico, 105, 113–14
and mosquito control, 28
native range of, 55, 113
in Oceania, 144, 145, 146
and parasite transfer, 242, 244, 245
as pest, 38
in Puerto Rico, 168
Poecilia sphenops (liberty molly), 135, 141, 152
Poecilia vittata (Cuban limia), 135, 141, 143, 152

Poecilia vivipara, 162, 163, 168
Poeciliopsis gracilis (porthole livebearer), 42, 55-56
Pomoxis spp. (crappies), 334
Pomoxis annularis (white crappie), 105, 120, 170
Pomoxis nigromaculatus (black crappie), 92, 105, 121
Potamotrygonidae (freshwater stingrays), 4
Protroctes oxyrhynchus (New Zealand grayling), 30
Pseudogobio rivulatus (river dodger), 38
Pseudorasbora parva, 29, 33, 38
?*Pterophyllum scalare* (angelfish), 79, 91
Pterygoplichthys multiradiatus, 53-54
Ptychocheilus lucius (squawfish), 242
Puntius gonionotus (tawes), 28, 38, 144, 147, 303
Puntius seale (barb), 144, 147

Rhinogobius similis, 38
Rhodeus sericeus (bitterling), 43, 50, 316
Rivulus marmoratus (rivulus), 162-63
Rutilus rutilus (roach), 38, 181, 201, 306

Salmonids (salmons and trouts)
  competition between, 347-50
  diseases of, lack of in New Zealand, 209
  early tranfers of, 22, 23
  impact of exotic fish on, 30
  impact on native fish, 184, 211-12
Salmo aguabonita (golden trout), 79, 82, 343
Salmo clarki (cutthroat trout), 14, 343
Salmo gairdneri (rainbow trout)
  and aquaculture, 23
  in Australia, 181, 182
  in Canada, 79, 82, 92
  early transfer of, 22
  in Hawaii, 134, 140-41, 143, 152
  impact of, 30-32
  impact of *S. trutta* on, 344
  in Mexico, 104, 106,
  mixed view of, 39
  native range of, 106
  in New Zealand, 201, 202-4, 205, 207, 209, 210, 211, 212
  and parasite transfer, 32, 33, 235-36, 253
  preadaptations of, 13
  in Puerto Rico, 164, 165
  in South Pacific, 147
Salmo letnica (Lake Ohrid trout), 4, 317-18
Salmo salar (Atlantic salmon)
  in Australia, 180, 181, 182
  in Canada, 94
  in New Zealand, 204, 209, 210
Salmo trutta (brown trout)
  in Australia, 180, 181
  in Canada, 78, 79, 82-86
  colonization by, 17, 323-24
  in continental United States, 43, 47, 343-44, 347-50
  in Hawaii, 134, 141
  impact of, 30-32, 314, 316, 343-44, 347-50
  mixed view of, 39
  native range of, 47
  in New Zealand, 201, 202-4, 209, 211
  and parasite transfer, 254
  in Puerto Rico, 165
  and sport fishing, 314, 316
Salvelinus alpinus (Arctic char or French alpine char), 93-94
Salvelinus fontinalis (brook trout)
  in Australia, 181, 182
  in Hawaii, 134, 139
  impact of *Salmo trutta* on, 82, 83, 85, 343-44, 347-50
  in Mexico, 104, 107
  mixed view of, 39
  native range of, 107
  in New Zealand, 201, 209, 211
  and parasite transfer, 236, 253
Salvelinus namaycush (lake trout), 202, 204, 209, 210-11, 343
Sandelia capensis (cape kurper), 31
Sardinella marquesensis (Marquesan sardine), 134, 138, 148
Scardinius erythrophthalmus (rudd), 43, 51, 201, 206, 210, 316
Scleropages jardini (Austrialian bonytongue), 178
Scleropages leichardti (spotted bonytongue), 178
Silurus glanis (wels), 241
Sorubim lima (duckbill catfish), 239
Stizostedion lucioperca (Zander or European pikeperch), 38
Symphysodon (discus), 236

Tilapia spp. (tilapias)
  aquaculture of, 23, 264, 274-80
  as forage fish, 340-41
  hybridization of, 32, 34-35, 275, 276, 283, 316

impact of, 124–25, 349–50
marketability of, 281
and parasite transfer, 33, 238, 246
preadaptations of, 17
predators of, 173
and reproduction control, 276
stunting of, 33
for weed control, 303, 304, 305
Tilapia aurea (blue tilapia)
aquaculture of, 274, 275
colonization by, 18, 19
in continental United States, 42, 60–62, 337, 339–40
impact of, 337, 339–40, 341–42
in Mexico, 122
mixed view of, 39
native range of, 60, 122
and overcrowding, 351
in Puerto Rico, 171, 172
and sport fishing, 316–17
weed control by, 304, 337
Tilapia galilaea (mango tilapia), 39, 304
Tilapia hornorum (Wami tilapia), 43, 62, 64, 171, 172, 275
Tilapia leucosticta, 39
Tilapia macrochir (longfin tilapia), 32, 137, 142, 152
Tilapia mariae (spotted tilapia)
accidental introductions of, 29
in Australia, 181, 183
colonization by, 18
in continental United States, 42, 62–63, 323
native range of, 62
as pest, 38
preadaptations of, 324–25
Tilapia melanopleura (= either *T. rendalli* or probably *T. Zilli*)
in Hawaii, 137, 142, 150, 152
in Mexico, 122
for weed control, 303, 304
Tilapia melanotheron (blackchin tilapia)
colonization by, 18, 19
in continental United States, 42, 63
in Hawaii, 137, 142, 152
mixed view of, 39
native range of, 63
Tilapia mossambica (Mozambique tilapia), 4
accidental introductions of, 29
aquaculture of, 275
in Australia, 181, 183–84
in continental United States, 42, 63–65, 316
in Hawaii, 132, 137, 142, 150–51, 152, 153
in Mexico, 105, 122
native range of, 63
native view of, 148
in Oceania, 132, 143, 144, 145–46
in Papua New Guinea, 187
as pest, 38
preadaptations of, 15, 18, 33
in Puerto Rico, 169, 171, 172, 173
and sport fishing, 316
for weed control, 303, 304, 337
Tilapia mossambica x *T. hornorum* (cherry snapper), 32, 147, 148
Tilapia nilotica (Nile tilapia), 26
aquaculture of, 274, 275
and commercial fishery, 27
hybridization of, 32
in Mexico, 122
mixed view of, 39
in Puerto Rico, 171, 172
in South Pacific, 144, 147
for weed control, 303, 304, 337
Tilapia rendalli (redbreast tilapia), 26
in Hawaii, 133
as pest, 38
in Puerto Rico, 171, 172, 173
for weed control, 28, 171, 173, 303, 304
Tilapia variabilis, 32
Tilapia zilli (redbelly tilapia), 26
colonization by, 18
in continental United States, 42, 65–67, 323, 337
in England, 29
in Hawaii, 137, 142, 150, 152
impact of, 32, 306, 337
in Mexico, 105, 122
native range of, 65, 122
in North Pacific, 144, 147
as pest, 38
in South Pacific, 145, 147
for weed control, 28, 303, 304, 305–6, 337
Tinca tinca (tench)
in Australia, 181
in Canada, 92
in continental United States, 43, 51–52, 316
early transfers of, 22
mixed view of, 39
native range of, 51
in New Zealand, 202, 204, 206, 210
and sport fishing, 316

Trachyistoma euronotus (hoarder), 31
Trichogaster leeri (pearl gourami), 38, 137, 141
Trichogaster pectoralis (sepat Siam), 38, 145, 147
Trichogaster trichopterus (blue gourami), 79, 91
Trichomycterus spp. (parasitic catfishes), 30
Trichomycterus rivulatus, 31
Trichopsis vittata (croaking gourami), 42, 68

Umbra krameri (European mudminnow), 38
Umbra pygmaea (eastern mudminnow), 29
Upeneus vittatus (goatfish), 137, 138–39

Xiphophorus couchianus (Monterrey platyfish), 114, 123
Xiphophorus helleri (green swordtail)
  in Australia, 181, 182, 183, 186
  in Canada, 90
  in continental United States, 42, 56
  in Hawaii, 135, 141, 143, 152
  hybridization of, 114, 123
  in Mexico, 105, 114
  native range of, 56, 114
  in Oceania, 144, 146
  as pest, 38
  in Puerto Rico, 168
Xiphophorus maculatus (southern platyfish)
  in Australia, 181, 182, 183, 186
  in continental United States, 42, 56–57
  in Hawaii, 135, 141, 152
  in Mexico, 105, 114
  native range of, 56, 114
  in Oceania, 144, 146
  as pest, 38
  in Puerto Rico, 168
Xiphophorus variatus (variable platyfish)
  in continental United States, 42, 57
  in Hawaii, 135, 142, 152
  in Mexico, 105, 114–15
  native range of, 57, 114

The Johns Hopkins University Press

DISTRIBUTION, BIOLOGY, AND MANAGEMENT OF EXOTIC FISHES

This book was composed in Times Roman type by Capitol Communication Systems, Inc., from a design by Susan P. Fillion. It was printed on 50-lb. Glatfelter paper and bound in Holliston Roxite A by Thomson-Shore Inc.